CW01267184

Chromosome banding

TITLES OF RELATED INTEREST

Cell movement and cell behaviour
J. M. Lackie

Chromosomes today, volume 9
A. Stahl, J. M. Luciani & A. M. Vagner-Capodano (eds)

Chromosomes today, volume 10
K. Fredga & M. D. Bennett (eds)

The eukaryote genome in development and evolution
B. John & G. Miklos

Gene regulation
D. S. Latchman

Genethics
D. Suzuki & P. Knudtson

The handling of chromosomes (6th edn)
C. D. Darlington & L. F. LaCour

The physical chemistry of membranes
B. Silver

Theoretical population genetics
J. Gale

Yeast biotechnology
D. R. Berry, I. Russell & G. G. Stewart (eds)

Chromosome banding

A. T. Sumner
MRC Human Genetics Unit, Edinburgh

London
UNWIN HYMAN
Boston   Sydney   Wellington

© A. T. Sumner, 1990
This book is copyright under the Berne Convention. No reproduction without permission. All rights reserved.

Published by the Academic Division of
Unwin Hyman Ltd
15/17 Broadwick Street, London W1V 1FP, UK

Unwin Hyman Inc.,
955 Massachusetts Avenue, Cambridge, Mass. 02139, USA

Allen & Unwin (Australia) Ltd,
8 Napier Street, North Sydney, NSW 2060, Australia

Allen & Unwin (New Zealand) Ltd in association with the
Port Nicholson Press Ltd,
Compusales Building, 75 Ghuznee Street, Wellington 1, New Zealand

First published in 1990

British Library Cataloguing in Publication Data

Sumner, A. T.
　Chromosome banding.
1. Organisms. Cells. Chromosomes. Analysis. Laboratory techniques
I. Title
574.87322
ISBN 0-04-445279-9

Library of Congress Cataloging-in-Publication Data

Sumner, A. T. (Adrian Thomas), 1940–
　Chromosome banding / A. T. Sumner.
　　p. cm.
Includes bibliographical references.
ISBN 0-04-445279-9
1. Chromosome banding. 2. Chromosomes. I. Title.
[DNLM: 1. Chromosome Banding–methods. 2. Chromosomes. QH 600 S956c]
QH600.S86 1990
574.87'322–dc20
DNLM/DLC
for Library of Congress　　　　　　　　　　　　　　　　　　90–12338
　　　　　　　　　　　　　　　　　　　　　　　　　　　　　　CIP

Typeset in 10 on 12 point Times by Computape (Pickering) Ltd, North Yorkshire and printed in Great Britain by Cambridge University Press

Preface

In the 20 years or so since modern chromosome banding methods first came into general use, they have become essential tools in fields as diverse as clinical genetics and evolutionary studies, and as a vital adjunct to the endeavours of molecular biologists in mapping genes on chromosomes, for example. If chromosome banding were only a useful tool for chromosome identification, it would be of immense importance, but in addition, the mechanisms of chromosome banding have proved to be a fascinating study in their own right. Perhaps most important of all, however, are the insights that banding has given us into the organization of chromosomes. Far from being rather uniform structures, chromosomes have now been shown to be highly differentiated into regions with distinctive functional properties.

Although several important review articles on different aspects of chromosome banding have appeared over the years, it has been impracticable for them to cover the subject comprehensively. In the present work my purpose has been to describe in depth, for all types of chromosome banding, the methods used to produce banding patterns (including practical details), the mechanisms involved, the practical applications of the methods and the functional significance of the different types of banding. It is clear that although much is known about these matters, a lot still remains to be discovered. I hope that this book will be not only a source of information on all aspects of chromosome banding, but also a stimulus for further research on the subject.

A. T. Sumner
Penicuik, Scotland

Acknowledgements

Many colleagues, both in Edinburgh and from other parts of the world, have helped me, through discussions, to clarify and crystallize many of the ideas presented in this book, without which it would undoubtedly be a less satisfactory work. Many have also generously supplied illustrations, without which the book could not have been regarded as complete; they are acknowledged individually in the figure legends. Most of the other figures have been prepared by Mr Norman Davidson and Mr Sandy Bruce, while Mr Andrew Ross and Mrs Liz Graham have prepared material for electron microscopy. Miss Sheila Mould has obtained all sorts of books and papers for me, many of them quite obscure, without which the bibliography would certainly have been far from complete. To all these I offer my sincere thanks. Finally, I should like to thank Dr Clem Earle of Unwin Hyman whose enthusiasm has been a great help in bringing this work to fruition.

Contents

			page	vii
Preface				

1	Origins and evolution of banding methods		1
	1.1	Definition of banding	1
	1.2	Origins and evolution of chromosome banding methods	2
	1.3	Modern era of banding	4
	1.4	Current state of chromosome banding	7

2	Classification and nomenclature of chromosome bands		9
	2.1	Introduction	9
	2.2	Classification of types of chromosome bands	10
	2.3	Classification of banding methods	12
	2.4	Chromosome band nomenclature	16

3	Structure and composition of chromosomes		21
	3.1	Introduction	21
	3.2	Constituents of chromosomes	21
	3.3	Assembly of chromosomal constituents into chromosomes	26
	3.4	Changes during chromosome condensation	32
	3.5	Effects of fixation on chromosomes	34

4	C-banding and related methods		39
	4.1	Introduction	39
	4.2	C-banding	40
	4.3	G-11 banding	59
	4.4	N-banding	63
	4.5	Methods for demonstrating heterochromatic bands using hydrochloric acid	65
	4.6	Silver staining of heterochromatin	67
	4.7	Concluding remarks	68

5 G-banding — 70

5.1 Definition of G-banding — 70
5.2 G-banding methods — 71
5.3 Characteristics of G-bands — 81
5.4 Mechanisms of G-banding — 86
5.5 Applications of G-banding — 93
5.6 Conclusion — 103

6 R- and T-banding — 105

6.1 Introduction and definitions — 105
6.2 Methods for R- and T-banding — 106
6.3 Nature of R- and T-bands — 112
6.4 Mechanisms of R- and T-banding — 114
6.5 Applications of R- and T-banding — 118

7 Q-banding — 122

7.1 Introduction — 122
7.2 Q-banding methods — 123
7.3 Nature of Q-bands — 132
7.4 Mechanisms of Q-banding — 137
7.5 Applications of Q-banding — 147

8 Banding with fluorochromes other than quinacrine — 155

8.1 Introduction — 155
8.2 Ethidium: a fluorochrome that shows no base specificity — 155
8.3 Banding with fluorochromes showing specificity for A+T-rich DNA — 157
8.4 Banding with fluorochromes showing specificity for G+C-rich DNA — 169
8.5 Banding with fluorochromes in combination — 174
8.6 Light-induced banding — 184
8.7 Concluding remarks — 185

9 Nucleolar organizers (NORs) — 187

9.1 Nature and organization of NORs and nucleoli — 187
9.2 Methods for selective staining of NORs — 191
9.3 N-banding (Giemsa staining of NORs) — 192
9.4 Silver staining of NORs (Ag-NOR staining) — 194

10 Kinetochores — 206

- 10.1 Introduction — 206
- 10.2 Staining methods for kinetochores — 209

11 Banding produced by treatment of living cells: condensation and replication banding — 220

- 11.1 Introduction — 220
- 11.2 Inhibition of condensation — 221
- 11.3 Replication banding: BrdU substitution — 232
- 11.4 Unclassified effects of agents which induce banding in living cells — 250
- 11.5 General discussion — 252

12 Banding with nucleases — 253

- 12.1 Introduction — 253
- 12.2 Digestion of chromosomes with non-specific nucleases — 253
- 12.3 Digestion of chromosomes with restriction endonucleases — 259
- 12.4 General remarks on banding with nucleases — 269
- 12.5 Note on nomenclature — 270

13 Banding demonstrated by immunocytochemical methods — 272

- 13.1 Scope of applications of immunocytochemical methods to chromosomes — 272
- 13.2 Antibodies to nucleic acids and nucleosides — 272
- 13.3 Antibodies to chromosomal proteins — 280

14 Chromosome banding polymorphisms, heteromorphisms and variants — 283

- 14.1 Introduction — 283
- 14.2 Mechanisms of formation of heteromorphisms — 284
- 14.3 Quantitative assessment of heteromorphic bands — 286
- 14.4 Human chromosome heteromorphisms — 288
- 14.5 Heteromorphisms in primates — 302
- 14.6 Heteromorphisms in other mammals — 303
- 14.7 Heteromorphisms in lower vertebrates — 305
- 14.8 Heteromorphisms in insects — 306
- 14.9 Heteromorphisms in plants — 307
- 14.10 Concluding remarks — 309

15 Banding and chromosome evolution — 311

- 15.1 Introduction — 311
- 15.2 Mechanisms of evolutionary change as seen by chromosome banding — 312
- 15.3 Use of banding to study chromosome phylogenies — 320
- 15.4 Evolution of sex chromosomes — 327
- 15.5 Concluding remarks — 331

16 Genome organization in the light of chromosome banding — 332

- 16.1 Introduction — 332
- 16.2 Heterochromatic bands — 333
- 16.3 Euchromatic bands — 336

Bibliography — 349

Index — 425

Tables

		page
2.1	Definition of banding methods, according to Paris Conference (ISCN 1985).	13
2.2	A banding alphabet.	14
2.3	Three letter code to describe banding techniques (from the Supplement to the Paris Conference Report, 1971, published by The National Foundation) (ISCN 1985).	15
2.4	Species for which standard karyotypes have been published.	16
2.5	Numerical description of size and intensity of variable bands (from the Supplement to the Paris Conference Report, 1971, published by The National Foundation) (ISCN 1985).	20
4.1	Some examples of types of DNA found in C-bands.	43
5.1	Proteolytic enzymes (other than trypsin) used to induce G-bands on mammalian chromosomes.	74
5.2	Properties of positive and negative G-bands.	86
7.1	Quinacrine fluorescence properties of regions of heterochromatin of known base composition, as identified by *in situ* hybridization.	134
8.1	Properties of the fluorochromes described in this chapter.	156
8.2	Fluorescence of chromosomal regions of known base composition, as identified by *in situ* hybridization, with Hoechst 33258, DAPI, dibutyl proflavine and daunomycin.	161
8.3	Staining properties of human heterochromatic segments with quinacrine-type and Hoechst 33258-type fluorochromes.	167
8.4	Fluorescence of chromosomal regions of known base composition, as identified by *in situ* hybridization, with chromomycin A_3, mithramycin and 7-amino-actinomycin D.	173
8.5	Combinations of fluorochromes used for chromosome banding.	175
8.6	Pairs of dyes used for chromosome banding which interact by energy transfer.	178
8.7	Species reported to have distamycin/DAPI-positive segments on their chromosomes.	182
9.1	Properties of some nucleolar proteins (after Sommerville 1986).	190
10.1	Molecular weights of kinetochore proteins recognized by CREST sera.	217
10.2	Classification of kinetochore proteins.	218

11.1	Agents causing differential inhibition of condensation of chromosomes.	222
11.2	Patterns of human heterochromatin decondensation induced by A+T-specific ligands (after Rocchi *et al.* 1979).	228
11.3	Human chromosomes showing lateral asymmetry after replication in the presence of BrdU.	249
11.4	Substances which induce banding by undetermined mechanisms when applied to living cells.	251
12.1	Restriction endonucleases used for chromosome banding in mammals.	260
12.2	Restriction endonuclease banding in organisms other than mammals.	264
13.1	Banding patterns produced on mammalian chromosomes by different anti-nucleoside antibodies.	274
13.2	Patterns of binding of anti-nucleoside antibodies to mammalian heterochromatin.	276
16.1	Distribution of different types of banding in different groups of eukaryotes.	343

1 Origins and evolution of banding methods

1.1 Definition of banding

Chromosome banding is a lengthwise variation in staining properties along a chromosome. This variation in staining properties is normally independent of any immediately obvious structural variation; thus the polytene chromosomes of Dipteran insects and certain other organisms do not display chromosome banding in the sense in which the term is used in this book, but instead have a banded structure that is visible without any staining. Chromosome banding may be produced in a variety of ways on chromosomes that show little or no structural differentiation along their length, and which are uniformly stained with many dyes. A single dye, either through differential binding or fluorescence, can often be used to produce a banding pattern on chromosomes. In many cases, a pretreatment is used followed by staining with a dye that would produce uniform coloration of the chromosomes if the pretreatment were omitted. Treatment of cells in culture, before fixation, can be used to induce changes detectable in fixed chromosomes: for example, base analogues can be incorporated into the chromosomal DNA, thereby labelling segments that were replicating DNA at a specific time. Certain types of banding depend on reagents of very high specificity, such as enzymes, antibodies and nucleic acid sequences, to label specific proteins and DNA sequences in the chromosome.

The term chromosome banding refers both to the process of producing banding patterns on chromosomes, and to the patterns themselves. As we shall see, there is great diversity in both banding processes and patterns, yet all have in common the objective of revealing a certain level of chromosome organization. As well as being of immense practical value in identifying chromosomes and parts of chromosomes, banding and the study of its mechanisms provides an insight into the highest level of chromosome organization below that of the whole chromosome. The object of this book is not only to describe the different classes of banding techniques that have been devised, together with an assessment of their practical value, but also to give an account of the implications of banding for chromosome organization.

1.2 Origins and evolution of chromosome banding methods

Chromosome banding has a respectable antiquity, the first observations of differential staining of a part of a chromosome having been made almost a century ago. In 1894 Metzner, using a mitochondrial stain, observed bodies at the centromeric region of *Salamandra* chromosomes that probably corresponded with kinetochores. During the next 40 years or so, several authors reported seeing similar chromosomal bodies in both plant and animal material (summarized by Levan 1946). Most of these observations were made on material fixed in a mixture containing osmium tetroxide, embedded, sectioned and stained with haematoxylin, and were thus very different technically from contemporary chromosome preparations made by squashing or spreading cells fixed with acetic acid. These earlier methods did not, therefore, lead to the methods used at present to stain kinetochores specifically.

Only five years after Metzner's observation of kinetochores, Ruzicka (1899) saw silver staining of nucleoli. During the early years of the century, the development of silver impregnation methods for neurocytology permitted more detailed observations on nucleoli, which were also stained by these methods, and the use of silver staining for studies of nucleoli has continued until recent times (reviewed by Schwarzacher & Wachtler 1983). Although silver staining of nucleolus organizing regions on chromosomes was not introduced into routine practice until relatively recently (Goodpasture & Bloom 1975, Howell *et al*. 1975), such methods are descended directly from those used to study interphase nucleoli over 90 years ago.

In 1928, Heitz introduced the concept that chromatin consists of two components, euchromatin and heterochromatin. Using a simple staining technique, he showed that in liverworts, and later in many other organisms, certain segments of chromosomes remained visible after telophase, and these segments he named heterochromatin, in contrast to the euchromatin which decondenses and becomes indistinguishable as discrete chromosomal bodies after telophase. For a brief review of the life and work of Emil Heitz, see Passarge (1979). The concept that chromosomal material is divided into euchromatin and heterochromatin has proved very valuable, and has been developed further from the original ideas of Heitz (Brown 1966). As we shall see in Chapter 4, modern banding studies have modified our concept of heterochromatin still more, providing greater insight into the molecular constitution of euchromatin and heterochromatin.

During the 45 years following the discovery of heterochromatin, the specific staining of heterochromatin by a variety of methods in many organisms (predominantly plants) was described (Levan 1945, 1946, Yamasaki 1959, 1971). Interesting and valuable though these observations were, the methods were not universally applicable, and with the development of modern techniques the older ones have been largely forgotten.

An alternative method of demonstrating heterochromatin was the phenomenon of so-called 'nucleic acid starvation' (Darlington & LaCour 1938, 1940). If the chromosomes of certain plants and animals were exposed to cold, certain segments appeared thinner than the remainder of the chromosome and took up less stain. Although the mechanism proposed for this phenomenon, that DNA synthesis was inhibited in these segments by the cold, has proved to be wrong (Ch. 11), these methods are important as the forerunners of a number of techniques whereby a banding pattern can be induced in chromosomes by treatment before fixation.

In 1956 Tjio & Levan demonstrated beyond all doubt that the human mitotic chromosome complement consisted of 46 chromosomes, and thus began the serious study of human cytogenetics, and of mammalian cytogenetics in general. Technical innovations were important for the development of human cytogenetics, and the methods of accumulating metaphase cells using colchicine, swelling the cells with a hypotonic solution, fixation in a mixture of alcohol and acetic acid, and spreading the chromosomes by air drying, are employed almost universally (although squashing is still commonly used for plant and insect chromosomes). These methods are amply described in several textbooks (Yunis 1974, Priest 1977, Rooney & Czepulkowski 1986, Macgregor & Varley 1988, Verma & Babu 1989) and will not be repeated in this book; it is, however, important to recognize that such methods are the basis of modern cytogenetics, and that modern banding techniques are designed for material prepared in such a way, in contrast to the diversity of preparative methods used in early studies.

The human chromosome complement comprises 23 pairs of chromosomes, but on purely morphological criteria they could only be classified reliably into seven groups by the Denver Conference of 1960 (ISCN 1985). Evidence soon accumulated that certain genetic defects were associated with particular classes of chromosomes, and it was apparent that a method for identifying individual chromosome pairs was needed. As well as the primary constriction, or centromere, some chromosomes in certain circumstances show non-centromeric secondary constrictions. Catalogues of sites where secondary constrictions may occur on human chromosomes have been prepared by Saksela & Moorhead (1962) and by Ferguson-Smith *et al.* (1962). However, secondary constrictions cannot always be demonstrated on particular chromosomes, and in any case several lack them, so that they had very limited value for identifying chromosomes. Of considerably greater value was the application of autoradiography to the problem of identifying chromosomes. Different parts of a chromosome replicate their DNA at different times during the S (synthetic) phase of the cell cycle. Replicating DNA can be labelled using radioactive (tritiated) thymidine, and under appropriate conditions an autoradiograph will show a pattern of silver grains over the chromosomes indicating which segments were replicating their DNA at a particular time. For a review of the applications of autoradiography in cytogenetics, see O. J. Miller (1970). In practice, autoradiography proved to be of limited use in identifying chromosomes, since many did not exhibit distinctive patterns, and the technique is inevitably slow.

1.3 Modern era of banding

In 1968 Caspersson *et al.* published their first paper describing the use of quinacrine mustard to stain chromosomes, and thereby ushered in a new era of chromosome banding. Within a few years the whole practice of cytogenetics was transformed by the introduction of a broad range of banding techniques that owed little to the methods used sporadically during the previous 70 years or so, and which were developed largely in ignorance of the previous work. The modern banding techniques have not only evolved independently, but also have so many features that distinguish them from the older ones that it is entirely proper to regard the period since 1968 as a distinct era. This era is characterized by techniques that are universal, have come into widespread use, and have valuable practical applications. The last characteristic is due to a large extent to the development of methods that demonstrate bands throughout the length of the chromosomes, at least in amniote vertebrates; these bands form distinctive patterns that permit the identification of individual chromosome pairs. This has proved invaluable in clinical cytogenetics, evolutionary studies and gene mapping, for example. The universality of modern banding techniques means that a technique developed for the chromosomes of one organism can be applied to another, plant or animal, vertebrate or invertebrate, with only minor modifications. Modern banding techniques also tend to have a rational basis, and the desire to understand the basis of different sorts of chromosome banding has led to extensive research into this aspect of chromosome organization.

Chromosome banding is now a complex fabric woven from many threads. As a result of several lines of evolution, a wide variety of techniques is available, some in everyday use for identifying chromosomes, others less often used, perhaps for particularly difficult cases; yet other techniques are primarily research tools, helping our understanding of chromosome organization. The first thread or line of evolution is that started by Caspersson *et al.* (1968, 1969a,b) in which a fluorochrome (quinacrine mustard) showing DNA base specificity was applied to chromosomes in the belief that there must be differences in DNA base composition along a chromosome. Variations in the intensity of fluorescence along chromosomes were indeed found, and such bands have become known as Q-bands. Subsequently, Hilwig & Gropp (1972) introduced the fluorochrome Hoechst 33258, which has a more clearly defined base specificity than quinacrine mustard and quinacrine, and since 1976 a variety of other base-specific fluorochromes have been used, singly or in combination (Ch. 8). A second thread began with the invention of *in situ* hybridization in which the sites of specific DNA sequences on chromosomes could be identified by hybridizing to them labelled complementary nucleic acid sequences. This thread has branched in no less than three directions, each of which has been of immense value. The first branch, *in situ* hybridization itself, has now been refined to the point where it can be used to locate a single gene on a chromosome (Szabo & Ward 1982). When Pardue & Gall (1970) used *in situ* hybridization to locate satellite DNA in mouse chromosomes, they noticed that the centromeres were more strongly stained with Giemsa dye. This observation of selective centromeric staining is the origin of techniques known as C-banding (Ch. 4), which are extensively used to demonstrate constitutive heterochromatin; indeed, in all except higher vertebrates, C-banding techniques are almost certainly the most widely used banding methods for both animal and plant chromosomes. A third branch from the stem of *in situ* hybridization has proved to be the most useful of all: if chromosomes are treated in a less drastic way than that required to produce C-banding, and then stained with Giemsa, a series of bands is revealed throughout the length of the chromosomes (Sumner *et al.* 1971a). These bands, known as G-bands after the dye mixture used to reveal them, have proved to be of immense value in the cytogenetics of higher vertebrates, and especially of man. The banding pattern is largely similar to that revealed by quinacrine in higher vertebrates, but G-banding has the advantage of producing permanent preparations and not requiring a fluorescence microscope. A wide variety of treatments has been used to produce G-bands, but the most popular have involved proteolytic enzymes, especially trypsin.

Although the French were among the pioneers of G-banding, they have to a large extent preferred what has become known as R-banding, in which the chromosomes are treated with a very hot buffer solution before Giemsa staining (Dutrillaux & Lejeune 1971). The pattern is the reverse of that given by G-banding, hence the name. R-banding is useful in certain circumstances, particularly for showing up the ends of chromosomes, which are often scarcely visible after G-banding, but the rest of the world has not found it to be as generally useful as G-banding.

In 1973 a method was published for staining nucleolus-organizing regions (NORs) with Giemsa, after suitable pretreatment (Matsui & Sasaki 1973). In practice, this method was little used for the study of NORs, although it has been employed for staining certain heterochromatic regions on plant chromosomes (Ch. 4). It was not until two years later that silver methods were introduced to reveal NORs (Goodpasture & Bloom 1975, Howell *et al.* 1975) and since then a large amount of work has been done on the location, heteromorphisms and activities of NORs on chromosomes of different organisms.

The evolution of methods for staining the kinetochores parallels that of NOR-staining methods. The first modern method for demonstrating kinetochores was the C_d method of Eiberg (1974), but this proved inconsistent in practice, and for routine use was superseded by silver staining (Denton *et al.* 1977). More recently, an immunocytochemical method has been introduced for kinetochores (Moroi *et al.* 1980), but unfortunately it does not work on routine chromosome preparations fixed with acetic acid.

Interest in the mechanisms of chromosome banding methods has led to the development of a number of methods which are mainly of research rather than immediate practical interest. Many of these methods were designed to reveal various features of chromosomal DNA, particularly base composition or sequence. Methods using base-specific fluorochromes have already been mentioned. Immunocytochemical methods were soon developed to detect specific DNA bases (Dev *et al.* 1972c), and more recently they have been used to show the distribution of Z-DNA in chromosomes. Light-induced banding (Mezzanotte *et al.* 1979) is another method designed to demonstrate differences in DNA base composition along chromosomes (Ch. 8). Most recently, restriction endonucleases have been used to induce banding patterns on chromosomes (Ch. 12). Restriction endonucleases cut DNA molecules only at specific sequences. Using different restriction enzymes, a variety of banding patterns has been demonstrated on chromosomes from a variety of species, but as has happened so often in the history of chromosome banding, the mechanisms of induction of these bands has turned out to be more complicated than expected.

Quite early during the modern era of banding it was observed that treatment of cells before fixation with certain substances could induce banding patterns without any special post-fixation treatment (Hsu *et al.* 1973). In particular it has been found that various substances which show base-specific interactions with DNA can inhibit the condensation of certain heterochromatic regions of chromosomes. Nevertheless, the treatment of living cells in culture to induce bands might have remained only of academic interest had not bromodeoxyuridine (BrdU) been used. This base analogue is incorporated into DNA instead of thymine. With appropriate treatment, regions of chromosomal DNA that have incorporated BrdU can be made to stain differently from unsubstituted DNA with a variety of dyes, especially Giemsa. Initially this method was used for studying the replication of whole chromosomes, and the identification of sister chromatid exchanges (Perry & Wolff 1974). If, however, the BrdU was supplied to the cells during only part of the DNA synthetic period of the cell cycle, patterns were produced on the chromosomes corresponding to the stage when particular chromosomal segments were replicating their DNA (Epplen *et al.* 1975). These replication patterns turn out to correspond closely with G-, Q- and R-banding patterns in higher vertebrates, but can also be induced in organisms that do not possess these other types of banding.

Chromosome banding has been of vital importance for mapping the sites of genes on chromosomes, and for identifying the sites of various alterations (such as translocations or deletions) associated with congenital abnormalities or with tumours. When banding techniques were first used, it was not possible to do such mapping with great precision. In a human haploid metaphase chromosome set, there are approximately 300 G-bands; if, as has been estimated, the human genome contains between 30 000 and 50 000 genes, then an average band would contain 100–150 genes. It is, therefore, unlikely that a genetic defect caused by the deletion of one or a few genes would be detectable on a banded metaphase chromosome. It was noticed very early on that less condensed chromosomes displayed more bands. More recently, deliberate attempts have been made to produce chromosome preparations with a large proportion of prophase cells (Yunis 1976), and preparations with up to 2000 bands per haploid set have been reported (Yunis 1981). As a result of this effort, small chromosome deletions associated with birth defects (Yunis 1976) or tumours (Yunis 1982) have been observed. 'High-resolution banding' is rapidly becoming an essential method for investigating chromosome abnormalities.

1.4 Current state of chromosome banding

In the previous section the development of a wide variety of chromosome banding techniques has been described. It is probable that further techniques will be developed in the future. Nevertheless, there is already a comprehensive set of methods that can be used for a variety of purposes.

Certain methods are used primarily for the identification of chromosomes, either to provide assurance that a normal set of chromosomes is present, or to identify a chromosome abnormality associated with a pathological state, or as part of the process of mapping genes to chromosomes, or for comparative purposes in evolutionary studies (Ch. 15). In higher vertebrates, G-, Q- and R-banding techniques, which can give a detailed pattern of bands throughout the length of the chromosomes, are used for this purpose. In all other organisms, in the chromosomes of which such bands cannot normally be induced at present, identification can often be made using patterns of C-bands or of other bands that represent subsets of heterochromatin.

Second, the use of C-banding and other techniques that stain heterochromatic regions of chromosomes has revealed a considerable amount of heteromorphism in chromosomes. That is, heterochromatic segments vary in size within and between individuals of a species. This has been an important field of study in its own right (Ch. 14).

Third, there are methods the results of which reflect the activity or behaviour of chromosomes. These methods are those for NORs (where silver staining is related to the activity of the ribosomal genes; see Chapter 9), and for kinetochores.

Fourth, there is a large category of techniques that are not in routine use for banding chromosomes. Such techniques include immunocytochemical methods for investigating the DNA composition of chromosomal segments, and the use of restriction enzymes to identify segments containing particular DNA sequences. Such methods are used primarily to identify and understand the fundamental chromosome organization underlying the bands.

The subject of chromosome banding is thus a very varied one, ranging from the highly practical to the esoteric. In the following chapters the techniques, applications and mechanisms of the different classes of chromosome bands will be described. For many purposes, clinical cytogenetics or evolutionary studies for example, it may be sufficient to apply a few standard techniques routinely without any immediate concern for how the techniques work or what underlying features of chromosomal organization they demonstrate. Nevertheless, attempts to understand what the bands represent are essential for our knowledge of chromosome organization and behaviour, both normal and abnormal.

2 Classification and nomenclature of chromosome bands

2.1 Introduction

With the development of a range of banding techniques at the beginning of the 1970s, the need for some sort of standard nomenclature became pressing, particularly in the field of human cytogenetics where the new techniques were adopted enthusiastically. Fortunately, the mechanism for establishing such a nomenclature for human chromosomes was already in being, and the report of the Paris Conference in 1971 was only one of a series, still continuing (ISCN 1985), that aims to codify all aspects of the nomenclature of human chromosomes. The Paris report was the first attempt to provide nomenclature for chromosome banding in any species, and thus many of its recommendations have been adopted for non-human species as well.

The nomenclature for chromosome banding proposed by the Paris Conference and its successors is described later in this chapter (Section 2.3). This nomenclature has been generally accepted and used, but it must be realized that it only addresses parts of the problem of classification of chromosome bands. It was (and still is) reasonable to give a distinctive title to each distinctive class of methods for putting bands on chromosomes, but the fact is that different banding techniques may demonstrate the same sort of band (in terms of chromosome organization), while a single technique often shows up two distinct sorts of bands. In some ways it is unfortunate and confusing that names such as G-bands, Q-bands and R-bands have come into use at all, since although they indicate a set of bands produced by a distinctive method, the use of such terms tends to obscure the existence of the biologically different types of bands that form a chromosome.

In this chapter, therefore, two different systems of classification of chromosome bands are presented: first, a classification based on the functional significance of different types of bands in the organization of the chromosome, and second, the classification of the Paris Conference and its successors (ISCN 1985), which is a system based primarily on the techniques used to produce the bands. In addition, the Paris Conference provided a system for designating individual bands, and subsequent reports have dealt with other aspects of the nomenclature related to chromosome banding as the need arose; these points, too, will be described.

2.2 Classification of types of chromosome bands

Four essentially different types of chromosome bands can be found in chromosomes, although some types are not present in all chromosomes (Sumner 1981, 1982). The first type corresponds to constitutive heterochromatin. The second type consists of those bands that occur throughout the length of chromosomes. The third and fourth classes of bands are nucleolar organizers and kinetochores.

2.2.1 Heterochromatic bands

These bands conform to the classical definition of heterochromatin in remaining condensed in interphase. The term 'heterochromatic bands' may, however, indicate a change in our concept of what heterochromatin is. In most cases it has not been shown that heterochromatic bands, as demonstrated by staining techniques, do actually remain condensed in interphase, and thus it has become the practice to describe as heterochromatin those chromosomal regions that stain distinctively with certain banding techniques. This is not quite correct. There are, in fact, reports of chromosomal segments that are heterochromatic by the classical criteria, but which have not yet been demonstrated to stain distinctively with any banding technique (John & King 1977, Camacho *et al.* 1984). Nevertheless, C-banding is generally regarded as demonstrating virtually all the constitutive heterochromatin in the chromosome complement of a species. Several other banding techniques reveal various subsets of heterochromatic bands; some of these involve distinctive pretreatments followed by Giemsa staining (Ch. 4), while others use various fluorochromes, singly or in combination (Ch. 8). Thus heterochromatic bands are also heterogeneous (Rocchi 1982).

Heterochromatic bands have been found in the chromosomes of all organisms where attempts have been made to demonstrate them, although their size and number vary greatly. Nearly all chromosomes possess heterochromatic bands although a few abnormal chromosomes appear to lack them (Levan *et al.* 1978).

It should be noted that heterochromatic bands represent only constitutive, and not facultative, heterochromatin. C-banding and other techniques that stain constitutive heterochromatin distinctively do not produce any characteristic staining reaction with facultative heterochromatin, such as the inactive X chromosome in female mammals. Nevertheless, the inactive X chromosome does stain more darkly than other chromosomes in certain mammals, particularly the mouse; this has been observed in both unbanded (Ohno & Hauschka 1960) and banded preparations (Buckland *et al.* 1971), and a special technique has been developed to accentuate this difference (Kanda 1973).

2.2.2 Bands occurring throughout the length of chromosomes

In certain organisms, mainly amniote vertebrates (reptiles, birds and mammals) a pattern of bands can be revealed throughout the length of the chromosomes, unlike the highly localized heterochromatic bands, nucleolar organizers and kinetochores. So far there has been no entirely satisfactory general name for these bands. A name which has much to commend it is 'euchromatic bands' (P. F. Ambros, personal communication). If the heterochromatic bands represent all the heterochromatin in the genome, then bands in other parts of the chromosome must necessarily be euchromatic. Nevertheless, certain of these bands (positive G- and Q-bands) have been regarded as 'intercalary heterochromatin' (Comings 1972). Nagl (1976) described the same material as 'functional heterochromatin', but it was regarded by Goldman *et al.* (1984) as 'ontogenetic euchromatin'. Clearly there have been a number of different opinions on the nature of this material, and it is not yet possible to give it a definitive title. Nevertheless, for the sake of convenience the term 'euchromatic bands' will be used in this book, to cover both the positively and negatively staining bands demonstrated throughout the length of chromosomes with a variety of techniques. Although 'euchromatic bands' must be regarded as only a provisional title, it is the author's opinion that this nomenclature corresponds best with the use of the terms heterochromatin and euchromatin by Heitz (1928). It also provides a useful contrast with the better established term 'heterochromatic bands'.

The principal techniques that demonstrate this class of bands are G-, Q- and R-banding. All produce a pattern of dark and light bands of various sizes throughout the length of the chromosome; G- and Q-banding produce essentially the same pattern, whereas that produced by R-banding is complementary, that is, it produces dark (positive) bands where there would be negative G- or Q-bands. It is, therefore, not possible to refer to positive bands or to negative bands as a class without referring to the technique used to demonstrate them, an unsatisfactory situation. However, recent developments in our understanding of the nature of these bands (Ch. 16) will almost certainly help in providing a more satisfactory descriptive nomenclature.

The pattern of euchromatic bands resembles that of DNA replication in the same chromosome (Ch. 11), although there seem to be minor differences (Sheldon & Nichols 1981a). It could be that replication bands are equivalent to euchromatic bands or they may represent a separate system of organization within the chromosome, which happens to coincide largely with the system represented by euchromatic bands.

The rarity of euchromatic bands in all organisms except higher vertebrates is something of a mystery (Ch. 16). It seems surprising if there should be a fundamental difference in chromosome organization between higher vertebrates and all other eukaryotes, and indeed with continuing technical improvements, bands have been demonstrated throughout the length of the chromosomes of certain plants, insects, fish and amphibia. So far these are only isolated examples (Ch. 5), but it could turn out that this supposedly fundamental difference is attributable entirely to technical factors.

2.2.3 Nucleolar organizers and kinetochores

The existence of nucleolar organizers and kinetochores as distinct regions of chromosomes is so well established that no further discussion is required at this point. Methods for staining paired bodies at the centromeres date back over 90 years to the first days of chromosome banding (Metzner 1894). Although it is not, in most cases, firmly established that the kinetochores themselves are being stained, there is no doubt that the bodies demonstrated are closely associated with the kinetochores (Ch. 10). The development of staining methods for nucleolar organizers is only slightly more recent (Ruzicka 1899), but their specificity has been thoroughly investigated and is firmly established (Ch. 9).

2.3 Classification of banding methods

As mentioned in Section 2.1, the need for a standardized nomenclature for chromosome banding was first met by the report of the Paris Conference in 1971. This report defined a band as 'a part of a chromosome which is clearly distinguishable from its adjacent segments by appearing darker or lighter' with various banding methods. Further, it was decided that 'The chromosomes are visualized as consisting of a continuous series of light and dark bands, so that by definition there are no "interbands"'. It is quite clear that this was the correct decision so far as euchromatic bands were concerned, such bands being either positively or negatively stained. However, this principle was irrelevant for the highly localized heterochromatic bands, and indeed (to quote the report of the Paris Conference) 'the C-band corresponds in position to the centromeric region'; in other words, it is a specific part of the chromosome, and the remainder of the chromosome is not regarded as a 'negative C-band'.

2.3.1 Nomenclature of banding methods

The Paris Conference quite clearly adopted a nomenclature for bands based on the method used to produce them. Thus C-bands were those stained by 'C-staining methods', Q-bands those stained by 'Q-staining methods' and so on. At that time, only four main classes of banding methods were available, and these were defined as in Table 2.1. Subsequently, T-bands, which are a subset of R-bands (Dutrillaux 1973), were added to the list.

Table 2.1 Definition of banding methods, according to Paris Conference (ISCN 1985).

Banding method	Definition
C	Those methods which demonstrate 'constitutive heterochromatin'. A C-band is a unit of chromatin stained by these methods
Q	Methods using quinacrine mustard or quinacrine to demonstrate bands along chromosomes
G	Techniques which demonstrate bands along the chromosomes using the Giemsa dye mixture
R	The 'reverse-staining Geimsa method' which gives patterns opposite in staining intensity to those obtained by the G-staining methods

2.3.2 A banding alphabet

Following the lead given by the Paris Conference report, it has become customary to refer to bands produced by different techniques by one or more letters. Most of these designations are unnecessary, some have been applied to unreproducible techniques, and the same letter has been used (by different people) to refer to essentially different phenomena. A process of natural selection has eliminated most of these terms, and only those that have proved useful have survived. Nevertheless, since all these terms are in the literature, and may be encountered from time to time, they are listed in Table 2.2. It must be emphasized that this list is probably incomplete, and does not include terms used by authors to denote specific techniques rather than a class of bands.

Apart from the terms recommended by the Paris Conference report and its Supplement (Section 2.3.1), a number of the terms given in Table 2.2 are worthy of further notice, and can be regarded as useful. C_d banding (Eiberg 1974) demonstrates 'centromere dots', paired bodies at the centromere which possibly represent the kinetochores (Ch. 10). G-11 banding (Bobrow *et al.* 1972b) was initially proposed as a method for staining a subset of C-bands in human chromosomes, but has achieved greater use for differentiating between rodent and primate chromosomes in hybrid cells (Section 4.3). The term thus denotes not only a staining method (Giemsa solution at pH 11), but also a distinctive staining pattern. Hy-banding (Greilhuber 1973) is a type of heterochromatin staining in plants; it cannot strictly be regarded as a subset of C-bands, as it stains (positively or negatively) regions of chromosomes that do not respond to C-banding techniques. Finally, N-banding (Matsui & Sasaki 1973) was introduced as a Giemsa staining method for nucleolar organizers. For this purpose it has been largely superseded by silver staining, but the method has also proved useful for staining certain heterochromatic bands, not associated with NORs, in plants and insects. As now used, N-banding refers to a staining technique rather than to a class of bands as described in Section 2.2, but it still remains a useful term.

Table 2.2 A banding alphabet.

Type	Description	Reference
A	Inconsistent bands produced on aged slides	Crossen 1974
C	**Heterochromatin staining**	ISCN 1985
C_d	**Centromere dots (kinetochores)**	Eiberg 1974
CT	Combination of C- and T-banding	Scheres 1974
D	Fluorescent bands with daunomycin or adriamycin (equivalent to Q)	C. C. Lin et al. 1975
D	Chromosomal regions showing DNase I hypersensitivity	Kerem et al. 1984
E	G-bands produced by enzymic digestion	Lejeune 1973
EM	Structural bands observed by electron microscopy	Bahr et al. 1973
F	All types of fluorescent bands (with suffix to indicate type and reagent)	Hsu 1976
G	**Giemsa-stained euchromatic bands**	ISCN 1985
G-11	**Differential staining produced with Giemsa solution at pH 11**	Bobrow et al. 1972b
H	Fluorescent bands with Hoechst 33258	Pimpinelli et al. 1976a
Hy	**A class of heterochromatic bands produced by acid treatment**	Greilhuber 1973
N	**Giemsa staining for nucleolar organizers**	Matsui & Sasaki 1973
O	Trypsin/orcein banding in plants (heterochromatic bands)	Lavania & Sharma 1979
Q	**Fluorescent bands with quinacrine**	ISCN 1985
R	**Reverse-Giemsa banding**	ISCN 1985
T	**Predominantly terminal bands**	Dutrillaux 1973
W	G- and C-bands stained with Wright's stain	de la Maza & Sanchez 1976

Recommended terms are printed in bold.

2.3.3 Subclassification of banding methods

The Supplement to the Paris Conference report (ISCN 1985) proposed a three-letter code to describe banding techniques, in which 'the first letter denotes the type of banding, the second letter the general technique and the third letter the stain'. This classification is reproduced in Table 2.3. The possibility of unlimited expansion of this scheme was envisaged in the statement that 'Any new triplet should be defined in the text of the publication in which it is first used' although few, if any, additions have been made in fact.

This three-letter code was described under the heading 'Description of heteromorphic chromosomes' and it is clear from both the context and the examples given that the code was for use in the description of heteromorphic chromosomes (Section 2.4.2). It has, however, come into more general use as a shorthand for describing the banding technique used in a particular study, whether or not heteromorphic chromosomes were involved. The use of this

Table 2.3 Three letter code to describe banding techniques (from the Supplement to the Paris Conference Report, 1971, published by The National Foundation) (ISCN 1985).

Code	Banding technique
Q	Q-bands
QF	Q-bands by fluorescence
QFQ	Q-bands by fluorescence using quinacrine
QFH	Q-bands by fluorescence using Hoechst 33258
G	G-bands
GT	G-bands by trypsin
GTG	G-bands by trypsin using Giemsa
GAG	G-bands by acetic saline using Giemsa
C	C-bands
CB	C-bands by barium hydroxide
CBG	C-bands by barium hydroxide using Giemsa
R	R-bands
RF	R-bands by fluorescence
RFA	R-bands by fluorescence using acridine orange
RH	R-bands by heating
RHG	R-bands by heating using Giemsa
RB	R-bands by BrdU
RBG	R-bands by BrdU using Giemsa
RBA	R-bands by BrdU using acridine orange
T	T-bands
TH	T-bands by heating
THG	T-bands by heating with Giemsa
THA	T-bands by heating with acridine orange

three-letter code in any circumstances is to be regretted. The usual scientific convention that methods should be described in sufficient detail for the work to be repeated by another should always be adhered to (even though 'sufficient detail' may only be a reference). Admittedly, in many cases it may be adequate merely to say that the chromosomes were G-banded, Q-banded etc., but particularly in studies on chromosome heteromorphisms, and certainly if chromosomes of a hitherto unstudied species are being investigated, it is necessary to give full experimental details. A three-letter code cannot provide such details as time and temperature of treatment and concentrations of reagents, all of which may influence the demonstration of particular banding patterns. If some method of denoting the type of banding is required in descriptions of heteromorphic chromosomes, a single letter (C, G, Q or R) should suffice, the method used having been described in detail elsewhere in the paper.

2.4 Chromosome band nomenclature

The report of the Paris Conference also introduced a system of designating chromosome bands in human chromosomes. Such a system is necessary for various purposes: designating break-points in rearranged or deleted chromosomes, and defining the location of genes on chromosomes, for example. Subsequently the system of nomenclature proposed for human chromosomes has been extended to other primates and various other mammals, for which standard karyotypes have been published (Table 2.4). A somewhat similar system was put forward for mouse chromosomes (Evans 1989). In principle, such systems could be applied to any species, but there is only a demand for them in those species in which intensive programmes of cytogenetical investigation and detailed gene mapping are carried out.

A special problem arises with certain heterochromatic regions of chromosomes, which are heteromorphic. A complete description of such bands requires some method of describing their size or staining intensity or both. The nomenclature for such cases is described in Section 2.4.2

2.4.1 System for numbering bands

The system for numbering chromosome bands proposed by the Paris Conference, and adopted as standard for human chromosomes starts by dividing

Table 2.4 Species for which standard karyotypes have been published.

Species	Reference
Mouse (*Mus musculus*)	Evans 1989
Chinese hamster (*Cricetulus griseus*)	Ray & Mohandas 1976
Deer mouse (*Peromyscus* spp.)	Committee 1977
Norway rat (*Rattus norvegicus*)	Committee 1973
Rabbit (*Oryctolagus cuniculus*)	Committee 1981
Ox (*Bos taurus*)	Ford *et al.* 1980
Sheep (*Ovis aries*)	Ford *et al.* 1980
	Committee 1985b
Goat (*Capra hircus*)	Ford *et al.* 1980
Pig (*Sus scrofa*)	Ford *et al.* 1980
	Gustavsson 1988
Horse (*Equus caballus*)	Ford *et al.* 1980
Cat (*Felis catus*)	Ford *et al.* 1980
Blue fox (*Alopex lagopus*)	Committee 1985a
Silver fox (*Vulpes fulvus*)	Committee 1985c
Cebus apella paraguayanus	Matayoshi *et al.* 1986
Orangutan (*Pongo pygmaeus*)	ISCN 1985
Chimpanzee (*Pan troglodytes*)	ISCN 1985
Gorilla (*Gorilla gorilla*)	ISCN 1985
Man (*Homo sapiens*)	ISCN 1985

*Full titles of the committees are given in the *bibliography*.

Figure 2.1 Diagrammatic representation of human chromosome bands, showing the numbering of the bands devised by the Paris Conference. Reproduced from ISCN (1985) by permission of Professor D. G. Harnden and S. Karger AG.

the short and long arms (already designated p and q respectively) into a number of regions, each defined as any area of a chromosome lying between two adjacent landmarks. Landmarks were defined as consistent and distinct morphological features important for identifying chromosomes. (Actually landmarks, like other bands, are features of staining rather than morphological features in the strict sense of the word.) The regions were numbered consecutively from the centromere to the telomere on each arm; within each region the individual bands were numbered in the same direction. Thus the complete designation of a band would consist of the chromosome number, a letter to indicate the short or long arm, a number for the region and a number for the band (e.g. 4q26). The complete system for the human karyotype is illustrated in Figure 2.1.

It was realized at the time of the Paris Conference report that bands might require to be subdivided, and provision was made for this simply by adding, after a full stop, further figures from 1 upwards to indicate subbands, and if necessary sub-subbands. Thus band 10p12 could be divided into 10p12.1, 10p12.2 and 10p12.3, and the last further divided into 10p12.31, 10p12.32 and 10p12.33. The wisdom of this provision for almost unlimited expansion became apparent with the development of high-resolution banding (Section 5.5.8) and a 1981 report of the standing committee on Human Cytogenetic Nomenclature (ISCN 1985) was concerned entirely with the numbering of bands in prophase chromosomes (Fig. 2.2).

The system proposed for numbering bands on mouse chromosomes (Evans 1989) does not differ in principle from that for human chromosomes, but uses letters instead of numbers to designate the main regions of the chromosomes. Also, since all mouse chromosomes lack short arms, no letter is necessary to distinguish short and long arms.

2.4.2 *Method for describing heteromorphic regions of chromosomes*

Certain heterochromatic regions of human chromosomes were observed to vary in their appearance between homologues and between individuals. Such chromosomes, and the variable bands themselves, are referred to as heteromorphic. Although of course these variable bands can be numbered according to the system described in Section 2.4.1, it is clearly desirable to be able to describe their size and staining intensity in detail. The system proposed in the 1975 Supplement to the Paris Conference report is shown in Table 2.5. Little comment is called for. In the absence of actual measurement, such a system is clearly required, and five grades of size and staining intensity is probably as many as can be distinguished by eye. It must be admitted, however, that some of the descriptive terms for intensity of quinacrine fluorescence are unhappily chosen: intense and brilliant really amount to much the same thing, and the use of pale to denote something which is really rather dark is misleading. Much more helpful terms could have been chosen. Nevertheless, the whole system of nomenclature proposed by successive standing committees

NOMENCLATURE 19

on Human Cytogenetic Nomenclature (brought together in ISCN 1985) has proved to be a sure foundation for the progress not only of human cytogenetics, but also the cytogenetics of other species, in the banding era.

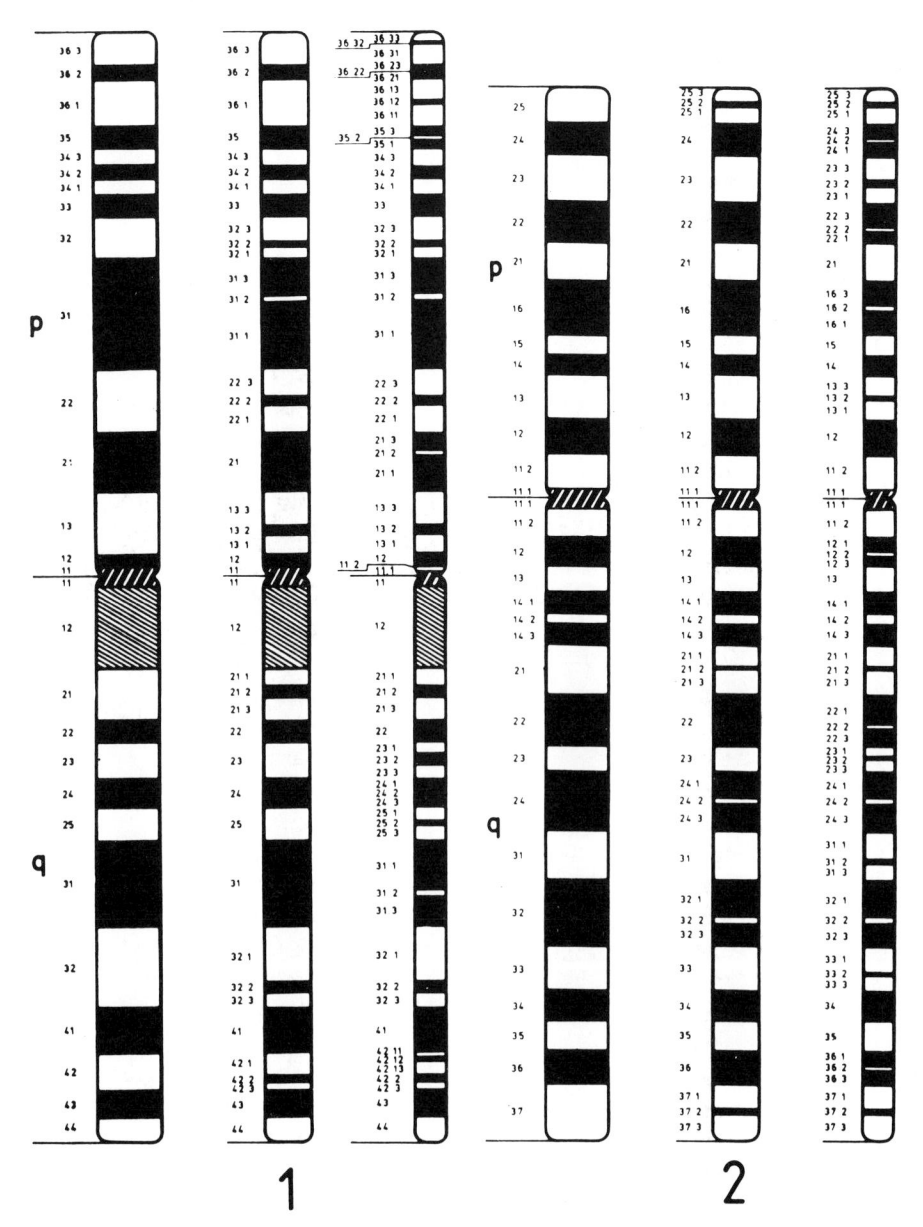

Figure 2.2 High-resolution banding patterns and their numbering system for human chromosomes 1 and 2 at the 400-, 550- and 850-band stage. Reproduced from ISCN (1985) by permission of Professor D. G. Harnden and S. Karger AG.

Table 2.5 Numerical description of size and intensity of variable bands (from the Supplement to the Paris Conference Report, 1971, published by The National Foundation) (ISCN 1985).

Size	Intensity
Q-banding	
1 Very small	1 Negative (no or almost no fluorescence)
2 Small	2 Pale (as on distal 1p)
3 Intermediate	3 Medium (as the two broad bands on 9q)
4 Large	4 Intense (as the distal half of 13q)
5 Very large	5 Brilliant (as on distal Yq)
C-banding	
1 Very small	
2 Small	
3 Intermediate	
4 Large	
5 Very large	

3 Structure and composition of chromosomes

3.1 Introduction

Banding is a phenomenon induced in chromosomes by treating them in certain ways, and therefore it is necessary for a full understanding of chromosome bands to understand the structure and composition of unbanded chromosomes. Since bands are not random, but form patterns characteristic of specific chromosomes, it is axiomatic that they must represent some feature or features characteristic of normal chromosome organization. It is the purpose of this chapter to describe those general features of chromosome organization that are necessary for a proper interpretation of banding.

Although some types of bands can be demonstrated on unfixed chromosomes, it is essential for the production of most types of bands that the chromosomes be fixed in a specific way. We shall therefore also consider the changes induced in chromosomes by fixation.

3.2 Constituents of chromosomes

Chromosomes are complexes of nucleic acids and proteins. DNA is, of course, the essential component of the chromosome, which may be regarded simply as a device for carrying the genetic information from a parent cell to its daughters. All the other components of the chromosomes may be regarded as subservient to this purpose, with the exception of a few which appear to be adventitious. These other components are RNA, at least some of which merely seems to be carried on mitotic chromosomes, and the two classes of chromosomal proteins, histone and non-histone. Both these classes can be divided further into a number of distinct categories, as will be described shortly. Evidence for the presence of carbohydrates in chromosomes is tenuous, and the same can be said of lipids, apart from the fragments of nuclear membranes that can often be seen adhering to the telomeric regions of isolated chromosomes (Daskal *et al.* 1978). Various ions are also components of chromosomes.

No attempt is made here to provide a comprehensive account of the various constituents of chromosomes, as the material is covered amply by a large number of books and reviews (e.g. Comings 1972, Bostock & Sumner

1978, Adolph 1988, van Holde 1989). Instead, only a brief outline will be given, emphasizing those aspects of the various chromosomal components that are most relevant to the understanding of different types of chromosome banding.

3.2.1 DNA

DNA normally occurs as a double-stranded polymer (although the strands must separate for replication to occur), the two strands being held together by hydrogen-bonding (Fig. 3.1) between complementary bases (adenine + thymidine, A + T, or guanine + cytosine, G + C). The double-stranded DNA molecule can be denatured (separated into single strands) by a variety of agents such as heat, acid or alkali. When returned to a favourable environment, the separated strands can come together and re-form the original double-stranded molecule (annealing). The ease with which DNA can be denatured, particularly by heat, depends strongly on its base composition, DNA rich in A + T denaturing more easily than G + C-rich DNA.

The properties of a DNA molecule are affected in another way by its base composition. DNA molecules can be fractionated by centrifugation in caesium chloride (and similar) density gradients, the molecules richer in G + C having greater buoyant density. If the DNA from a eukaryote species is treated in this way, the distribution of DNA on the gradient often shows one or more satellites (Beridze 1986) as well as the main band. Satellite DNAs are therefore fractions that differ in density, and therefore in base composition, from the norm for that species. Satellite DNAs usually consist of a short sequence of DNA bases repeated very many, often millions of, times with little variation. The unit of repetition may vary from two up to a few hundred bases. With very short units of repetition, it is not surprising

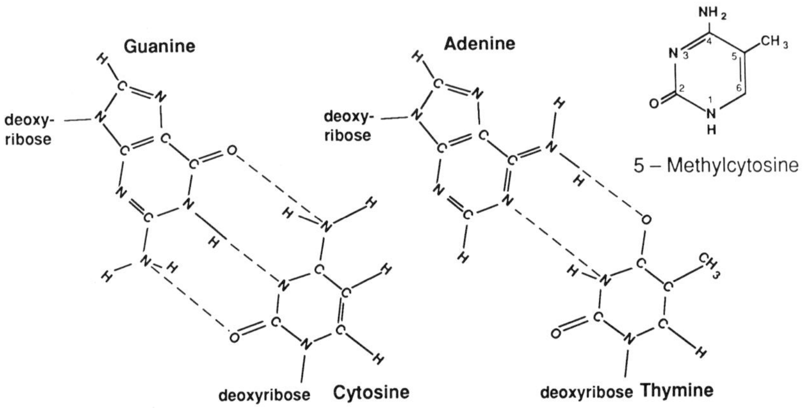

Figure 3.1 Structural formulae of the four main bases in eukaryotic DNA, showing the hydrogen bonding (broken lines) between guanine and cytosine and between adenine and thymine. The modified base 5-methylcytosine is also shown.

that a skewed base composition can occur, although not all repetitive DNA fractions have base compositions notably different from that of main band DNA. Another consequence of a short unit of repetition is that the base composition may differ markedly between the two strands of DNA (Flamm *et al.* 1969); indeed, in the highly repetitive DNA of some species a particular base may be absent from one strand (Southern 1970).

The usual method of studying the sequence complexity of DNA is by observing the rate of reassociation of the denatured (separated) single strands. The rate of reassociation is proportional to the concentration of the DNA sequences, so clearly if total DNA from an organism is denatured and allowed to anneal, DNA sequences present in a large number of copies will anneal faster than those present only as a single copy. Experiments of this sort indicate the existence of four classes of DNA (Kao 1985). (1) Inverted repeats anneal virtually instantaneously, occur in thousands of copies per genome, and their function is obscure. (2) Highly repetitive or rapidly annealing DNA has already been mentioned; it consists of relatively short sequences repeated a large number of times. It commonly, but probably not exclusively, occurs in large blocks, and is often equated with satellite DNA. (3) The intermediate fraction, or moderately repetitive DNA, consists of sequences repeated between 100 and 100 000 times, and appears to comprise a number of families of sequences, generally of considerably greater length than those of highly repetitive DNA. Moderately repetitive sequences appear to be scattered throughout the genome rather than being concentrated in blocks. (4) Unique DNA sequences are those that are present only once in the haploid genome. This class includes the majority of genes, of course, but by no means all unique DNA consists of genes.

As mentioned above, DNA contains four bases; one of these, cytosine, is often modified after a new DNA molecule has been synthesized, to form 5-methyl cytosine (5-MeC) (Fig. 3.1). Although this is a post-synthetic modification, the methylation of particular sites in DNA molecules is heritable, methylation occurring predominantly at sites that contained 5-MeC before replication. A large body of evidence indicates that methylation of cytosine is associated in some way with repression of gene activity (Cooper 1983, Razin & Cedar 1984).

The two complementary strands of native (double-stranded) DNA are wound round each other to form a double helix. The common form of the double helix, not only in aqueous solution, but probably also in cells (Hamilton *et al.* 1959, Hanlon *et al.* 1972) is a right-handed helix, generally B-DNA, although recent research suggests that the form of the DNA helix is considerably more variable than was once supposed, being influenced by base sequence and DNA-protein interactions (van Holde 1989). In recent years it has been shown that DNA may also, in certain circumstances, form a left-handed helix, Z-DNA (Rich *et al.* 1984, Jovin *et al.* 1983). In solution, the Z-form of DNA is generally only stable in strong salt solutions, reverting to the B-form if the salt concentration is lowered. However, there seem to be two situations in which Z-DNA is stable at physiological salt concentrations: if the DNA contains methylated cytosine, and if the DNA is supercoiled. Both these situations are biologically significant, and Z-DNA has indeed been detected in chromosomes (Section 13.2.4).

3.2.2 RNA

Metaphase chromosomes have a higher ratio of RNA to DNA than interphase chromatin (Bostock & Sumner 1978), but the values obtained for this ratio are highly variable, and there is no doubt that a substantial amount of non-chromosomal RNA can become attached to chromosomes during isolation procedures (Mendelsohn 1974).

RNA is generally produced and processed in the nucleus, and much of it is exported to the cytoplasm. On the whole, therefore, it does not appear to be a permanent constituent of chromosomes. The main classes of RNA are ribosomal, transfer, messenger, heterogeneous nuclear and small nuclear RNAs. Some RNA, probably ribosomal RNA, becomes attached to the surface of chromosomes and is carried through mitosis in this way (Pailweletz & Risueño 1982).

3.2.3 Histones

The histones are a group of small, basic proteins that are highly conserved in evolution, are confined to nuclei and chromosomes, and are present in a quantity roughly equal to that of DNA. They are readily extracted with dilute acids or strong salt solutions. Five types of histones are normally present in eukaryotic chromosomes: H1, which is lysine-rich; H2A and H2B, slightly lysine-rich; and H3 and H4 which are arginine-rich. H1 differs from the other histones in having a larger molecule, and consisting of several different but closely related fractions. Although the histones were formerly believed to have an important role in the control of transcription, it is now clear that they are structural proteins (Section 3.3.1). All the histones are subject to modification by methylation, acetylation, phosphorylation or ADP-ribosylation, and in addition H2A and H2B may also combine with the polypeptide ubiquitin (Wu *et al.* 1986). It has been suggested that such modifications might be involved in functions such as transcription or chromosome condensation, but conclusive evidence is lacking as yet.

3.2.4 Non-histone chromosomal proteins

All chromosomal proteins other than histones tend to be lumped together as non-histone chromosomal proteins (NHCP), although this term covers several distinct classes of proteins, the members of which show a wide range of molecular sizes and amino acid compositions. There are at least several hundred different NHCPs (Peterson & McConkey 1976) although only about 15–20 major ones (Elgin & Weintraub 1975). Together these make up the bulk of mitotic chromosomes, being up to three or four times as abundant as DNA (Bostock & Sumner 1978).

Several classes of NHCPs can be recognized in interphase nuclei, although probably not all of them are present in chromosomes. These classes include structural proteins, contractile proteins, enzymes, DNA-binding proteins, proteins concerned with regulation of gene activity, hormone receptor proteins and the HMG (high mobility group) proteins (van Holde 1989). Few members of any of these classes have been characterized adequately, and it would be inappropriate to attempt to give more than a sketchy account here. Contractile proteins such as actin and myosin have often been reported as components of chromosomes (Comings & Harris 1975, Douvas et al. 1975), and their presence is attractive, as providing a mechanism for chromosome condensation. Nevertheless, it has been claimed that such proteins are artefacts, absorbed during isolation, and indeed this is a perennial problem in the study of chromosomal proteins. Many enzyme activities have been associated with interphase chromatin (Elgin & Weintraub 1975). However, it is clear that several of these enzymes would not be required during mitosis, and while it is possible that the enzymes may simply be held on the chromosomes in an inactive state, it seems equally possible that they could be lost at mitosis. The same sort of uncertainties apply to proteins concerned with gene regulation, although specific proteins could be required for the overall suppression of gene activity that occurs during cell division. DNA-binding proteins that affect DNA conformation could conceivably be very important in mitotic chromosomes, but again detailed information is lacking. Finally, the HMG proteins (Johns 1982) are a group of small proteins rich in both basic and acidic amino acids, and which may be functionally diverse. Certain of them (HMG 14 and 17) may be concerned in maintaining the structure of transcriptionally active chromatin (Einck & Bustin 1985), while HMG 1 may modulate chromosome condensation (Kohlstaedt et al. 1987).

3.2.5 Inorganic ions

Extraction of inorganic ions, especially divalent cations, from chromosomes causes them to swell, and this swelling can be reversed if the appropriate ions are restored (Hadlaczky et al. 1981a, Zelenin et al. 1982). At a finer level, the presence or absence of divalent cations affects the appearance of chromatin fibres (Ris 1975). There is thus good evidence that certain inorganic ions have an important structural function in chromosomes. Potassium and chloride are the most abundant ions in chromosomes; there is a moderate amount of sodium, and relatively small quantities of magnesium and calcium (Cameron et al. 1979, Warley et al. 1983).

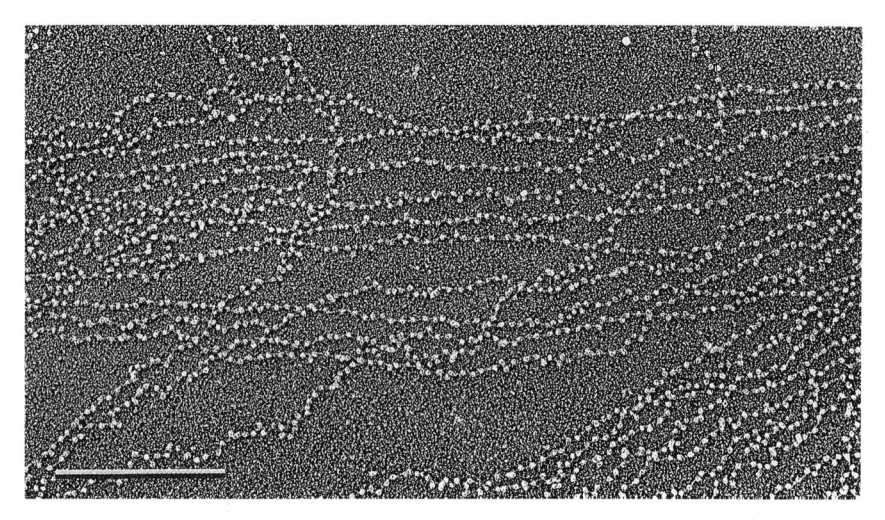

Figure 3.2 Electron micrograph of chromatin fibres from chicken erythrocytes showing nucleosomes as 'beads on a string'. The bar represents 0.5μm. Reproduced from Zentgraf *et al.* (1987) in *Electron microscopy in molecular biology*, 81–100. (J. Sommerville & U. Scheer, eds) Oxford University Press.

3.3 Assembly of chromosomal constituents into chromosomes

The major problem in the construction of chromosomes is how to pack an enormous length of DNA into a manageable volume. The amount of DNA in the human haploid genome (and in that of most other mammals) is approximately 3pg. This is equivalent to about one metre of linear DNA molecule, which has to be packed into a chromosomal length of the order of 100μm at metaphase. The ratio of the total length of DNA to that of the chromosomes is known as the packing ratio, and in the example just given is about 10 000. Although we have now a fairly clear understanding of the lower orders of packing, the situation regarding the higher orders of chromosome organization, which may be of greater relevance to the interpretation of banding, is much less clear.

3.3.1 Nucleosomes

A variety of evidence has shown that the first level of packing of DNA is into bodies called nucleosomes (Pederson *et al.* 1986, van Holde 1989). These consist of 146 base pairs (bp) of DNA wrapped round an octameric histone bead which contains two molecules each of histones H2A, H2B, H3 and H4. These beads are connected to each other by a short length of linker DNA, the whole appearing as 'beads on a string' in electron micrographs (Fig. 3.2.). The total length of DNA per nucleosome, including linker, is variable but is in the region of 200 bp. The DNA not wrapped around the histone bead is in some way compacted by histone H1, and certain HMG proteins may also be bound to the linker (Einck & Bustin 1985). The diameter of nucleosomes with DNA wrapped round them is about 10 μm, and there seems little doubt that the 10 nm fibres seen by electron microscopy are made up of strings of nucleosomes. The packing ratio for DNA organized into nucleosomes is about seven (Griffith 1975). In highly active genes, histones are still present but apparently not organized into nucleosomes, but less active genes appear to retain the nucleosomal structure (Yaniv & Cereghini 1986).

3.3.2 *Solenoids*

Fibres of 10 nm are seen in electron micrographs of chromosomes prepared by methods that use chelating agents, which remove divalent cations. More often the chromosome fibres are in the region of 20–30 nm in diameter. These thicker fibres are produced by the coiling of the thinner, 10 nm nucleosomal fibre into a 'solenoid' (Widom & Klug 1985, Walker & Sikorska 1987); as well as divalent cations, these solenoids apparently require histone H1 to stabilize them (Finch & Klug 1976). The formation of solenoids results in a total packing ratio for the DNA of between 40 and 50 : 1 (Finch & Klug 1976). An alternative suggestion for the packing of nucleosomal fibres into 30 nm fibres by the formation of 'superbeads' no longer attracts much support, although the superbeads may represent an intermediate stage in the coiling of the 10 nm fibre into the 30 nm solenoid (Walker & Sikorska 1987).

3.3.3 *Higher-order packing of chromosome fibres*

The DNA of a human haploid genome, packed into solenoids 30 nm in diameter, would still be about 2 cm in length, and it is clear that it must be compacted very much further. Over the years there have been several suggestions as to how this compaction might be accomplished, although so far none has achieved unquestioned acceptance. In a general sense it is probably variations along the chromosome in the higher-order organization that are responsible for many aspects of chromosome banding.

When considering how the 30 nm fibres might be arranged in a chromosome, it is necessary to bear in mind certain points. The first is that the chromosomal material is arranged in a definite, fixed order. This is shown by the essential constancy of banding patterns within a species (Burkholder & Comings 1972, Nesbitt & Donahue 1972), and also by the substantial evidence for a fixed order of genes on chromosomes. Second, there is now an overwhelming body of evidence that chromosomes are uninemic; that is, the unreplicated chromosome contains only a single DNA molecule, and not two or more arranged side by side (Comings 1972). Third, the chromosomes contract substantially during prophase, so that any proposed structure must be able to change in length.

Since the two lowest orders of packing of DNA into chromosomes are achieved by coiling, it is reasonable to suggest that the remainder of the packing might also be accomplished in the same way (Bostock & Sumner 1978). It is indeed true that a spiral structure has been observed in the chromosomes of many species of plants (Manton 1950, Kaufman *et al.* 1960), and can be induced in the chromosomes of certain animals by appropriate treatment (Ohnuki 1965, Rudak & Callan 1976). Some eletron micrographs of chromosomes also show some sort of spiral structure (Fig. 3.3). However, although a structure based on a hierarchy of coils would meet satisfactorily all the requirements mentioned in the previous paragraph, the majority of chromosome preparations show no sign of a spiral structure, and indeed may show structures apparently incompatible with a spiral arrangement.

In fact, electron micrographs of whole chromosomes usually show no regular structure above that of the 30 nm fibre (Fig. 3.4). The fibres appear to form a tangle, and at the periphery of the chromosome they loop back into the main mass of fibres; free ends are not seen. Observations of such electron micrographs of chromosomes, and careful measurements of the quantity of 30 nm fibres within them, led DuPraw (1970) to propose his 'folded-fibre' model of chromosome organization. The essentials of this model were that the chromosome is composed of nothing but chromatin fibres, that there is a single fibre per chromatid, and that this fibre is folded in

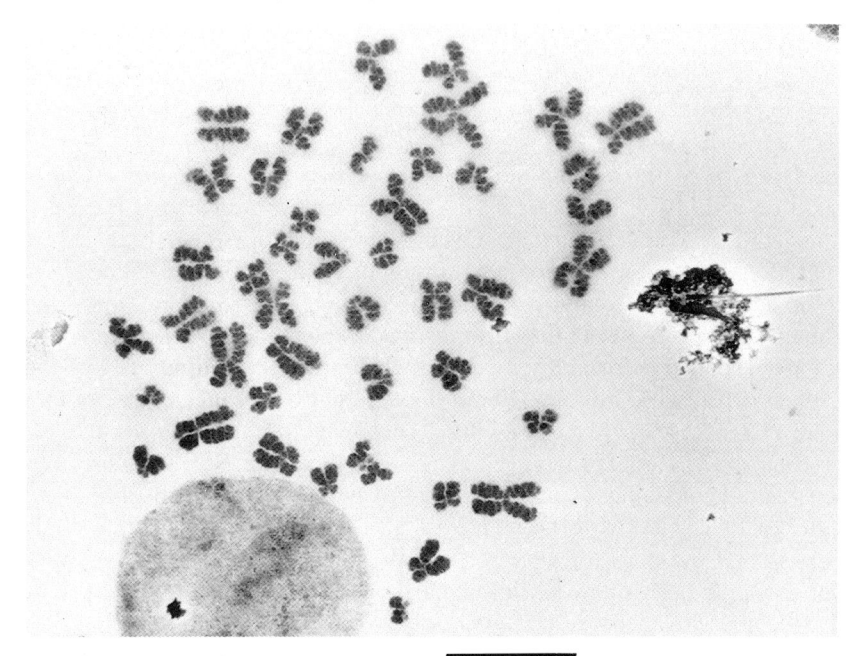

Figure 3.3 Electron micrograph of human chromosomes, showing spiral structure of chromatids.

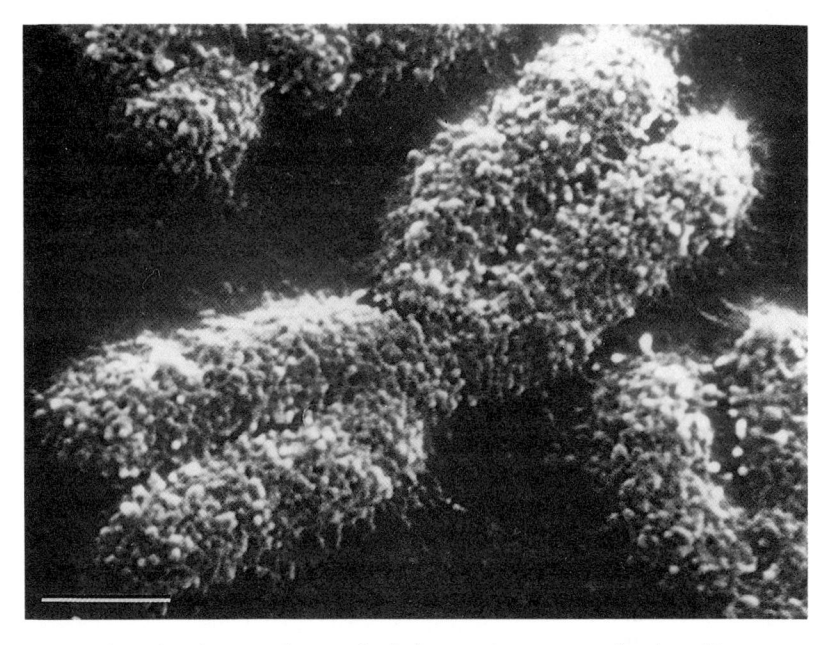

Figure 3.4 Scanning electron micrograph of a human chromosome, showing a fibrous appearance without any other substructure. The bar represents 1 μm.

various ways to form a typical chromosome. The method of folding would vary at different stages during the mitotic cycle, and even between homologous chromosomes in different cells. Although such a model is consistent with the morphological appearance of chromosomes in electron micrographs, it does not appear to be consistent with the longitudinal organization that can be demonstrated by banding.

Probably the most widely accepted model of chromosome structure today is that based on a core or scaffold from which radiate loops of chromatin fibres. Although such structures had been predicted and were illustrated many years ago (Bostock & Sumner 1978), it was not until the publications of Laemmli and his colleagues (Adolph *et al.* 1977, Paulson & Laemmli 1977) that this type of chromosome organization became widely accepted. In essence, when histones are extracted from isolated chromosomes, and the resulting material spread on an aqueous surface, objects are found which have the general shape of a chromosome, but usually a more open structure, from which radiate numerous loops of DNA (Fig. 3.5). The central structure has been named the scaffold. A core-like structure in conventionally fixed metaphase chromosomes can be stained with silver after certain pretreatments (Howell & Hsu 1979). A chromosome core is obviously a very satisfactory method for maintaining linear order in chromosomes, and the scaffold model has much to commend it. Apart from this, it is not certain that the model has much direct relevance to banding, which could well be

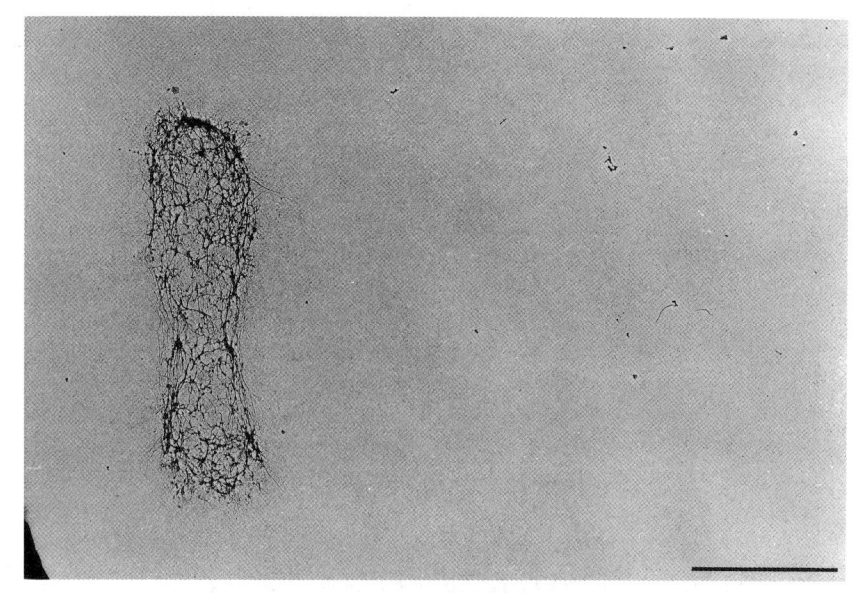

Figure 3.5 Transmission electron micrograph of a dehistonized CHO chromosome, showing a central chromosome-shaped scaffold surrounded by a halo of DNA fibres. The bar represents 5 µm. Reproduced from Hadlaczky et al. (1981b) by permission of the authors and Springer-Verlag.

mediated by proteins released during the histone extraction procedures (which also remove non-histones). Nevertheless, some sort of pattern analogous to banding has been observed on chromosome scaffolds, but has not been investigated in detail (Hadlaczky et al. 1981a).

Nearly all the chromosomal DNA is, according to the scaffold model, arranged in large loops surrounding the scaffold itself. These loops appear to be equivalent to the domains described for interphase nuclei, which are regarded as fundamental units of chromatin organization (Pienta & Coffey 1984). These domains average about $20\mu m$ of DNA, but of course they will be greatly condensed when complexed with proteins to form a 30 nm fibre (Section 3.3.2). Electron micrographs of cross-sections of chromosomes that have been slightly swollen in certain buffers (Fig. 3.6) show fibres radiating from a central mass (Marsden & Laemmli 1979, Adolph 1981). These radiating fibres are presumably the condensed loops of DNA. In the Radial Loop model (Marsden & Laemmli 1979), chromosomes are composed of loops of nucleoprotein fibres radiating from a central axis.

Some form of radial loop/scaffold model appears to be the best explanation of chromosome structure at the present time, but it does not yet offer an adequate explanation of all relevant chromosomal phenomena. In its simplest form, it does not take account of the sporadic observations of spiral structure in chromosomes, but it appears that the chromosome axis or scaffold may assume a spiral form (Mullinger & Johnson 1980, Rattner &

Lin 1985, Boy de la Tour & Laemmli 1988). One objection to the scaffold model has been that no such structure is visible in sectioned chromosomes (Okada & Comings 1980); however, it seems likely that the scaffold is a diffuse structure rather than a solid rod, and is therefore unlikely to be conspicuous in sections (Earnshaw 1988, Paulson 1989). A greater difficulty seems to be that chromosome breaks can be induced simply as a result of a double-strand break in DNA (Obe *et al.* 1982); the organization of the chromosome fibre into discrete loops suggests that a break in one would not affect the integrity of the chromosome as a whole. This observation would be much more easily explained if the chromosome were organized as a hierarchy of spirals. Certain observations on the condensation of chromosomes (Section 3.4) do not at present fit either model well, and serve to emphasize that current ideas of chromosome structure must be regarded as provisional.

3.3.4 Distribution of chromosomal constituents along chromosomes

A simple explanation of chromosome banding might be that the constituents of chromosomes are not uniformly distributed along chromosomes. Evidence on this point is conflicting. Caspersson *et al.* (1969b) and Sumner *et al.* (1973)

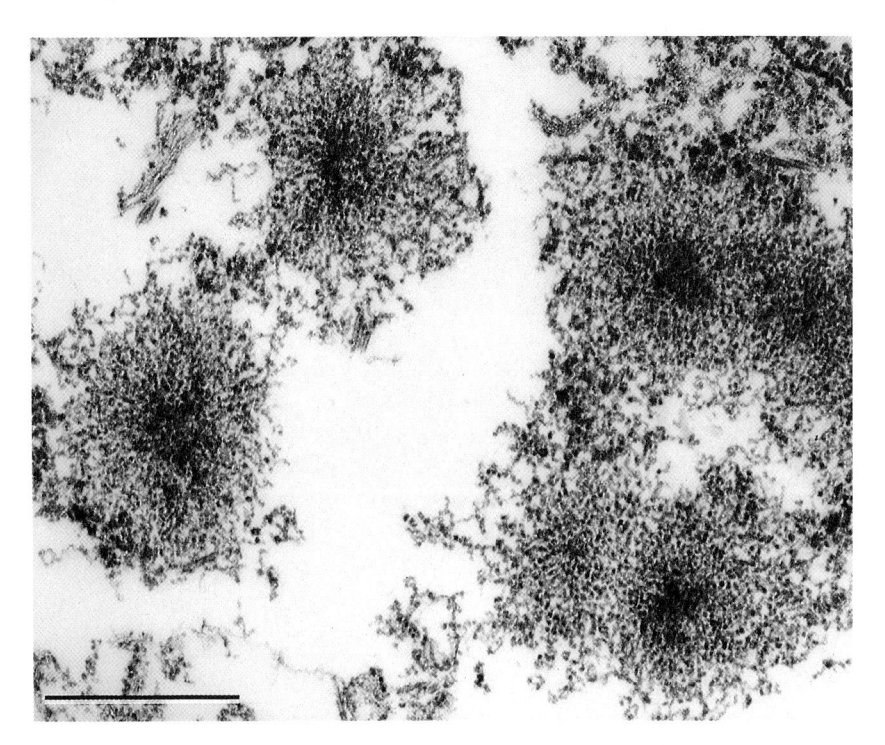

Figure 3.6 Transmission electron micrograph of CHO chromosomes swollen before fixation in buffer containing 1 mM $MgCl_2$ according to Adolph (1981), to show radial fibre structure. The bar represents 1 μm.

found that DNA appears to be uniformly distributed along chromosomes. On the other hand, McKay (1973) and Yunis & Sanchez (1973) found a pattern of DNA distribution on mouse chromosomes that corresponded to G-bands, while Prescott *et al.* (1973) found a greater concentration of DNA in heterochromatic regions of kangaroo rat (*Dipodomys ordii*) chromosomes. Bahr & Larsen (1974) found a correspondence between distribution of dry mass and banding patterns in electron micrographs of whole chromosomes, although many such micrographs do not show any obvious pattern. Zatsepina *et al.* (1988) showed that a structural pattern corresponding to G-banding could be induced in mammalian chromosomes by hypotonic treatment. In the chromosomes of rye, DNA was shown not to be uniformly distributed, and its distribution was similar to that of the chromomeres of pachytene chromosomes (Heneen & Caspersson 1973). Finally, it should be noted that there is a close correspondence between G-band patterns of mitotic chromosomes, and the patterns of chromomeres on meiotic prophase chromosomes at pachytene (Ferguson-Smith & Page 1973, Okada & Comings 1974, Luciani *et al.* 1975).

If chromosome bands are an inherent part of chromosome organization, it would not be surprising that they should be represented as varying amounts of DNA or protein along the chromosome. What can be said with confidence, however, is that any such variations must generally be small, otherwise it would have been unnecessary to devise special banding techniques. The question still remains whether such techniques merely exaggerate a pre-existing structural pattern, or whether they demonstrate some other form of differentiation that is not expressed structurally. Much of this book is concerned with this problem.

It is difficult to explain why some investigators find that DNA and other chromosomal constituents are uniformly distributed, while others find that their distribution is related to banding patterns. Differences in the species of origin of the chromosomes, or in the methods of preparing them, may explain some of the differences. More detailed studies than those carried out hitherto, using a variety of techniques, could well yield valuable information on this point. A further possibility is that different appearances may be produced by chromosomes at different stages of contraction, the more extended prophase chromosomes perhaps showing a non-uniform distribution of material, while metaphase chromosomes exhibit a uniform degree of contraction throughout. This point will be considered in more detail in the next section.

3.4 Changes during chromosome condensation

Information on the changes that occur during the condensation of interphase chromatin to form metaphase chromosomes is unfortunately rather meagre and fragmentary, and it is not possible to give a satisfactory account of the process. Although there are clearly many differences, both structural and chemical, between interphase and metaphase chromosomes, it is often not known precisely when the transitions occur. This is unfortunate, since it has been recognized for many years that the more extended prophase chromosomes show many more euchromatic bands than fully condensed metaphase chromosomes (Seabright et al. 1975, Sumner 1976); indeed it has been proposed that banding is a manifestation of the condensation process (Sumner 1976). Clearly, a good understanding of the processes involved in chromosome condensation might be of great importance for understanding chromosome banding.

Studies with the light microscope indicated that the condensation of chromosomes was due to the tighter coiling of the chromonema, a chromosomal fibre about $0.2\mu m$ in diameter (Manton 1950). Although electron micrographs have been published which claim to show a similar coiled chromonema in metaphase chromosomes (Ris & Korenberg 1979, see also Sect. 3.3.3), such a structure is not apparent in most electron micrographs of chromosomes. Golomb & Bahr (1974b) describe what appears to be a completely different pattern of condensation of chromosome fibres. At the earliest stages of condensation, configurations appear consisting of parallel fibres with accumulations of tortuous looping fibres at the distal parts of each chromatid. Subsequently such accumulations of fibres are seen all along the chromatids, and are regarded as equivalent to the chromomeres seen by light microscopy. In fully condensed chromosomes, the longitudinal fibres are no longer visible and only the tortuous looping fibres are seen. Similar appearances have been described in prophase chromosomes (Yunis & Bahr 1979), and comparable observations on prematurely condensed chromosomes have been made by Gollin et al. (1984). The process of condensation described by Golomb & Bahr (1974b) appears to be more compatible with the production of something resembling a banding pattern than a process of progressively tightening the coiling of the chromonema. There may, however be a closer relationship between the two processes than might appear at first sight, since it has been claimed that banding patterns are related to patterns of chromosome coiling (Kato & Yosida 1972, Goradia & Davis 1977, Rønne 1977c), and indeed that the former is derived from the latter (Takayama 1976).

Chromosome condensation is associated with an increase in non-histone proteins, phosphorylation of histone H1, conversion of protein sulphydryl groups to disulphide cross-links, cessation of RNA synthesis, and accumulation of RNA on the chromosomes (Bostock & Sumner 1978). Only two of these changes seem to have a direct connection with the process of condensation. A substantial amount of phosphorylation of H1 takes place at the end of interphase and beginning of prophase (Bradbury et al. 1973, Gurley et al. 1973, 1974), although the nature of the connection with chromosome condensation is not yet established (Wu et al. 1986, van Holde 1989).

A substantial body of evidence suggests that in decondensed chromatin, free protein sulphydryls are predominant, while in condensed chromatin and metaphase chromosomes, disulphide cross-links are formed (Sumner 1983). This change apparently occurs in both histone H3 and in non-histone proteins (Ord & Stocken 1966, Sadgopal & Bonner 1970). A role for disulphide bonds in chromosome condensation is easy to imagine; possibly the disulphide bonds stabilize a structure which has been condensed by some other means.

It has previously been noted (Section 3.2.5) that increasing concentrations of inorganic cations can cause condensation of isolated chromosomes. Although there is some evidence that their concentration changes during the cell cycle (Cameron *et al.* 1979, Warley *et al.* 1983), there is little evidence that a change in concentration of inorganic ions causes chromosome condensation in intact living cells (Rao & Johnson, 1974, Henry *et al.* 1980).

3.5 Effects of fixation on chromosomes

Chromosome preparations to be used for banding are normally fixed. Although it is possible in some cases to demonstrate banding on unfixed chromosomes (Buys *et al.* 1984b), fixation is generally employed for a number of reasons. Historically, methods of preparing chromosomes for microscopical examination have involved fixation. Methods of preparing good quality unfixed whole metaphase plates are even now not widely available, whereas good quality metaphase preparations can be made routinely from appropriately fixed material. Perhaps most importantly, fixation appears to be necessary if certain types of chromosome bands are to be produced (Hancock & Sumner 1982, Sumner 1982). In this context, fixation means the use of a mixture of acetic acid with either alcohol or water; other types of fixative have not been shown so far to be compatible with the induction of bands. It is clearly necessary for a proper understanding of banding methods to know the nature of the substrate on which those methods act. Fixation with acetic acid is in fact a somewhat destructive process, and indeed acetic acid fails to fix many constituents of cells (Baker 1958). Thus the fixed chromosomes to which banding techniques are applied are different in composition and organization from living chromosomes.

Apart from embedding and sectioning pieces of tissue, the classical way of making metaphase chromosome preparations is by squashing. Fixation is commonly in a 3 : 1 mixture of ethanol and acetic acid, the material then being transferred to 45% acetic acid for squashing on a slide under a coverslip. There are problems in squashing plant material because of the rigid cell walls, and thus more drastic treatment is needed. Hydrochloric acid treatment, hot or cold at various concentrations, is usually used to break down the cell walls (e.g. Filion 1974, Döbel *et al.* 1978), but more recently cellulase and pectinase have been used (Friebe 1977, Tanaka & Ohta 1982). In some cases all or part of the staining procedure is also carried out before squashing (Greilhuber 1973).

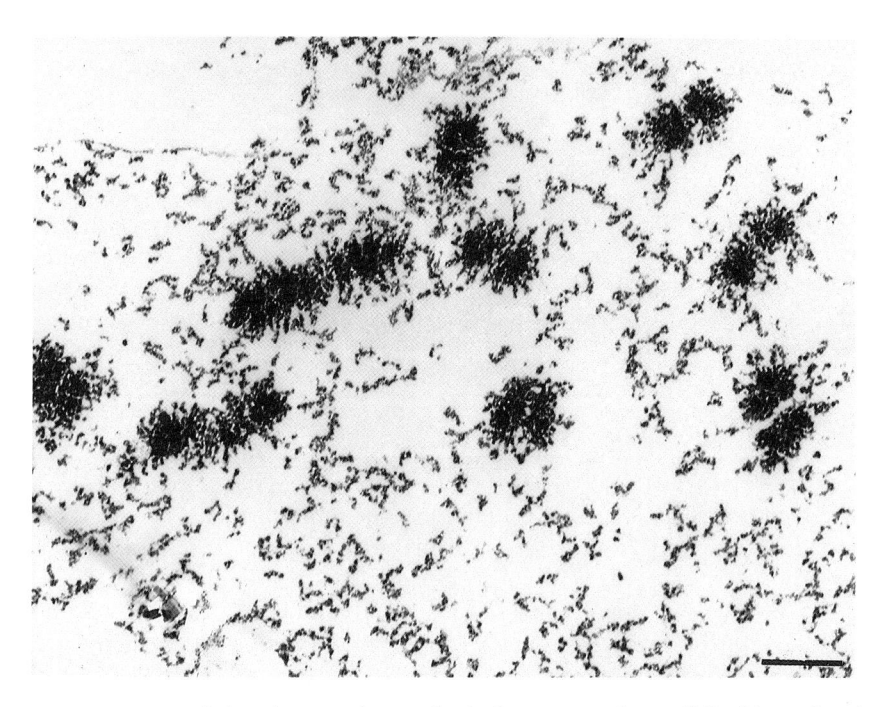

Figure 3.7 Transmission electron micrograph of a human metaphase cell fixed in methanol-acetic acid, showing good preservation of chromosome structure but damaged membranes and extensive extraction of cytoplasm. The bar represents 1 μm.

The methods generally used for preparing mammalian chromosome spreads, and increasingly chromosome spreads from other organisms, are completely different. Isolated cells in suspension are swollen with a hypotonic solution, fixed in suspension with a 3 : 1 mixture of methanol and acetic acid, and a drop of the fixed suspension dropped on to a slide. As the fluid spreads across the slide and dries, the cells burst, leaving nuclei and chromosome spreads which in the best preparations are essentially devoid of cytoplasm. This basic 'air-drying' procedure was introduced for human lymphocyte cultures (Moorhead et al. 1960), but is obviously applicable to any types of cells that can be obtained in suspension. Cultured cells can be brought into suspension by treatment with trypsin, and cells from solid tissue by mechanical disruption (Harnden 1974). The technique has also been extended to plant material, in which the cell walls are removed enzymically, and the resulting protoplasts treated in the same way as animal cells in suspension (Mouras et al. 1978).

Acetic acid has little fixative action on proteins, swells tissues and cells, and leaves them in a soft condition, but is a powerful precipitant of nucleic acids (Baker 1958). These properties no doubt contribute greatly to the value of fixatives containing acetic acid in preparing chromosomes by either the squash or air-drying techniques; only the chromosomes are fixed whereas all the surrounding material, if not actually dissolved by the fixative, is left in a condition in which it can easily be squashed or dispersed. Electron micrographs of metaphase cells which have been fixed in methanol-acetic acid show very dense chromosomes, but little other material remaining in the cell (Fig. 3.7).

It has long been recognized that acetic acid can extract histones, one of the major components of chromosomes, from cells (Baker 1958), and more recently the extent of extraction of proteins during fixation has been investigated in some detail (Dick & Johns 1968, Brody 1974, Sivak & Wolman 1974, Pothier *et al.* 1975, Retief & Rüchel 1977, Burkholder & Duczek 1980a, 1982a, b, Hancock & Sumner 1982). Different studies have used different starting material (e.g. isolated chromatin, isolated nuclei or isolated chromosomes) some of which is not a precise model for the fixation of chromosomes. As well as the more commonly used methanol-acetic acid, 45% aqueous acetic acid and 3 : 1 ethanol-acetic acid have been tested. It is therefore not surprising that estimates of the amount of histone extracted during fixation vary from 7% to 90%. Nevertheless, there does appear to be a consensus that histone H1 is most susceptible to extraction, while the other (core) histones may be retained in substantial amounts. Those histones not extracted during fixation are apparently firmly bound to the chromosomes and are not readily removed by further acid treatment (Pothier *et al.* 1975). The distribution of histones remaining in fixed chromosomes tends to be inhomogeneous and may resemble banding patterns (Pothier *et al.* 1975, Turner 1982, Section 13.3.1).

Some non-histone chromosomal proteins are also extracted during fixation (Sivak & Wolman 1974, Burkholder & Duczek 1980a, 1982a, b), but

Figure 3.8 SDS-polyacrylamide gel electrophoresis patterns of (b) proteins remaining in chromosomes after fixation with methanol-acetic acid, and (c) proteins extracted from chromosomes by methanol-acetic acid. Lane (a) contains molecular weight markers. Reproduced from Hancock & Sumner (1982) by permission of the authors and The Faculty Press.

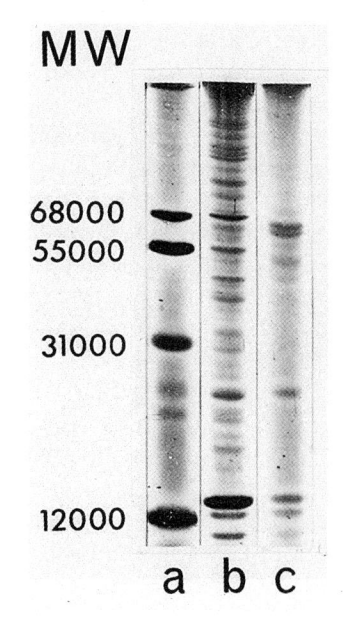

this has been investigated much less thoroughly than the loss of histones. It appears, however, that there is some degree of specificity in this process, the gel electrophoresis pattern of the extracted proteins (Fig. 3.8) differing markedly from that of the proteins remaining in the chromosomes (Hancock & Sumner 1982).

Losses of DNA during fixation with alcohol-acetic acid fixatives are small. Earlier work suggested that such losses were trivial (Sandritter & Hartlieb 1955), but more recent measurements show DNA losses of between 7% and 15% during fixation of isolated nuclei (Franke et al. 1973, Matsui 1974). The DNA remaining in nuclei and chromosomes after fixation is altered from the native state. Acid treatment might be expected to denature and depurinate the DNA, and introduce nicks into it. Rigler (1966) showed that the majority of the DNA is denatured when human chromosomes are squashed in aqueous mixtures of acetic acid. However, methanol-acetic acid fixation causes little denaturation of DNA in chromosomes (Mace et al. 1972, Sumner et al. 1973, Kurnit 1974, Mezzanotte et al. 1989).

Depurination of DNA also occurs during fixation with acetic acid (Sederoff et al. 1975, Holmquist 1979) but the amount is apparently small. There is too little depurination to produce a visible Feulgen reaction, or to cause denaturation of DNA. However, standard C-banding methods require a certain degree of depurination (Section 4.2.5) and this can sometimes be provided solely by fixation. Holmquist (1979) pointed out that depurination could in principle be prevented by using anhydrous fixation, but even with alcohol-acetic acid mixtures this is difficult to arrange because of the water content of the material being fixed. Depurinated sites can be converted to nicks by alkali, but there seems to be no evidence that acetic acid fixation itself nicks DNA. However, chromosomal DNA can become nicked before fixation, apparently by an endogenous nuclease (Stickel & Clark 1985). Nicking of DNA also occurs during prolonged storage in methanol-acetic acid (Nishigaki et al. 1988), and after the chromosomes have been spread on the slides (Mezzanotte et al. 1988).

Acetic acid fixation also affects the interactions between DNA and proteins. Treatment of chromosomes with aqueous or methanolic acetic acid destroys the nucleosomal structure of the chromatin fibres (Kuo 1982, Hutchison & Weintraub 1985), and causes irreversible changes to the interactions between DNA and proteins. This is not simply a matter of extraction of proteins; in fact the proteins are more tightly bound to the DNA than before fixation (Kuo 1982).

To summarize this section, chromosomes fixed with mixtures containing acetic acid retain most of their DNA, although it suffers slight depurination. The DNA is predominantly double-stranded after alcohol-acetic acid fixation, but is partly denatured by aqueous acetic acid. Once the chromosome preparations are spread on slides, further denaturation (Mezzanotte et al. 1989) as well as nicking, can occur. Losses of protein are more extensive, and appear to involve specific non-histones and all types of histones. Some of the histone is retained, however, although H1 is more labile than the others. The interactions between the DNA and protein are changed, and the nucleosomal structure is destroyed. Acetic acid fixation does not appear to produce gross morphological changes in chromosomes (Cowden & Curtis 1973); indeed, a number of studies have made use of acetic acid fixation in the preparation of chromosomes for electron microscopy (Ford et al. 1968, Ris 1978, Harrison et al. 1981, 1985a, b, Mullinger & Johnson 1983, Welter & Hodge 1985). Not only is the gross structure comparable to that of chromosomes prepared in other ways, but chromatin fibres of 20–30 nm diameter are preserved.

One point which cannot be over-emphasized is the great variability of chromosome preparations fixed with acetic acid. We have already seen that according to different investigators the amount of histone extracted during fixation varies from almost none to almost all. A similar degree of variability in chromosome preparations is familiar to practising cytogeneticists; some chromosome preparations spread beautifully, others poorly; some band easily, others with difficulty. It is a common observation that poor banding is associated with poor spreading, and both are often associated with the retention of cellular protein round the chromosomes. Although a systematic study of these points has yet to be made, it is a reasonable deduction that poor banding is associated with inadequate extraction of certain chromosomal components, presumably proteins. Chromosomes derived from different tissues also differ in morphological appearance and in 'bandability'. Thus, although we know in a general way the sort of modifications that occur when chromosomes are fixed, these modifications take place to differing extents in different organisms, in different individuals, in different tissues, and on different parts of the same slide. Both practical cytogenetics and the study of chromosome banding mechanisms would benefit if the fixed chromosome were a standardized object, but it is only realistic to accept that is not. This variable, unstandardized object is the basic material for all the work to be discussed in the remainder of this book.

4 C-banding and related methods

4.1 Introduction

In this and the following chapters the various different classes of banding techniques will be considered in turn, and in general according to a common plan. First, the methods will be described, with most detail given for the method or methods that are regarded as most useful. This will be followed by a consideration of the characteristics of the bands demonstrated, the mechanism of action of the methods, so far as they are known, and finally a section on applications. It would not be appropriate to devote a single chapter to methods for heterochromatic bands, another to methods for euchromatic bands and so on, partly because a single method can reveal more than one type of band, but also because of the great variety of methods which can be used to demonstrate the different classes of bands. For these reasons the present approach of describing different classes of methods in each chapter has been adopted.

There are, nevertheless, some topics to which banding techniques have made vital contributions, and it is felt that these topics should be discussed separately. Thus can a more coherent view be presented of the contribution of a variety of classes of banding methods to such subjects as heteromorphism (Ch. 14) and chromosome evolution (Ch. 15).

The present chapter will cover methods for demonstrating heterochromatic bands, of which the principal is C-banding. Other methods for demonstrating subsets of C-bands are included; some are minor variants of C-banding, while others (G-11, and certain methods for plant chromosomes, for example) are based on different principles. Also included here is N-banding; although introduced as a method for showing nucleolar organizers (Matsui & Sasaki 1973), it has been used widely for demonstrating certain heterochromatic bands in insects and plants. Fluorescence methods that demonstrate certain heterochromatic bands will, however, be dealt with elsewhere (Ch. 7 & 8). Yet other methods that reveal heterochromatic bands are described in Chapters 11, 12 and 13.

4.2 C-banding

It is not easy to define what is meant by a C-banding method. C-banding methods are undoubtedly those that produce C-bands, which in turn are most satisfactorily defined as those bands produced by C-banding methods. Although C-bands are held to represent constitutive heterochromatin, it appears that in some cases there are heterochromatic bands not demonstrated by C-banding methods (John & King 1977, Camacho *et al.* 1984), which in some cases can be stained using other methods (Section 4.4 & 4.5). Moreover, there are few species of which it can be said that bands revealed by C-banding methods conform to the classical definition of heterochromatin (Heitz 1928); that is, they remain condensed in interphase. In most cases the matter simply has not been studied.

If we wish to regard C-bands as those produced by C-banding methods, it must be understood that they are not a homogeneous set of methods (Section 4.2.1). However, when several completely different methods give similar results on a species that is well known cytogenetically (e.g. man, *Vicia faba*), it is reasonable to regard all as C-banding methods. Many methods have not been so tested, and in this chapter such methods are also regarded as C-banding if they resemble established C-banding methods in their principles.

4.2.1 C-banding methods

Modern C-banding methods can take as their evolutionary origin the classic observation of Pardue & Gall (1970), who carried out *in situ* hybridization of satellite DNA to mouse chromosomes, and found stronger staining of the centromeric regions. The essentials of this method were treatment with 0.2_N hydrochloric acid, digestion with RNase to remove any RNA attached to the chromosomes, denaturation of the DNA with 0.07_N sodium hydroxide, and incubation of the preparation with the labelled complementary nucleic acid in saline-citrate ('2 × SSC', 0.3_N NaCl + 0.03_N tri-sodium citrate). With the omission of the complementary nucleic acid, this procedure was soon used for C-banding human chromosomes (Arrighi & Hsu 1971, Chernay *et al.* 1971). Various simplified and modified methods using sodium hydroxide treatment have been described over the years for both plant and animal chromosomes. Hydrochloric acid treatment was found by most workers to be unnecessary. Sodium hydroxide treatment proved to be destructive, and lower concentrations were proposed (Gagné *et al.* 1971, Alfi & Menon 1973), and some authors used saline or buffer brought to a moderately alkaline pH (Chuprevich *et al.* 1973, Dallapiccola & Ricci 1975). The use of an alcoholic sodium hydroxide solution has also been described (Hansen-Melander *et al.* 1974). The subsequent saline treatment was also modified, involving increases in concentration to 6 × or 12 × SSC (Gagné *et al.* 1971, Sinha *et al.* 1972), and prolonged incubation (Sinha *et al.* 1972, Voiculescu *et al.* 1972), usually at 60–65°C. Extremes included 12 × SSC at 100°C for one minute (Alfi & Menon 1973), Hanks' solution at room temperature (Chuprevich *et al.* 1973), or in some cases no saline incubation at all (Dallapiccola & Ricci 1975, Noda & Kasha 1978). Although methods using sodium hydroxide have now been almost entirely superseded, they provided some valuable data on C-banding patterns, and led directly to less destructive methods. Moreover, during their development, a number of significant observations were made. Alfi & Menon (1973) found that best results were obtained when slides had been aged for two to three weeks after preparation. Chuprevich *et al.* (1973) and Dallapiccola & Ricci (1975) showed that the number of C-bands demonstrated varied with the length of incubation in saline, while Voiculescu *et al.* (1972) and Noda & Kasha (1978) reported that their methods showed only certain C-bands, and not others which had been described using other techniques.

The solution to the destructiveness of sodium hydroxide was the substitution of barium hydroxide (Sumner *et al.* 1971a). This modification proved so successful that the BSG (Barium hydroxide/Saline/Giemsa) technique (Sumner 1972) can be regarded as the standard C-banding method for virtually all plant and animal chromosomes. This method, which will be described in detail in the next section (4.2.2), consists of treatment with 0.2N hydrochloric acid, 5% barium hydroxide at 50°C, 2 × SSC at 60°C, and Giemsa staining. Although many variants have been published, important modifications have been few. The preliminary hydrochloric acid treatment has been omitted for both insect (Gallagher *et al.* 1973) and plant (Vosa & Marchi 1972a) chromosomes. Phosphate buffer has occasionally been used instead of 2 × SSC (Stack & Clarke 1973, Döbel *et al.* 1978), and this stage has even been reduced to nothing more than a short wash in deionized water (Tanaka & Taniguchi 1975). In general, however, the 2 × SSC treatment has been found best.

Various authors have obtained C-banding without any alkaline treatment, using only hot saline treatment. It was originally supposed, by analogy with *in situ* hybridization, that the purpose of the alkaline treatment in C-banding was to denature the chromosomal DNA, and that of the subsequent saline treatment to anneal it. Logically therefore, any treatment that would denature DNA should suffice. Both Yunis *et al.* (1971) and Polani (1972) used a hot (85 + °C) saline treatment, followed by longer treatment at 62–65°C. Treatment at two temperatures proved to be unnecessary. Takagi (1971) used distilled water at 92°C, and Chamla & Ruffié (1976) Hanks' solution at 91°C, but more commonly temperatures of 60–65°C have been used, with a variety of salt solutions (Eiberg 1973, McKenzie & Lubs 1973, Fiskesjö 1974). Friebe (1977), using a technique of this type on plant chromosomes, commented that the C-bands demonstrated were not necessarily the same as those shown using a barium hydroxide method.

C-banding has been induced by enzyme treatment, although such techniques are probably of more interest in revealing the mechanism of

C-banding than as practical methods. Production of C-bands by deoxyribonuclease (DNase) actually has quite a respectable antiquity (see also Section 12.2), having been used on sectioned material by Yamasaki (1961). Alfi *et al.* (1973), who used the enzyme on human chromosome spreads, commented that their method was not suitable for routine diagnostic use. Nevertheless, Kato *et al.* (1974) used DNase to demonstrate heterochromatic bands in chromosomes of the Indian muntjac which were different from the C-bands shown by the BSG technique. It is presumed that in all these cases DNase I was used, although the authors do not state this. Reference must also be made at this point to the use of restriction endonucleases to reveal certain classes of C-bands, a subject considered fully in Section 12.3. Trypsin has also been used to produce C-bands (Ray & Hamerton 1973, Merrick *et al.* 1973). Lavania & Sharma (1979) used acetic-orcein to stain plant chromosomes after trypsin digestion, and unnecessarily named the result 'O-bands'. Özkinay & Mitelman (1979) published a method which could be used to demonstrate G-bands or C-bands or a combination of the two, depending on the length of trypsin treatment.

A variety of other methods, which can only be described as miscellaneous, has been used to produce C-bands. Dev *et al.* (1972a) used incubation in a formamide solution. A similar method was used by Marshall (1975) to induce C-bands in the chromosomes of mice, but not of rat, man or hamster, after trypsin G-banding. McKay (1973) and Döbel *et al.* (1978) found urea solutions to be efficacious.

This survey of C-banding methods indicates that a wide variety of methods has been used to demonstrate C-bands on chromosomes. If some of the more historical work is included, the variety becomes even wider (Levan 1945, 1946, Yamasaki 1959, 1965). It is not surprising that during the early years of the modern development of C-banding, in the early 1970s, numerous methods should evolve only to be eliminated by stringent natural selection. Nevertheless, it is clear that some methods have been devised for particular purposes, such as use on difficult material. A constantly recurring theme is that not all C-bands react in the same way, and certain methods have been proposed because they show either a subset of C-bands in a particular species, or a different set of bands from those demonstrated by a 'conventional' C-banding method. These observations serve to emphasize the point that C-bands are not a homogeneous class. We shall see that different C-bands contain different types of DNA (Section 4.2.3, Table 4.1), and later sections of this chapter will describe certain special methods that have been designed to reveal specific bands within the general class of C-bands. The theme is continued in later chapters, particularly those dealing with fluorochrome methods (Ch. 7 & 8). These facts must obviously have a bearing on our interpretation of the nature of C-bands, and of the mechanisms behind the methods used to demonstrate them.

Table 4.1 Some examples of types of DNA found in C-bands

Species	Type of DNA	Location	Reference
Plants:			
Monocotyledons			
Scilla siberica	G+C-rich satellite	Centromeric and telomeric C-bands	Timmis *et al.* 1975
Scilla spp.	G+C-rich satellites	Telomeric and interstitial but not centromeric C-bands	Deumling & Greilhuber 1982
Secale cereale	Satellite DNA	Mainly telomeric C-bands	Appels *et al.* 1978
Dicotyledons			
Vicia faba	Light satellite DNA	Heterochromatic bands	Cionini *et al.* 1985
Animals:			
Nematoda			
Parascaris spp.	Highly repetitive DNA	C-bands	Goday & Pimpinelli 1984
Insecta (Orthoptera)			
Atractomorpha similis	Satellite-1	Telomeric C-bands	Miklos & Nankivell 1976
Insecta (Diptera)			
Drosophila melanogaster	Highly repetitive DNA	Centromeric heterochromatin	Jones & Roberston 1970
D. hydei	Highly repetitive DNA	Mostly in X chromosome; little in heterochromatic Y	Hennig 1972
D. nasuta	A+T-rich satellite	Heterochromatic regions of all chromosomes	Ranganath *et al.* 1982
D. nasutoides	Four A+T-rich satellites	Heterochromatin of chromosome 4 but not in other heterochromatin	Wheeler *et al.* 1978
D. virilis	Three A+T-rich satellites	Centromeric heterochromatin of all chromosomes	Gall *et al.* 1971
Rhynchosciara	Two A+T-rich satellites	Centromeric heterochromatin of all chromosomes	Eckhardt & Gall 1971
Sarcophaga bullata	Highly repetitive DNA	C-bands of C & E chromosomes but not in other chromosomes	Samols & Swift 1979
Amphiba			
Plethodon cinereus	G+C-rich satellite	Centromeric heterochromatin	Macgregor & Kezer 1971
Aves			
Coturnix coturnix japonica	G+C-rich satellite	C-bands of microchromosomes	Brown & Jones 1972
Gallus domesticus	Two satellites	C-bands of microchromosomes and W chromosome	Stefos & Arrighi 1974

Table 4.1 (*cont.*)

Species	Type of DNA	Location	Reference
Mammalia (Marsupialia)			
Macropus rufogriseus	Two satellites	C-bands of all autosomes and centromeres of sex chromosomes; not in other C-bands of sex chromosomes	Dunsmuir 1976
Mammalia (Primates)			
Cercopithecus aethiops	α-satellite	Centromeric heterochromatin	Kurnit & Maio 1973
	β-satellite	Concentrated in C-bands	Kurnit & Maio 1974
	γ-satellite	Centromeric C-bands	
Saimiri sciureus	A+T-rich satellite	In telomeric but not centromeric C-bands	Lau *et al.* 1977
Great apes and man	Several satellites with complex relationships	In most but not all C-bands	See text
Mammalia (Rodentia)			
Akodon molinae	No satellite	C-bands present	Catala *et al.* 1981
Cavia porcellus	Satellites I, II and III	Centromeric and telomeric C-bands but not on C-banded Y	Duhemel-Maestracci *et al.* 1979
Cricetulus griseus	Repetitive DNA	Not in C-banded long arm of X, entire Y, and centromere of 10; present in all other C-bands	Arrighi *et al.* 1974
Dipodomys ordii	HS-β satellite (G+C-rich)	Most centromeric C-bands	Prescott *et al.* 1973
	HD satellite	C-banded short arms; centromeres that lack HS-β	
Mus musculus	A+T-rich satellite	Centromeric C-bands	Pardue & Gall 1970, Jones 1970
Neotoma spp.	Highly repetitive DNA	Not concentrated in C-bands	Mascarello & Hsu 1976
Peromyscus eremicus	G+C-rich satellite	C-banded short arms of chromosomes	Hazen *et al.* 1977
Mammalia (Artiodactyla)			
Bos taurus *Capra hircus* *Ovis aries*	G+C-rich satellites	Centromeric C-bands	Kurnit *et al.* 1973, 1978
Mammalia (Cetacea)			
Balaenoptera spp.	Satellite $\rho = 1.702$	Centromeric C-bands of a few chromosomes	Árnason *et al.* 1978
	Satellite $\rho = 1.710$	Terminal C-bands	
		Many centromeric C-bands lack satellite	

4.2.2 BSG method

The BSG (Barium hydroxide/Saline/Giemsa) method was introduced by Sumner (1972), and has become, with only trivial modifications, the standard method for producing C-bands on plant and animal chromosomes. The procedure is as follows:

1. Standard chromosome preparations are treated with 0.2N hydrochloric acid for 1 h at room temperature, followed by a rinse with deionized or distilled water.
2. The slides are placed in a freshly prepared 5% aqueous solution of barium hydroxide octahydrate ($Ba(OH)_2.8H_2O$) at 50°C for about 2 – 5 min, followed by thorough rinsing with deionized or distilled water.
3. Slides are incubated for 1 h at 60°C in 2 × SSC (0.3 M sodium chloride containing 0.03 M tri-sodium citrate), followed again by a rinse with water.
4. The chromosome preparations are stained with Giemsa. Gurr's Giemsa R66 (BDH), 1 ml in 50 ml of buffer, pH 6.8, made with Gurr's buffer tablets, is recommended, and the slides should be stained for about 45 min.
5. After a final rinse with distilled water the slides are blotted, allowed to dry thoroughly, and mounted in a suitable mountant.

Results of the application of this method to the chromosomes of animals and plants are illustrated in Figures 4.1 and 4.2.

A number of comments on the method are called for. The BSG method was designed for human chromosomes from lymphocyte cultures fixed in methanol-acetic acid (3:1), but works satisfactorily with any chromosome preparation fixed conventionally in a fixative containing acetic acid. C-banding can also be induced in unfixed preparations (Sumner, unpublished), but not on chromosomes fixed in aldehydes (Hancock & Sumner 1982).

In the original description of the BSG method it was noted that the hydrochloric acid treatment could be omitted, particularly with good preparations in which very little cytoplasm surrounded the chromosomes. Many workers have omitted this stage and obtained completely satisfactory results (e.g. Gallagher et al. 1973, Vosa & Marchi 1972a). Nevertheless, acid treatment was held to improve the consistency of the technique, and since it has no deleterious effect on good preparations, it is convenient to include this stage as a routine, regardless of the quality of individual preparations.

The barium hydroxide treatment is the critical stage of the procedure, and this is the stage that should be varied if required, to produce optimal results. Treatment which is too short or at too low a temperature may produce a complete G-banding pattern, while excessive treatment can cause swelling and distortion of the chromosomes, with pale staining and progressive loss of the C-bands. Although treatment at 50°C is generally recommended,

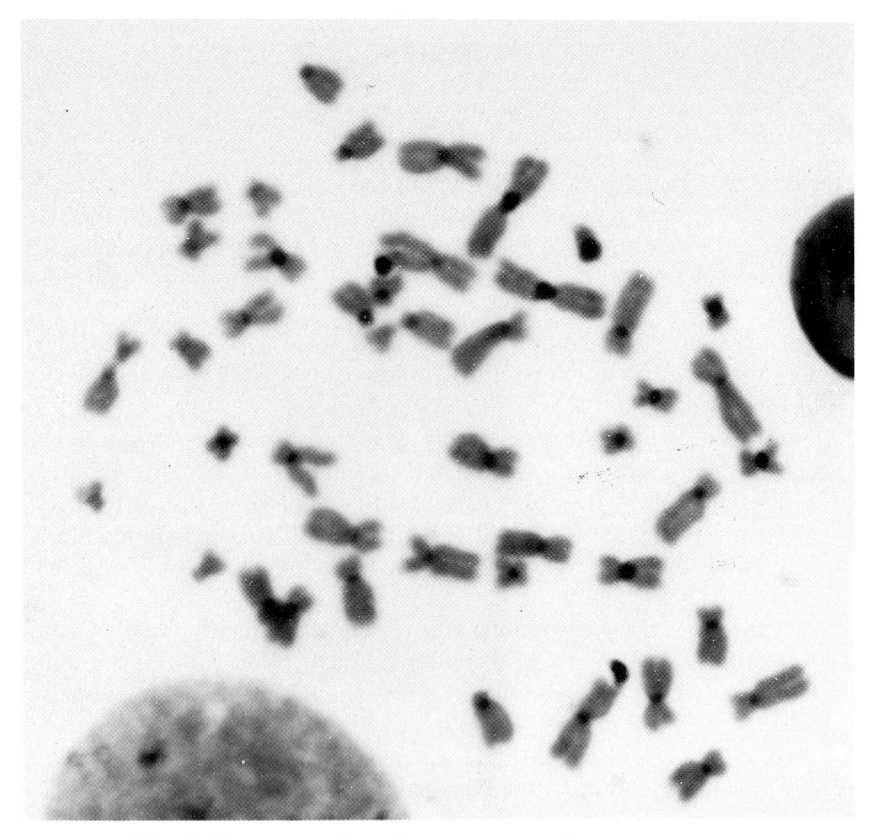

Figure 4.1 C-banded human metaphase chromosome spread.

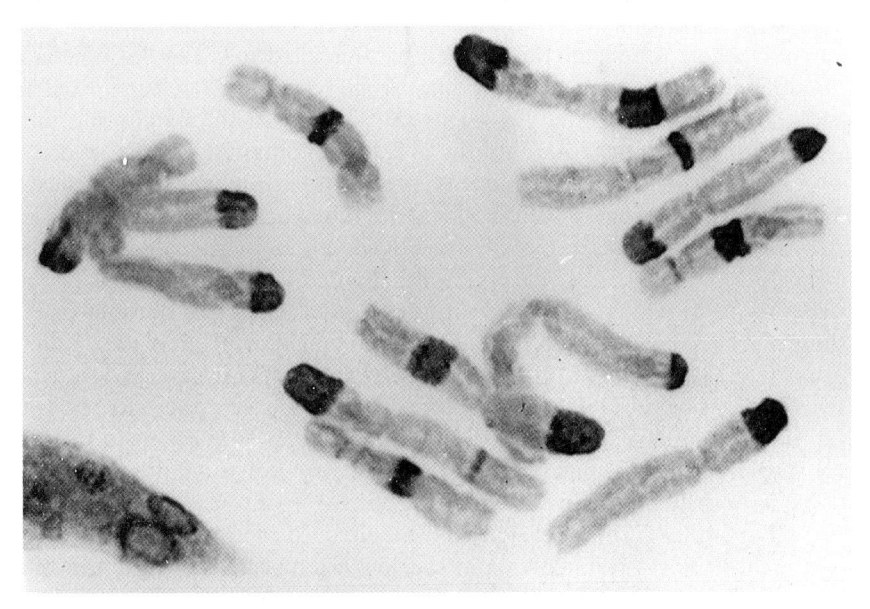

Figure 4.2 C-banded metaphase chromosome spread from the plant *Scilla sibirica*. Reproduced from Vosa (1973) by permission of the author and Springer-Verlag.

Chandley & Fletcher (1973) found that 37°C was preferable for meiotic chromosomes. The best length of time for the barium hydroxide treatment must be found by trial and error, and varies from one laboratory to another according to their routine methods of chromosome preparation, and also from one tissue or species to another.

Regardless of the amount of care taken in making up the barium hydroxide solution, a scum of barium carbonate forms on the surface. To minimize this process, it is best to warm the water to the required temperature first, and add the barium hydroxide crystals only a few minutes before use; the solution should be discarded after a short time. Although it is usual to wash the slides rather well after barium hydroxide treatment, any scum left on the slides dissolves away in the $2 \times$ SSC, and indeed satisfactory C-banding can be obtained without any washing. There is no reason to suppose that any reputable commercial Giemsa mixture will not be satisfactory for C-banding, although the concentration and staining time may need to be varied.

Finally, a note about mounting C-banding preparations (and indeed any chromosome preparation stained with Giemsa or similar mixtures). I use DPX (BDH Chemicals Ltd), a polystyrene-based mountant dissolved in xylene, and any mountant of this type is suitable. The Giemsa stain is very stable after mounting, so long as the mountant has not become acidic with age. It is unnecessary to soak the slides in xylene before mounting, and they must not be dehydrated in alcohols, as this will dissolve out the Giemsa stain.

4.2.3 *Nature of C-bands*

C-bands are usually equated with constitutive heterochromatin. The situation is not, however, as simple as this. It has already been mentioned (Section 4.2.1) that certain C-banding techniques only show some C-bands, and not others which can be demonstrated by different techniques. Although this may be dismissed as a technical triviality, it indicates that any given C-banding technique does not necessarily show up all the constitutive heterochromatin in an organism. Unfortunately there are very few cases in which the chromosomal distribution of heterochromatin has been investigated using independent criteria, such as those described by Schmid (1967). It has been confirmed in the fruit fly *Drosophila melanogaster* (Hsu 1971) and in the field vole *Microtus agrestis* (Arrighi *et al.* 1970) that the C-banded material corresponds to the heterochromatin observed in pre-banding cytogenetical studies. Baliček *et al.* (1977) showed that during prophase the C-bands of human chromosomes contracted less than the euchromatic regions; this is consistent with the C-bands being heterochromatic and therefore having to undergo less condensation. However, not all animals show complete correspondence between C-bands and heterochromatin. In the grasshopper *Cryptobothrus chrysophorus* several pairs of chromosomes have terminal blocks of heterochromatin, as defined by their condensation at meiotic prophase, but the heterochromatic blocks on chromosome M5 never C-band, although those on the other chromosomes always do (John & King 1977). In plants it has been widely accepted that differentiated regions of chromosomes induced by cold treatment (Section 11.2.1) are heterochromatic, but in *Adoxa moschatellina* there are cold-sensitive regions that do not C-band, as well as those that do (Greilhuber 1979). Evidently even when allowances are made for technical difficulties, there are some organisms, but probably not all, in which there is not a complete correspondence between C-banding and constitutive heterochromatin defined by other criteria. As long as this possibility is kept clearly in mind, it is permissible to regard heterochromatin as equivalent to C-bands in most cases. It would, however, be wrong to redefine constitutive heterochromatin solely in terms of reactivity to C-banding methods. On the other hand, there is no reported case of C-banded material that is not heterochromatic.

A similarly complicated situation occurs with the DNA content of C-bands. When Pardue & Gall (1970) accidentally discovered C-banding, they found that the C-band regions were the almost exclusive sites of the highly repetitive mouse satellite DNA. Subsequently, highly repetitive DNA has been located in C-band regions in a wide variety of species (Table 4.1), so often indeed that C-bands are commonly regarded as sites of highly repetitive DNA even in the absence of direct evidence. Although this assumption is frequently correct, there are several situations in which highly repetitive DNAs have not been found in C-bands. In addition it must not be assumed that where highly repetitive DNAs have been located in C-bands, the C-band DNA is exclusively of this type. In the Chinese hamster, for example, the long arm of the X chromosome, the entire Y, and the centromeric region of chromosome 10 are all C-banded, but do not appear to contain a high proportion of repetitive DNA (Arrighi *et al.* 1974). In Balenopterid whales, many centromeric C-bands appear to contain no satellite, although other centromeric and telomeric C-bands do so (Árnason *et al.* 1978). *Drosophila nasutoides* has four DNA satellites which all hybridize to the large heterochromatic chromosome 4, but not with other constitutive heterochromatin (Wheeler *et al.* 1978). It is always possible, of course, that lack of *in situ* hybridization to a particular C-band might be due to technical factors. Failure to isolate the repetitive DNA specific to a particular C-band is always a possibility. A consideration of the distribution of satellite DNAs among the C-bands of human chromosomes illustrates how complex these matters can be. Gosden *et al.* (1975b) localized satellites I, II, III and IV to several C-bands, including all the major ones and some minor ones. Each satellite showed a distinctive pattern. Manuelidis (1978) isolated another satellite which hybridized to the C-bands of chromosomes 7, 10 and 19, where specific satellites had not previously been reported. Repeated DNA fractions specific to the Y chromosome have also been found (Cooke *et al.* 1982). For a summary of this work, see Miklos & John (1979) and Singer

(1982). Although the C-bands of some human chromosomes have not been shown to contain specific types of DNA, progress so far suggests that it may only be a matter of time before repetitive DNA fractions have been located in all human C-bands.

An important question is whether C-bands, where they have been shown to contain highly repetitive DNAs, contain such types of DNA exclusively. Miklos & John (1979) claim that as much as 20% of the human genome consists of C-bands, while only 4% of human DNA is satellite. There can be no doubt, however, that their figure for the amount of C-band material is much too high. Considering the difficulties in making accurate measurements of both the amounts of C-band material (Section 14.3) and the amounts of satellite, it is unlikely that any such comparison will give a definitive result. Nevertheless, there is a much closer correspondence between the amount of C-band material and the amount of satellite in the mouse. Kram et al. (1972) found that in Drosophila melanogaster the simple highly repeated DNA sequences were interspersed with more complex and presumably less highly repeated sequences. Mouse satellite DNA appeared to be organized in a similar way. An ingenious approach to this problem was made by Kuo & Hsu (1978), who made RNA transcripts from C-banded chromosomes of Peromyscus eremicus. In these transcripts, non-repetitive sequences were covalently linked to repetitive ones, suggesting that the C-bands contained both classes of DNA. The fact that certain human C-bands contain more than one sort of satellite DNA (Gosden et al. 1975b) clearly indicates that they are not homogeneous. Although the evidence so far is sparse and unsatisfactory, there are certainly indications that non-repetitive DNA fractions may be included in C-bands with satellite DNAs.

Information on the protein composition of C-bands is much less comprehensive than that for DNA composition. Several authors have described various proteins in Drosophila species which bind preferentially to highly repetitive DNA, and which in some cases have been located in heterochromatin (Hsieh & Brutlag 1979, Will & Bautz 1980, Schmidt & Keyl 1981, James & Elgin 1986, Viglianti & Blumenfeld 1986). Similarly, Strauss & Varshavsky (1984) have described an HMG-like protein from African green monkey that binds preferentially to the satellite DNA of this species, and which may, by its interactions with nucleosomes, promote the formation of a particularly compact chromatin structure. Other observations have described particular histone subfractions or modifications in constitutive heterochromatin. Blumenfeld et al. (1978) found that in Drosophila spp., those fractions of H1 most closely associated with satellite DNAs were phosphorylated. Halleck & Gurley (1980) in Peromyscus spp., and Holmgren et al. (1985) in Drosophila spp. both found a correlation between compaction of heterochromatin and a high degree of phosphorylation of H2A. On the other hand, histones H2B, H3 and H4 appear to be unacetylated in Peromyscus heterochromatin (Halleck & Gurley 1981); however, experimentally-induced hyperacetylation of these histones did not affect

C-banding or condensation of heterochromatin (Halleck & Schlegel 1983). Study of the proteins of C-bands is obviously inhibited by the fact that there is no method of isolating them comparable in technical simplicity with that for satellite DNAs. Nevertheless, there is already evidence both for the presence of distinctive proteins in C-bands, and for specific types of modifications of histones, and further efforts in this field should give valuable results.

The DNA of C-bands is replicated late in the S phase (Bostock *et al.* 1972, Citoler *et al.* 1972, Sperling & Rao 1974), which is as expected since heterochromatin is generally late replicating (Schmid 1967, Lima-de-Faria & Jaworska 1968). Similarly, C-bands do not in general appear to contain active genes. There is, indeed, good reason to believe that satellite DNAs, which are often the predominant component of C-bands, are not transcribed and may not in fact be transcribable because of their sequences (Bostock 1980). Nevertheless, this is not an absolute rule. Several genes have been described in the heterochromatin of *Drosophila melanogaster*, although the density of genes is much lower than in euchromatin (Hilliker *et al.* 1980), and it is possible to argue that these genes could be contained in minute segments of euchromatin, microscopically undetectable, within the heterochromatin. Nagl & Schmitt (1985) found that repetitive DNA could be transcribed in the plant *Vicia faba*, while Sperling *et al.* (1987) reported that constitutive heterochromatin could be transcribed in the vole *Microtus agrestis*. In newts it has been shown that transcription of satellite DNA occurs in meiosis, at the lampbrush chromosome stage (Varley *et al.* 1980). This is particularly interesting in view of the suggestion that the function of C-bands is chiefly in meiosis, rather than in somatic cells (see below). In *Drosophila melanogaster*, Pimpinelli *et al.* (1985) have described a situation in which a specific type of heterochromatin counteracts the effect of a maternal-effect mutation which reduces the probability of an egg developing into an adult. This heterochromatin exerts its effect only during the early embryonic stages when the genome is transcriptionally inactive. Although these examples are few in number, it is clear that in certain situations the DNA of heterochromatin can be transcribed, and that it can also interact with genes in other ways.

The heterogeneity and heteromorphism of C-bands have been referred to previously in various places. We have seen that they vary in the type of DNA they contain, not only between species, but between chromosomes within the same species, and even within the same C-band (Table 4.1, Ch. 14). They also vary in their staining reactions. This subject cannot be described adequately until several other banding techniques have been described in later chapters, but a few examples are given here. There are two separate points to be made. First, different regions of heterochromatin may stain with different intensities with the same C-banding method (Camacho *et al.* 1984, Herrero & Garcia de la Vega 1986); it has, of course, already been mentioned that some regions of heterochromatin do not C-band at all. Second, not all C-bands respond in the same way to other banding techniques. The situation is so complicated that one can do little more than quote a few random examples. A general review of this field, with particular reference to the heterochromatin of human chromosomes, has been written by Rocchi (1982). Among plants Vosa (1976a) showed that in the M chromosome of *Vicia faba*, all seven heterochromatic bands could be distinguished from one another by using appropriate staining techniques. In the chromosomes of several species of *Scilla*, C-bands show differing responses to staining with the fluorochromes quinacrine, DAPI and chromomycin (Deumling & Greilhuber 1982). Species of *Allium* also show variations in the fluorochrome staining properties of their C-bands (Loidl 1983). Among animals, the heterochromatin of the chromosomes of *Drosophila* species has been studied intensively. In the *melanogaster* species group, there are complex relationships between patterns and intensities of C-banding and quinacrine fluorescence: C-bands may be quinacrine positive or negative, or partly positive and partly negative (Lemeunier *et al*. 1978). In *D. nasutoides*, C-bands may show bright fluorescence with Hoechst 33258, but dim fluorescence with quinacrine, or *vice versa* (Wheeler *et al*. 1978). Gatti *et al*. (1976) recognized no less than 14 types of heterochromatin in a group of *Drosophila* species, based on their response to Hoechst 33258 and quinacrine before and after fixation. More examples could be given, from many other groups of organisms, but my object has been to point out the heterogeneity of heterochromatin, and not to catalogue it. It must be added that this heterogeneity, although widespread, is not universal. For example, the C-bands of mouse chromosomes (except the Y) all contain the same satellite DNA and respond in the same way to a variety of banding techniques.

C-bands are also heteromorphic. This topic will be considered in detail in Chapter 14, after various other types of bands have been described that also show heteromorphism. Suffice it to say here that in virtually all species that have been investigated adequately, there may be differences in the size of C-bands on homologous chromosomes both within and between individuals. Differences between the chromosomes of closely related species are also often due to differences in the amount of C-band material they contain (Ch. 15). Within a species differences in the amount of C-band material result in variations in the total amount of nuclear DNA. There is a constant amount of euchromatic DNA, but the heterochromatic DNA is variable in amount and additional to the euchromatin. This has been established for human chromosomes by careful measurements of their DNA content (Geraedts *et al*. 1975, Sumner 1977a). It has also been shown that variations in genome size between certain closely related species are due essentially to differences in the size of C-bands, as for example in grasshoppers of the genus *Atractomorpha* (Miklos & Nankivell 1976), rodents of the genus *Peromyscus* (Deaven *et al*. 1977), and plants in the *Scilla siberica* group (Greilhuber & Speta 1978). In other groups, however, there is no clear correlation between genome size and C-band size (e.g. *Scilla*, Greilhuber (1977a); certain grasshoppers, King & John (1980)).

4.2.4 Functions of C-bands

Many functions have been proposed for C-bands; these include involvement in chromosome folding, recognition of homologous centromeres and meiotic pairing, structural protection of centromeres and ribosomal genes, a role in the phenomenon of affinity (in which sets of chromosomes from each parent in cells of a hybrid tend to segregate separately) (Walker 1971, Yunis & Yasmineh 1971), acting as a 'bodyguard' against clastogenic agents (Hsu 1975) and determination of nuclear volume (Cavalier-Smith 1978). In fact, there is very little evidence for any of these suggestions, and good evidence that C-band material does not play a part in meiotic pairing. In *Drosophila melanogaster*, in which it is possible to construct chromosomes having various amounts of either euchromatin or heterochromatin, it is found that it is the euchromatin and not the heterochromatin that is essential for proper meiotic pairing and segregation (Yamamoto & Miklos 1978, Yamamoto 1979).

The evidence which is available strongly suggests that the function of heterochromatin should be looked for mainly in the germ line, and not in somatic cells. In the polytene chromosomes of Diptera, the heterochromatin is usually under-replicated, suggesting that it is not very important in these chromosomes (Endow & Gall 1975). In a variety of organisms (e.g. *Parascaris*, Goday & Pimpinelli (1984); *Cyclops*, Beermann (1977)) the heterochromatin is completely eliminated in somatic cells, but is always retained in cells of the germ line. The germ line function of heterochromatin that is most clearly established is an effect on chiasma formation at meiosis. Crossing-over rarely occurs in heterochromatin itself (John 1976), but perhaps of greater significance is the fact that blocks of heterochromatin affect chiasma formation in adjacent euchromatic segments of chromosomes. John & Miklos (1979) have tabulated several examples from both animals and plants where additional heterochromatin affects chiasma frequency, usually by increasing the number of chiasmata. In the Australian grasshopper *Atractomorpha* the chiasma frequency is clearly correlated with the amount of heterochromatin; the greater the amount of heterochromatin, the lower the number of chiasmata (Miklos & Nankivell 1976). Not only is the frequency of chiasmata altered, but also their position, since the telomeric heterochromatin in particular causes a substantial reduction in chiasma formation in the adjacent euchromatin. A similar situation has been described in *Drosophila melanogaster* (Yamamoto & Miklos 1978). Blocks of heterochromatin do not always have the same effect on chiasmata, however. In species of *Allium*, chiasmata are found preferentially adjacent to C-bands, but where C-bands are absent, the chiasmata are much less localized (Loidl 1979, 1982). An effect of C-bands or heterochromatin on meiotic recombination is thus the most clearly established effect of this material, although the details of this effect may vary, and it is not known how widespread this effect is. This is not the only effect of heterochromatin in the germ line, however; in maize it has also been reported to cause preferential segregation of certain chromosomes, and neocentromere formation (Rhoades 1978).

4.2.5 Mechanism of C-banding

Because C-banding methods were derived from methods for *in situ* hybridization, it was at first believed that C-bands were produced by an analogous mechanism. That is, the chromosomal DNA was denatured and the most highly repetitive DNA, in the C-bands, renatured most rapidly, and was stained most strongly. Mace *et al.* (1972), using antiserum specific for single stranded DNA, did in fact report that after C-banding treatment the DNA was denatured except in the C-bands. Other considerations make a mechanism based solely on denaturation and selective annealing of DNA unlikely, however. As we have seen (Table 4.1) C-bands do not invariably contain highly repetitive DNA, and in the absence of repetitive DNA a different mechanism would be required. In any case, it appears that denaturation of chromosomal DNA by alkaline treatment is largely irreversible (Kurnit 1974, Holmquist 1979) although a small amount of annealing can occur in certain circumstances (Raap *et al.* 1986). Further evidence for an alternative mechanism comes from observations that C-banding can be produced by reagents whose action is essentially on proteins, and not on DNA (McKay 1973, Merrick *et al.* 1973, Ray & Hamerton 1973). It seems, therefore, that a mechanism based on differential annealing of DNA cannot explain the production of C-bands satisfactorily.

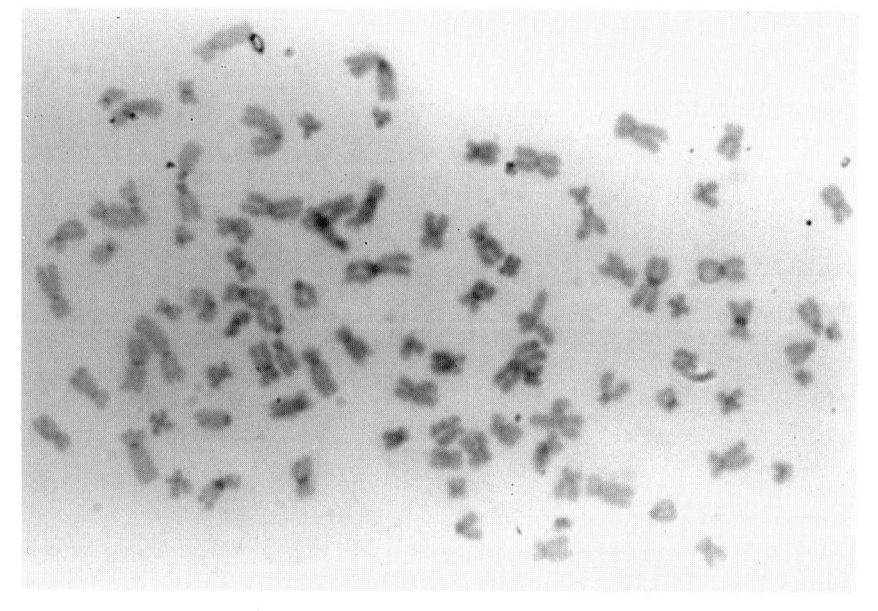

Figure 4.3 Human metaphase chromosome spread treated for C-banding, but stained with methylene blue to show distribution of DNA.

It was soon established that C-banding involved the preferential removal of DNA from non-C-banded regions of chromosomes (Fig. 4.3), leaving a greater amount of DNA in the C-bands (Comings *et al.* 1973, Pathak & Arrighi 1973). It should, nevertheless, be noted that Kongsuwan & Smyth (1978) claimed that C-bands were formed in chromosomes of *Lilium* without differential loss of DNA, although substantial uniform losses occurred. Differential loss of DNA is apparently not in general accompanied by a similar differential loss of proteins (Gendel & Fosket 1978), although there is a general loss of proteins from chromosomes. Burkholder & Duczek (1980b, 1982b) showed that the extraction involved a specific set of proteins.

The chemical processes responsible for loss of DNA during C-banding have been described by Holmquist (1979). Each of the three stages investigated, acid treatment, alkaline treatment, and incubation in warm saline, has a specific part to play in the process. The preliminary treatment with 0.2N hydrochloric acid depurinates the DNA, but does not break the depurinated sites. This supplements depurination which takes place during methanol-acetic acid fixation (Sederoff *et al.* 1975). Extraction of the depurinated DNA occurs during the final stage of incubation in warm saline, as a result of β-elimination at the depurinated site, leading to chain breakage (Fig. 4.4). This process can be inhibited by reducing the free aldehydes at apurinic sites with borohydride, which also prevents C-banding. The alkaline treatment causes breakage of apurinic sites, as well as denaturing the DNA irreversibly (Kurnit 1974, Sederoff *et al.* 1975, Holmquist 1979). As a result the DNA is reduced to smaller fragments which are more easily lost from the chromosomes.

If the mechanism just described for the extraction of DNA during C-banding is accepted, the question must be asked why DNA is extracted preferentially from non-C-banded regions of chromosomes. There are evidently three possibilities: first, that the C-band DNA is depurinated less than the rest; second, that it is denatured less in alkali, or anneals more rapidly, thus rendering it less extractable; or third, that it may be inherently less extractable in warm salt solution. Obviously a combination of these factors could operate. There is indeed evidence that DNA in condensed chromatin is depurinated more slowly than that in dispersed chromatin (Duijndam & van Duijn 1975) and cytological evidence of this is provided by the existence of Feulgen banding of chromosomes (Section 4.6). As described above, there is little evidence for selective annealing of C-band DNA. If one accepts the conclusion of Kurnit (1974) and of Holmquist (1979) that annealing does not occur after alkaline treatment, the situation is still the same: that the alkaline treatment does not appear to act differentially on euchromatin and heterochromatin. The third possibility is that DNA in C-bands is more tightly bound to the chromosomal material and is therefore less readily extracted in warm saline, regardless of its degree of depurination. We have already seen that there is evidence for distinctive types of protein in C-bands (Section 4.2.3), so that there is a clear possibility that there might be distinctive

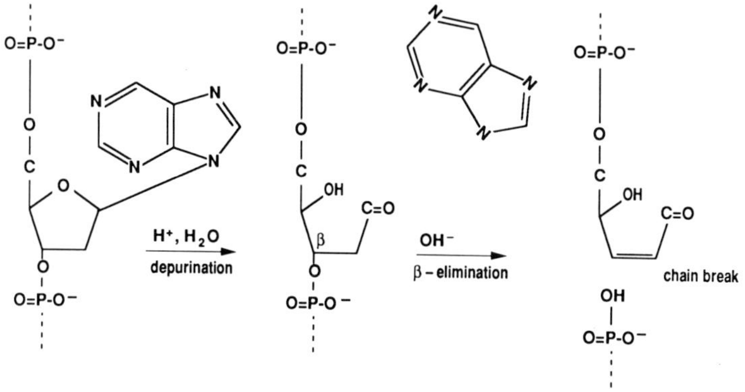

Figure 4.4 The process of β-elimination in DNA. Acid treatment causes depurination, leaving the deoxyribose residue in the aldehyde form. The process of β-elimination in the presence of alkali or heat leads to DNA chain breakage. Reproduced from Holmquist (1979) by permission of the author and Springer-Verlag.

interactions between DNA and proteins in C-bands. There is certainly evidence that there are proteins in chromatin that bind more strongly to highly repetitive DNAs (Maio & Schildkraut 1969, Musich et al. 1977, Hsieh & Brutlag 1979) in both mammals and *Drosophila*, but it is not yet known whether such proteins occur in all C-bands, regardless of the type of DNA they contain. Burkholder & Weaver (1977) showed that C-bands were protected against digestion with DNase by the presence of non-histone proteins, and concluded that the protection was largely due to the tighter binding of non-histones to DNA, which restricted access of the enzyme. Access of the reagents used in conventional C-banding is less likely to be restricted in this way, but the stronger binding of DNA to proteins in C-bands is a highly plausible mechanism for the differential extraction of DNA from chromosomes.

In summary, then, differential annealing of denatured DNA does not appear to be part of the mechanism of C-banding. The important features of the process seem to be a progressive attack on the DNA molecules, rendering them more labile to extraction, and the stronger binding of DNA to proteins in C-bands, so that C-band DNA of whatever type is less readily extracted. Although this is a plausible explanation, more evidence in support of it would be very welcome, especially as there is evidence that alternative mechanisms might be involved in at least some cases. It has already been mentioned that in one species C-bands can apparently be demonstrated without differential extraction of DNA (Kongsuwan & Smyth 1978). Gendel & Fosket (1978) claimed that C-bands could be demonstrated in *Allium cepa* after total extraction of DNA; although total protein was uniformly distributed along the chromosomes, the protein in the C-bands had a greater affinity for Giemsa. There is an interesting parallel here with the mechanism of N-banding (Sections 4.4 and 9.3).

Figure 4.5 Unstained transmission electron micrograph of a C-banded mouse chromosome, showing the preferential retention of material in the centrometric heterochromatin. Micrograph kindly provided by Dr V. J. Goyanes.

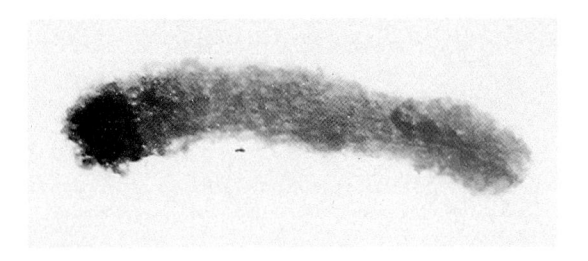

4.2.6 Electron microscopy of C-bands

A number of observations have been made on C-bands using electron microscopy, which may be conveniently considered together. These observations cover the greater resistance to destruction of C-bands, the size of chromatin fibres in C-bands compared with those in euchromatin, and the surface structure of C-banded chromosomes.

The greater resistance of C-bands to C-banding procedures has been observed by several authors (Comings *et al.* 1973, Burkholder 1975, Goyanes 1985, Jack *et al.* 1985). The C-bands remain compact whereas the remainder of the chromosomal material has a much looser structure (Fig. 4.5). The looser structure in the chromosome arms appears to result not merely from extraction of material but also from a greater dispersion of the remaining material. Similar appearances have been described by Ris (1978) and Ris & Korenberg (1979) in chromosomes treated with various solutions to extract H1 histones. The same authors observed that the C-bands contained thick, 50 nm fibres, whereas the remainder of the chromosomes consisted of 20 nm fibres. Similar observations on differences in fibre size between euchromatin and heterochromatin have been made by Comings & Okada (1976) and by Weith (1983).

Observations of the surface structure of C-banded chromosomes show that the C-bands appear as elevations above the level of the rest of the chromosome (Ross & Gormley 1973, Blakey & Filion 1976). The elevation of the C-bands is due solely to the presence of Giemsa dye, since no elevations are present before staining, or after extraction of the dye. Ross & Gormley (1973) studied each stage of the BSG procedure, and found that there were characteristic changes during each part of the process. The initial acid treatment made the chromosomes stand out more clearly, apparently because of the removal of surrounding proteinaceous material. Subsequent treatments with barium hydroxide and $2 \times SSC$ produced a 'collapsed' appearance (Fig. 4.6) in which only the outlines of the chromatids were raised, and the interior appeared almost empty. These structures were then rebuilt during staining.

4.2.7 Applications of C-banding

In most eukaryotes, heterochromatic bands are the principal class of bands that can be demonstrated, since G-, R- and Q-bands throughout the length of the chromosomes are largely confined to higher vertebrates. Thus C-banding is valuable for the identification of chromosomes, particularly in insects and plants. The characteristics that aid identification are the size and position of the C-bands; they may be located pericentromerically, interstitially or terminally. In a few species, aberrant C-bands have been described. Usually C-bands appear as a solid block of material occupying the whole width of the chromosome, but in *Scilla mischtschenkoana* C-bands have been described which appear as clusters of narrow bands or dots (Greilhuber & Speta 1978), while in *Triturus marmoratus* C-bands which occupy only part of the width of a chromatid have been reported (Herrero & Gosálvez 1985).

Chromosomes which lack C-bands completely are rare, and possibly may not occur in a normal karyotype. The best authenticated case is the CM (C-minus) chromosomes in a mouse tumour cell line, described by Levan *et al.*

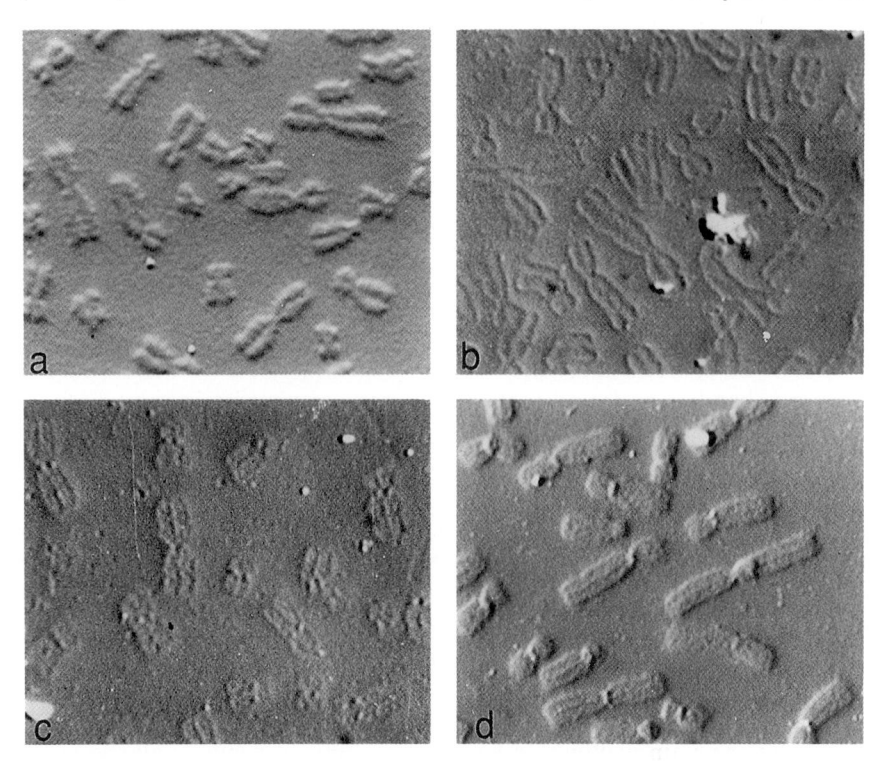

Figure 4.6 Surface topography of human metaphase chromosomes at successive stages during the BSG C-banding technique (a) fixed and treated with 0.2N HCl; (b) 0.2N HCl followed by barium hydroxide; (c) treatment with 0.2N HCl, barium hydroxide, and 2 × SSC at 60° C; (d) complete BSG treatment including Giemsa staining. Reproduced from Ross & Gormley (1973) by permission of the authors and Academic Press.

(1978). In deciding whether a chromosome truly lacks a C-band, it is important to bear in mind that different C-banding techniques may not show identical patterns of bands, and that small bands in particular may be lost easily by excessive treatment. At least a small band is usually present at the centromere of chromosomes, although this may actually represent staining of the kinetochores (Ch. 10). Absence of C-banding has also been reported for double minutes (Levan *et al.* 1976), minute chromosome-like bodies that sometimes occur in tumour cells, are made of amplified DNA, and which lack centromeres. (For a review of double minutes, see Cowell 1982.) It should be stressed that the presence or absence of C-bands, whether in double minutes or, for example, in suspected dicentric chromosomes, cannot be taken as evidence for the presence or absence of a functional centromere; other techniques, specific for the kinetochores, are necessary to establish this (Ch. 10).

C-banding has proved to be useful in the identification of meiotic chromosomes, even in species such as mammals which show good G-banding patterns on mitotic chromosomes. It has turned out to be almost impossible to obtain satisfactory G- or Q-banding on mammalian meiotic chromosomes, and in these circumstances C-banding is valuable to identify bivalents at diakinesis, using both the centromere position (as shown by the position of the C-band) and the size of the C-band (Chandley & Fletcher 1973).

C-bands can be demonstrated in interphase nuclei as satisfactorily as on chromosomes. Hsu *et al.* (1971) used C-banding to show the distribution of centromeric heterochromatin in interphase nuclei of the mouse, and found that the pattern of distribution varied and was apparently characteristic for each type of cell. Fussell (1975) used C-banding of interphase nuclei of *Allium cepa* (onion) to study the orientation of chromosomes within the nucleus, making use of the facts that in plant meristems the orientation of daughter nuclei is fixed, and that onion chromosomes have centromeric and telomeric C-bands. She found that the chromosomes maintain the telophase configuration throughout interphase, with the centromeres on one side of the nucleus and the telomeres at the opposite side.

There are, of course, many species in which C-banding alone is inadequate for unequivocal identification of all the chromosomes, and in which it has not so far been practicable to induce euchromatic bands. In such cases it is often useful to combine C-banding with one of the fluorescence techniques described in Chapters 7 & 8. By this means different subclasses of heterochromatic bands may be revealed, producing greater differentiation between chromosomes. Some of the techniques to be described later in this chapter can also be used in the same way.

C-bands have been shown to be heteromorphic in virtually all species in which the matter has been studied in sufficient depth. The subject of banding heteromorphisms will be considered in detail in Chapter 14, but it is appropriate to summarize here some of the areas in which heteromorphisms are of interest. It has been shown that the heteromorphisms are stable, and are inherited in a typical Mendelian fashion (Robinson *et al.* 1976). They are thus good markers for such diverse applications as paternity testing (Baliček *et al.* 1978) and gene mapping. Attempts have been made to associate heteromorphisms, particularly the more extreme variants, with mental retardation in man (Podugolnikova *et al.* 1984), reduced reproductive fitness (Erdtmann 1982), and with a tendency to develop certain types of tumours (Atkin & Brito-Babapulle 1981, Berger *et al.* 1985). The evidence will be considered in more detail in Chapter 14, but it should be mentioned here that some of the evidence linking C-band heteromorphisms with clinically observed defects is contradictory, and much of it is not very strong. In the mouse it has been shown that different inbred strains show characteristic patterns of C-band variation (Davisson 1989). Differences in patterns of heteromorphisms may also occur between wild populations of the same species (John & King 1977, 1983), and between different human populations and races (Lubs *et al.* 1977, Berger *et al.* 1983, Zanenga *et al.* 1984).

The subject of banding in relation to karyotype evolution will be considered fully in Chapter 15, but it is appropriate to mention here that differences in C-bands between species occur quite frequently and may be substantial. For example, a major means of karyotype evolution in the rodent genus *Peromyscus* has been the development of short arms which are entirely C-banded (Greenbaum & Baker 1978). When the sheep and goat are compared, it is found that the formation of metacentric chromosomes by Robertsonian fusion is accompanied by loss of most of the C-band material (Evans *et al.* 1973).

4.3 G-11 banding

G-11 banding was introduced almost simultaneously by Bobrow *et al.* (1972b) and by Gagné & Laberge (1972). In principle the method consists simply of staining chromosome preparations with a Giemsa solution at a pH of about 11. In human chromosomes this procedure resulted in the specific magenta staining of the centromeric heterochromatin of chromosome 9, while everything else was stained blue (Fig. 4.7). In practice the method has proved somewhat fickle, and the regions of heterochromatin stained vary from one metaphase to another, and from one slide to another. Both Bobrow *et al.* (1972b) and Gagné & Laberge (1972) observed that several other sites of heterochromatin, particularly on the acrocentric NOR-bearing chromosomes, were also stained magenta, and that two large magenta bodies, believed to be the centromeric heterochromatin of chromosome 9, could be seen in interphase nuclei.

G-11 banding has only been demonstrated in a few species (only the great apes and man so far as the author is aware), and the method might have remained of limited value, especially considering its unreliability, had it not been observed that it could be used to distinguish between human and mouse chromosomes in somatic cell hybrids (Bobrow & Cross 1974). It has

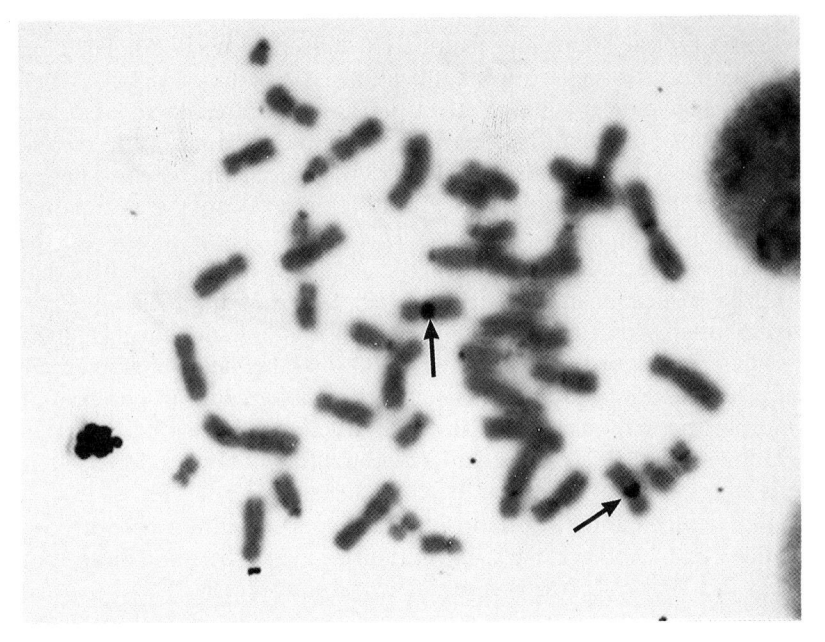

Figure 4.7 G-11 banding of a human metaphase chromosome spread, showing strong staining of the large blocks of centromeric heterochromatin on chromosome 9 (arrows), and some small centromeric blocks elsewhere. Note that the large blocks of heterochromatin on chromosomes 1, 16 and Y are not picked out by this method.

subsequently been found that this is a general method for distinguishing the chromosomes of higher primates from those of rodents, and it has become very valuable in somatic cell genetics.

4.3.1 Giemsa-11 methods

In the original technique described by Bobrow *et al.* (1972b), standard chromosome preparations were simply stained in alkaline Giemsa, made by diluting a stock solution 1:50 with distilled water adjusted to pH 11 with a suitable alkali. The length of staining had to be determined by trial and error, with 10–20 min being optimal. Too short a period of staining resulted in an overall blue colour, and too long a period in overall magenta. Staining at 37°C reduced the amount of precipitate that formed on the slides. Gagné & Laberge (1972) made up their Giemsa solution by diluting a stock solution 1:50 with 0.1% disodium phosphate $12H_2O$ adjusted to pH 11.6 with sodium hydroxide. Subsequently a pH of 11.5 was recommended for the Bobrow method by Pearson & van Egmond-Cowan (1976) who also advised using five-day-old slides, while Bobrow & Cross (1974) suggested leaving the slides for a week at room temperature, or two days in a dry oven at 60°C, before staining. The latter authors stated that the slides could subsequently be destained and then Q-banded or G-banded.

In a series of experiments designed to elucidate the mechanism of G-11 banding, and to improve its reliability, Wyandt et al. (1976) found that choice of a suitable Giemsa solution was important; azure B plus eosin Y in appropriate proportions also gave excellent differentiation. Their method involved incubation for 2–3 min in an alkaline phosphate or borate-carbonate buffer at precisely pH 11.6 at 37°C. Giemsa stock solution was added at a dilution of 1:50 and incubation continued for 6–7 min. The pre-incubation gave uniform staining over the entire slide, and the use of buffer resulted in better chromosome morphology and differentiation than unbuffered alkali.

Methods subsequently described for G-11 banding have involved variations on the themes already described. Friend et al. (1976) soaked specially prepared slides for 1–2 h in distilled water to age them. They emphasized the use of a fresh staining solution (alkaline Giemsa solutions precipitate rapidly), and pointed out that the density of cells on the slides was critical, overall blue staining being obtained with too many cells. Minimal cytoplasmic background was also essential to give good colour differentiation, and methods were described for reducing the amount of cytoplasm. Alhadeff et al. (1977) used a pre-incubation in distilled water for 2 h at 60°C, and prepared their staining solution half an hour in advance. The method of Buys et al. (1984a) is similar to that described by Wyandt et al. (1976), with a pre-incubation in alkaline buffer before staining; the difference lies in the staining solution, which is a defined mixture of azure B, eosin Y, and an oxidized methylene blue solution of unknown composition.

4.3.2 Mechanisms of Giemsa-11 banding

Little is known of the mechanism of G-11 banding. The use of an alkaline treatment suggests an analogy with C-banding. However, it cannot simply be that DNA is extracted from the chromosome arms, and that for some reason only the remaining C-bands stain magenta. If the technique is applied to mouse chromosomes, it is the arms that are stained magenta, and the centromeric regions blue (Bobrow & Cross 1974). The nature of the substrate that stains magenta with alkaline Giemsa remains a mystery, but clearly it may be found in heterochromatin in some species and in euchromatin in others. Bühler et al. (1975) pointed out that the pattern of G-11 staining was correlated with the distribution of DNA satellite III, but obviously such a correlation does not occur in rodents, and therefore this does not help to explain the staining mechanism.

Wyandt et al. (1976) showed that formation of an azure-eosin complex (involving either azure A or azure B) was necessary to produce the magenta staining with Giemsa-11, a conclusion supported by the observations of Buys et al. (1984a). However, formation of such complexes is a normal part of the process of staining chromosomes with Giemsa (e.g. in G-banding, Ch. 5), and indeed of Romanowsky-Giemsa staining in general (Wittekind 1983), and no explanation has been proposed for the differential G-11 staining. The work of Wyandt *et al.* (1976) indicated that differential denaturation of DNA was not involved.

In passing, it may be mentioned that a similar colour differentiation has been obtained with Giemsa staining after a C-banding procedure in plant chromosomes (Sato *et al.* 1979). In this case, some C-bands were magenta, while other C-bands and the chromosome arms were blue.

4.3.3 Applications of Giemsa-11 banding

Most of the possible applications of G-11 banding were described by Bobrow *et al.* (1972b) in their original paper; these included the study of heteromorphisms, identification of human chromosome 9 in meiosis, and recognition of this chromosome in non-dividing nuclei, particularly those of spermatozoa. Banding heteromorphisms will be considered in detail in Chapter 14, but a few observations may be noted here. Gagné *et al.* (1973) observed that the G-11 stained material in human chromosome 9 consists of several segments. Donlon & Magenis (1981) observed that the G-11 stained material was only part of the C-band of chromosome 9, while Magenis *et al.* (1978) found considerable variability in the G-11 stained region of human chromosome 1. The value of G-11 staining for identifying chromosome 9 in

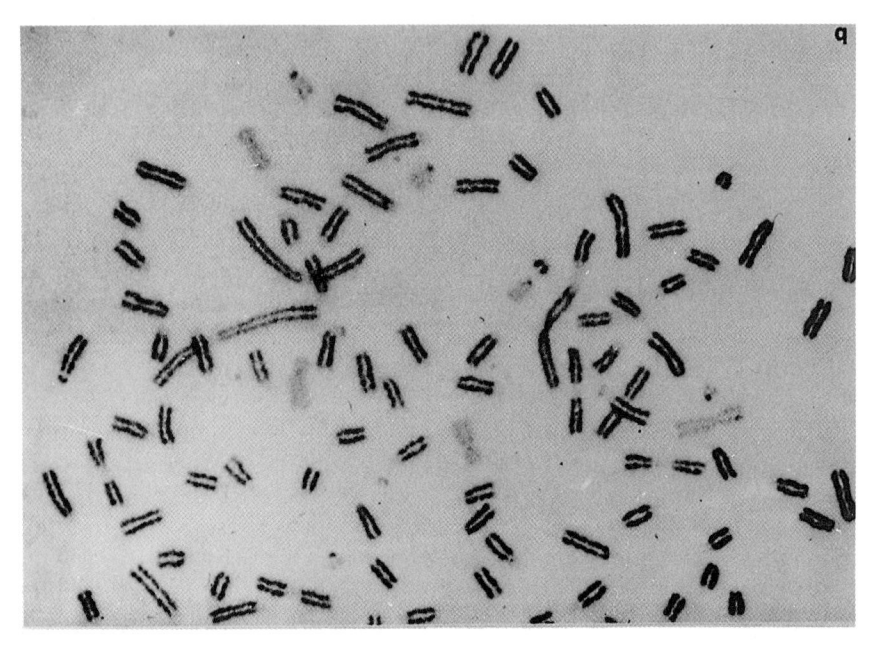

Figure 4.8 G-11 staining of a mouse-human hybrid cell. The mouse chromosomes appear dark (magenta) with pale (blue) centromeres, while the human chromosomes are pale throughout except for small magenta centromeric bands in some cases. Reproduced from Buys *et al.* (1984) by permission of the authors and Springer-Verlag.

human male meiosis was shown by Page (1973), who also described the distinctive structure of this region of heterochromatin.

One of the most interesting applications of G-11 staining was to study aneuploidy for chromosome 9 in human spermatozoa (Pearson *et al.* 1974). Although most sperm heads showed only a single magenta body, as expected, about 2% showed two magenta bodies. In such cases it was believed that non-disjunction had occurred, and that these spermatozoa were disomic for chromosome 9. However, subsequent direct observations of disomy in human spermatozoa have failed to confirm such a high rate of non-disjunction (Martin *et al.* 1983, Brandriff *et al.* 1985).

The chief use of the G-11 technique is now undoubtedly for the distinction of rodent and primate chromosomes (Fig. 4.8) in somatic cell hybrids (Bobrow & Cross 1974, Friend *et al.* 1976, Buys *et al.* 1984a). Since the method can be combined with other banding techniques to identify individual chromosomes (Bobrow & Cross 1974, Buys *et al.* 1984a), the G-11 technique has become extremely valuable in somatic cell genetics. The range of species whose chromosomes stain magenta or blue has not yet been established, but the magenta colour is obtained with both mouse and Chinese hamster chromosomes, and the blue colour with chimpanzee and human chromosomes (Friend *et al.* 1976). Not only can primate chromosomes be distinguished easily amongst a set of rodent chromosomes, but in some cases chromosomes can be seen which show both colours (Friend *et al.* 1976). Such chromosomes are interpreted as interspecific translocations, although it must always be borne in mind that certain segments of primate chromosomes stain magenta, and that the centromeres of mouse chromosomes stain blue.

4.4 N-banding

Strictly speaking, N-banding is a method for staining the nucleolus-organizing regions (NORs) of chromosomes, and as such will be considered in detail in Section 9.3. However, the method has been used to stain certain regions of heterochromatin that are not associated with NORs, in various organisms, and this application is properly described here.

The original N-banding method was introduced by Matsui & Sasaki (1973), and comprised successive treatments with hot trichloracetic acid and hot hydrochloric acid (see Section 9.3.1 for details). These treatments were believed to extract nucleic acids and histones respectively. This interpretation was supported by various experiments, and it was concluded that Giemsa was staining acidic proteins which were left after the extraction procedures. With this method Matsui & Sasaki (1973) showed that the sites of N-bands on the chromosomes of several mammalian species corresponded to those of nucleolar organizers, although in the mouse the N-banding technique stained centromeric heterochromatin. Subsequently, the same group (Funaki et al. 1975) modified the technique, using a 1M phosphate solution (pH 4.2) at 96°C to extract nucleic acids and histones. With the modified technique they showed a correspondence between N-bands and NORs for 27 species of animals and plants. Meanwhile, Faust & Vogel (1974) had questioned whether the N-banding technique was actually staining NORs, and suggested that it was staining adjacent blocks of heterochromatin.

A number of authors have used one or other of the original N-banding techniques on a variety of species, and have demonstrated patterns of bands many of which do not correspond to nucleolar organizers, and which in many cases are also different from C-bands. This includes a substantial amount of work on various species of cereals (Gerlach 1977, Jewell 1981, Endo & Gill 1984, Schlegel & Gill 1984), *Drosophila* species (Pimpinelli et al. 1976a, Hägele 1977), *Chironomus* (Hägele 1977) and the locust *Schistocerca* (Hägele 1979). It was noted that different patterns of N-bands could be produced by varying the length of treatment in trichloracetic and hydrochloric acids (Hägele 1979), or by altering the conditions of treatment with hot phosphate (Jewell 1981).

As with other banding techniques, N-bands may be defined as those bands produced by N-banding methods. These in turn may be defined as one of the two original methods (Matsui & Sasaki 1973, Funaki et al. 1975) which demonstrate nucleolar organizers in a variety of species (but not in all), and which apparently act by extracting nucleic acids and histones. The stained bands would then represent acidic proteins. However, the definition of N-bands becomes clouded when the term is applied to bands produced by substantially different techniques. Armstrong (1982) treated roots of wheat plants with the complete Feulgen technique for DNA before making squash preparations of chromosomes, and then treated the squashes with a modified BSG (C-banding) technique, in which the barium hydroxide treatment was at room temperature. The banding pattern produced was essentially the same as that described by Gerlach (1977) using a conventional N-banding technique. Armstrong therefore referred to these bands as N-bands, although he recognized that his method was more akin to C-banding. An even more extreme variant of N-banding was described by Fox & Santos (1985), and applied to various species of grasshoppers. In this method, squash preparations of chromosomes were treated with a hot mixture of formamide and $1 \times$ SSC, and then stained with one of a variety of DNA-specific dyes. Clearly the technique differs greatly from other N-banding techniques, yet the result is similar in that some of the bands represent nucleolar organizers while others do not. It should also be noted that this method does not rely on complete extraction of DNA.

N-banding can thus be used to demonstrate a distinctive set of heterochromatic bands, as well as nucleolar organizers. Although the nature of the original methods implied that N-bands were regions rich in acidic proteins that were resistant to extraction, it cannot yet be said that this is true of all bands revealed by N-banding methods. It has still to be determined what the common characteristics of N-bands are. Nevertheless, these methods are a valuable addition to the armamentarium of techniques available to workers on plant and insect chromosomes.

4.5 Methods for demonstrating heterochromatic bands using hydrochloric acid

A number of methods for demonstrating certain types of heterochromatin, mainly in plant chromosomes, depend on the hydrolysis of the chromosomes, as in the Feulgen reaction for DNA. Indeed, most of the published methods use the complete Feulgen reaction, including staining with Schiff's reagent, although staining with orcein or carmine has also been used.

The earliest method of this sort appears to be that described by Yamasaki (1956, 1959). In this method, plant tissue (*Cypripedium debile*) was first treated with a mixture of hot acetic and hydrochloric acids, transferred to alcoholic acetic acid, and the complete Feulgen reaction carried out, after which squash preparations were made. Takehisa (1968, 1969) used a similar method, this time on chromosomes of *Vicia faba*. More recently, a conventional ethanol-acetic acid fixation has been employed before performing the Feulgen reaction. Ennis (1975) used this procedure on the chromosomes of the beetles *Chilocorus* spp., while Tempelaar *et al.* (1982) and G. E. Marks (1983) investigated various plants.

Yamasaki (1971) omitted Feulgen staining, and reduced the procedure to heating in a solution of orcein in a mixture of acetic and hydrochloric acids, followed by squashing in aceto-orcein. Then in 1973, Greilhuber (1973, 1975) introduced his Hy-banding method. This technique, which the author regarded as a development of the earlier ones, involved conventional fixation in ethanol-acetic acid and treatment with hot hydrochloric acid. Staining was in aceto-orcein before squashing. The abbreviation Hy-banding refers to the author's belief that hydrolysis by the hydrochloric acid is the essential feature of these methods.

It is characteristic of bands produced by these techniques that they may be darker or paler than the euchromatic parts of the chromosome, both types sometimes occurring together on the same chromosome (Takehisa 1969). Greilhuber (1973) refers to Hy+ bands and Hy− bands. These bands are not normally equivalent to the whole of the heterochromatin revealed by other methods, such as cold treatment (Section 11.2.1) or C-banding, and thus appear to represent a distinct class or classes of heterochromatin (Greilhuber 1973, 1975). Such methods may also reveal heterochromatic bands not demonstrated by other methods (Takehisa 1968).

The production of Hy-bands and Feulgen bands is believed to depend on differences in extractability; in fact, Greilhuber (1973) showed that such bands were visible by phase contrast microscopy after hydrolysis, without

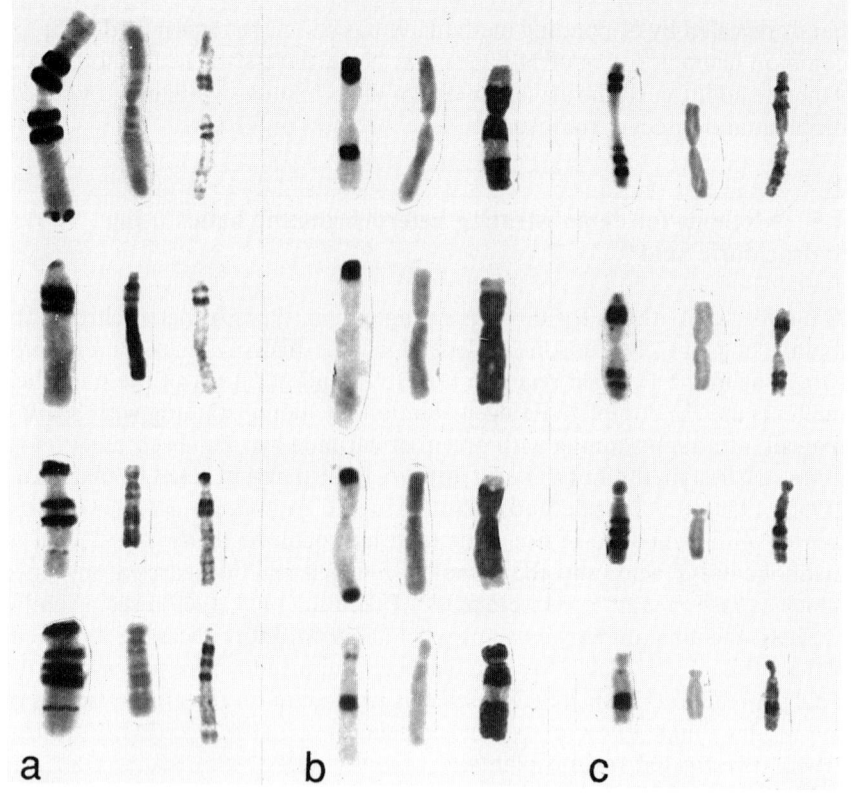

Figure 4.9 Feulgen banding of plant chromosomes: (a) *Fritillaria lanceolata*, (b) *Scilla sibirica*, and (c) *Anemone blanda*. For each species, the left-hand column illustrates conventionally C-banded chromosomes, the middle column chromosomes stained with Feulgen after short hydrolysis, and the right-hand column chromosomes stained with Feulgen after prolonged hydrolysis. Reproduced from Marks (1983) by permission of the author and the Company of Biologists Ltd.

staining. Such an explanation seems highly plausible in principle, as differences in the rate of hydrolysis and extraction of DNA in condensed and dispersed chromatin are well known (Duijndam & van Duijn 1975, Mello & Vidal 1980). Feulgen banding is very dependent on the length of hydrolysis, as shown by the work of G. E. Marks (1983) on the chromosomes of three species of plants (Fig. 4.9). In *Anemone blanda*, short hydrolysis gave uniform staining, and long hydrolysis gave dark staining of heterochromatic bands. Short hydrolysis of the chromosomes of *Scilla siberica* again resulted in uniform staining, but longer hydrolysis produced palely stained heterochromatic bands. Finally, in the chromosomes of *Fritillaria lanceolata*, heterochromatic bands were stained more weakly than euchromatin after short hydrolysis, but more strongly after prolonged hydrolysis. All these results can be explained in terms of differential rates of hydrolysis and extraction of DNA, and serve to emphasize that different classes of DNA are being demonstrated. The results of Yamasaki (1973) on *Cypripedium* and of Ennis (1975) on *Chilochorus* suggest that Feulgen positive bands are equivalent to quinacrine negative bands and *vice versa*, and Greilhuber (1975) suggested that there was a similar relationship between Hy-bands and Q-bands. However, G. E. Marks (1983) found the opposite relationship in the species he studied. Assuming that the intensity of quinacrine fluorescence is an indication of DNA base composition (Section 7.4), it is clear that Hy-bands or Feulgen bands do not represent regions of any particular base composition, and indeed there appears to be no evidence that base composition alone affects the rate of hydrolysis of DNA. The cause of differential staining of bands after hydrolysis must be sought in the proteins associated with the DNA. Yamasaki (1971) showed that the densely stained heterochromatic bands of *Cypripedium* were unstained by the fast green method for histones, whereas in *Trillium*, the weakly stained heterochromatic bands stained as strongly with fast green as the euchromatin. Whether such a difference is sufficient to explain the phenomena described in this section is not certain, but as with C-banding (Section 4.2.5) distinctive types of protein interacting with the DNA seems to be the most likely explanation of this type of band.

4.6 Silver staining of heterochromatin

Since silver staining has been used to demonstrate chromosomal components such as DNA (Korson 1964) and proteins (Black & Ansley 1964), and chromosomal structures such as cores (Howell & Hsu 1979), nucleolar organizers (Ch. 9) and kinetochores (Ch. 10), it hardly seems surprising that silver has also been used to demonstrate heterochromatic bands. These methods are, however, few and of little practical significance. They are included because of the insight which they may provide into the nature of heterochromatin and its differentiation as chromosome bands.

Utsumi & Takehisa (1974) used the ammoniacal silver method described by Black & Ansley (1964) to stain the chromosomes of the plant *Trillium kamtschaticum*, in the belief that heterochromatic bands differed in protein content from euchromatic regions. They found that certain C-bands were stained with silver, and that the silver staining was prevented by prior treatment with cold hydrochloric acid, showing that histones were being stained. They suggested that the composition of the histones differed from that of euchromatin on the basis of the colour of the silver staining, in accordance with the observations of Black & Ansley (1966). Howell & Denton (1974) also applied an ammoniacal silver technique to chromosomes, and obtained specific staining of the heterochromatin of human chromosome 9. These authors omitted the formaldehyde pre-treatment required for histone staining and made no claims about the specificity of their method, although they drew a parallel with the Giemsa-11 method

(Section 4.3). Two years later, however, the same authors, believing that the silver was reacting with a non-histone protein, treated chromosomes with a hot alkaline solution to extract non-histone proteins and obtained negative silver staining of heterochromatic bands (Howell & Denton 1976).

These reports of silver staining do not provide very specific information about proteins in heterochromatic bands. Nevertheless, they do add to the small amount of direct evidence that heterochromatic bands contain distinctive proteins.

4.7 Concluding remarks

This survey of methods for staining heterochromatic bands has highlighted a number of points about them. C-banding in particular has almost come to be regarded as representing all heterochromatin, yet this is clearly not so in many cases. There are relatively few species in which attempts have been made to correlate the bands revealed by the newer staining methods with heterochromatin as defined by more classical methods, such as failure to decondense at telophase, and the induction of constrictions by cold treatment in some plants. However, it does seem safe to say that all C-bands are heterochromatic, but that some heterochromatin is not stained by C-banding methods. Other methods that stain heterochromatic bands (including some fluorescence methods to be described in Ch. 7 & 8) generally show up a more restricted range of bands, but may nevertheless reveal bands not demonstrated by C-banding. This leads to the observation that heterochromatic bands are in fact very heterogeneous, something which is obscured by the wide use of C-banding. When other staining techniques are used, as well as analysis of the type of DNA, it becomes clear that a large number of types of heterochromatin exist.

The heteromorphism of heterochromatic bands, a subject to be considered in detail in Chapter 14, is also very characteristic, and should perhaps be considered with the concept of the inertness of heterochromatin. How else can the genome tolerate substantial fluctuations in the size and composition of some of its components unless those parts have no important function? It is true that variations in heterochromatin are often associated with speciation (Ch. 15), but it is not known whether these variations are cause or effect. Certainly within a species it has proved remarkably difficult to find functions for heterochromatin. Apart from the absence, or at least the very low density, of genes, most of the functions that have been suggested remain without supporting evidence. There are only indications that perhaps heterochromatin has some function in the germ line, and in particular that the size and position of blocks of heterochromatin may regulate the position of chiasmata. So far this appears to be the only function of heterochromatin for which there is widespread evidence.

Finally, a word about the nature of heterochromatic bands. Years of research have established that they often contain highly repeated DNA sequences (but apparently not always), and that they may contain distinctive proteins which are bound particularly strongly to the DNA. These observations help to explain a good deal about the staining properties of heterochromatic bands. Their content of highly repetitive DNA sequences would explain the virtual absence of genes, but at present it cannot be said that a knowledge of the organization of heterochromatin, in terms of biochemistry and molecular biology, has provided any significant clue to their function. It would be easy to dismiss them as 'junk', having no reason for existence, or as parasitic material whose end is its own perpetuation. These are negative hypotheses, however, which are unlikely to lead to advances in our understanding of the functions of heterochromatin. In any case, we already know that a few functions or effects can be ascribed to heterochromatin, although in a few organisms only. The ready availability of methods for staining heterochromatic bands, and for distinguishing them with a battery of techniques, must help the quest for the functions of heterochromatin.

5 G-banding

5.1 Definition of G-banding

G-bands may be defined as a system of alternating dark and light bands throughout the length of the euchromatic parts of chromosomes. This brief definition requires some amplification and comment. First, the definition above applies equally to R-bands (Ch. 6), although G-banding and R-banding produce complementary patterns. It could be said that positive G-bands are the same as positive Q-bands, but this similarity only applies to euchromatic bands, since there are striking quinacrine fluorescence patterns in heterochromatin that are not revealed by G-banding. Second, comparison with Q-banding raises the point that distinctive patterns of Q-banding can be demonstrated in chromosomes of most organisms that have been studied. These Q-bands are generally heterochromatic, and the euchromatic bands which we are considering in this chapter are almost entirely confined to the amniote vertebrates: reptiles, birds and mammals. G-banding methods do not, in general, produce distinctive patterns on the chromosomes of other organisms.

We are, in fact, forced back to the circular definition that G-bands are those bands produced by G-banding techniques. The G-banding pattern of human chromosomes has been defined by the Paris Conference of 1971 (ISCN 1985) and this arbitrary, but very necessary, decision can be taken as the standard against which a G-banding method can be assessed. G-bands would therefore be those bands produced by a method which produces the internationally agreed G-banding pattern on human chromosomes. Ultimately, no doubt, it will be possible to define G-bands in terms of chromosome organization, but at present, although there are a number of clues (Section 5.3) it is not yet possible to define G-bands in this way.

The term G-banding arose because of the general use of Giemsa's dye mixture to stain such bands, after appropriate treatment of the chromosomes. However, when plant cytogeneticists refer to Giemsa banding, they are in fact referring to heterochromatic bands (C-bands), since as already mentioned, G-bands cannot normally be produced in plants. Finally, it should be noted that according to the decision of the Paris Conference (ISCN 1985), the chromosomes consist of a series of light and dark bands, rather than G-bands and interbands. Thus it is appropriate to refer to positive and negative G-bands. Nevertheless, a considerable number of cytogeneticists appear to regard the darkly stained bands obtained with a G-banding method as the G-bands, while the pale G-bands, which are stained darkly by R-banding methods (Ch. 6), are described as R-bands. According to the Paris Conference these should be G-positive or R-negative, and G-negative or R-positive, respectively. The description of bands may thus become somewhat confusing, and it could well be that in time it will become the convention to regard chromosomes as consisting of alternating G- and R-bands.

5.2 G-banding methods

5.2.1 Review of G-banding methods

G-banding methods have evolved along two principal independent lines, which have occasionally converged, plus a substantial number of miscellaneous methods which have no clear ancestry. One of the principal lines of evolution is from C-banding methods, using incubation in salt solutions, whereas the other line involves the use of proteolytic enzymes. It is a curious fact that while G-banding methods continue to be less reliable than one would wish, and new methods are still being published, some of the earliest methods devised continue to be the most widely used. The greatest development of new methods took place between 1971 and 1973, although more recently there has been the introduction of 'high-resolution' banding, the induction of bands on elongated prophase chromosomes. This latter topic is considered in Section 5.2.4.

In late 1970 and early 1971, several cytogeneticists were experimenting with C-banding techniques, which then consisted of alkaline treatment followed by 'annealing' in a hot solution of sodium chloride and citrate (SSC), followed by Giemsa staining. It was soon discovered that a milder alkaline treatment resulted in a pattern of bands throughout the length of human chromosomes (Drets & Shaw 1971, Schnedl 1971a). Indeed, the best results were obtained if the alkaline treatment was omitted altogether, leading to the ASG technique (Sumner *et al.* 1971a) which has continued to be one of the most widely used G-banding techniques (Section 5.2.2). This method consists of incubation in $2 \times$ SSC at 60°C for 1 h, followed by staining with Giemsa. Several variants of this have been published, some involving only minor variants of time and temperature, while others have used different salt or buffer solutions (Chaudhuri *et al.* 1971, Bhasin & Foerster 1972, Bosman & Schaberg 1973, Meisner *et al.*. 1973b, Beil & Limon 1975). Little would be gained by listing all the papers describing methods of this sort, many of which should be regarded as minor technical variants. It should be pointed out, however, that some workers continued to find useful treatment with sodium hydroxide before incubation in buffer (Crossen 1972, Murtagh 1977). Others have found treatment with dilute acid before incubation in $2 \times$ SSC to be beneficial (Bath & Gendel 1973, de la Maza & Sanchez 1976).

Kato & Moriwaki (1972), in an attempt to understand the mechanism of banding, tested no less than 73 reagents for their ability to produce G-bands on mammalian chromosomes. They found that acids were ineffective, but that salts could induce bands, particularly if the solutions were alkaline. Other classes of substances which Kato & Moriwaki (1972) found to induce G-bands were strong alkalis, which act in only a few seconds; protein denaturants such as urea and guanidine hydrochloride; and various detergents. Several of these substances were used in G-banding techniques published at about the same time or subsequently. For example, Berger (1972a) used dilute sodium hydroxide without any other pretreatment, but no doubt the alkaline treatment would be too destructive and not easily controlled. Urea, however, has proved more popular and has been used by several authors (Berger 1972b, Kato & Yosida 1972, Shiraishi & Yosida 1972). Detergents have also been quite widely used. Sodium dodecyl sulphate (SDS) has probably been used most (Yosida & Sagai 1972, Lee *et al.* 1973), but sodium deoxycholate and a variety of commercial products have also been applied successfully (Lee *et al.* 1973, Stephen 1977).

Among the earliest G-banding methods was the 'Giemsa-9' technique (Patil *et al.* 1971), in which chromosome preparations were simply stained with a Giemsa solution adjusted to pH 9, without any prior treatment. With the correct staining time and pH, good G-bands were obtained, although this simple method did not catch on. It is nevertheless important as it leads to a series of observations that G-bands can be induced using a simple Giemsa solution without any pretreatment (Seleznev 1973, Yunis & Sanchez 1973, Walther *et al.* 1974, Mäkinen *et al.* 1975, Steiniger & Mukherjee 1975, Lichtenberger 1983). Although these methods differ in detail, in general they involve the use of relatively fresh chromosome preparations, and relatively dilute Giemsa solutions. Ionic strength and pH are also important factors in producing G-bands by direct staining (Mäkinen *et al.* 1975, Bignone *et al.* 1983). As with C-banding, it is normal practice to leave preparations a few days before attempting to G-band them, but this is not required, and indeed seems undesirable, for direct staining. In spite of their simplicity, these methods have had limited application, and it may be supposed that, as with so many G-banding methods, they are not sufficiently reliable for routine use. A recent variant of this approach involves treatment of fresh chromosome preparations in an oven at 100°C and 98% relative humidity before staining (den Nijs *et al.* 1985). The technique was designed to obtain good quality G-bands from chromosomes of leukaemic cells, from which it is notoriously difficult to obtain good preparations.

There is a small group of G-banding methods which have in common the use of chemical reagents that are believed to affect chromosomal proteins, although the reagents themselves are diverse. These methods are of value in understanding the mechanisms underlying G-banding, although they have not been adopted as practical banding techniques. Kato & Yosida (1972) obtained good G-banding using a mixture of urea, SDS, and 2-mercapto-ethanol, which had previously been used for extracting protein from chromatin. Utakoji (1973) treated chromosomes with permanganate to produce G-bands, an effect attributed to the oxidation of disulphides and sulphydryls in chromosomal proteins. Rønne & Sandermann (1977) also used reagents which had previously been employed to extract proteins from chromatin: ammonium sulphate solution followed by either trichloracetic or perchloric acid. The quality of bands was stated to be superior to that produced by the ASG method.

Mention of methods using reagents which affect chromosomal proteins leads naturally to a consideration of the most important class of such methods, that is, the use of proteases to produce G-bands. Although the first such method was published earlier in 1971, using pronase (Dutrillaux et al. 1971), it was the publication later in the year of a method using trypsin (Seabright 1971) which popularized such methods. So successful has Seabright's method become that it is almost certainly the most widely used and quoted banding method in the world. This method, described fully in Section 5.2.3, consists simply of treatment of standard metaphase chromosome preparations with a trypsin solution for several seconds, followed by staining in Leishman's solution (like Giemsa, one of the Romanowsky dye mixtures). Many variants of Seabright's method have been published, usually with apparently unimportant variations in enzyme concentration, buffers, times or temperatures.

Various workers have tried alternatives to trypsin, and a wide variety have proved effective, although apparently showing no superiority to trypsin, which is readily available and is probably the most economical protease. Seabright herself reported that other proteases could be used, and several others are listed in Table 5.1. Neumann et al. (1980) tested the effects of pretreatments with various detergents on trypsin banding. Such pretreatments led to sharper bands, and on some chromosomes extra bands could be identified. This improvement has apparently not been found valuable in practice.

Alternatives to the usual Romanowsky stains (Giemsa, Leishman, or Wright's stain) have been tried following trypsin treatment, but have not been found to have any advantage: Tabuchi et al. (1979) used a methyl green-pyronin mixture, while Scheres et al. (1982) employed an alkaline solution of basic fuchsin. Others have introduced techniques involving only a single stage, by mixing the trypsin and Giemsa solutions (Sperling & Wiesner 1972, Sun et al. 1974). Little simplification is obtained in this way, but the development of bands is more controllable (5–25 min, rather than seconds) and moreover can be watched under a microscope.

In view of the success of both the ASG technique, using pretreatment in hot $2 \times$ SSC, and of the trypsin methods, it is perhaps not surprising that attempts have been made to combine the two. Gallimore & Richardson (1973) treated chromosome preparations first with $2 \times$ SSC followed by trypsin, while Bulatova & Radjabli (1974) used these reagents in the reverse order. The former method was devised for rat chromosomes, and the latter

Table 5.1 Proteolytic enzymes (other than trypsin) used to induce G-bands on mammalian chromosomes.

Proteolytic enzyme	Reference
Bromelain	Seabright 1972
α-chymotrypsin	Finaz & de Grouchy 1971
Collagenase	Trusler 1975
Ficin	Seabright 1972
Pankreatin	Müller & Rosenkranz 1972
Papain	Seabright 1972
Pepsin	Rønne et al. 1977
Pronase	Dutrillaux et al. 1971
Thrombin	Lee et al. 1978
Viokase	Kozak et al. 1977

for avian chromosomes, and indeed Gallimore & Richardson's method seems to be especially valuable for banding chromosomes from a variety of mammals (Short et al. 1974, Seuanez et al. 1976a).

Finally, patterns similar to G-bands have been observed on Feulgen stained chromosomes. Yunis & Sanchez (1973) observed banding without any pretreatment, but the contrast of the Feulgen bands was so low that even using an appropriate filter and high contrast photographic methods the bands were much less clear than with Giemsa. Rodman (1974) and Rodman & Tahiliani (1973) obtained much clearer bands in chromosomes of mouse and man, but used a pretreatment consisting of alkaline buffer followed by prolonged exposure to 12 ×SSC. Although in general the pattern of bands resembled that shown with Giemsa staining, a distinctive feature was the relatively dense Feulgen staining of the telomeric regions of many chromosomes.

G-banding techniques have always proved somewhat variable in their results, and attempts have been made to improve the slide preparation with the intention of making the results more reproducible. Islam & Levan (1987) varied the composition of the fixative, while van Prooijen-Knegt et al. (1982) treated glass slides with aminoalkylsilane to improve spreading. Improved banding was claimed with both procedures.

5.2.2 ASG method

The ASG (Acetic/Saline/Giemsa) method (Sumner et al. 1971a) consists of two stages. Conventional methanol-acetic acid fixed, air-dried chromosome preparations are treated as described below. The use of a fixative containing acetic acid is the 'Acetic' part of the title of the method.

1 Chromosome preparations are treated for 1 h at 60°C in 2 × SSC (0.3M sodium chloride containing 0.03M tri-sodium citrate).

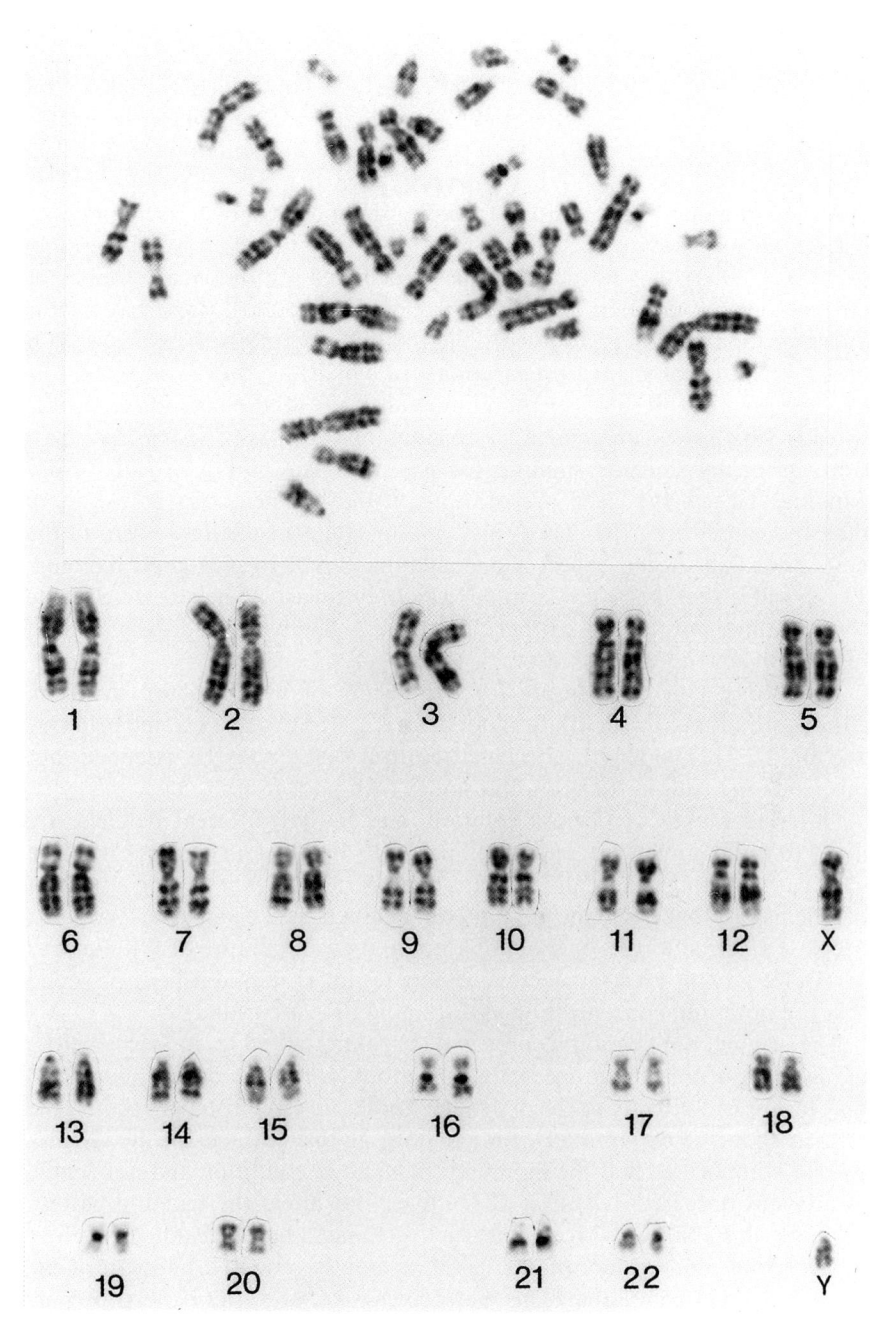

Figure 5.1 G-banded human metaphase chromosome spread and karyotype. Reproduced from Sumner et al. (1971) *Nature New Biology* **248**, 55 by permission of the authors. Copyright © 1971 Macmillan Magazines Ltd.

2   After rinsing in distilled or deionized water, the slides are stained with Giemsa for 45 min (Gurr's Giemsa R66 stock solution (BDH Ltd.), 1 ml diluted to 50 ml with buffer, pH 6.8, made from Gurr's buffer tablets).
3   The slides are rinsed again with water, blotted dry, and mounted.

A human metaphase spread prepared in this way is shown in Figure 5.1.

Like all G-banding techniques, the ASG technique is not invariably successful. (This fact no doubt helps to account for the large number of different techniques that have been proposed.) The most important factor affecting the result is undoubtedly the quality of fixation. It is essential to obtain clean chromosome preparations, in which the chromosomes are as free as possible of any surrounding material. Metaphase chromosome spreads which are surrounded by remains of cytoplasm, visible by phase contrast or by Giemsa staining for example, will not give satisfactory banding. To achieve clean preparations, it may be necessary to use more than the usual three changes of methanol-acetic acid fixative. Even on the best slides one should expect some metaphases not to give good banding. These will either be a few around which cytoplasm persists, or else the chromosomes will be too strongly contracted. The best banding is seen on reasonably extended chromosomes.

No artifice can compensate for the results of unsatisfactory fixation, although treatment with dilute alkali before 2 × SSC may be some help (see Section 5.2.1). The length of incubation in 2 × SSC may be extended, but this tends to result in fuzzy chromosomes with poor morphology.

Different makes of Giemsa solution, and indeed different batches, are likely to differ somewhat, and it will probably be necessary to experiment a little with staining times, and perhaps concentrations, when using a new lot of Giemsa. Other Romanowsky mixtures, such as Leishman's, Wright's, McNeal's Tetrachrome, or the simple azure B-eosin mixture of Wittekind *et al.* (1982) will all give satisfactory results (Sumner, unpublished) provided that the optimum concentration and staining time are found.

The original publication recommended soaking the slides in xylene before mounting, but this is not necessary. The mountant (DPX in my laboratory) may be applied directly to the dry slides. Under no circumstances should the stained slides be dehydrated with alcohols, as this extracts all the Giemsa dye. It is important that the mountant be in good condition and not acidic; acidity not only causes fading of the dyes, but alters the banding pattern (Sumner, unpublished). Mountants such as Canada balsam cannot therefore be used, but modern mountants based on polystyrene and known organic solvents should be reliable. This is not always so, and if G-banded preparations appear to be fading, it is wise to obtain a fresh supply of mountant. It is not known whether faded preparations can be restained successfully. Rønne *et al.* (1977) recommended the use of Spurr's resin, an embedding medium for electron microscopy, as a mountant for G-banded chromosomes. They claimed that this resulted in more stable preparations than those mounted in resins dissolved in organic solvents. Since Spurr's resin is reputed to be carcinogenic it cannot be recommended, but it may be that other polymerizable resins might prove to be effective substitutes.

It is often stated that chromosome preparations should be left for a few days (say, three days to a week) before banding (e.g. Pearson & van Egmond-Cowan 1976). There is undoubtedly an effect of ageing on the quality of banding, but the rate of ageing is certainly affected by environmental factors such as temperature and humidity. Zuelzer et al. (1973) and Salamanca & Armendares (1975) obtained G-bands on some slides up to 10 years old, although the quality was inferior to that of fresher preparations. Perfectly good G-banding has been obtained on slides several months old, but this cannot be relied upon. A few days' ageing definitely seems to be best.

5.2.3 Seabright's trypsin method

The original trypsin G-banding method consisted of the following stages (Seabright 1971):

1 After fixing and making chromosome preparations in the standard way, slides are placed horizontally and flooded with 0.25% trypsin (Difco) in isotonic saline for 10–15 s. The slide is then rinsed twice in isotonic saline. At this stage the slide may be examined by phase contrast microscopy to assess the action of the enzyme. No bands are visible without staining, but the phase difference should be reduced compared with untreated chromosomes, and the chromosomes should appear slightly swollen (Seabright 1972). Obvious distortion indicates excessive treatment.
2 Stain for 3–5 min with Leishman's stain (BDH Ltd.) diluted 1:4 with Gurr's buffer, pH 6.8.
3 Rinse well with buffer, pH 6.8. Blot dry, place in xylene, and mount in a neutral medium.

Several of the more general points about G-banding in the previous section (5.2.2) also apply here. Good quality chromosome preparations are always desirable. The remarks about mounting are also relevant here. Trypsin banding was claimed to be less sensitive to the age of the slide, even slides up to three years old being useable (Seabright 1971). Later, however, a treatment with hydrogen peroxide was recommended for very new or very old slides (Seabright 1973). The effect on new slides could be that of accelerated ageing, but it is not clear how hydrogen peroxide acts on old slides. I have no information that this treatment has been widely adopted.

The main variable in Seabright's technique is the length of trypsin treatment. Although 10–15 s was originally specified, the length of treatment required is much more variable than that, depending not only on the age and quality of the chromosome preparations, but also on the batch of trypsin used. Seabright herself pointed out that the more refined the trypsin, the more precise and constant its effect. Nevertheless, the chief disadvantage of this method is the necessity to determine the optimum conditions of treatment for each batch of slides and for each lot of trypsin. The great popularity of the technique is proof that these are not considered serious disadvantages.

The original method prescribed Leishman staining, but as with the ASG technique it is probable that any Romanowsky mixture will be effective, subject to appropriate adjustment of concentration and staining time. Probably most trypsin G-banding methods have specified Giemsa rather than Leishman (e.g. Wang & Fedoroff 1972).

In a straightforward popularity contest, Seabright's trypsin method would probably receive more votes than the ASG method. This probably implies that the trypsin method is more suitable for routine use. In other words, it is regarded as quick, simple and reliable. Nevertheless, both methods have their advantages and disadvantages. So far as I am aware there are no figures on the reliability of the two methods, and I suspect that at least on good quality chromosome preparations the success rate of the two methods is not significantly different. The author can hardly be considered entirely unbiased, but his view that chromosome morphology is retained better after the ASG procedure has independent support (Pearson & van Egmond-Cowan 1976). Trypsin tends to produce a fuzzy outline to the chromosomes, but on the other hand the bands themselves may appear sharper. For routine use, good chromosome morphology may be less important than obtaining the clearest banding pattern.

5.2.4 *High-resolution banding*

Even during the earliest days of G-banding it was observed that more elongated chromosomes showed more bands. More detailed banding patterns would be useful to locate more precisely chromosome breakpoints in translocations, for example, or to identify small deletions which might be associated with genetic diseases (Sections 5.5.2 & 5.5.4). It is therefore not surprising that observations were soon made of G-banding patterns on prophase chromosomes. In a standard chromosome preparation made from cells arrested in metaphase with colchicine, a small proportion of the mitotic cells have chromosomes still in prophase. Such cells were used in earlier studies, and a number of workers have studied high-resolution banding patterns without using any special treatment of the cells (Patterson & Petricciani 1973, Yunis & Sanchez 1975, Francke & Oliver 1978). Recently, Cheung *et al.* (1987) have obtained high-resolution banding by careful control of such parameters as colcemid concentration, fixation and slide preparation. The use of Ohnuki's medium as a hypotonic treatment before fixation (Bigger & Savage 1975) or a prolonged colcemid treatment in the cold can probably also be regarded as situations in which minimal special treatment was applied. High-resolution banding has also been obtained on prematurely condensed chromosomes (Unakul et al. 1973, Röhme & Heneen 1982). In all these situations standard G-banding methods were applied.

The number of spreads in which elongated chromosomes occur naturally is very low, and it is not possible to arrest cells in prophase as can be done in metaphase with spindle poisons. Efforts to obtain chromosome preparations with a high mitotic index and a high proportion of elongated chromosomes have, therefore, been directed towards synchronizing the cells in a culture, and then harvesting them at a time after removal of the agent that induces synchrony, such that as many cells as possible will be in prophase. It is clear that some of the reagents used to obtain a high level of prophase cells also induce banding themselves – indeed, this has been done deliberately – and should properly be considered in Section 11.3.5. For the sake of completeness, however, such methods will also be mentioned here. More recently, reagents have been added during the final stages of culture that apparently have the effect of inhibiting normal condensation. These have either been used alone or in conjunction with synchronization techniques.

Yunis (1976) used methotrexate (amethopterin) to block cell cultures at the beginning of the S phase of the cell cycle. The block was released by changing to medium free of methotrexate and rich in thymidine, and culturing for a precisely determined time, at which a large number of cells were found in prophase. Viegas-Péquignot & Dutrillaux (1978) used a blockade with a high concentration of thymidine for the same purpose, while Rønne (1984) proposed the use of fluorouracil for synchronization. For leukaemic cells, Rapoza et al. (1984) found that cold synchronization gave superior results to the other methods they tested for producing elongated chromosomes. It appears that in each of these cases the synchronizing agent does that and nothing else. When a reagent such as bromodeoxyuridine (BrdU) is used for synchronization (Dutrillaux & Viegas-Péquignot 1981) the situation is more complicated, however. BrdU appears to block the cell cycle in the middle of S phase, and is therefore incorporated into those parts of the chromosomes that replicate early. Incorporation of BrdU into chromosomes induces banding patterns which depend on the portion of the S phase during which this modified nucleotide is incorporated into the DNA. Thus these bands are *replication bands* (Section 11.3) indicating the pattern of chromosome replication, and not G-bands in the strict sense. As we shall see (Section 5.3) the two types of bands may be similar but not necessarily identical. (Replication bands may also be similar to R-bands (Ch. 6) depending on whether early or late replication patterns have been induced.) In fact, BrdU has been used to release from blockage in the cycle of cells synchronized with methotrexate (Yunis 1982), thymidine (Scheres et al. 1982) or fluorouracil (Rønne 1984), with the deliberate intention that it should induce replication bands.

Another stratagem for inducing elongated chromosomes is to add an agent to the culture which inhibits chromosome condensation evenly throughout the chromosomes, and among the chemicals used are certain dyes that bind to DNA without showing any base preference. Matsubara & Nakagome (1983) used acridine orange for 30 min, and Ikeuchi (1984) proposed the use of ethidium bromide for 2 h before harvesting the cells. These treatments appeared merely to elongate the chromosomes, and not to induce or alter G-banding patterns. Ethidium has also been used as a supplement to methods involving synchronization of cells to produce prophase chromosomes (Rønne 1985). A different approach to inhibiting condensation was used by Kao et al. (1983). On the basis that chromosome condensation involved formation of disulphide bonds, they added 2-mercaptoethanol, a reagent known to break disulphide bonds, to the hypotonic treatment immediately before fixation.

Obtaining a good yield of cells in prophase is only part of the process of achieving high-resolution banding. Very good spreading of the chromosomes is necessary if they are to be analysable, and this not only involves high quality fixation, but possibly also special methods of making the slides. To ensure that the chromosomes are not inhibited from spreading by surrounding cytoplasm, extra changes of fixative may be necessary (Scheres et al. 1982). Leaving cells in fixative overnight before making slides is believed to result in more elongated chromosomes (Rønne et al. 1979). Many workers have used standard methods for spreading chromosomes on slides, although Yunis (1981, 1982) dropped the cell suspension from a great height on to a cold slide held at an angle and wetted with water or acetic acid. The fixative was then evaporated rapidly from the slide by blowing.

After all these complex manipulations, G-banding of elongated chromosomes appears to be somewhat easier than on metaphase chromosomes. Yunis (1976) used direct staining with Wright's stain, without any pretreatment, while others have used standard trypsin-Giemsa methods (e.g. Francke & Oliver 1978, Kao et al. 1983, Matsubara & Nakagome 1983).

It is not proposed to pick out any one method as being particularly reliable or popular. Although the banding of extended chromosomes has been studied for some 15 years, methods for their study are still being actively developed. In any case, it may be necessary to try more than one method on any particular sample to ensure optimal results.

Finally, a note on nomenclature. The report of the Standing Committee on Human Cytogenetic Nomenclature (ISCN 1985) refers to High-Resolution Banding, and this is a preferable term to prophase banding. Several authors have referred to chromosomes of different degrees of contraction as being in prophase, prometaphase or metaphase. However, these phases are defined by factors other than the degree of contraction of the chromosomes, such as whether the nuclear membrane is disrupted, and by the arrangement of the chromosomes on the spindle (Rieger et al. 1968). Since extra-chromosomal structures are swept away during the preparation of a chromosome spread, it is obviously not possible to say with any accuracy at what stage of the cell cycle the chromosomes are. It is better to describe the degree of extension of the chromosomes by the number of bands that can be recognized. This, after all, is the purpose of high-resolution banding. ISCN (1985) illustrates human karyotypes with approximately 400, 550 and 850 bands, while Yunis (1981) has claimed 2000 bands.

5.3 Characteristics of G-bands

This section is concerned with the question of what the differences are in chromosome organization between positive and negative G-bands. As we shall see later, nearly all R-bands (Ch. 6) and the majority of Q-bands (Ch. 7) in higher vertebrates are equivalent to G-bands, as are replication bands. A complete consideration of the properties and relationships of euchromatic bands cannot be made until all the different types of banding have been described, and a full discussion of these matters is deferred to Section 16.3.

G-bands can be demonstrated easily and regularly on chromosomes of higher, amniote, vertebrates (reptiles, birds and mammals), but only sporadically in other animals and in plants. Reasons have, therefore, been sought for the absence of G-bands and other euchromatic bands on the chromosomes of virtually all organisms except amniotes. Greilhuber (1977b) proposed that plant chromosomes did not show G-bands because they were much more contracted at metaphase than those of mammals, and that because of the limits of resolution of the light microscope, any bands present would be too close to be resolved. However, both Anderson *et al.* (1982) and Bennett *et al.* (1983) have shown that the density of DNA in metaphase chromosomes of a variety of plants and mammals was rather similar, whether measured as amount of DNA per cubic micrometre, or as amount per micrometre of chromosome length.

It must be asked whether G-bands are really absent from the chromosomes of all organisms except higher vertebrates, or whether the failure to demonstrate them is simply a consequence of technical factors. This is a very real possibility; in those lower organisms in which satisfactory G-banding has been obtained, considerable technical effort has often been required to get good results. It has already been noted (Section 5.2.2) that satisfactory G-banding can only be obtained in chromosomes which are free, or almost so, of surrounding cytoplasmic material. Since the standard methods of preparing plant and insect chromosomes are usually squashing, it is to be expected that the chromosomes will still be surrounded by a lot of cytoplasmic material, and thus could not be G-banded purely for technical reasons. Careful study of the literature indicates that this could well be so. It must be admitted that many of the examples of G-banding claimed in, for example, fish, insects and plants are of such poor quality that they are not worthy of serious consideration. In recent years, however, good quality G-banding has been obtained in various amphibia (Stock & Mengden 1975, Stock 1984, Cuny & Malacinski 1985) and fish (Blaxhall 1983, Wiberg 1983). In all cases chromosome preparation was done using methods similar to those used for mammalian chromosomes, rather than squash techniques. Similarly, Steiniger & Mukherjee (1975) used this type of preparation for mosquito chromosomes, and obtained G- and Q-bands of reasonable quality. Lester *et al.* (1979) obtained quite good G-bands in housefly chromosomes. Bigger (1975) also showed some G-banding on butterfly chromosomes, and Brum-Zorrilla & Postiglioni (1980) on spider chromosomes. Among plants, there have been a few reports, unfortunately not so far confirmed, of quite detailed G-banding, especially in prophase chromosomes (Drewry 1982, Murata & Orton 1984, Wang & Kao 1988). It should be noted that in certain insects, treatment of the chromosomes with a G-banding technique gives a fairly detailed banding pattern, but only in certain chromosomes (Rees *et al.* 1976, Webb 1976, Webb and Neuhaus 1979). Although certain of these bands are regarded by the authors as being equivalent to G-bands in mammalian chromosomes, it is not in fact clear that they are anything other than a subclass of heterochromatic bands.

It was established in the early days of the study of euchromatic bands that their patterns do not change during embryonic development and differentiation (Burkholder & Comings 1972, Nesbitt & Donahue 1972) and that similar banding patterns are found in chromosomes from different tissues (Caspersson *et al.* 1972). The matter has not been explicitly investigated with high-resolution banding techniques, but the ability to use these techniques to identify consistent chromosomal alterations in human neoplasms, for example (Yunis 1983), implies that alterations in banding patterns due to other causes are absent. It would, nevertheless, be valuable to confirm that no changes in high-resolution banding patterns occur during development and differentiation. For the present, however, all the evidence indicates that G-banding and related patterns are a fixed, intrinsic feature of the chromosomes.

G-banding patterns are, of course, only comparable when chromosomes are compared at a similar degree of contraction. Elongated prophase chromosomes have many fine G-bands, while many fewer but larger G-bands are visible on metaphase chromosomes. In maximally contracted chromosomes, differentiation may be absent; in effect, one band occupies the whole chromosome (Sumner 1976). This striking reduction in the number of bands is achieved by fusion, two adjacent positive bands swallowing up the pale band between them (Seabright *et al.* 1975, Yunis *et al.* 1978, Röhme & Heneen 1982, Sen & Sharma 1985). To a limited extent, pale G-bands may also fuse with the elimination of the intervening dark band. Observations of banded chromosomes with the electron microscope do not reveal any extra bands (Gormley & Ross 1972), and it is concluded that a genuine fusion of bands occurs (Sumner 1976). These observations do not support the alternative possibility that as the chromosome contracts, the bands are brought too close together to be resolvable by light microscopy. It may be concluded that G-bands are a reflection of the process of chromosome contraction. They do not represent a fixed part of the genome, but increase during prophase to extend over more and more of the chromosome.

Additional evidence for a connection between G-bands and the process of chromosome condensation comes from a comparison with the patterns of pachytene chromomeres. The chromomeres of meiotic prophase chromosomes are evidently the regions which condense first (Section 3.4), and a resemblance to patterns of G-bands would imply that the latter were also the first-condensed regions of mitotic chromosomes. In fact, several authors have reported that G-banding patterns resemble those of pachytene chromomeres, in Chinese hamster (Okada & Comings 1974, Jhanwar & Chaganti 1981), the mouse (Jagiello & Fang 1980), and in man (Ferguson-Smith & Page 1973, Luciani et al. 1975, Hungerford & Hungerford 1978, 1979). Satisfactory G-banding of pachytene chromosomes is not practicable, and the comparisons between pachytene chromomere patterns and mitotic banding patterns are subjective and not always convincing. A more objective approach was used by Ambros & Sumner (1987). When chromosomes are doubly stained with the fluorochromes DAPI and chromomycin A_3, which are believed to bind to DNA according to its base composition (Ch. 8), DAPI produces a banding pattern equivalent to G-bands. It was found that most pachytene chromomeres were DAPI positive (that is, equivalent to positive G-bands), although a few, particularly at the ends of chromosomes, were chromomycin positive and thus equivalent to negative G-bands. The observations described in this paragraph, therefore, reinforce the conclusion that G-bands are a manifestation of chromosome condensation, as well as emphasizing the resemblance between meiotic and mitotic chromosomes.

It was established early on that there is a correlation between patterns of euchromatic bands and the sequence of DNA replication in chromosomes. More specifically, it was shown that positive Q-bands (which are generally equivalent to G-bands, see Chapter 7) generally replicate their DNA late in the S phase (Ganner & Evans 1971, Calderon & Schnedl 1973). In these early studies, the time of replication of DNA was assessed by autoradiography of tritiated thymidine, and thus the resolution was inadequate for any detailed comparison. However, the introduction of methods for studying chromosome replication by the incorporation of bromodeoxyuridine (BrdU) meant that replication patterns could be obtained with as good a resolution as the banding patterns (Section 11.3). Using such methods, it was established that negative G-bands (positive R-bands) were early replicating, while positive G-bands were late replicating (Section 11.3.4). Although subsequent work has upheld and reinforced this general conclusion, it has been found that patterns of replication and of G-banding do not necessarily agree in all details. As Epplen et al. (1975) pointed out, 'banding and replication patterns are independent phenomena', although there is clearly a strong linkage between them.

Suggestions that G-banding patterns were related to the distribution of different classes of DNA along the chromosomes were made at the time that

G-bands were first discovered (Sumner *et al.* 1971a), although the finding that G-bands could be induced by proteolytic enzymes made it unlikely that differences in DNA were the direct cause of G-banding. In fact, there appear to be several differences in the quality of DNA between positive and negative G-bands. Differences have been reported in overall DNA base composition, in the type of DNA sequence, and in the type and distribution of genes.

Evidence from a variety of sources indicates that the DNA in positive G-bands is relatively rich in the bases adenine and thymine (A + T-rich), while that in the negative G-bands is relatively rich in guanine and cytosine (G + C-rich). Direct evidence comes from the use of base-specific fluorochromes (Ch. 8) and from studies of immunostaining with antibodies to different DNA bases (Ch. 13). Indirect evidence for the non-homogeneous distribution of DNA bases along chromosomes comes from the study of DNA replication. As we have just seen, positive G-bands correspond in general to late replicating regions of chromosomes, and negative G-bands to early replicating regions. There is abundant evidence that the early replicating DNA is relatively G + C-rich, and the late replicating DNA A + T-rich in a variety of mammals (Tobia *et al.* 1970, Bostock & Prescott 1971a, b, c, Holmquist *et al.* 1982). It must be emphasized that this shift in base composition is not due to the late replication of A + T-rich satellite DNA in heterochromatin, but refers to non-satellite sequences in euchromatin. Although, as pointed out above, patterns of replication and of G-banding are not perfectly concordant, the differences are not sufficient to affect the conclusion that negative G-bands should be G + C-rich and positive ones A + T-rich. Yet there is a serious difficulty here that has not been resolved. The reduction in number and increase in extent of bands as chromosomes contract during prophase has already been remarked. If in fact the positive G-bands come to include an increasing part of the chromosomal DNA as the chromosome contracts during prophase, it is clear that they cannot comprise a fixed class of DNA. Evidently it is important that accurate measurements of the amount of DNA in positive and negative G-bands at different stages of chromosome contraction should be made. It may be that existing ideas on chromosome condensation and on mechanisms of banding need some fundamental revisions.

The observations that late replicating regions of chromosomes in mammals (positive G-bands) contain A + T-rich DNA, while early replicating regions contain G + C-rich DNA, may have some relevance to the first question considered in this section, the absence of G-bands from the chromosomes of all organisms except higher vertebrates. It has been shown in recent years that the main band DNA of higher vertebrates consists of a few major components which differ in their base composition (Thiery *et al.* 1976, Cuny *et al.* 1981, Bernardi *et al.* 1985). Each component is arranged in a number of very large DNA segments, the isochores, which, it has been suggested, may actually be equivalent to G-bands. The interesting thing is that amphibia, fishes and the other eukaryotes studied had much more homogeneous DNA, not separable into A + T-rich and G + C-rich components. If, therefore, differences in base composition are essential to the formation of G-bands, one would only expect to find G-bands in higher vertebrates. We shall return to this point when considering the mechanism of G-banding (Section 5.4) and also in Section 16.3.

Not surprisingly, efforts have been made to characterize in more detail the DNA and types of DNA sequence in positive and negative G-bands. It was suggested many years ago that negative G-bands were richer in transcribed genes. This conclusion was derived from the observation that those human trisomies that are compatible with live birth involve chromosomes which have a relatively small proportion of negative G-bands (Ganner & Evans 1971), and this concept was extended by Hoehn (1975), Wahrman *et al.* (1976) and Korenberg *et al.* (1978). More direct methods have tended to confirm that negative G-bands are richer in genes. Yunis *et al.* (1977) and Yunis & Tsai (1978) used *in situ* hybridization to locate sequences complementary to messenger RNA, and found they were preferentially located in the negative G-bands. On the other hand, moderately repetitive DNA sequences were concentrated in positive Q-bands (equivalent to G-bands) (Sanchez & Yunis 1974). Attempts to map DNase-sensitive sites on chromosomes, which appear to correspond to sites of potentially transcribable genes (Section 12.2.1) have given more equivocal results. Kerem *et al.* (1984) found that in different species of mammals, DNase-sensitive sites usually corresponded to negative G-bands, but that not all such bands were DNase-sensitive. However, Adolph & Hameister (1985) found that the pattern of DNase-sensitive sites did not correspond completely to either G- or R-banding. It is clear that much more work is required on this technique. Goldman *et al.* (1984) separated early and late replicating fractions of DNA (which they equated with negative and positive G-bands respectively) and concluded that constitutive, housekeeping genes were in negative G-bands, whereas tissue-specific genes were in late-replicating, positive G-bands. Switching on a tissue-specific gene by replicating it early would not produce any microscopically visible changes in the replication pattern on chromosomes. Bernardi *et al.* (1985) similarly found genes in both the heavy (G + C-rich) and light (A + T-rich) fractions of mammalian main band DNA. In humans, sufficient genes have now been mapped with sufficient precision to be able to state that the majority are in negative G-bands (Rodionov 1985, Bickmore & Sumner 1989). Intermediate repetitive DNA sequences also appear to be non-randomly distributed in chromosomes, the long intermediate sequences being concentrated in dark G-bands, and the shorter ones in pale G-bands (Manuelidis & Ward 1984, Korenberg & Rykowski 1988).

Current views on the nature of positive and negative G-bands are summarized in Table 5.2. It will be noticed that no mention is made of the proteins of G-bands, a subject that has been largely neglected, in spite of the implication that proteins are involved in G-banding, since G-bands can be induced by

Table 5.2 Properties of positive and negative G-bands.

Positive G-bands	Negative G-bands
Positive Q-bands	Negative Q-bands
Negative R-bands	Positive R-bands
Pachytene chromomeres	Interchromomeric regions
Early condensation	Late condensation
Late replicating DNA	Early replicating DNA
A + T-rich DNA	G + C-rich DNA
Tissue-specific genes	'Housekeeping' genes
Long intermediate repetitive DNA sequences (LINEs)	Short intermediate repetitive DNA sequences (SINEs)

proteases. Sumner (1974) proposed, on the basis of indirect evidence, that positive G-bands were relatively rich in protein disulphide bonds, while negative G-bands were relatively rich in protein sulphydryl groups that were not cross-linked. There is a considerable body of evidence that disulphide cross-links are involved in chromatin condensation (see Sumner 1983 for a brief summary), and this idea would have fitted well with the suggestion that G-bands represent centres of condensation. However, sensitive fluorescence cytochemistry of sulphydryl and disulphide groups failed to demonstrate any differentiation along chromosomes (Sumner 1984). Nevertheless, there is evidence from other sources that sulphydryl and disulphide groups may be connected with G-banding. Not only can G-bands be induced by various reagents that affect these groups (Kato & Yosida 1972, Utakoji 1973), but such reagents can also be used to give improved high-resolution banding (Kao *et al.* 1983).

5.4 Mechanisms of G-banding

The object of this section must be to try to explain how the G-banding procedures described in Section 5.2 interact with the characteristic features of positive and negative G-bands described in Section 5.3 to produce the characteristic patterns along chromosomes illustrated in Figure 5.1. In spite of years of research, we still lack a satisfactory hypothesis of G-banding. The following discussion will assume that, although the methods by which G-banding can be induced are diverse, their mechanisms are essentially similar. The strongest evidence that this is so is that all these methods produce a common pattern of bands. Nevertheless, the possibility cannot be excluded that there may be important differences in the mode of action of different reagents.

The first practical G-banding methods were derived from C-banding methods, and attempts were made to explain them in a similar way (Sumner *et al.* 1971a). It was proposed that positive G-bands contained DNA sequences that annealed more readily than those in negative G-bands, and thus stained more strongly as a result. No concrete evidence was available for this suggestion, and indeed the idea that C-banding was the result of denaturation and differential annealing of DNA was soon abandoned (Section 4.2.5). More importantly, the introduction of G-banding using proteolytic enzymes was strong evidence that action on chromosomal proteins is required to produce G-bands. There is, however, little doubt that the substrate for Giemsa staining of chromosomes is DNA, and therefore it has been usual to try to interpret G-banding mechanisms in terms of DNA-protein interactions.

It is convenient to start with a description of the morphological changes that occur during the G-banding of chromosomes. In Section 3.3.4 the question was considered whether the chromosomal constitutents were uniformly distributed, or whether a structural banding pattern was already present. Some authors reported uniform distribution of DNA, while others described a pattern of DNA distribution corresponding with that of the G-bands. Similarly, there is no consistency between reports on the distribution of the chromosomal constituents after G-banding. Gormley & Ross (1972, 1976, Ross & Gormley 1973) found that pretreatment, whether for the ASG or the trypsin method, produced what they described as a collapsed appearance, a ridge of material surrounding a rather empty looking chromatid (Fig. 5.2a). Only after staining was a banded appearance seen (Fig. 5.2b), consisting of ridges across the chromosome, which now showed a fibrous surface structure. Schuh et al. (1975) obtained similar results with trypsin G-banding. Ross & Gormley (1973) found that removal of the Giemsa dye eliminated the banded structure of the chromosomes, but Schuh et al. (1975) found that the bands were retained even after extraction of the dye. These differences could be due both to the use of different G-banding methods, and

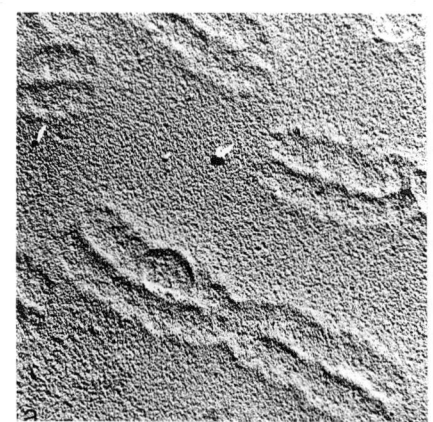

Figure 5.2 Surface topography of human metaphase chromosomes treated by the ASG technique for G-banding: (a) collapsed appearance after treatment with hot 2 × SSC; (b) banded appearance after staining. The bar represetns 1 μm. Reproduced from Gormley & Ross (1972) by permission of the authors and Academic Press.

to different methods of observation. Rather different results have been obtained by other workers, using trypsin G-banding (Burkholder 1974, 1975, Hozier et al. 1981) They reported that after trypsin treatment alone, without Giemsa staining, structural bands are visible. If the banding pretreatment alone can induce structural bands, one need look no further for an explanation of G-banding; however it is clear that although such a phenomenon occurs in some cases, Giemsa staining is still necessary in other cases to produce a banding pattern. This argument, in fact, resembles that over whether there is a pre-existing structural banding pattern in the fixed but otherwise untreated chromosome. The existence of structural patterns in some cases does not explain the production of G-bands in the cases in which structural bands cannot be seen.

Several workers have suggested a strong connection between chromosome coils and G-bands (Kato & Yosida 1972, Hatami-Monazah 1974, Goradia & Davis 1977, Johnson et al. 1981). Takayama (1976) investigated this point further, and found that by repeated trypsin treatment, chromosome coiling could be converted progressively to banding. Some coils moved closer to each other to form bands, while other coils became further apart. According to this hypothesis, banding would be the result of redistribution of material within the chromosome.

A pattern of banding visible without staining could alternatively result from selective loss of material from certain parts of the chromosome. If no banding pattern is visible until the chromosome is stained, there might be no substantial loss of material from the chromosome. Several attempts have been made to measure losses of material from chromosomes during G-banding, but as so often, conflicting results were obtained. The need for some loss or reorganization of material during banding is strongly indicated by the fact that cross-linking the chromosome structure with aldehyde fixatives prevents banding (Takayama 1974, Hancock & Sumner 1982, van Duijn et al. 1985). Comings et al. (1973) and Sumner et al. (1973) concluded that there was negligible loss of DNA or protein during various G-banding procedures, and this was confirmed for DNA by Schmiady et al. (1975). On the other hand, Franke et al. (1973) reported substantial losses of DNA and protein in a procedure similar to, but less drastic than the ASG method. Crossen (1973) and Matsui & Sasaki (1975) also reported loss of DNA. Staining with protein-specific fluorochromes also supports the view that there are substantial losses of chromosomal proteins during G-banding (Utakoji & Matsukuma 1974, Lee & Bahr 1983, Sciorra et al. 1985); there are indications that the pattern of fluorescence after the G-banding procedure may correspond to that of the G-bands (Matsukuma & Utakoji 1976). Clearly most of the evidence is that some DNA and protein are lost during G-banding procedures. In a comprehensive series of biochemical studies, Burkholder & Duczek (1980a, 1982a, b) showed that a remarkably consistent set of chromosomal proteins was extracted by a variety of different G-banding procedures, and also that C- and R-banding methods extracted

Figure 5.3 Structural formulae of Azure B and Eosin Y.

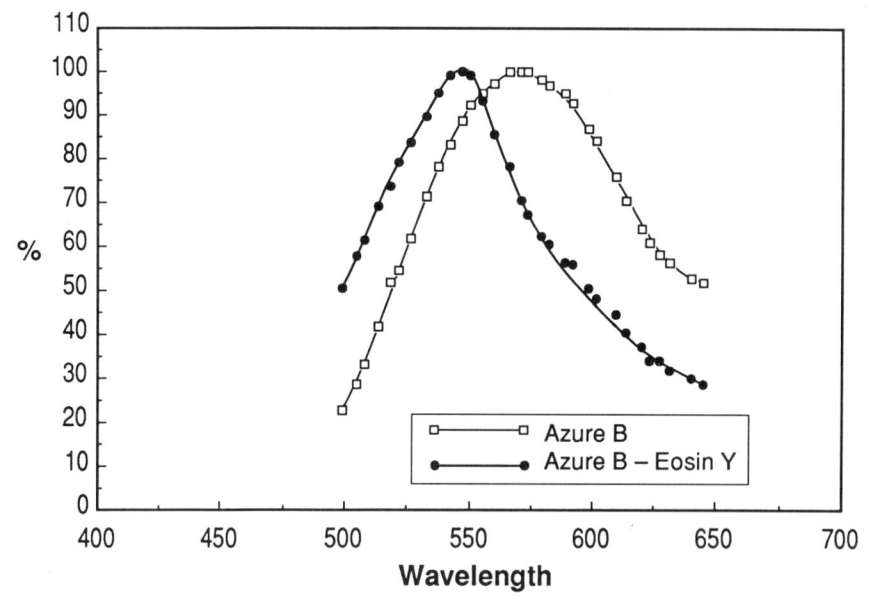

Figure 5.4 Absorption spectra of human lymphocyte nuclei stained with purified Azure B and with Azure B-Eosin Y.

different sets of proteins. This is strong evidence for the involvement of specific protein extraction in G-banding, although so far the chromosomal locations and functions of extracted and residual proteins are not known.

If specific proteins are extracted from chromosomes during the G-banding process, how does extraction give rise to bands? Evidently not by inducing a morphological differentiation along the chromosomes, since as we have already seen, such differentiation is frequently undetectable. Nor do the G-banding pretreatments appear to produce differential availability of binding sites on chromosomal DNA. Incubation in $2 \times$ SSC or trypsin actually reduces the number of DNA phosphates available for dye binding (van Duijn et al. 1985) but the reduction is uniform throughout the chromosomes. In general, no banding is visible until the chromosomes are stained with Giemsa (or some other related Romanowsky dye mixture). There must be some special property of Romanowsky dye mixtures that can give rise to bands. Fortunately these dyes have been studied extensively in recent years, not only in relation to chromosome banding, but also in relation to their original field of application in haematology. Romanowsky dyes are mixtures of oxidation products of methylene blue with eosin Y (Fig. 5.3); however, Wittekind (1983) and Wittekind *et al.* (1982) have shown that in haematology all the differential staining patterns known as Romanowsky staining can be produced by a mixture of purified azure B and eosin Y. The same mixture is also completely adequate for G-banding (Sumner, unpublished). In practice, traditional Romanowsky dye mixtures are used for banding, and it is quite likely that other dyes in these mixtures also take part in the staining. In the following discussion, therefore, this fact will be recognized by the use of the generic term thiazine dye to cover all those dyes derived from methylene blue which can be found in Romanowsky mixtures.

The fundamental observation about the staining of chromosomes (banded or unbanded) with Giemsa is that the colour is magenta or purple (Sumner & Evans 1973) and not the blue colour obtained when chromosomes are stained with thiazines alone (Figure 5.4). Although attempts have been made to explain the purple colour by stacking of thiazine molecules (Comings 1975a) this does not explain adequately the absorption spectrum of Giemsa-stained chromosomes. In fact, it is clear that the magenta colour is due to a complex of thiazine dye and eosin which is formed *in situ* (Sumner & Evans 1973). The complex consists of two thiazine molecules and one of eosin, and appears to be the same as the precipitate which forms in diluted Giemsa solutions (Sumner 1980). Formation of the complex takes place in two stages, and requires a solid substrate. First, thiazine dye molecules bind ionically to the substrate, which is DNA in the case of chromosomes, and subsequently eosin binds to the thiazines to produce the magenta colour (Sumner & Evans 1973). The thiazine-eosin complex is not bound ionically, but accumulates in hydrophobic sites (Sumner & Evans 1973, Sumner 1980; see also the theoretical studies of Curtis & Horobin 1982). As a point of both theoretical and practical importance, it should be noted that the intensity of staining with Giemsa is much greater than that normally obtained with thiazine dyes alone (Sumner 1980, van Duijn *et al.* 1985). It is the intensity of staining that makes Giemsa such an attractive dye for staining small objects such as chromosomes. From the theoretical point of view, it may be inferred that the amount of dye bound is not dependent on the number of binding sites available on the DNA, but rather that the complex accumulates as a precipitate at hydrophobic sites. The suggestion made by Marshall & Galbraith (1984) that the magenta colour is simply the result of overlapping spectra of independently bound thiazines and eosin is not tenable in the case of chromosomes, since a negligible amount of eosin is bound to chromosomes in the absence of thiazines (Sumner & Evans 1973).

Knowledge of the nature and properties of the magenta-coloured complex produced in chromosomes by staining with Giemsa or other Romanowsky-type dye mixtures should enable us to make predictions about the differences between the staining and non-staining regions of G-banded chromosomes. Sumner & Evans (1973) suggested that, since the first stage of staining is the attachment of thiazines to the DNA, followed by attachment of eosin to the thiazine, the distance apart of the thiazine molecules was critical for the formation of the magenta colour. It was, therefore, postulated that the amount of Giemsa staining would be exquisitely sensitive to the local concentration of DNA, and might distinguish regions of chromosomes whose degree of condensation differed too little from that of adjacent regions to be distinguished without staining. This hypothesis was consistent with the idea that G-bands were a manifestation of chromosome condensation (Section 5.3) and that in some cases variations in the degree of condensation along chromosomes could be seen without staining. However, direct measurements of Giemsa uptake compared with thiazine binding to DNA in nuclei in which the DNA had been partially blocked or extracted showed that Giemsa binding is not especially sensitive to DNA concentration, and indeed may be less sensitive than other dyes (Sumner 1980). Although there can be little doubt that there is a relationship between patterns of chromosome condensation and G-banding, it does not seem that the degree of condensation can be the direct cause of G-banding.

Giemsa binding is generally held to show no DNA base specificity, and in contrast to most fluorescence banding methods, explanations of G-banding have been based on rather vague DNA-protein interactions and not on the properties of the DNA alone. Recently, however, van Duijn *et al.* (1985) have put forward a hypothesis in which the established facts on variation of DNA base composition along chromosomes, on chromatin structure, and on the nature of Giemsa staining are combined to explain not only the G-banding process itself but also related phenomena such as chromosome collapse and swelling. The complete hypothesis is rather complex, and it must be emphasized that although it is consistent with many known facts, it is so far essentially untested. It is proposed that during fixation, the methanol acts to shrink and denature the nucleosomes irreversibly. The H1 histones may also be denatured. When hot saline treatment is used for G-banding it is postulated that rearrangements of the histones and DNA occur. Polyarginine, and by implication the arginine-rich core histones, bind more strongly to G + C-rich DNA, and could thus prevent the insertion of the Giemsa complex between such DNA and the nucleosomes. In A + T-rich regions of DNA the interactions between histones and DNA would not be strong enough to prevent Giemsa binding. The action of trypsin is explained largely in terms of digestion of histone H1, which again would permit the interactions between core histones and DNAs of different base composition as described above for hot saline treatment.

Various aspects of the hypothesis of van Duijn *et al.* (1985) have been foreshadowed by previous authors. For example, Comings & Avelino (1975) showed that the number of DNA phosphate groups available for staining was reduced by saline incubation, and postulated that chromosomal proteins were denatured so as to cover the DNA more effectively and block staining. Burkholder & Weaver (1977), in a study of the digestibility of chromatin by DNase, concluded that proteins were more tightly bound to DNA in positive G-bands than in negative ones. Ris & Korenberg (1979) produced structural bands using salt solutions that extracted histone H1, which resembles the way in which van Duijn et al. (1985) proposed that trypsin might act. McKay (1973) argued that banding treatments produced chromosome collapse by extracting divalent cations, and that the introduction of Giemsa dye molecules could reconstitute the chromosome structure.

Two essential features of the explanation of G-banding offered by van Duijn et al. (1985) are the retention of histones in the chromosomes (although this need not be complete), and the existence of variations in DNA base composition along the chromosomes. In spite of claims to the contrary, it is now clear that substantial amounts of histones are retained in acetic acid-fixed chromosomes (Section 3.5). The observation that extraction of histones from fixed chromosomes has no effect on G-banding (Comings & Avelino 1974) was not supported by the experiments of van Duijn et al. (1985). In fact, Pothier et al. (1975) have shown that, once fixed, histones are not readily extracted from chromosomes by the normal procedures such as dilute acid. There is no doubt that in mammals there are variations in base composition along chromosomes that correspond to banding patterns. The relationship between G-banding and DNA base composition is discussed further in Section 16.3; the occurrence of G-banding in a number of organisms that do not appear to show any significant variation in base composition along chromosomes makes this aspect of the hypothesis of van Duijn et al. (1985) unlikely, although there may well be differences in the binding of DNA to protein between positive and negative G-bands that are unrelated to base composition. At present we must conclude that there is still no satisfactory hypothesis to explain the mechanism of G-banding.

Before leaving this subject, a few small points should be mentioned. One of these is the persistent reporting that G-banding can be obtained by the use of thiazine dyes alone, in the absence of eosin (Comings 1975a, Löber et al. 1976, Wyandt et al. 1980). Other reports indicate that adequate G-banding cannot be produced by thiazines alone (Sumner & Evans 1973, Meisner et al. 1974) and it remains the universal practice to use Romanowsky-type mixtures for routine G-banding. Nevertheless, the observation of bands using thiazines alone cannot be dismissed as an aberration, and it was noted by Comings (1975a) that the ability of thiazines to produce banding depended on their degree of methylation, which is related to their degree of hydrophobicity, which is apparently one of the most important factors determining the selectivity of Giemsa. It should be noted that Curtis & Horobin (1975) regarded Bernthsen's methylene violet, a water insoluble dye related to the thiazines, as an essential component of Romanowsky mixtures for banding.

In fact, the theoretical studies of Curtis & Horobin (1982) indicated that banding could be produced by any dye showing the appropriate hydrophobic and certain other structural features. Dyes other than thiazines could be used for banding, but it was pointed out that factors other than the structure of the dye must influence the result of the staining. Horobin & Walter (1987) explained differential Giemsa staining of blood smears on the basis of different rates of penetration of the dyes into different structures. It is not clear, however, how this concept might apply to G-banded chromosomes, where the most strongly stained regions are probably also the most densely textured, and are at any rate the most resistant to the actions of proteases and DNase (Section 12.2) (Burkholder & Weaver 1977).

The role of inorganic ions in G-banding has often been considered, but its importance is uncertain. At one time it was suggested that chelation of ions bound to chromosomes was a common feature of diverse G-banding methods (Dev *et al.* 1972b); indeed, it was proposed that this was the mode of action of trypsin, rather than proteolysis, although this was later refuted (Lundsteen *et al.* 1974, Korf *et al.* 1976). However, Meisner *et al.* (1973b), who produced G-banding by prior incubation in a caesium chloride solution, obtained evidence suggesting that the banding pattern was the result of competitive binding of caesium and Giemsa. Beil & Limon (1975) used cacodylate buffer to produce G-banding, and believed that an important effect of this buffer was to extract magnesium from the chromosomes.

5.5 Applications of G-banding

G-banding is one of the three principal methods for demonstrating euchromatic bands, the other two being R-banding and Q-banding. As already noted, replication banding (Section 11.3) sometimes has a slightly different pattern, and represents a different phenomenon, and its distinctive applications will be considered separately. The use of G-, R- and Q-banding is, however, essentially interchangeable so far as the study of euchromatic bands is concerned. Nevertheless, G-banding is undoubtedly the most widely used of the three, because experience has shown it to have advantages over the other methods. Compared with R-banding methods, G-banding generally gives clearer and more detailed patterns, and is probably more reliable for routine use. G-banding also gives more detailed patterns than Q-banding, and while Q-banding is very simple to perform, and is probably as reliable as any banding method, it has the disadvantage of being a fluorescence method. This means that permanent preparations cannot be made, the preparations fade during observation and photography, and special equipment (a fluorescence microscope) is necessary. These factors have combined to make G-banding the pre-eminent method for studying euchromatic bands, but it should be made clear that a large amount of important work has also been done using R-banding and Q-banding. The following account is, therefore, really a description of the applications of methods for euchromatic bands in general, although applications to evolutionary studies are described in Chapter 15. Distinctive applications of G-banding will, of course, be referred to in this chapter, while applications in which R-banding or Q-banding are especially advantageous will be referred to in Chapters 6 and 7 respectively.

The applications of G-banding can be grouped together under the heading of identification. In contrast, other banding techniques may be used for studying heteromorphisms (C-banding and Q-banding; see Ch. 4, 7 & 14), patterns of replication or condensation (Ch. 11), or the location of particular chromosome organelles such as nucleolar organizers and kinetochores (Ch. 9 & 10). In such cases either an additional banding method has to be used for identification, or the ability to identify the chromosome is a useful adjunct to the primary purpose of the method.

5.5.1 *Chromosome identification*

The first and most important application of G-banding, from which all else is derived, is the identification of individual chromosomes, which enables one

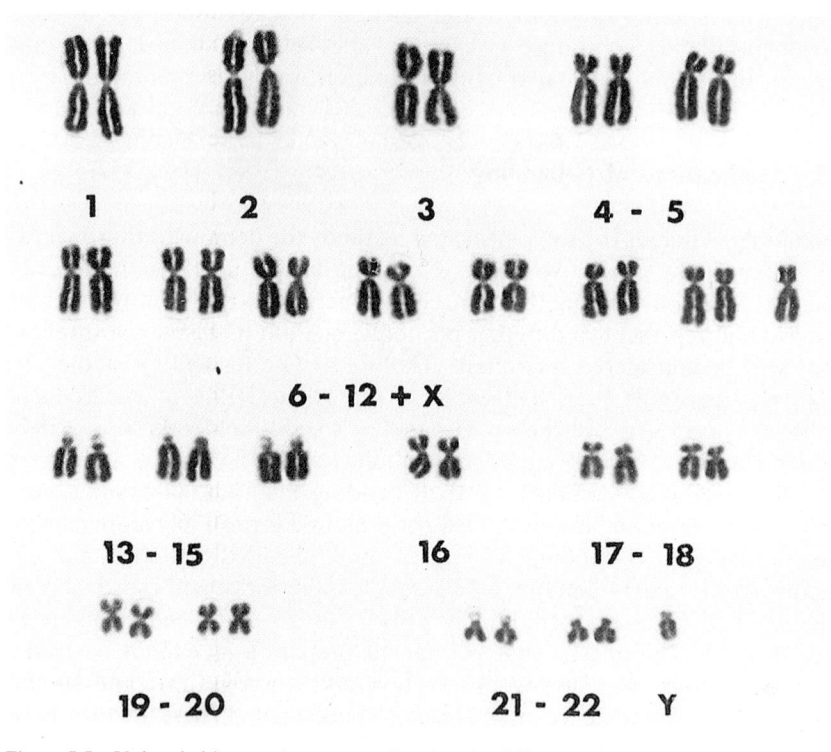

Figure 5.5 Unbanded human karyotype, showing the difficulty of distinguishing many of the chromosomes without banding. Reproduced from Bostock & Sumner (1978) by permission of the authors and Elsevier Science Publishers.

to prepare precise and detailed karyotypes. A couple of examples will suffice. In Figure 5.5 is an unbanded human karyotype, in which the 46 chromosomes are grouped into seven classes on the basis of their sizes and the position of their centromeres. In group A, it is actually possible to distinguish pairs 1, 2 and 3 with some confidence on purely morphological criteria. In group E, chromosome 16 can be distinguished from 17 and 18. Most chromosomes, however, are not distinguishable morphologically from others in the same group. In group C, for example, although there is a noticeable difference in size between the largest (number 6) and the smallest (12), there are no differences large enough to distinguish any chromosome from the next in size. Although several attempts were made to distinguish chromosomes on the basis of size and centromere index, this was not really satisfactory, as many of the differences were too small for confidence. The situation was completely transformed by the discovery of banding, which provided a distinct pattern for each chromosome pair, even those closely resembling each other in size and shape. Indeed, in more recent years chromosomes have been identified by banding before measuring their DNA content, so that the relative amounts of DNA in each chromosome can now be measured with confidence (Mendelsohn *et al.* 1973, Bosman *et al.* 1977b).

The positive identification of all the human chromosomes was an event of great importance, but not such a difficult problem as that of certain other species, such as the mouse and ox, in which all the autosomes are acrocentric, and are continuously graded in size. Although the differences in size between the largest and smallest are considerable, there was no means of identifying individual unbanded chromosomes. With the discovery of banding, 'the miserable mouse chromosomes suddenly became wonderful' (Hsu 1974), and each pair could be identified (Fig. 5.6). Definitive karyotypes have now been published for many species of higher vertebrates, and for a number of mammals the patterns and numbering of the chromosomes have been agreed upon and promulgated by international committees of experts (Table 2.4). Of course, G-banding alone is not sufficient to define a karyotype, and other methods are essential, especially for heterochromatic bands, but in all cases G-banding has been one of the chief tools for defining the karyotype.

The identification of chromosomes using G-banding has been almost entirely restricted to mitotic metaphase chromosomes fixed in the conventional way with methanol-acetic acid. In spite of the well-established resemblance between patterns of G-bands and of pachytene chromomeres, it has, in general, proved almost impossible to obtain G-banding on meiotic chromosomes. It is, therefore, worth noting that Drwinga & Pathak (1982) successfully G-banded prematurely condensed mouse diplotene bivalents.

Another field for which there is great potential is the application of G-banding to morphologically classified cells (Knuutila & Teerenhovi 1989). Cells are first identified by immunofluorescence, then fixed in methanol-acetic acid, and then G-banded using a variant of a standard procedure. Use

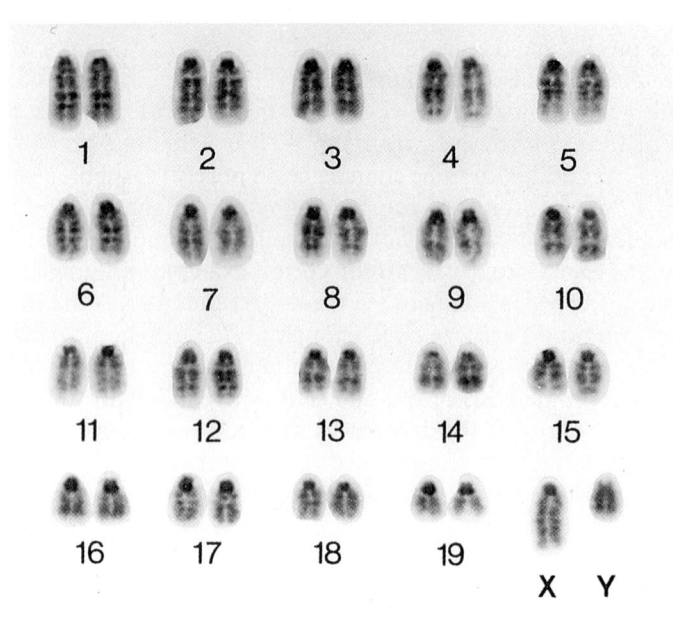

Figure 5.6 G-banded karyotype of mouse chromosomes, showing the especial value of banding for distinguishing chromosomes in a species in which they are all morphologically similar. Reproduced from Buckland *et al.* (1971) by permission of the authors and Academic Press.

of such methods permits the association of a chromosome abnormality with a particular cell type in a mixed population, for example.

5.5.2 Chromosome abnormalities: aneuploidy, breakage and rearrangement

One of the main uses of banding methods in the early days was for the identification and better characterization of chromosome abnormalities associated with various human genetic diseases. Hoehn (1975) listed some achievements in this field resulting from improved identification of human chromosomes using banding methods (and see also the review by Lawler & Reeves 1976). These include confirmation of established syndromes and identification of new syndromes involving trisomy or specific deletions or additions of parts of chromosomes (see the appendix in Hoehn 1975 for a comprehensive list of references up to that date). Banding has been important in determining the types of aneuploidy associated with fetal loss at various stages of gestation in humans (Boué *et al.* 1985, Hassold 1986). Specific trisomies causing characteristic abnormalities in domestic animals have also been reported (Gustavsson 1980).

Since banding patterns remain unaltered when parts of chromosomes are translocated or inverted, G-banding, especially in combination with other methods, is invaluable for the detailed description of such events. Among the simpler examples are the identification of the chromosomes involved in Robertsonian translocations, for example in patients with Down's syndrome with 46 chromosomes (D. A. Miller *et al.* 1971, Jacobs *et al.* 1974), or in mice with different numbers of chromosomes (Gropp *et al.* 1972, Capanna *et al.* 1975) or in domestic sheep (Bruère *et al.* 1974, 1978). Many more complex situations have, of course, been analyzed, but it is only possible to give one or two examples. Banding techniques were used to study the chromosomes in mice carrying genetically defined paracentric inversions; corresponding inversions were found in the chromosomes (Davisson & Roderick 1973). In a survey of structural rearrangements in human chromosomes (Jacobs *et al.* 1974), 58 Robertsonian translocations, 53 reciprocal translocations, and 10 inversions were analyzed, and it was concluded that the distribution of points of breakage and exchange were non-random. Non-random distribution of chromosome breaks was also observed in the chromosomes of patients with Fanconi's anaemia, which is characterized by increased chromosome breakage (von Koskull & Aula 1973). These authors found that all the breaks were in weakly stained G-bands, and that certain sites were broken particularly frequently. The tendency for breaks to occur in negative G-bands was also observed by Nakagome & Chiyo (1976) for structural rearrangements. Sites of induced chromosome breakage are concentrated in pale G-bands as well, whether the breakage is induced by radiation (San Roman & Bobrow 1973) or by chemicals (Morad *et al.* 1973). However, Buckton (1976) used sequential G- and R-banding (Sehested 1974) on the same chromosomes, and claimed that the actual site of breakage was at the interface between the dark and light bands. This demonstrates the importance of using more than one banding method in such cases. It must be pointed out, however, that the interface between dark and light bands, whether G- or R-bands, is not sharp but diffuse; moreover, if the positive G-bands fuse together and increase in size as the chromosomes condense (Section 5.3), there can be no fixed interface. Further, as Savage (1977a, b) has pointed out, there is in many cases (particularly chromosome-type exchanges) an uncertainty in defining the actual breakpoint, which could in fact occur in any one of three adjacent bands. The breakpoint is commonly said to occur in a light band, although it could equally well have occurred in either of the adjacent dark bands. Use of R-banding as well as G-banding does not resolve this problem, although it may suggest a different location of the breakpoint.

5.5.3 *Chromosomes of cultured cells*

Cultured cell lines normally show differences in the morphology and number of their chromosomes compared with those of the animal from which the cells were derived. Using banding techniques, it is possible to work out how the chromosomes in the cell line have been derived from those of the parental animal, whether the cultured cells are more or less diploid or are wildly aneuploid, and how the karyotype changes during culture. The Chinese hamster line Don is almost diploid, differing from the animal by only one extra band (Chen 1985). This is an unusual case, however, and the Chinese hamster line CHO, although having a nearly diploid number of chromosomes, has in fact many chromosomes modified by translocation, deletion, or pericentric inversion (Deaven & Petersen 1973). In spite of this, most of the original genome is still present, and the modified karyotype is relatively stable (Worton et al. 1977). Similar studies have been carried out on cell lines derived from mice (e.g. Farber & Liskay 1974), while in human cell lines banding was used to study chromosome changes during prolonged culture (Shade et al. 1980, Steel et al. 1980).

5.5.4 Chromosome banding and cancer

One of the most valuable applications of banding to the study of chromosomal alterations has been in cancerous cells. It was recognized for many years that the karyotypes of cells from leukaemias and solid tumours were often abnormal, sometimes grossly so, but with one exception no consistent pattern of abnormality was found until intensive studies were made using banding methods. The one exception, the Philadelphia (Ph 1) chromosome, found in most cases of chronic myeloid leukaemia, is itself of interest, since although it was recognized before the introduction of banding, it required the use of banding methods to establish its true nature. This chromosome had been supposed, for various reasons, to be a partly deleted no. 21 chromosome. Studies using banding showed not only that it was a deleted no. 22 chromosome, but that the deleted part was translocated to another chromosome, usually no. 9 (Rowley 1973). Banding techniques were also used to demonstrate the existence of a chromosome abnormality in Burkitt's lymphoma (Manolov & Manolova 1972, Zech et al. 1976). Many more examples could now be quoted, but to do so would be tedious and merely repeat information already collated elsewhere. Suffice it to say that the involvement of specific types of chromosome alterations in certain specific types of cancers in now well established, and that this has been established entirely through the use of chromosome banding methods. Mitelman (1984) concluded from all this work that there was only a limited number of chromosomal breakpoints involved in human cancer and leukaemia, and that the genes of importance for transformation of a normal cell to a cancerous one are located at such sites. The presence of oncogenes at such sites, and their activation by the translocation process, is at present the subject of intense study (Haluska et al. 1987). For those wishing to know more about what is now a very large field, see the reviews by LeBeau & Rowley (1984), Sheer (1986) and Teyssier (1989), and the books by Sandberg (1980) and Heim & Mitelman (1987). Although papers in this field appear in many journals, it is perhaps worth noting that there is one (Cancer Genetics and Cytogenetics) which is largely devoted to this field. It should be emphasized that the association of particular chromosome abnormalities with specific types of cancer not only gives insights into possible mechanisms of carcinogenesis. Evidence is now accumulating that there are correlations between types of chromosomal changes, and the prognosis for a particular disease, with the practical implication that the therapy could be adjusted on the basis of the cytogenetical findings (Sandberg 1977, Third International Workshop on Chromosomes in Leukaemia 1981).

The chromosomal changes associated with cancer described in the previous paragraph are apparently restricted to the cancer cells themselves. There are, however, certain cancers that are associated with constitutional chromosome anomalies that occur in all cells of the body. The two best known tumours of this type are retinoblastoma and Wilms' tumour. Each of these may occur either as a result of an autosomal dominant mutation, or as a result of a small chromosomal deletion (Rowley 1980). If the cancers result from a deletion, other abnormalities are usually present also (mental retardation in the case of retinoblastoma, and aniridia in the case of Wilms' tumour). The use of banding techniques was again essential to establish the presence of a chromosomal deletion in these conditions (Orkin 1984, Sparkes 1984).

5.5.5 Homogeneously staining regions

Biedler & Spengler (1976) used G-banding to investigate strikingly long marker chromosomes which they had found in two types of cultured cells: antifolate-resistant Chinese hamster cells and cultured human neuroblastoma cells. In both cases the long marker chromosomes proved to have long segments that did not band, but instead showed a uniform and intermediate level of staining with Giemsa after treatment which produced typical G-banding patterns elsewhere on the chromosomes. These distinctive segments (Fig. 5.7) were therefore named 'homogeneously staining regions' or HSRs (Biedler & Spengler 1976), or alternatively 'extended chromosomal regions' (ECRs) (Stark *et al.* 1989). During the past ten years or so a considerable amount of work has been done which shows that HSRs, together with double minutes (minute chromosome-like bodies which lack centromeres), are a manifestation of DNA amplification (see reviews by Cowell 1982, Schimke 1982, 1984, George 1984, Hamlin *et al.* 1984, Stark *et al.* 1989). HSRs have been reported in cultured cells resistant to a variety of drugs, the resistance being conferred by increased production of the target enzyme for the drug, which in turn is the result of gene amplification (Caizzi & Bostock 1982, Wahl *et al.* 1982, Meltzer *et al.* 1985). Where HSRs occur in tumour cell lines, it has been shown that oncogenes have been amplified (Alitalo *et al.* 1983, Emanuel *et al.* 1985, Schwab *et al.* 1985). A HSR has also been reported in several normal wild mouse populations (Fig. 5.7b) (Traut *et al.* 1984), although no evidence for gene amplification is available in this case.

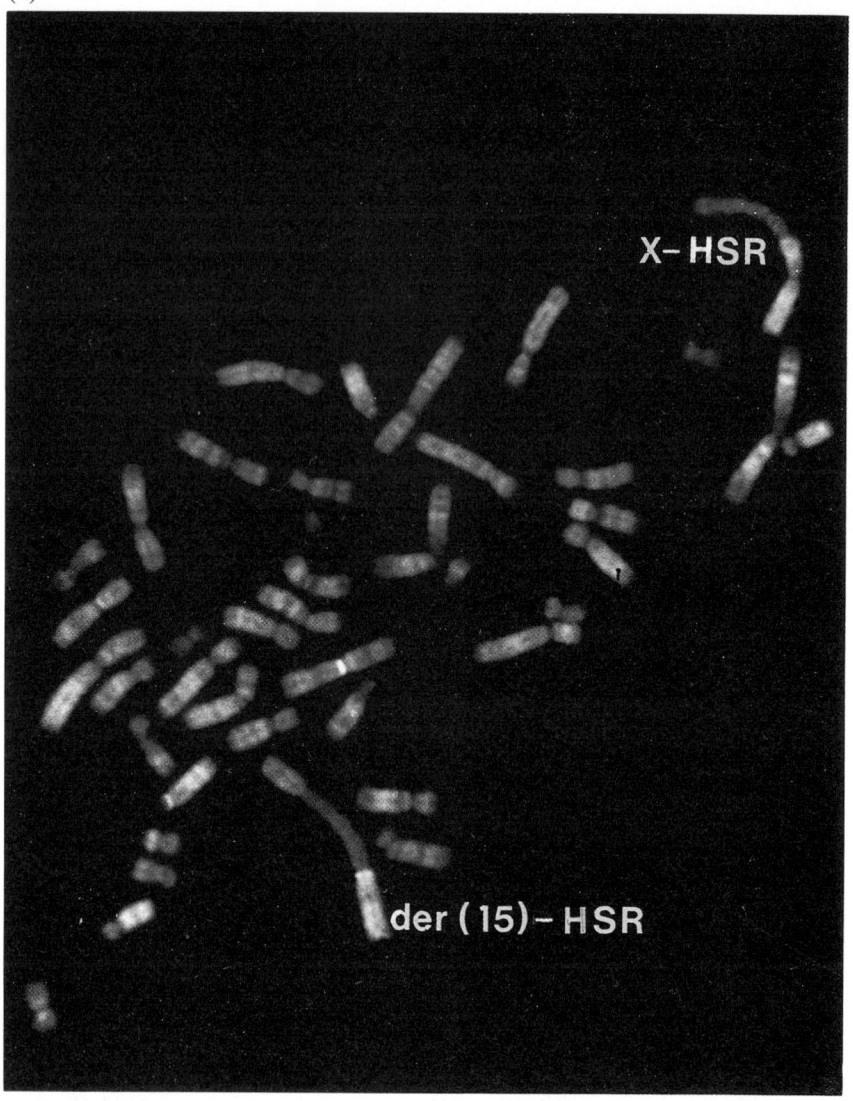

Figure 5.7 Homogeneously staining regions. (a) Q-banded metaphase from the human melanoma cell line MeWo. Reproduced from Holden et al. (1985) by permission of the authors and publisher. Copyright 1985 by Elsevier Science Publishing. (b) G-banded normal chromosome 1 (left) and chromosome 1 with an HSR (right) from a wild mouse. Reproduced from Traut et al. (1984) by permission of the authors and S. Karger.

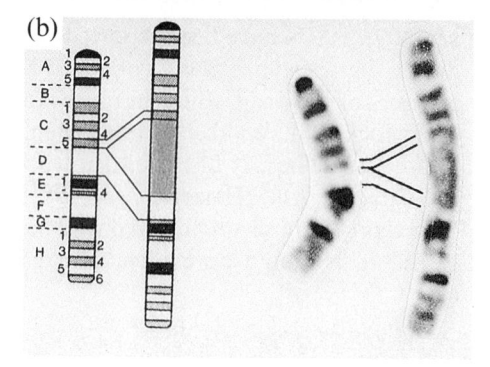

HSRs are not necessarily homogeneous, and although it is convenient to use the original term to describe them, fine regular banding structures can often be seen if appropriate banding techniques are used (Barker et al. 1980, Bostock & Clark 1980, George & Francke 1980). HSRs often, but not always, show some C-band positivity (Barker et al. 1980, Bostock & Clark 1980, Traut et al. 1984). A particularly complicated case was described by Holden et al. (1985); the HSR showed weak uniform fluorescence with quinacrine, but was banded with Giemsa, G-banding, C-banding, and distamycin/DAPI. Multiple copies of both nucleolar organizers and centromeres were also present. Fine banding patterns in HSRs may perhaps seem readily explicable in terms of DNA amplification, since the amplified material contains DNA other than the amplified gene (Bostock & Clark 1980, Caizzi & Bostock 1982, Montgomery et al. 1983, Holden et al. 1985), and the unit of amplification is large in molecular terms (Bostock & Clark 1980). However, there may be as many as 300 copies of the amplified genes in certain HSRs (Bostock & Clark 1980, Stark & Wahl 1984), considerably more than the number of fine bands that may be visible. The relationship between the unit of DNA amplification and the repeating unit of the banding pattern is thus not clear. In this connection, it is reasonable to speculate that the intermediate level of Giemsa staining reported for G-banded HSRs could result from a repeated pattern of dark and light G-bands too fine to resolve by light microscopy.

The ability to induce HSRs by treating cultured cells with appropriate drugs has permitted studies of their development. Biedler & Spengler (1976) reported that HSRs only appeared in cell lines showing a very high degree of resistance to the drug. Chromosomes from cells with lesser degrees of resistance show other changes: abnormal banding patterns (Biedler et al. 1980); breaks and rearrangements of the chromosome on which the HSR later appears (Andrulis et al. 1983); an additional chromosome of the type on which the HSR appears; and the appearance of single unpaired small chromatin bodies (Kopnin et al. 1985). Some HSRs are stable, but in other cases the HSR and the drug resistance are lost in the absence of the drug. The length of the HSR is related to the degree of drug resistance; as the latter declines, the HSR gets smaller (Biedler et al. 1980). It has been proposed that double minutes are formed from the breakdown of HSRs (Balaban-Malenbaum & Gilbert 1980), but this does not always appear to be so, and double minutes are also associated with the process of amplification.

HSRs are not necessarily formed at the original sites of the sequence which is amplified. In Chinese hamster cells resistant to methotrexate, the amplified genes for dihydrofolate reductase (dhfr) occurred in HSRs which were most often on the long arm of chromosome 2. However, HSRs have also been seen on certain other chromosomes, and on chromosome 2 they occur at a variety of different sites (Biedler et al. 1980, Biedler 1982). The site of the dhfr gene has apparently not yet been established, but clearly all these HSRs cannot be at the site of the unamplified gene. In human neuroendocrine tumour cell lines, the HSRs containing the amplified c-*myc* genes were on the X chromosome, and not in the normal position of the gene on chromosome 8 (Alitalo *et al.* 1983).

5.5.6 *Gene mapping*

G-banding has an important part to play in gene mapping, simply because it is essential to identify the chromosome or part of a chromosome on which the gene is located. Since the main effort in gene mapping at the present time is directed towards mapping human genes, and to a lesser extent those of other mammals, the patterns of euchromatic bands are invaluable for identification.

There are essentially three main methods for assigning genes to specific chromosomes: family studies, somatic cell hybridization, and *in situ* hybridization; for more information on the first two methods see the reviews by Shows *et al.* (1982) and Kao (1983). In family studies, it is necessary that the gene of interest and a marker chromosome segregate together. The most obvious sort of marker is a heteromorphic block of heterochromatin. Sometimes, however, the absence of a gene is associated with a small deletion. In this way Ferguson-Smith *et al.* (1973) located the human acid phosphatase gene to the short arm of chromosome 2. In somatic cell hybridization, banding is used to identify the few remaining human chromosomes (to give the most commonly used example) in a human-rodent cell hybrid. The presence of particular human chromosomes can then be correlated with the presence of particular human enzymes or other distinctive gene products. To locate the position of a gene more precisely on a particular chromosome, it is necessary to use human cells carrying translocations for the chromosome of interest. In such a case, the presence or absence of the gene product can be correlated with the presence or absence of part of the chromosome, and using a number of different translocations, it is possible to map the gene with a fair degree of precision, limited only by the nature of the available translocations. Obviously the use of banding is necessary not only to identify the chromosomes but to locate the breakpoints of the translocations.

5.5.7 *High-resolution banding*

The philosophy behind the development and use of high-resolution banding (HRB) is that almost anything that can be done with conventional banding can be done better with high-resolution banding. HRB is not, of course, entirely a routine technique, nor is it really suitable for simple identification. The elongated chromosomes required for HRB can rarely be prepared without overlapping, often to a considerable extent, and the distinctive patterns which permit identification of metaphase chromosomes at a glance are less obvious when the chromosomes are more extended and the bands more numerous. HRB is, however, particularly advantageous in the precise delineation of breakpoints, and the identification of small deletions.

It is in the study of human neoplasia that HRB appears to have had most success. Yunis (1983) has gone so far as to suggest that the malignant cells of most tumours can be shown to have characteristic chromosomal defects when studied with HRB, a view generally supported by Testa (1984) for leukaemias. More precise identification of the chromosome abnormalities is already leading to the delineation of different subclasses of certain leukaemias, which have different chromosome alterations and may have different prognoses. In routine clinical cytogenetics the advantages of HRB are perhaps fewer. This may be due partly to the frequently poor quality of chromosome preparations from cancer cells, so that any improvement would be worthwhile; lymphocyte chromosome preparations are generally of a higher standard, so that the use of HRB offers less improvement. The advantages and disadvantages of HRB for routine diagnosis are discussed by Barnes & Maltby (1986) and Hoo (1986). A striking application of HRB is the field which has been named 'microcytogenetics' by de Grouchy & Turleau (1986), as a result of which specific small chromosome deletions have been found in a number of clinical conditions believed to have a genetic origin.

5.6 Conclusion

G-banding is undoubtedly the most widely used banding technique. This is due to the relative ease and reliability of performing the techniques, and the permanence and ease of observation of the results. It is essentially a method of identifying chromosomes, in contrast to certain other methods which only draw attention to certain features of chromosomes such as heterochromatin, but which are of limited value for identification. The fields of application have just been reviewed (Section 5.5), but are worth restating briefly: defining karyotypes, locating breakpoints in clinical conditions, in cancer cells, and in cultured cells; in evolutionary studies, and in gene mapping. In all these fields the use of G-banding, as well as other techniques for demonstrating euchromatic bands, has been indispensable. The discovery of homogeneously staining regions must be singled out for especial mention, as a chromosomal phenomenon the discovery of which was wholly attributable to the use of banding.

It seems paradoxical that in spite of its immense value, relatively little is known about certain aspects of G-banding. It is a fortunate chance that although at present G-banding is only applicable to a limited group of organisms, that group happens to include the mammals, and particularly man, the species in which we are most interested. If this were not so, G-banding would be little more than a curiosity instead of the essential tool it has become. We know relatively little about the nature of positive and negative G-band material, and the mechanisms by which these bands are demonstrated, although knowledge of these matters is increasing steadily. These deficiencies are, however, more than counterbalanced by the immense practical value of G-banding.

6 R- and T-banding

6.1 Introduction and definitions

In May 1971, Dutrillaux & Lejeune published a short paper describing a new technique of analysing the human karyotype. In this technique, subsequently referred to as *dénaturation ménagée* ('controlled denaturation') (Dutrillaux *et al.* 1972), chromosome preparations were heated in a phosphate buffer at 87°C, and then stained with Giemsa. The stained chromosomes showed a pattern of bands (Fig. 6.1) which was complementary to that obtained with G- and Q-banding; bands strongly stained with Giemsa by this method were equivalent to those showing pale G-banding or weak quinacrine fluorescence. As a matter of historical fact this was the first published method for euchromatic bands that did not involve fluorescence; typical G-banding methods were not published until later in 1971. The report of the Paris Conference (ISCN 1985) in naming the technique noted that: 'One of the techniques using the Giemsa reagent, however, gives patterns which are opposite in staining intensity to those obtained by the G-staining methods ... the resulting bands [are called] R-bands.'

R-banding was a French discovery, and it is in France that the technique has been most extensively used. Cytogeneticists in other countries have not in general adopted R-banding as their routine method. One serious disadvantage of the original method was that the staining was very pale, so that phase contrast microscopy was required to see it adequately (Dutrillaux & Lejeune 1971). Although later developments with Giemsa staining have remedied this problem to a large extent, an alternative principle of R-banding soon appeared, using the fluorochrome acridine orange (Bobrow *et al.* 1972a). Apart from giving much better contrast than the original Giemsa staining method, the acridine orange methods have revealed colour heteromorphisms in certain segments of human chromosomes (Verma & Lubs 1975a).

As Bobrow *et al.* (1972a) pointed out, one of the advantages of R-banding patterns is that the ends of the chromosomes are strongly stained. This is a great help, for example, in studying translocations. However, the interstitial bands stain (or fluoresce) with the same intensity as the terminal bands, which could be confusing in certain situations. Dutrillaux (1973) therefore introduced some new techniques in which the intensity of most of the R-bands was reduced, although the terminal bands were still strongly stained. Because of their terminal position, these were named T-bands. As with R-bands, variants using Giemsa or acridine orange were described.

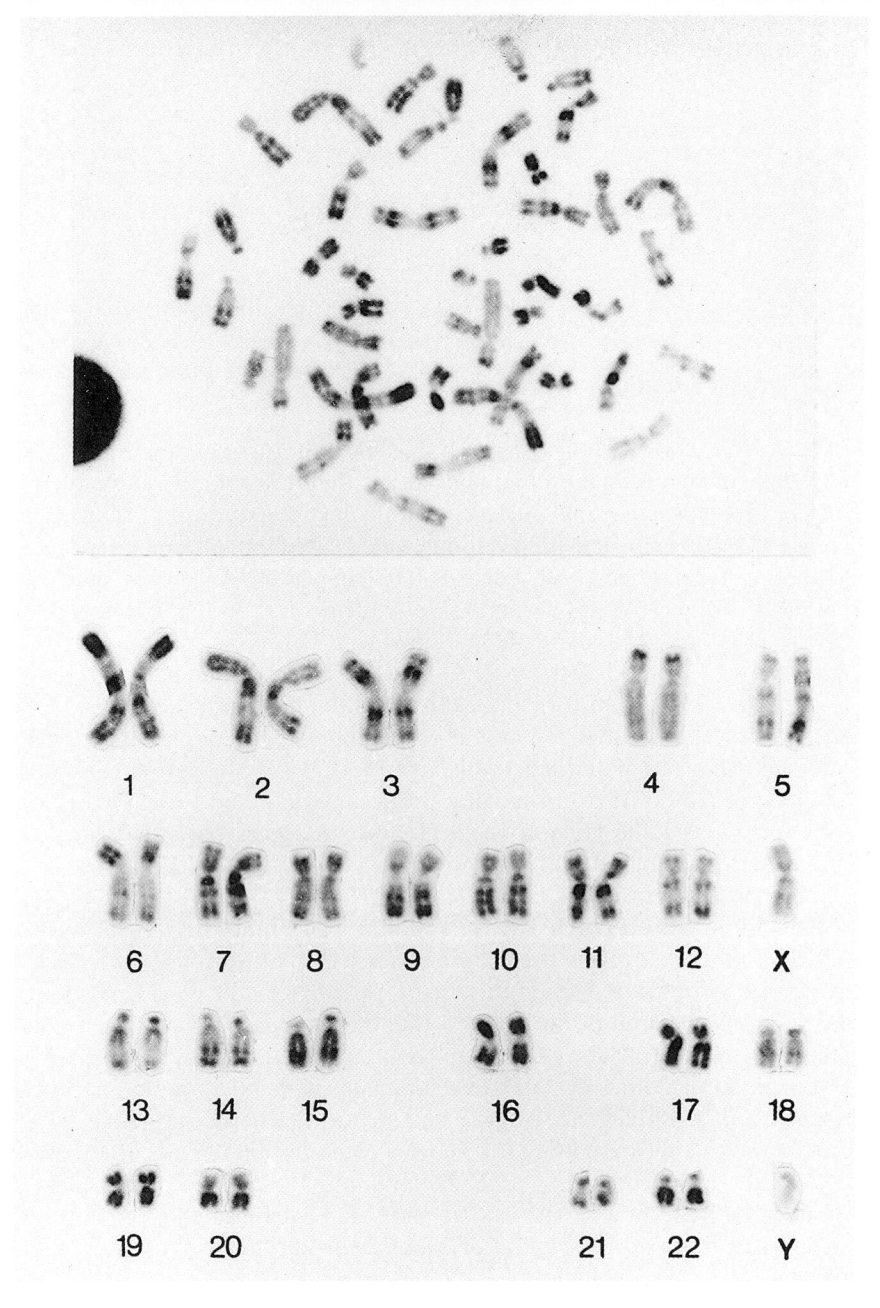

Figure 6.1 Giemsa-stained, R-banded metaphase and karyotype of human chromosomes. Reproduced from Bostock & Sumner (1978) by permission of the authors and Elsevier Science Publishers.

6.2 Methods for R- and T-banding

Although R-banding has proved less popular than G-banding, there are still a number of different techniques for producing R-banding. They may be divided into two classes: those that use Giemsa or, in a few cases, another dye that can be viewed by transmitted light (Section 6.2.1), and those that use acridine orange fluorescence (Section 6.2.2). As will be seen, although the results are, in principle, similar, the acridine orange methods are not in general simply the substitution of a fluorochrome for Giemsa. Section 6.2.3 deals with T-banding methods.

As with G-banding (Ch. 5), R-banding has been used for high-resolution banding. Although in a few cases conventional R-banding methods have been applied to prophase chromosomes accumulated by cell synchronization methods (Drouin *et al.* 1988a), most reports have used BrdU or some similar agent. As explained in Section 5.2.4, such methods really demonstrate replication bands (Section 11.3) and not R-bands *sensu stricto*, and will not be considered in this chapter.

6.2.1 R-banding with Giemsa and similar methods

The original R-banding method described by Dutrillaux & Lejeune (1971) consisted of the following steps:

1 After culturing peripheral blood cells in the usual way, and treating with hypotonic solution, the cells are first fixed in Carnoy's fixative (ethanol, 6 volumes; chloroform, 3 volumes; acetic acid, 1 volume) for 35 min, followed by 20 min in Clark's 3:1 mixture of ethanol and acetic acid, and chromosome spreads made in the usual way.
2 After drying, the chromosome preparations are treated for 10–12 min in 20 mM phosphate buffer at pH 6.5 and 87°C.
3 Slides are stained for 10 min with Giemsa, diluted in the same buffer.

Because the staining was rather pale, phase contrast microscopy was used to observe the banding.

To make the method more reliable, and to permit its application to fibroblast chromosomes, Carpentier *et al.* (1972) experimented with a number of modifications to the original method. Their main alteration was the use of Earle's solution at pH 6.5 instead of a simple phosphate buffer, and they commented on the desirability of leaving the slides for a certain time (hours, or perhaps days or weeks) between making the chromosome preparations and carrying out the R-banding treatment. These improvements permitted observation without phase contrast, which, however, was still recommended for photography. Dutrillaux & Covic (1974) studied the technical variables of R-banding in much more detail. Briefly, they found the optimal pH for R-banding was between pH 5.1 and 7.5, and the optimal temperature between 85° and 89°C. At pH 6.5, R-banding was obtained with an incubation time between a few minutes and four hours. The relationship between the age of the preparations and the length of incubation was also studied.

Other methods are variants on the same principles. Holmberg & Jonasson

(1973) incubated chromosome preparations for 2 h at 95°C in a 2M sodium chloride solution buffered with tris buffer (pH 7.0) and containing 0.01M magnesium sulphate. Such a complex solution is unnecessary; the method of Sehested (1974), which has been widely used, involves incubation in 1M sodium dihydrogen phosphate (which has a pH of 4.0–4.5) at 88°C for 10 min. Both these methods use conventional hypotonic treatment of the cultures, followed by fixation in methanol-acetic acid (3:1). As an alternative to Giemsa staining, Stahl & Vagner-Capodano (1974) used the dye pseudo-isocyanine after the original R-banding pretreatment of Dutrillaux & Lejeune (1971). This dye gave strong staining and could also be observed by fluorescence, but does not seem to have been used subsequently.

A completely different approach to R-banding was that used by Kanda (1976) and by Scheres (1974, 1976a,b). These authors used what were essentially modifications of the BSG method for C-banding (Section 4.2.2). It is characteristic of these methods that C-bands are strongly stained, as well as R-bands. The method of Kanda (1976) in fact used the BSG technique with reduced acid and alkaline treatment, and Giemsa staining after the incubation in $2 \times$ SSC. Scheres (1974) omitted the treatment with hydrochloric acid, and later (Scheres 1976a) found that incubation in $2 \times$ SSC was also dispensable, so that his method consisted finally of incubation in saturated barium hydroxide solution at 60°C for 10 min, followed by staining. Instead of Giemsa, Scheres (1974, 1976a) used the dye 'Stains-All' (4, 5, 4',5'-dibenzo-3,3'-diethyl-9-methyl-thiacarbocyanine bromide) dissolved at 0.005% in a 1:1 mixture of formamide and water. Subsequently, Scheres (1976b) showed that a variety of divalent (alkaline earth) hydroxides could be used to produce R-banding, but that hydroxides of monovalent cations either destroyed the chromosomes or produced G-bands. It was also shown that an alkaline basic fuchsin solution could be used to stain R-bands after barium hydroxide solution (Scheres 1976c). Unfortunately, although the original papers illustrate banding of excellent quality, these latter methods do not seem to have been adopted to any significant extent.

6.2.2 *R-banding with acridine orange*

The main deficiencies of the early R-banding methods using Giemsa were poor contrast due to the low level of staining, and poor definition of bands. These problems were overcome by the use of acridine orange instead of Giemsa. This metachromatic fluorochrome has been used following a wide variety of pretreatments, and produces green to yellow fluorescence in positive R-bands and orange to red fluorescence in negative R-bands (Fig. 6.2). Bobrow *et al.* (1972a) were apparently the first to use acridine orange in this way, and they found that the same pattern of fluorescence could be obtained using several different methods. These methods included the standard R-banding treatment (Dutrillaux & Lejeune 1971), a G-banding method ($2 \times$ SSC at 60°C, Sumner *et al.* 1971a), and mild proteolytic

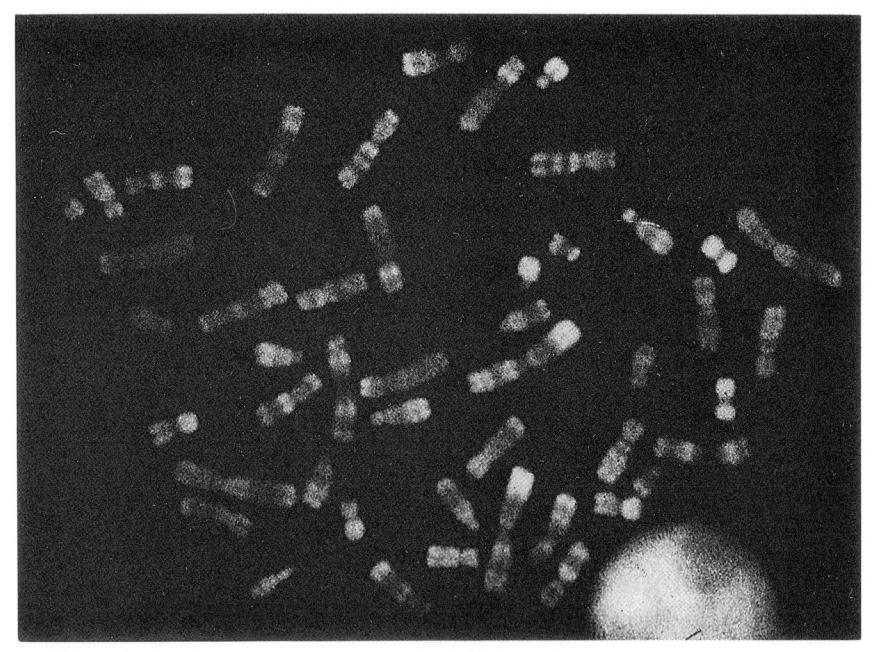

Figure 6.2 Fluorescent acridine orange R-banded human metaphase chromosome spread. Reproduced from Verma *et al.* (1977) *J. Hered.* **68**, 262–3 by permission of the authors and Oxford University Press.

treatment. The method recommended by Bobrow *et al.* (1972a) was as follows:

1. Incubate the chromosome preparations in M/15 phosphate buffer, pH 6.5, for 30 min at 85°C.
2. Stain in 0.01% acridine orange in the same buffer for 5 min.

Overheating or overstaining produced too much red colour, which could be removed by further rinsing; insufficient heating or excessive rinsing produced uniform yellow-green fluorescence. Lubs *et al* (1973) modified this method in detail, and incubated slides in phosphate buffer (pH 6.5) at 85°C for 6 min, and then stained in 0.5% acridine orange.

Wyandt *et al.* (1974) used a completely different type of pretreatment to produce R-bands. The chromosome preparations were first incubated in a solution of ribonuclease, and then treated in a solution of sodium hydroxide (0.07–0.10M) containing 10% formalin, for 2½ min at room temperature. Staining was in a 0.014% solution of acridine orange in McIlvaine's buffer, pH 6.0, for 5 min. In the method described by Fox & Santos (1985), slides were incubated for an hour in a solution containing 50% formamide and 1 × SSC. The chromosomes were then stained with 0.05% acridine orange in McIlvaine's buffer, pH 7.0, for 5 min. Certain authors (Lubs *et al.* 1973,

Wyandt et al. 1974) described the use of their acridine orange R-banding techniques in a double staining method following Q-banding. In the method described by Niikawa & Kajii (1975), the initial stage of Q-banding was an essential part of the method. Following observation of the Q-bands, the slides were destained in methanol-acetic acid, and then incubated in Hanks' solution, pH 5.1 for 13 min at 85°C. Staining was in a 0.005% acridine orange solution in phosphate buffer, pH 6.8, for 10 min. Only chromosomes in cells which had been photographed with quinacrine fluorescence showed red and green bands with acridine orange, whereas other cells showed uniform green coloration.

The age of the chromosome preparations is to be regarded as important for all banding techniques, but this aspect appears to have been given particular attention in the case of acridine orange R-banding. Fresh slides tend to show uniform green fluorescence, whereas slides which are too old fluoresce uniformly red (Wyandt et al. 1974). Niikawa & Kajii (1975) proposed a method of storing chromosome preparations which permitted the induction of good acridine orange R-bands on slides at least 18 months old. The slides were stored in a vacuum dessicator containing silica gel; storage in an air-tight box over silica gel was not adequate, although it reduced the rate of ageing of the slides. There can be little doubt that this method of storage would be beneficial for slides bearing chromosome preparations intended for almost any banding technique.

Photographic procedures for acridine orange banding have also been described in some detail. Apart from the problems associated with photographing any fluorescent object through the microscope (low light intensity, fading of fluorescence), it is also necessary to ensure a clear distinction between the red and green colours of fluorescence. Although black-and-white photography has been employed successfully, the use of colour transparency film has been particularly valuable for acridine orange banding (Verma & Lubs 1975a). As well as providing a permanent record of the colour, the transparency can be used as a negative to produce black-and-white prints of superior quality. By printing through red or green filters, the distribution of the separate colours of fluorescence on the chromosome can be analysed (Castleman & Wall 1973). Verma & Lubs (1975a) and Verma & Dosik (1976) give advice on the choice of photographic materials to achieve the best results.

6.2.3 T-banding

Dutrillaux (1973) introduced simultaneously two distinct methods for producing T-bands (Fig. 6.3), either of which could be used in conjunction with Giemsa or acridine orange staining. For both methods it was recommended to leave the chromosome preparations for a few days before banding. The first method involves incubation at 87°C in a solution consisting of 94 ml of distilled water and 3 ml of phosphate buffer at pH 6.7. After several minutes,

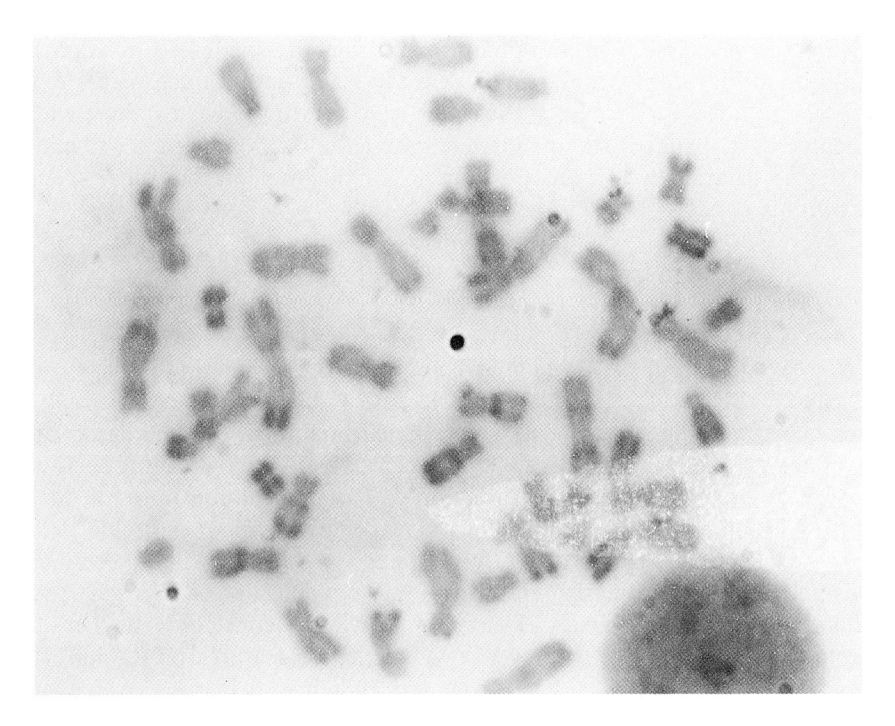

Figure 6.3 T-banded human metaphase.

3 ml of a Giemsa stock solution are added, and incubation continued for a further 5–30 min. If acridine orange staining is required, the Giemsa is washed out with alcohols, and after rinsing with distilled water, the chromosome preparations are stained for 20 min in acridine orange solution (5 mg acridine orange in 100 ml phosphate buffer, pH 6.7). The slides are finally rinsed and mounted in the same buffer.

The second method consists of incubation in Earle's solution, phosphate-buffered saline, or phosphate buffer, all at pH 5.1 and 87°C, for 20–60 min. The chromosomes are then either stained with Giemsa, or with acridine orange as described above. Optimal incubation conditions were investigated in detail by Dutrillaux & Covic (1974).

Later methods for T-banding are classified as CT-banding, as they reveal C- and T-bands simultaneously. The first such method was described by Scheres (1974). Chromosome preparations more than two weeks old are required if T-bands are to be demonstrated. Slides are incubated in a saturated aqueous solution of barium hydroxide for 40 min at 60°C, followed by $2 \times$ SSC for 1 h at 60°C, and finally stained in either Giemsa (in phosphate buffer, pH 6.8) or in a solution of Stains-All in a mixture of formamide and water. The same author later described the use of basic fuchsin in this technique (Scheres 1976c). The method for CT-banding described by Chamla & Ruffié (1976) is more akin to Dutrillaux's original method in principle. Chromosome preparations are incubated in Hanks' solution at 91°C for 30 min, cooled with tap water, and stained with a 4% Giemsa solution for 15 min. If the slides have previously been stained with Giemsa or R-banded, the incubation time should be reduced to 10 min. Slides up to a few weeks old are suitable for this method, but older slides show only C-bands.

6.3 Nature of R- and T-bands

From the definition of R-bands as being opposite in staining intensity to G-bands (ISCN 1985) it is clear that R- and G-bands must be essentially two aspects of the same phenomenon, euchromatic bands. The nature of R-bands may therefore be largely, but not entirely, found by reading Section 5.3 in the previous chapter on G-banding. The information found in that section is summarized in Table 5.2. The reader of this chapter should read 'negative R-bands' for 'positive G-bands' and *vice versa*. It will be seen that positive R-bands are early replicating, late condensing, G + C-rich, and contain housekeeping genes. These points will be considered further in Section 16.3. There are, however, two distinctive features of R- and T-banding which will be considered rather more extensively in this section. One feature is the terminal segments which are stained distinctively with R-banding, and especially with T-banding methods. The other feature is certain regions of human chromosomes which show colour heteromorphism with acridine orange R-banding (Verma & Lubs 1975b).

The ends of chromosomes have long been recognized as having distinctive properties. The telomeres have been defined as regions of DNA at the ends of linear chromosomes that are required for replication and stability of the chromosomes (Blackburn & Szostak 1984). Defined in this way, telomeres are very small, consisting of perhaps only a few hundred base pairs of DNA, plus associated proteins. Structures of this size are, of course, far too small to be resolved by light microscopy and obviously cannot be T-bands. There are, however, larger blocks of DNA with distinctive properties at the ends of chromosomes. Blocks of heterochromatin often occur at the ends of chromosomes, but terminal blocks of heterochromatin are not particularly common in mammals, and T-bands can be demonstrated in the absence of terminal heterochromatic regions. T-bands (using the term to include terminal R-bands) cannot therefore be regarded as a subclass of heterochromatin. There is, in many organisms, evidence for telomere-associated DNA sequences, which are repeated but quite complex, and are adjacent to the telomeres themselves (Blackburn & Szostak 1984). So far such sequences have not been identified in mammalian chromosomes, but it is tempting to suppose that distinctive types of DNA may be responsible for T-banding. Certainly there is evidence from *in situ* hybridization experiments for the location of particular types of DNA at the end of chromosomes, but the

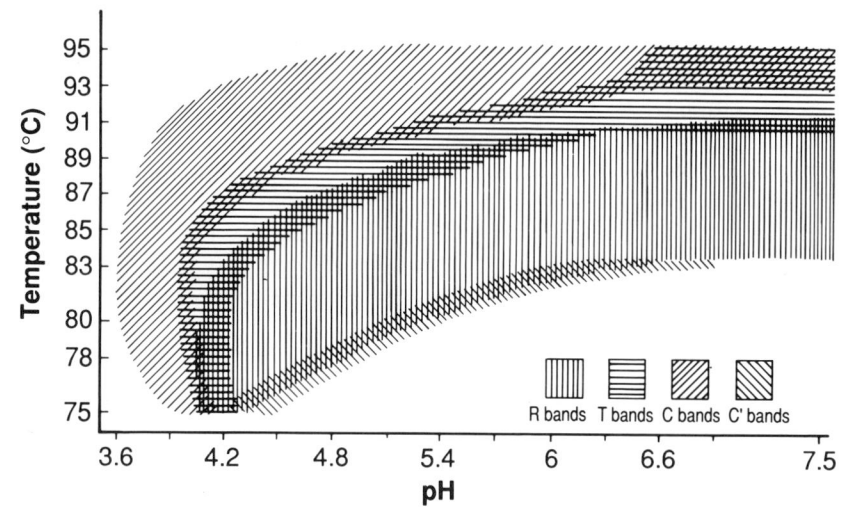

Figure 6.4 Diagram showing the relationships between temperature and pH for production of R- and T-bands. Reproduced from Dutrillaux & Covic (1974) by permission of the authors and Academic Press.

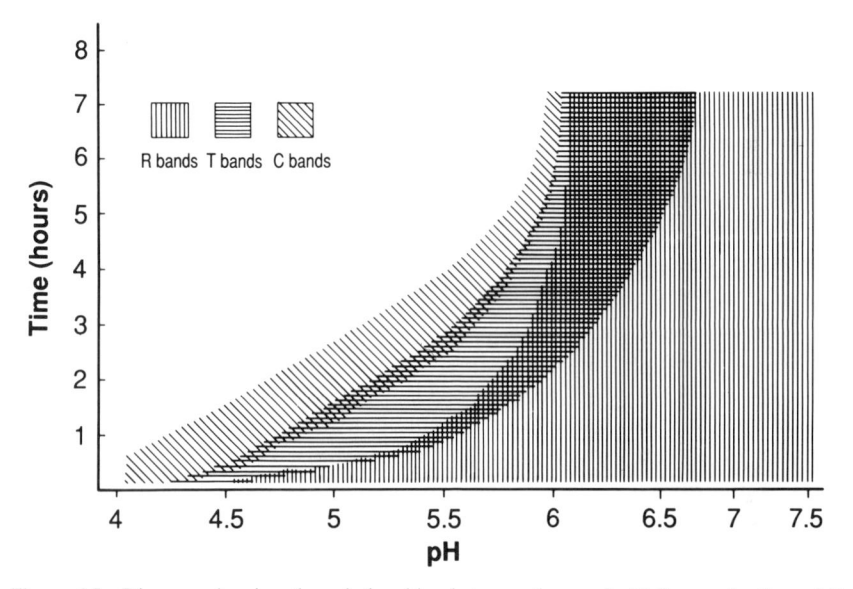

Figure 6.5 Diagram showing the relationships between time and pH for production of R- and T-bands. Reproduced from Dutrillaux & Covic (1974) by permission of the authors and Academic Press.

sequences involved have not been adequately characterized (Hsu *et al.* 1972, Sanchez & Yunis 1974). These regions also show distinctive staining properties with certain fluorochromes (Ambros & Sumner 1987), indicating that they may contain G + C-rich DNA (Ch. 8).

The regions on human chromosomes that show colour heteromorphism with acridine orange R-banding are apparently the secondary constrictions on the short arms of the acrocentric chromosomes, the satellite stalks (Verma & Lubs 1975b, Verma *et al.* 1983). These are the regions that contain the nucleolar organizers, although it seems likely that the acridine orange heteromorphisms involve surrounding heterochromatic material rather than the NORs themselves; however, the resolution of the light microscope is scarcely adequate to settle this point. It should be noted that other regions of heterochromatin show distinctive colours with acridine orange R-banding (for example, in the muntjac, Verma *et al.* 1979), but such regions do not show colour heteromorphism.

6.4 Mechanisms of R- and T-banding

It is appropriate to start a consideration of the mechanisms of R- and T-banding by referring to the work of Dutrillaux & Covic (1974), who studied the effects of several factors (pH, ionic strength and temperature of the incubation medium, length of treatment and age of the chromosome preparations) on the production of R-bands and T-bands (Figs 6.4 & 6.5). They concluded that G-, R-, T- and C-bands represent successive stages in the destruction of chromosome structure, a view supported by the work of Kanda (1976) who studied a completely different type of R-banding method. Other authors have also described the factors required to produce R-bands rather than other types of bands. Sehested (1974) showed that G-bands were produced by incubation at a higher pH than that used for R-banding. Eiberg (1973) kept time and temperature of incubation constant, and investigated the effect of a wide variety of salts and pH values as well as the age of the chromosome preparations. Again a more acid pH was preferable for producing R-bands, but the most important factor appeared to be the age of the chromosome preparations, those one to six months old being best for R-banding, while new slides were better for G-banding. Scheres (1976b) compared the effects of hydroxides of monovalent and divalent cations on chromosomes, and showed that whereas hydroxides of monovalent cations either induced weak G-banding or destroyed the chromosomes completely, hydroxides of the divalent alkaline earth metals induced R-banding. He postulated that R-banding resulted from the strong denaturation of stabilized chromosomes; in his own experiments, the divalent cations were responsible for the stabilization. In the original method of Dutrillaux &

Lejeune (1971), the high incubation temperature would cause denaturation, while the acid pH would stabilize the chromosomes. In the method of Wyandt et al. (1974), the chromosomes are denatured with sodium hydroxide, at the same time being exposed to the stabilizing effects of formaldehyde. Scheres (1976b) did not state clearly what chromosomal components might be denatured and stabilized, but he probably had in mind DNA or nucleoprotein. Experiments such as these suggest, therefore, that limited destruction or extraction of chromosomal material is required for R-banding to occur.

The concept of limited destruction of chromosomal material during R-banding is supported by structural studies. Scheres et al. (1980) carried out Nomarski-optical studies on chromosomes R-banded with barium hydroxide, and observed ridges corresponding to the R-bands, even in unstained chromosomes. They interpreted this finding to indicate that R-bands resulted from loss of material from the negative R-band regions. Burkholder (1981) used electron microscopy to study the structural changes induced by R-banding with a hot sodium phosphate solution. He showed that there is a concentration of chromatin in the positive R-band regions, although the overall electron density of the chromosomes was markedly reduced. It was also noted that the chromatin fibres tended to be aggregated, a conclusion supported by the SEM studies of Jack et al. (1986).

Biochemical studies, in contrast to the structural studies, indicate that there is rather little extraction of material from the chromosomes during R-banding. Comings et al. (1973) found that only about 3% of both DNA and proteins were lost during Dutrillaux & Lejeune's original R-banding procedure. Burkholder & Dvczek (1980b, 1982a,b) also reported only small losses of protein during R-banding, less, in fact, than during any other banding technique they investigated. They found, however, that after incubation in hot Earle's solution, it became more difficult to dissolve the chromosomal proteins, and such aggregated proteins failed to enter the electrophoresis gel (Burkholder & Duczek 1982a,b). This provides an interesting and probably significant parallel with the electron microscopical observations of chromatin aggregation during R-banding (Burkholder 1981, Jack et al. 1986).

An alternative hypothesis to explain R-banding is based on the idea that denaturation of DNA is important. This hypothesis, and the proposal that R-banding involves selective extraction of chromosomal components, should not be regarded as mutually exclusive; on the other hand, the involvement of DNA denaturation is based primarily on observations with acridine orange R-banding, and not with Giemsa. An understanding of the staining properties of acridine orange is therefore essential to a proper interpretation of R-banding. Darzynkiewicz (1979) has reviewed the subject and his conclusions may be summarized as follows. Acridine orange (AO) is a non-specific metachromatic cationic dye, which can, however, be made highly specific for nucleic acids by using an appropriate concentration of the dye in a certain ionic environment. AO has two main modes of binding to nucleic acids: by intercalation between nucleic acid base pairs (Fig. 6.6), which results in green fluorescence; and externally, to the nucleic acid phosphates. When bound externally, the flat AO molecules can aggregate by stacking on each other (Fig. 6.6) and when aggregated, the fluorescence of AO is red. Intercalation can only occur in double-stranded nucleic acids, whereas external binding and aggregation occurs preferentially on single-stranded nucleic acids. Thus AO fluorescence can, in appropriate conditions, be used to distinguish between DNA (green fluorescence) and RNA (mainly single-stranded, therefore red fluorescence), or between double-stranded, native DNA (green) and denatured, single-stranded DNA (red). As with almost any dye, the binding of AO to nuclear DNA can be restricted by nuclear proteins. Bearing these points in mind, is it possible that acridine orange R-banding is simply the result of differential denaturation of chromosomal DNA?

Many authors have investigated the effects of DNA denaturing agents on nuclei and chromosomes, assessing the results not only with acridine orange (Rigler 1966, Comings *et al.* 1973, Sumner *et al.* 1973), but also with other suitable dyes (Sumner *et al.* 1973, Dreskin & Mayall 1974) or with ultraviolet

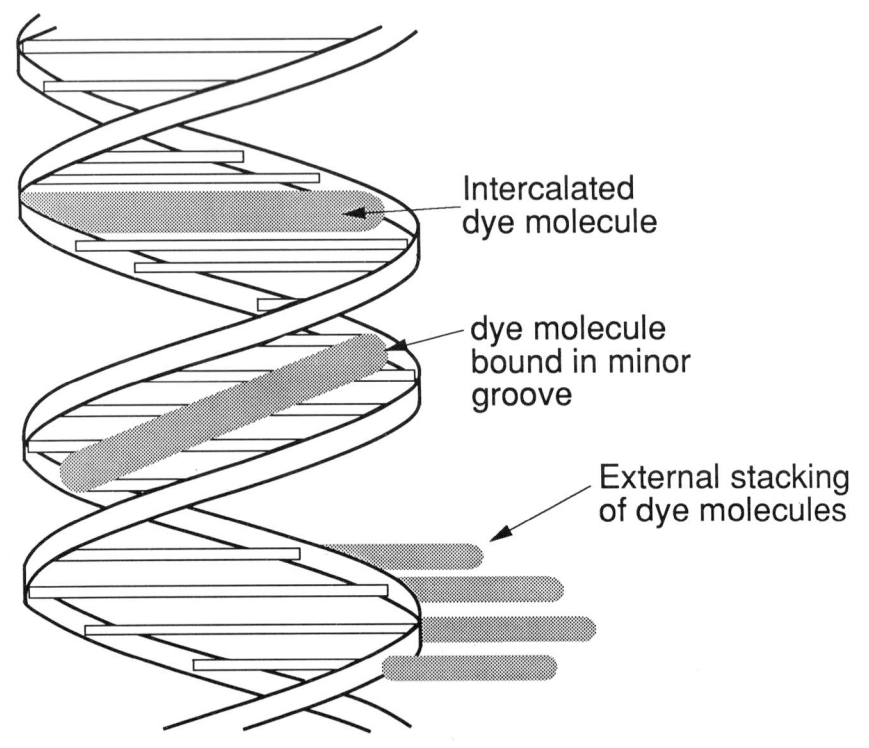

Figure 6.6 Diagram illustrating different modes of binding of dye molecules to DNA: intercalation, binding in the minor groove, and external binding with stacking.

absorption (Kernell & Ringertz 1972). Without going into details, it may be stated that the results obtained correspond to the effects of the denaturing agents on isolated DNA; fixed but otherwise untreated chromosomes and nuclei contain native DNA, which can be denatured by appropriate treatment with alkali, heat, or formamide. Many authors have used formaldehyde to prevent renaturation of the chromosomal DNA before staining (e.g. Rigler 1966), although Darzynkiewicz (1979) states that this is unnecessary, and that formaldehyde merely serves to lower the temperature of denaturation. Other evidence (see Section 4.2.5) also suggests that rapid renaturation of chromosomal DNA does not occur. Evidence, independent of AO fluorescence, that an R-banding treatment can denature chromosomal DNA comes from the work of Bernheim *et al.* (1984), who were able to carry out *in situ* hybridization to chromosomes that had been R-banded but not otherwise denatured.

The denaturation temperature of DNA depends upon its base composition, $G + C$-rich DNA denaturing at a higher temperature than $A + T$-rich DNA. The G- and R-bands on the chromosomes of higher vertebrates apparently consist of regions of different base composition, positive R-bands being $G + C$-rich, and negative R-bands $A + T$-rich (Table 5.2). In principle, therefore, it should be possible to denature negative R-bands at a lower temperature than positive R-bands, and the former would show red AO fluorescence, and the latter green. This is, of course, exactly the pattern which is found with acridine orange R-banding. Both de la Chapelle *et al.* (1973) and Bobrow (1974) showed that at an appropriate temperature patterns of red and green bands were obtained, while at lower temperatures the chromosomes were wholly green, and at higher temperatures wholly red. Interestingly, Bobrow (1974) found that the same result was produced whether or not formaldehyde was included in the incubation medium, but that the colour changes occurred at lower temperatures in the presence of formaldehyde. There is, therefore, a satisfactory body of evidence consistent with the hypothesis that acridine orange R-banding is due to differential denaturation of chromosomal DNA according to its base composition. It must be pointed out, however, that both DNA denaturation and AO binding can be affected by chromosomal proteins, although it does not appear necessary to invoke such factors to explain the generality of acridine orange R-banding. Nevertheless, there are some situations which suggest that 'all may not be as simple as it looks' (Bobrow 1974). Chief among these is the observation that brief trypsin treatment, which can hardly be expected to denature DNA, can produce acridine orange R-bands (Bobrow *et al.* 1972a, Bobrow & Madan 1973). This suggests that proteins may be influencing the pattern of AO fluorescence, but it would also be necessary to explain the red fluorescence in the absence of denaturation. Bobrow (1974) also failed to produce consistent R-band patterns with another dye, ethidium, that is sensitive to DNA denaturation, although this may simply have been due to failure to find the correct staining conditions.

If, as seems plausible, the production of AO R-banding is due to differential denaturation of DNA, can R-banding with Giemsa be explained in the same way? Experiments on the binding of Giemsa to double- and single-stranded DNA are inconsistent (Sumner & Evans 1973, Ahnström & Natarajan 1974), and single thiazine dyes are reported to bind equally well to native and denatured DNA (Comings & Avelino 1975). Furthermore, since the same treatment that results in G-banding with Giemsa produces R-banding with acridine orange (Bobrow & Madan 1973), one is forced to the conclusion that there are, in fact, important differences between the mechanism of R-banding with Giemsa and with acridine orange. Both Comings & Avelino (1975) and van Duijn *et al.* (1985) propose that the binding of proteins to DNA is modified by the R-banding pretreatment, so that binding of the dye is largely restricted to positive R-bands. It should be noted that, according to van Duijn *et al.* (1985), the formation of a magenta thiazine-eosin complex (Section 5.4) is not required for R-banding, in which the staining is blue. Comings & Avelino (1975) also attempted to explain AO R-banding in terms of blocking of dye binding, but their hypothesis does not explain adequately the inconsistencies referred to above. It is certainly not possible to conclude that the definitive mechanism for any type of R-banding has yet been elucidated. While differential denaturation of DNA is a plausible explanation for AO R-banding, it could equally well turn out that all types of R-banding involve rearrangements of chromosomal proteins on the DNA, thereby blocking dye binding at certain sites (negative R-bands). A combination of the two mechanisms is not inconceivable.

Very little has been said about T-banding in this section, because little effort has been made to investigate the mechanism of T-banding independently of that of R-banding. It must be supposed that T-bands are a subset of R-bands that are in some way more resistant. It has already been noted (Section 6.3) that the ends of chromosome arms tend to contain G + C-rich DNA, and this factor alone might explain T-banding with acridine orange, but not, in our present state of knowledge, T-banding with Giemsa.

6.5 Applications of R- and T-banding

R-banding produces variations in staining intensities along chromosomes which are the inverse of those produced by G-banding. Like G-banding, R-banding has only been reported for the chromosomes of higher vertebrates, particularly mammals. R-banding is, in fact, another method of staining euchromatic bands, and in principle, has similar applications to G-banding (Section 5.5). Nevertheless, R-banding has been much less used than G-banding, and a number of factors are responsible for this. It is general experience that R-banding patterns, whether with Giemsa or acridine orange, are less sharp and less detailed than typical G-banding patterns (Hsu 1974). Moreover, R-banding methods had a reputation for unreliability (Verma & Lubs 1975a, Kanda 1976), although efforts have been made to improve the situation (Bernheim & Berger 1981). Added to this were the problems that, with the early methods, phase contrast microscopy was required after Giemsa staining, while with AO R-banding, fluorescence microscopy was necessary with its attendant disadvantages. In spite of these limitations, which must now be regarded as only minor, R-banding has been used extensively in France, its country of origin, and to a much lesser extent elsewhere, in clinical cytogenetics, and also in evolutionary studies (Ch. 15). For the benefit of those working with human chromosomes, R-banded karyotypes have been published, showing up to 1300 bands (Drouin *et al.* 1988a). Some studies on chromosomes of domestic animals, using R-banding, have been summarized by Gustavsson (1980).

It is clear that although R-banding can be applied satisfactorily to problems similar to those which can be investigated with G-banding (Section 5.5) or Q-banding (Section 7.5) it has not proved as popular as either of these methods as the sole or principal technique in an investigation. There are, however, certain situations in which R-banding, and in one case T-banding, have distinctive contributions to make. These are described in the following sections.

6.5.1 *R-banding in combination with other methods for better delineation of breakpoints*

A combination of two different banding methods, especially on the same set of chromosomes, may be expected to give more information than either method used alone. This is true especially if an attempt is being made to determine the exact site of chromosomal breakpoints. The fact that a break may, for example, be in a negative G-band implies that it should be in a positive R-band, and the combination of the two methods should therefore permit one to locate the breakpoint with greater confidence. To this end, a number of double-staining methods have been described. Sehested (1974) showed that G- or R-banding could be produced simply by altering the pH of the incubation medium. Buckton (1976) used this method to study the breakpoints induced in human chromosomes by X-rays. Using sequential G- and R-banding, she concluded that the breaks occurred preferentially at the interfaces of the dark and light bands, in contrast to earlier results suggesting that such breaks were concentrated in (positive) R-bands (Holmberg & Jonasson 1973).

A sequential method for Q-banding and AO R-banding was published by Niikawa & Kajii (1975), who commented that the technique should be useful for the study of structural chromosome anomalies, although no examples were given. Similarly, Verma & Lubs (1976a) described sequential Q- and AO R-banding on the same preparations, and reported that sequential G- and AO R-banding could also be obtained, although the results were less good than after Q-banding.

6.5.2 Staining of terminal regions of chromosomes

It is characteristic of R-banded, and especially T-banded chromosomes, that the ends of the chromosomes are strongly stained. Whereas it is difficult to be sure where the ends of G-banded chromosomes are, there can be little doubt after R- or T-banding. Not only is it helpful to see how long normal chromosomes actually are, but the distinctive staining of the chromosome ends is also valuable for the interpretation of a variety of chromosomal rearrangements. Thus Bobrow *et al.* (1972a) used acridine orange R-banding, and Dutrillaux (1973) used T-banding, to show that certain translocations were in fact reciprocal; the terminal regions involved in the translocations were shown, by these methods, to be still present although now exchanged. Berger *et al.* (1981) described a situation in a child with acute non-lymphocytic leukaemia, who appeared by G-banding to have a terminal deletion of chromosome 11, and by R-banding a terminal deletion of chromosome 17; this discrepancy was interpreted as an 11;17 translocation.

Deletions of terminal regions of chromosomes are shown well by these methods (Wyandt *et al.* 1974), and they have been particularly useful in the study of ring chromosomes. Wyandt *et al.* (1974) and Chamla & Ruffié (1976) showed the presence of some terminal material in ring chromosomes using AO R-banding and CT-banding respectively, although Wyandt *et al.* (1974) showed, using densitometric traces of the banding patterns, that some terminal material had been lost. These studies would scarcely have been possible with G-banding, as the relevant regions of the chromosomes would have been stained only feebly.

6.5.3 Acridine orange R-band colour heteromorphisms

Verma & Lubs (1975a) described an interesting heteromorphism in the secondary constriction region of the short arms of the human acrocentric chromosomes, using acridine orange R-banding. These regions show colours varying from red, through orange and yellow, to green, as well as varying in size from very small to very large. These variations were shown to be inherited in a strictly Mendelian manner (Verma & Lubs 1976b). Comparisons with other banding techniques indicated that the acridine orange heteromorphism gave more information than Q-band heteromorphism for these regions, and that there was no consistent relationship between the brightness of the quinacrine fluorescence and the colour of the acridine orange fluorescence (Verma & Lubs 1975b, Verma *et al.* 1977). Similarly, no relationship could be found between the colour of acridine orange fluorescence on the short arms of the human acrocentric chromosomes, and the sizes of NORs seen by silver staining (Verma *et al.* 1983). It should be noted that all the work on this system of heteromorphism has come from one laboratory, although Niikawa & Kajii (1975) reported a heteromorphism for brightness, but not colour, on the short arms of the human acrocentric chromosomes. This heteromorphism was found useful for determining the origin of the extra chromosome in an abortus with trisomy 22. The acridine orange colour heteromorphisms will be considered in more detail in Chapter 14.

7 Q-banding

7.1 Introduction

Q-bands are defined as the fluorescent patterns produced on chromosomes after staining with quinacrine mustard or quinacrine (ISCN 1985). Q-banding methods were introduced by Caspersson *et al.* (1968), and are thus the oldest of the modern banding techniques, as well as being one of the simplest and almost certainly the most versatile. Moreover, Q-banding was the first method, and still one of the very few, to be designed on a rational basis (Caspersson 1973). Caspersson argued that the base composition in chromosomal DNA must be inhomogeneous, and that such inhomogeneities should be detectable using compounds that would react specifically with certain bases. Quinacrine mustard was chosen as a suitable compound, because of the known guanine-specificity of the nitrogen mustards, and because of the greater sensitivity of fluorescence methods. This principle proved to work in practice, and the fact that the mechanism of formation of Q-bands appears to be quite different from that originally envisaged does not in any way detract from Caspersson's foresight and vision.

A summary of the achievements in cytogenetics due to Q-banding is impressive. The first paper (Caspersson *et al.* 1968) described quinacrine fluorescence of both plant and mammalian chromosomes, and although the latter scarcely foreshadowed the developments that were to come, it was quickly established that quinacrine mustard was valuable for demonstrating heterochromatic regions of chromosomes (Caspersson *et al.* 1969a, b, Zech 1969, Vosa & Marchi 1972b). A particularly exciting and valuable discovery was the brilliant quinacrine fluorescence of the distal part of the long arm of the human Y chromosome (Zech 1969). This not only permitted the identification of the Y chromosome in mitotic metaphase cells (Zech 1969, George 1970), but also in meiotic cells (Pearson & Bobrow 1970a, b), in interphase nuclei (Caspersson *et al.* 1970b, Pearson *et al.* 1970), and in sperm heads (Barlow & Vosa 1970). It was soon recognized that Q-banding patterns could be demonstrated throughout the length of the chromosomes of man (Caspersson *et al.* 1970a, 1971a) and these became the basis for the standard description of the banded human karyotype (ISCN 1985). Similar patterns were quickly reported for the chromosomes of other species of mammals (O. J. Miller *et al.* 1971, Hansen 1972a). Caspersson *et al.* (1971c) used Q-banding to identify the chromosomes of each species in mouse-human hybrid cells, and Q-banding was also used to show that euchromatic banding patterns did not vary from one tissue to another (Caspersson *et al.* 1972) or during development (Nesbitt & Donahue 1972). Another valuable finding of those early years was the close correlation between banding and time of replication (Ganner & Evans 1971, Calderon & Schnedl 1973). Cytogeneticists also applied Q-banding to species other than plants and mammals: several early studies showed Q-banding patterns on *Drosophila* chromosomes (Adkisson *et al.* 1971, Ellison & Barr 1971, Zuffardi *et al.* 1971).

Many other discoveries could be listed, and it is still quite astonishing how much was accomplished using this method in the space of only two or three years. During this period, of course, C- and G-banding methods were also introduced, and those who were working in the field at that time will recall the intense excitement generated by the new methods, and by Q-banding in particular. A large number of papers poured out, often on successive pages of the same journal, describing applications which had scarcely been a dream only a few months earlier. Although Q-banding is now only one of a large number of banding methods, it remains important, not only for the many pioneering discoveries made with this technique, but because it is still a valuable practical technique in many fields. Like all fluorescence methods, Q-banding suffers from the impermanence and faintness of the image, and also from the requirement for special microscopes, which, however, should now be routinely available in cytogenetics laboratories. These disadvantages are greatly outweighed by the advantages: the simplicity and reliability of the technique, and above all, the versatility of Q-banding.

7.2 Q-banding methods

Q-banding methods are simple. Chromosome preparations are simply stained in a solution of the dye, mounted in an appropriate aqueous fluid, and viewed with a fluorescence microscope. None of the parameters of Q-banding appears to be critical, and a reasonable result can generally be obtained without difficulty. However, as with any other banding method, good chromosome preparations are required for the best results.

Many fluorochromes have been tested to see if they will produce chromosome bands. They may be divided into three classes: those that produce Q-banding patterns, those that produce slightly or very different patterns from Q-banding, which will be considered in the next chapter, and those that produce no banding at all without prior treatment of the chromosomes. The number of fluorochromes that has been found useful for Q-banding is quite limited, and these are illustrated in Figure 7.1. The first compound used for Q-banding was, of course, quinacrine mustard (QM) (Caspersson *et al.* 1968); by comparison, quinacrine (Q) was held to give inferior results and to show weaker and less stable fluorescence (Caspersson *et al.* 1968, 1969a). As well as these two compounds, Caspersson *et al.* (1969a) also tested propyl quinacrine mustard (PQM) (Fig. 7.1) and a number of monofunctional nitrogen mustard derivatives. The latter were found to give similar results to

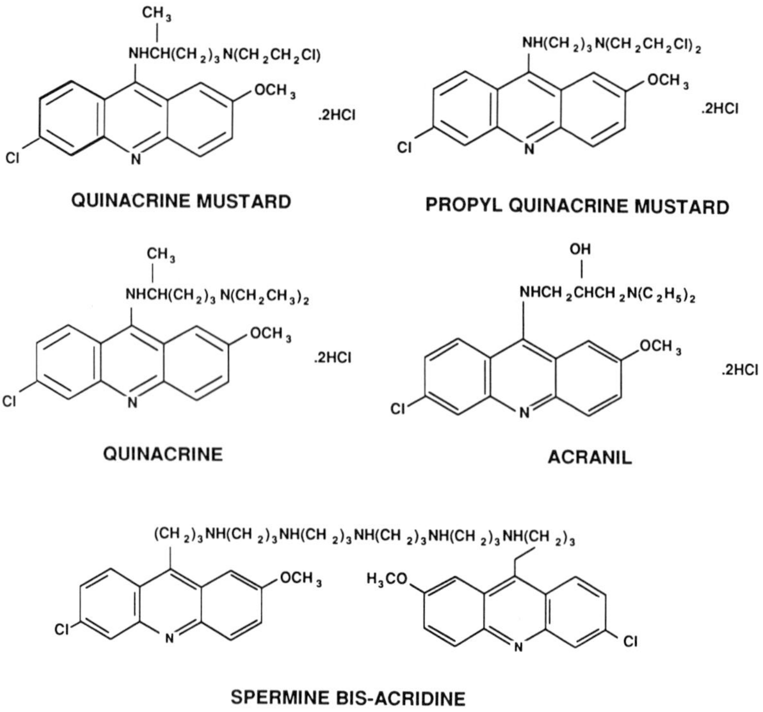

Figure 7.1 Structural formulae of quinacrine, quinacrine mustard, propyl quinacrine mustard, Acranil, and spermine bis-acridine.

QM, but with lower fluorescence intensities. PQM was reported to give results 'entirely the same' as those produced by QM (Caspersson et al. 1969a), and although it was neglected for several years, it has more recently been adopted and extensively used by workers in the USSR (Mikelsaar et al. 1978, Ibraimov et al. 1982). Several other quinacrine analogues have been tested, often with a view to establishing the distinctive molecular features required for banding (Limon et al. 1975, Tsou et al. 1975, Comings et al. 1978). Although many such compounds can give satisfactory results, the fact that they are not commercially available but must be specially synthesized has precluded their adoption for routine cytogenetics. Elsewhere, commercially available compounds related to quinacrine have been tested successfully (e.g. Salamanca et al. 1972); of these, Acranil (Fig. 7.1) possibly deserves further investigation because of its reportedly good resistance to radiation (Moscetti et al. 1971). It may be noted that Vosa et al. (1972) prepared extracts from roots of several species of plants, and found that some of them gave results strictly comparable to those obtained with quinacrine on various plant chromosomes and interphase human Y-bodies. It is, however, difficult to discern any practical advantages in the use of such extracts.

In spite of Caspersson's original remarks of the relative merits of quinacrine and quinacrine mustard (see above), it is now generally believed that there is little practical difference between Q and QM (Pearson & van Egmond-Cowan 1976), and most workers now use quinacrine, which is more readily available, cheaper, and not being a nitrogen mustard should also be safer to handle. Nevertheless, the quest has continued for improved stability of fluorescence and quality of banding, and van de Sande *et al.* (1979) reported on a spermine *bis*-acridine compound, $(CMA)_2S$. This compound, *bis*-N,N'''(6-chloro-2-methoxy-acridin-9-yl) spermine, is essentially a dimeric analogue of quinacrine (Fig. 7.1), and compared with quinacrine shows greater stability of fluorescence, stronger binding and sharper definition of bands. This compound can be strongly recommended, and is now available commercially.

7.2.1 Q-banding with quinacrine mustard and quinacrine

Quinacrine mustard was first used as an aqueous solution containing $50\mu g$/ml of QM. Chromosome preparations were stained for 20 min at 20°C, the slides were washed three times in MacIlvaine's phosphate-citrate buffer, pH 4.1, and then mounted in the same buffer under a sealed coverslip (Caspersson *et al.* 1969a). Rubber solution is particularly good for sealing coverslips: it is easily applied from a tube, sets in a few minutes, and can be peeled off easily if it is desired to remove the coverslip.

Significant variations from this original procedure have been rare. Various authors have used higher concentrations of QM in the staining solution (George 1970, Hansen 1972a, b), but there is no apparent advantage in doing so. Sometimes the solvent for the dye has been varied. Adkisson *et al.* (1971) dissolved QM in MacIlvaine's buffer at pH 4 rather than in water. Comings (1971) used QM dissolved in MacIlvaine's buffer, pH 7.0, containing 25% methanol, and reduced the staining time to 5 min. Vosa & Marchi (1972b) went further and dissolved QM is absolute alcohol, again restricting the staining time to 5 min. As is so often the case, no explanation was given for using these variant methods, but it may be supposed that the use of an alcoholic solvent might preserve the chromosomal morphology better than an aqueous one.

Although Caspersson's original method specified mounting in buffer at pH 4.1, higher pH values have been preferred subsequently. For human chromosomes, MacIlvaine's phosphate-citrate buffer at pH 7.0 was recommended for optimal results (Caspersson *et al.* 1970a), and Hansen (1972b) used the same pH buffer for mounting chromosomes from the pig. Vosa & Marchi (1972b) used a still higher pH, 7.8, remarking that such a buffer solution gave better enhancement of the fluorescent bands than mounting in distilled water. It seems clear that selection of the optimal pH for the mounting medium is necessary to obtain the best delineation of the bands, but that the other steps of the staining procedure can be varied within wide limits without seriously affecting the results.

Staining with quinacrine, rather than with quinacrine mustard, has generally been done with more concentrated solutions, perhaps because Q is cheaper and more easily available than QM. It should be noted that quinacrine is also sold as a chromosome stain under the name of Atebrin, and as a drug under the name of Mepacrine, quinacrine having been used originally in large quantities as an anti-malarial treatment. Quinacrine has commonly been used as a 0.5% solution in distilled water, with staining for 5 or 6 min, followed by washing in tap water (Pearson & Bobrow 1970b, Pearson *et al.* 1971, Evans *et al.* 1971). The period of washing has been varied from a rinse up to 3 min., and mounting in distilled water is commonly used. Several authors have used a buffer at about pH 5.5 for mounting (e.g. Breg 1972, Calderon & Schnedl 1973), and C. C. Lin *et al.* (1975) have shown that a pH value between 4.5 and 5.5 was the best compromise for sharpness of bands and brightness of fluorescence.

Variations on the above procedure, which is the method recommended for general application, have been few. Vosa & Marchi (1972b) used 0.5% quinacrine in absolute alcohol, which as already remarked, may perhaps better preserve chromosome morphology. Instead of using a high concentration of quinacrine, van Prooijen-Knegt *et al.* (1982) recommended staining for 2 h in a 0.002% solution of quinacrine in MacIlvaine's citrate-phosphate buffer at pH 4.1, and mounting in the staining solution. The dye concentration in the mountant is sufficiently low not to cause significant background fluorescence, and the method is claimed to have the advantage that the quality of Q-banding is more uniform across the slide. The intensity of fluorescence is comparable to that obtained with standard staining methods.

The most serious disadvantage of quinacrine as compared with QM is its greater tendency to fade when illuminated. None of the now standard additives which inhibit fading of fluorochromes appears to have any effect on the fading of quinacrine (Gill 1979, Sumner unpublished). Although various workers have used mountants containing glycerol (Pearson *et al.* 1970) or sucrose (Ellison & Barr 1971) there has been no suggestion that fading is reduced in such mountants. If fading of quinacrine fluorescence does prove to be a serious problem, then it will be necessary to use a more stable fluorochrome such as QM, or (CMA)$_2$S (Section 7.2.2).

7.2.2 Use of spermine bis-acridine[(CMA)$_2$S]

Dimeric analogues of quinacrine bind to DNA much more strongly than monomeric quinacrine (LePecq *et al.* 1975), show enhanced contrast between weakly and strongly fluorescent bands, and show negligible fading of fluorescence (van de Sande *et al.* 1979). Until recently is was necessary to prepare spermine and spermidine *bis*-acridines in the laboratory (see Sumner, 1989, for a description of the synthesis), but one of these compounds, either of which may be used interchangeably, is now available commercially (Molecular Probes Inc., Eugene, Oregon, USA).

The staining technique for (CMA)$_2$S, whether made with spermine or spermidine, is generally similar to the procedure with quinacrine.

1 A 0.005% solution is used: 5mg of (CMA)$_2$S is dissolved in 2ml of methanol, and the solution diluted to 100 ml with 10mM disodium hydrogen phosphate, adjusted to pH 6.5 with 0.2M hydrochloric acid.
2 Chromosome preparations are stained for 10 min.
3 Slides are washed for 2–3 min in running tap water, and mounted in deionized water or phosphate buffer, pH 6.5.

Definition of bands is very sharp, even on slides several years old.

7.2.3 Quinacrine staining of interphase nuclei and sperm heads

Certain regions on human chromosomes – the centromeric region of no. 3, short arms of the acrocentric chromosomes and part of the Y – fluoresce particularly strongly with QM and Q (Caspersson et al. 1970a), and can be detected in interphase nuclei (Caspersson et al. 1970b). The brightly fluorescent region of the Y chromosome is not only the largest and brightest, and therefore most easily detected in interphase nuclei, but also happens to be of the greatest practical interest, since it can be used to determine the chromosomal sex of the individual by examining uncultured, interphase cells (Section 7.5.3).

The actual staining methods for interphase nuclei do not, in most cases, differ at all from those used for metaphase chromosomes; the preparation, and in particular the fixation, of the cells varies, however. It is not necessary to fix interphase nuclei in methanol-acetic acid to demonstrate fluorescent Y-bodies, although this fixative has been employed in some cases. For amniotic fluid cells, buccal smear cells, fibroblasts and spermatozoa simple fixation in methanol, or occasionally ethanol, is perfectly satisfactory (Pearson et al. 1970, Barlow & Vosa 1970, George 1971, Červenka et al. 1971). For buccal smears, alcohol-ether fixation has also been used successfully (Pearson et al. 1970, George 1971). For hair roots, however, which are an easily accessible source of material for Y-body staining, fixatives containing acetic acid are recommended, presumably as much to macerate the tissue as to provide satisfactory preservation (Červenka et al. 1971, Schwinger et al. 1971, Korf et al. 1975). For a comprehensive review of these procedures, see Schwarzacher (1974).

It is less easy to distinguish the fluorescent Y-body when it is surrounded by other fluorescent material in a nucleus than it is to recognize the bright fluorescence of a metaphase Y chromosome against a dark background. Procedures have therefore been published for improving the contrast between the interphase Y-body and its surroundings. Schwinger et al. (1978) found that hypotonic treatment of amniotic fluid cells enhanced the visibility of the Y-body, as a result of the swelling, and therefore weaker fluorescence, of the rest of the nucleus. Intensely fluorescent sites on autosomes were also rendered more easily visible by hypotonic treatment. Hegde *et al.* (1978) applied a similar principle to human sperm heads, using dithiothreitol to decondense them before staining. Newburger & Latt (1979) used a completely different principle to enhance the visibility of the Y-body. After quinacrine staining, nuclei were counterstained with 7-aminoactinomycin D. This substance quenches the quinacrine fluorescence of most chromosomal regions, by the process of energy transfer (Section 8.5), but does not affect regions such as the Y-body which show brilliant fluorescence. Similarly, Klasen & Schmid (1981) used counterstaining with distamycin A to improve the visibility of the quinacrine-stained Y-body in human interphase nuclei and spermatozoa, while Hollander *et al.* (1976) used ethidium bromide as a counterstain.

7.2.4 Q-banding in combination with other banding methods

It has been noted in earlier chapters that double staining of the same set of chromosomes successively with more than one banding technique may give extra information not readily obtained by using the techniques independently. It was soon discovered that Q-banding can be used in this way. If human chromosomes are first G-banded, then destained and Q-banded, both G- and Q-band patterns can be demonstrated on the same chromosomes (Evans *et al.* 1971). Compared with straight Q-banding, the pattern of bands is the same except in certain regions of heterochromatin which show brighter fluorescence when stained with quinacrine after G-banding. Other workers to report sequential G- and Q-banding include Krajca & Wray (1977) and Lichtenberger (1983). Krajca & Wray (1977) used the two techniques on isolated chromosomes, and pointed out that while G-banding is inconsistent on isolated chromosomes, it does appear to stabilize subsequent quinacrine fluorescence. Lichtenberger (1983) claimed that Q-banding could be performed before or after G-banding, but this has not been general experience. G-banding after Q-banding is either completely unsuccessful (Evans *et al.* 1971), or C-banding is obtained instead (Chen 1974). This is no doubt connected with photochemical damage to DNA that occurs during irradiation of quinacrine-stained nuclei and chromosomes (Bosman *et al.* 1977b). There is indeed no problem about C-banding chromosomes that have already been Q-banded and examined (Rowley & Bodmer 1971, Martin & Rowley 1983).

As described in the previous chapter (Section 6.5.1) Q-banding and acridine orange R-banding can also be performed successfully on the same chromosome preparations (Niikawa & Kajii 1975, Verma & Lubs 1976a). The staining methods involved in these cases were essentially standard. Prior exposure of the Q-banded chromosomes to irradiation apparently had no adverse effects on the subsequent R-banding, and indeed it appeared to be an essential part of the procedure of Niikawa & Kajii (1975). They found that only those chromosomes previously photographed for Q-banding showed the red and green banding patterns when restained with acridine orange.

Finally, it may be noted that the fluorochrome Hoechst 33258 (Section 8.3.1) has been combined with quinacrine staining to distinguish between human and mouse chromosomes in hybrid cells (Carlin & Rao 1982); Hoechst 33258 is used to label the mouse chromosomes by their bright centromeric fluorescence. Although Hoechst 33258 produces patterns similar to Q-bands along chromosomes, such banding is less satisfactory than Q-banding with quinacrine or QM. Carlin & Rao, therefore, restained the chromosome preparations with QM to identify the human chromosomes which had been picked out by the Hoechst 33258 staining. The prior staining with Hoechst 33258 was reported to produce no alterations in the Q-banding pattern.

7.2.5 Observation and recording of Q-banding

Because of the nature of fluorescence, special equipment and procedures are required for the study of Q-banded chromosomes, and indeed of chromosomes banded with any fluorescent dye (for example, acridine orange R-banding (Section 6.2.2), and many other fluorescence methods described in Chapter 8). Fluorescence is produced by excitation with light of a shorter wavelength, the fluorescence itself being emitted at longer wavelength than that of the exciting radiation. Because the excitation of fluorescence is generally rather an inefficient process, and the amount of the fluorochrome

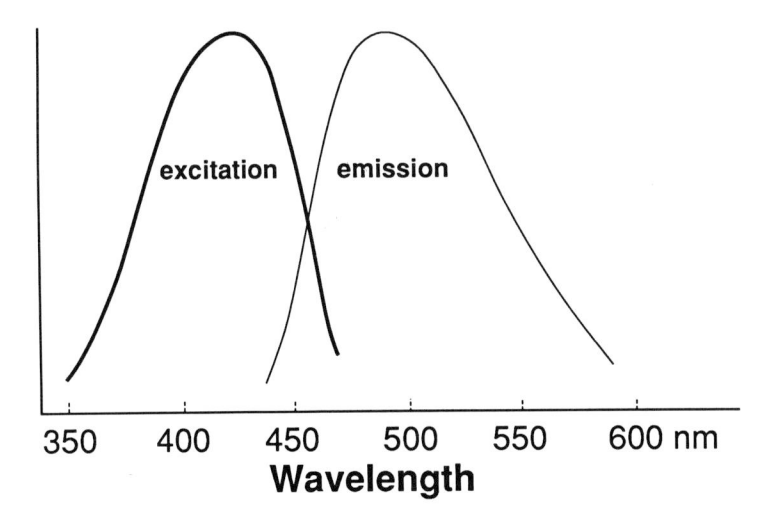

Figure 7.2 Excitation and emission spectra of deoxyribonucleoprotein stained with quinacrine. Spectra of chromosome preparations stained with quinacrine mustard were practically the same. Reproduced from van der Ploeg & Ploem (1973) *Histochemie* **33**, 61–70 by permission of the authors and Springer-Verlag.

is generally very small, the amount of light produced is very low, and sensitive means are required for its detection. To produce stronger fluorescence, a more powerful source of exciting radiation may be used, but this in turn is more likely to cause photochemical destruction of the fluorochrome, and consequently fading of the fluorescence. The great intensity of the exciting radiation, and the weakness of the emitted fluorescence, mean that the exciting and emitted light must be effectively separated. For satisfactory results it is therefore necessary to use an efficient fluorescence microscope and a sensitive recording system. Detailed direct visual observation of fluorescently banded chromosomes is not normally practicable because the fluorescence fades too rapidly.

There are several general accounts of the principles and practice of fluorescence microscopy (Ploem 1977, Rost 1980, Ploem & Tanke 1987) to which the reader may refer for background information. To observe Q-banded chromosomes, and preparations banded with other fluorochromes (Ch. 8) it is necessary to use a fluorescence microscope of maximum efficiency. This implies the use of epi-illumination, combinations of exciter and barrier filters which match the spectral characteristics of the fluorochrome used as closely as possible, and objectives specially designed for fluorescence work. Such systems are now available from all the major microscope manufacturers, and further comments on these points are made by Sumner (1989). The excitation and emission spectra of quinacrine bound to deoxyribonucleoprotein are illustrated in Figure 7.2; it should be noted that spectra of fluorochromes bound to DNA usually differ somewhat from those of the free fluorochrome in solution.

Because they are normally mounted in an aqueous mountant, Q-banded chromosome preparations are not permanent. Although storage overnight in a refrigerator is probably acceptable, the quality of banding deteriorates if they are kept any length of time, and in general it is best to examine Q-banded preparations on the day that they are made. Since quinacrine fluorescence also fades quite rapidly during observation, it will be clear the Q-banding must be recorded photographically.

Older papers dealing with the problems of photographing Q-banded chromosomes are those of Breg (1972) and Davidson (1973). Although it might be supposed that for photographing weak fluorescence, the photographic emulsion should be as fast as possible, the coarse grain of fast films is usually not acceptable. In addition, faster films tend to have lower contrast. In recent years, however, the situation has improved with the introduction of relatively fast films with much finer grain than formerly, such as Kodak T-max. Films of this type are eminently suitable for fluorescence photomicrography.

The sensitivity of photographic films to low levels of light can also be improved by pre-exposure, and this can be useful for fluorescence, especially if photographic cytofluorimetry is contemplated (van der Ploeg *et al.* 1976). The reason for this is that below a certain threshold the amount of light

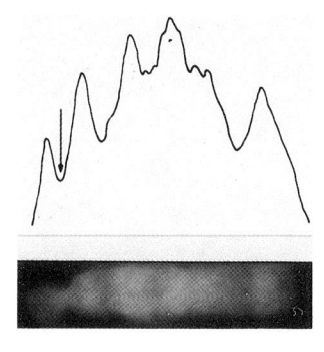

Figure 7.3 Intensity profile of a Q-banded chromosome (Z4) from the Chinese hamster cell line CHO. Profile reproduced from Sumner et al. (1981) by permission of the authors and Springer-Verlag.

falling on a photographic emulsion does not increase its density, while above the threshold, the optical density of the developed emulsion is proportional to the logarithm of the exposure time. The purpose of pre-exposure is to expose the emulsion to sufficient light to pass this threshold; all the light from the fluorescent chromosomes will then go towards producing a useful image. Use of pre-exposure was reported to bring the exposure time for Q-banded human chromosomes down from between 40 and 70 s to only 10–15 s (van der Ploeg et al. 1976). Not only is this an actual saving of time, but it also helps to record the image before serious fading has occurred.

Although photographic recording is generally adequate for detailed examination of Q-banded chromosomes, a further process, the production of intensity profiles, has sometimes been used. A scan along the image of a chromosome on a photographic negative is made with some type of densitometer, and a trace is produced representing the variation in optical density due to the bands (Fig. 7.3). It should be noted that optical densities recorded are not directly proportional to the intensity of fluorescence in the bands (van der Ploeg et al. 1977), although fluorescence intensities can be calculated if the negatives are prepared under standardized conditions. Nevertheless, brighter fluorescence will give greater optical density on the negative, and the positions of the bands will be defined.

The use of profiles of Q-banding patterns of chromosomes was introduced at the very beginning (Caspersson et al. 1969b). Banding profiles provide a more objective means of assessing patterns of bands than direct observations of the chromosomes. Caspersson et al. (1971a), when describing the Q-banding patterns of the 24 different human chromosomes, averaged the profiles obtained from no less than 5000 chromosomes, and pointed out the value of banding profiles not only for demonstrating the reproducibility of the patterns for any one particular type of chromosome, but also for illustrating the differences between different types of chromosomes. Profiles are also valuable when comparisons are made between the Q-banding pattern and the distribution of other chromosomal components (Caspersson et al. 1969b, Golomb & Bahr 1974a, Sumner et al. 1981). Not only can minor variations be demonstrated that are not easily seen by direct examination of the chromosomes, but it is also possible to compare different profiles objectively using a computer. A further application of banding profiles is in the quantification of Q-band heteromorphisms (Schnedl *et al.* 1977a). Although such heteromorphisms are generally assessed visually, this is not satisfactory as such judgements are necessarily subjective (see also Section 14.3).

7.3 Nature of Q-bands

There are two types of Q-bands in chromosomes: heterochromatic and euchromatic. At the outset of studies on Q-banding, it was observed, particularly in certain plant chromosomes, that certain restricted regions of the chromosomes showed particularly bright fluorescence. It was shown for the chromosomes of *Vicia* and *Trillium* that these brightly fluorescent regions corresponded to sites of heterochromatin as defined by cold treatment (Section 11.2.1) (Caspersson *et al.* 1968, 1969a,b). Such correlations between sites of intense quinacrine fluorescence and heterochromatin were soon extended to other species. In *Allium carinatum*, intensely fluorescent chromosomal segments were shown to correspond to heterochromatic segments visible as darkly stained regions on prophase chromosomes, and to interphase chromocentres (Vosa 1971). In *Drosophila* spp. it was reported that most (but not all) of the brilliantly fluorescent regions were at known sites of heterochromatin (Adkisson *et al.* 1971). In the related species *Samoaia leonensis* brightly fluorescent regions were shown to be late replicating (Ellison & Barr 1972). Late replication is widely accepted as a property of heterochromatin (Section 4.2.3) and similar correlations have now been reported in many species. One such species is man, in which various brightly fluorescent segments had already been reported as late replicating, and the brilliantly fluorescent distal portion of the long arm of the Y chromosome had been described as negatively heteropycnotic (Caspersson *et al.* 1970a). The general statement that regions of especially brilliant fluorescence are heterochromatic is valid and has not been questioned.

It is equally clear that by no means all regions of heterochromatin show either especially bright or indeed distinctive quinacrine fluorescence. The relationships are, in fact, quite complicated. With the general application of C-banding to chromosomes, it became apparent that three types of heterochromatin could be recognized on the basis of Q-banding properties. All such regions were C-band positive, but could show either intense or reduced quinacrine fluorescence (Fig. 7.4), or no differentiation with quinacrine (Vosa & Marchi 1972a, b). More than one type of heterochromatin, as defined by Q-banding, could occur in the same species, as for example in *Scilla sibirica* (Vosa 1973). A similar situation was described for human chromosomes by Gagné *et al.* (1971). The C-bands on no. 3, the D group and the Y chromosomes are all quinacrine positive, whereas the large C-bands on chromosomes 1, 9, and 16 all show very weak quinacrine fluorescence.

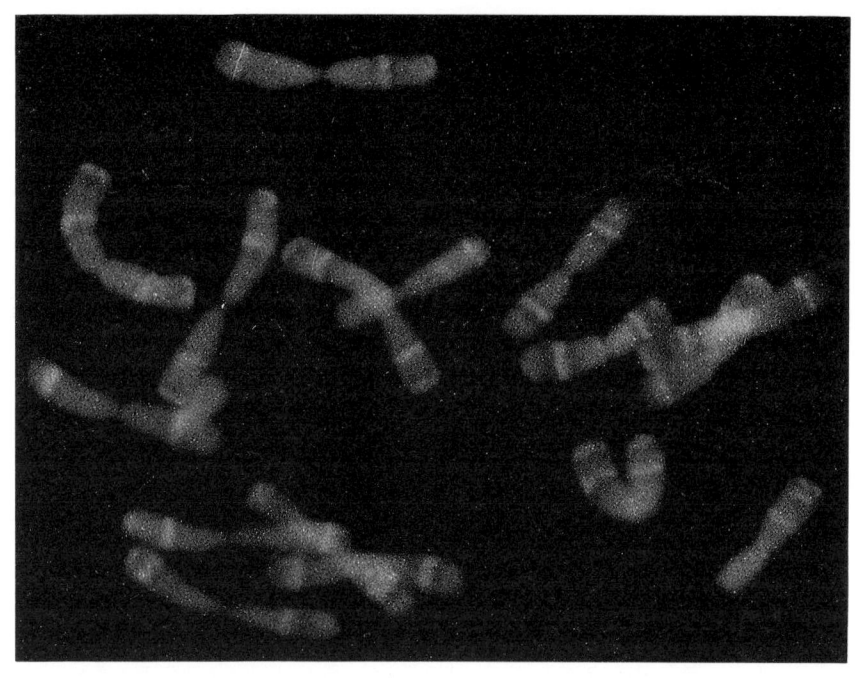

Figure 7.4 Q-banded chromosomes from the plant *Allium flavum*, showing regions of enhanced and reduced quinacrine fluorescence. Reproduced from Vosa (1973) in *Chromosome identification: Nobel symposium 23*, T. Caspersson & L. Zech (eds), 156–8 by permission of the author. Copyright The Nobel Foundation 1973.

The nature of the material in C-bands has already been discussed (Section 4.2.3). With the knowledge that C-bands generally contain satellite DNAs, often of distinctive base composition, it might be wondered whether quinacrine positive C-bands contain a particular class of DNA. To anticipate the next section of this chapter (7.4) it was found that quinacrine fluorescence was enhanced by A + T-rich DNA, and quenched by G + C-rich DNA. It was therefore a reasonable hypothesis that chromosomal segments showing brilliant quinacrine fluorescence contain A + T-rich satellite DNAs. Evidence to support this was obtained from species of Dipteran flies (Ellison & Barr 1972, Ranganath *et al.* 1982) and from *Vicia faba* (Cionini *et al.* 1985). The centromeric heterochromatin of ox (*Bos taurus*) chromosomes contains G + C-rich satellite DNA, and scarcely fluoresces with quinacrine (Schnedl 1972), again in accordance with the hypothesis, which is also supported by observations on several other species (Table 7.1). The situation in certain other mammals is different, however. Mouse centromeric heterochromatin contains A + T-rich satellite DNA, but shows rather weak quinacrine fluorescence (Rowley & Bodmer 1971), while in the Indian muntjac, the centromeric heterochromatin apparently contains DNA with a base composition similar to that of the rest of the chromosome, yet again shows weak fluorescence (Comings 1971). The chromosomes of the plant *Ornithogalum*

Table 7.1 Quinacrine fluorescence properties of regions of heterochromatin of known base composition, as identified by *in situ* hybridization.

Species	Satellite DNA composition	Quinacrine fluorescence	
Plants			
Allium cepa	G+C-rich	−ve	(Barnes *et al.* 1985)	(Vosa 1970)
Scilla sibirica	G+C-rich	−ve	(Timmis *et al.* 1975, Deumling 1981)	(Vosa 1973)
Secale cereale	Slightly G+C-rich	No differentiation	(Appels *et al.* 1978)	(Sarma & Natarajan 1973, Vosa 1974)
Vicia faba	A+T-rich	+ve	(Cionini *et al.* 1985)	(Vosa & Marchi 1972b, Cionini *et al.* 1985)
Insecta			
Drosophila melanogaster	A+T-rich	+ve	(Peacock *et al.* 1977)	(Holmquist 1975a, Gatti *et al.* 1976)
Drosophila nasuta	A+T-rich	+ve	(Ranganath *et al.* 1982)	(Ranganath *et al.* 1982)
Drosophila virilis	A+T-rich	+ve	(Gall *et al.* 1971)	(Adkisson *et al.* 1971)
Rhynchosciara hollaenderi	A+T-rich	+ve	(Eckhardt & Gall 1971)	(Stocker *et al.* 1978)
Amphibia			
Triturus sp.	A+T-rich	+ve	(Barsacchi-Pilone *et al.* 1986)	(Schmid *et al.* 1979)
Birds			
Coturnix coturnix japonica	G+C-rich	−ve	(Brown & Jones 1972)	(Comings & Wyandt 1976)
Gallus domesticus	G+C-rich	−ve	(Stefos & Arrighi 1974)	(Stahl & Vagner-Capodano 1972)
Mammals			
Cavia porcellus	One A+T-rich, two G+C-rich	−ve	(Duhamel-Maestracci *et al.* 1979)	(Natarajan & Raposa 1974)
Dipodomys ordii	G+C-rich	−ve	(Prescott *et al.* 1973)	(Bostock & Christie 1974)
Mus musculus	A+T-rich	−ve	(Pardue & Gall 1970, Jones & Robertson 1970)	(Rowley & Bodmer 1971)
Peromyscus eremicus	G+C-rich	−ve	(Hazen *et al.* 1977)	(Jalal *et al.* 1974)
Bos taurus	G+C-rich	−ve	(Kurnit *et al.* 1973)	(Hansen 1972a, Schnedl 1972)
Saimiri sciureus	A+T-rich	+ve	(Lau *et al.* 1977)	(Jalal *et al.* 1974)
Homo sapiens (Y chromosome)	A+T-rich	+ve	(Manuelidis 1978, Frommer *et al.* 1984)	(Zech 1969)

montanum contain both A + T-rich satellite DNA, and brilliantly fluorescent Q-bands (Barsanti *et al.* 1981). However, while different populations of this species have chromosomes containing similar amounts of satellite DNA, the amount of Q-banded material varies widely between populations. It will, therefore, be clear that there is no simple and general relationship between quinacrine fluorescence and DNA composition of heterochromatin. Although such a relationship holds in some species, there are others where it apparently does not hold. Further consideration of this point must be deferred until Section 7.4.

Like other types of heterochromatin, the heterochromatin revealed by quinacrine fluorescence shows a good deal of heteromorphism. This will be considered in more detail in Chapter 14. Most of this variation is in size, but in certain human chromosomes there is also variation in brightness. The short arms and the satellites of the human acrocentric autosomes (13–15, 21, 22) independently show variation (Fig. 14.2) from very dim to very bright fluorescence (Caspersson *et al.* 1970c, Evans *et al.* 1971, Buckton *et al.* 1976, Robinson *et al.* 1976, Verma & Dosik 1980, Ibraimov & Mirrakhimov 1985). Similar heteromorphisms have also been found in chromosomes of the great apes (Dutrillaux *et al.* 1973a, C. C. Lin *et al.* 1973, Lejeune *et al.* 1973, D. A. Miller *et al.* 1974), but not in any other species. Little is known about the nature of these heteromorphic regions, or the factors responsible for their variation in brightness.

In describing the Q-banding patterns of the human chromosomes, Caspersson *et al.* (1971a) wrote: 'In addition to these few very brightly fluorescing spots, all the chromosomes of the human metaphase plate show a faint fluorescence pattern along their whole length, ...' This statement essentially marks the recognition of a system of euchromatic bands throughout the length of chromosomes, although such bands had, in fact, been seen in Chinese hamster chromosomes in the original paper on quinacrine fluorescence of chromosomes (Caspersson *et al.* 1968). The characteristics of such bands have already been described in detail in Section 5.3 and summarized in Table 5.2, and will be discussed further in Section 16.3. To generalize, positive euchromatic Q-bands correspond to positive G-bands and pachytene chromomeres, show late DNA replication and early chromosome condensation, contain relatively A + T-rich DNA, and tissue-specific rather than 'housekeeping' genes. The negative Q-bands have the opposite characteristics. Quinacrine fluorescence has been important in developing these concepts. As just stated, it was through the use of quinacrine fluorescence that such bands were discovered. Further pioneering work demonstrated a correlation between positive Q-bands and late replication (Ganner & Evans 1971, Calderon & Schnedl 1973), while the idea that positive Q-bands contain A + T-rich DNA owes much to investigations on the mechanism of Q-banding that will be described in the next section (7.4.).

As with G-bands (Section 5.3) and R-bands (Ch. 6), euchromatic Q-bands are largely confined to higher vertebrates, where they have been demonstrated in all the chromosomes of all species to which the technique has been applied. Reports of this type of Q-banding in lower vertebrates, invertebrates and plants are scarce, although of course, quinacrine fluorescent heterochromatic bands are very widely distributed. In Anuran amphibia, Schmid (1978) reported that no banding patterns could be produced with quinacrine in the euchromatic parts of metaphase chromosomes, but that a great many Q-bands could be demonstrated along the arms of the highly extended prophase chromosomes. The lack of Q-bands on metaphase chromosomes was attributed to the high degree of contraction of Anuran chromosomes at metaphase, which is 1.5 to 3 times greater than the contraction of human chromosomes. Thus the bands would be too close together to be resolved at metaphase. However, Schmid (1978) noted that the amphibian prophase chromosomes showed density differences along their length. It is not clear, therefore, whether this alone might account for the 'banding' of the prophase chromosomes, in which case it would be merely a structural phenomenon and not true banding.

Among plants, Q-bands have been reported throughout the length of chromosomes of species of *Lilium* (Holm 1976, Kongsuwan & Smyth 1977, 1980) (Fig. 7.5), and in *Paris* (Smith & Ingram 1986). As in the Amphibia, there is some uncertainty about the exact nature of such Q-bands. Holm (1976) stated that the quality of the Q-banding in *Lilium* was highly dependent on the quality of preservation of the chromosomes. There was also a suggestion, not substantiated, that the pattern of Q-bands might be correlated with the distribution of Feulgen-stained DNA along the chromosomes.

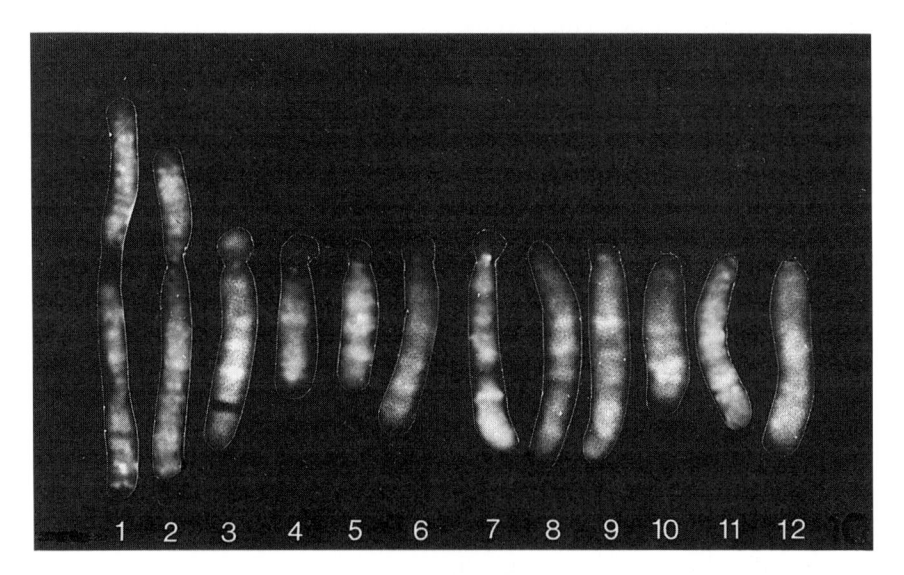

Figure 7.5 Q-banded karyotype of *Lilium longiflorum*, showing bands throughout the length of the chromosomes. Reproduced from Holm (1976) by permission of the author and the Carlsberg Laboratory.

There is evidence that these Q-bands in *Lilium* spp. may not correspond to those in mammals. First, the patterns appear to be quite different in closely-related species (Kongsuwan & Smyth 1977), and second, do not seem to correspond to patterns of DNA replication (Kongsuwan & Smyth 1980). Clearly only more research, both on *Lilium* and other plant species, will resolve the problem of whether euchromatic Q-bands equivalent to those of mammals occur in plants. For further discussion of this important question, see Section 5.3.

7.4 Mechanisms of Q-banding

Caspersson *et al.* (1968), when they introduced quinacrine mustard as a fluorochrome for chromosomes, did so in the belief that, as an alkylating agent, its most important binding sites on DNA would be guanine residues. It was, however, recognized that QM would also bind to DNA by intercalation, and that the pattern of QM fluorescence on chromosomes might be affected by preferential accessibility of the DNA. The discovery that quinacrine, the non-alkylating analogue of QM, could produce identical patterns (Section 7.2) indicated, however, that specific binding to guanine was unimportant for the production of bands. Instead, evidence began to appear that regions of chromosomes showing bright quinacrine fluorescence contained DNA rich in adenine and thymine (A + T-rich DNA). In the Drosophilid fly *Samoaia leonensis*, Ellison & Barr (1972) found brilliant quinacrine fluorescence in certain chromosomal regions which, during replication, incorporated thymine but not deoxycytidine. They deduced, therefore, that such chromosomal regions were very A + T-rich, and that the bright quinacrine fluorescence was related to this. This hypothesis was largely supported by other observations of correlations between brightness of quinacrine fluorescence and base composition of defined regions of chromosomes (Table 7.1), and also by studies on the effect of solutions of DNA of known base composition on quinacrine fluorescence (Section 7.4.2). Nevertheless, it was clear that if it were a rule that the brightness of quinacrine fluorescence reflected the underlying base composition of the chromosomal DNA, there were exceptions to this rule which would have to be explained. Moreover, it was not clear that there were necessarily sufficiently large differences in DNA base composition along chromosomes to account for Q-banding patterns. This applied particularly to euchromatic Q-bands, and indeed it has been suggested that Q-banding in euchromatin and heterochromatin might be somewhat different phenomena, which might need to be explained in rather different ways.

Finally, before starting on a detailed discussion of Q-banding mechanisms, the possibility that patterns may be influenced by local DNA concentrations should be noted. Caspersson *et al.* (1968, 1969a, b) were at pains to establish that the Q-banding patterns they described in plant chromosomes were independent of DNA distribution, which was uniform throughout the chromosome except for the primary and secondary constrictions. On the other hand, however, Golomb & Bahr (1974b) showed that the mass distribution along human chromosomes showed a pattern resembling that of the Q-bands. We have already noted the possibility that Q-bands throughout the length of amphibian prophase chromosomes (Schmid 1978) could simply represent variations in DNA concentration due to the chromosome structure (Section 7.3). It is not surprising that several other cases could be cited, particularly of heterochromatic Q-bands having a greater DNA concentration than the euchromatin (e.g. Wake & Ward 1975), since this is a common property of heterochromatic segments (Section 3.3.4). There is, however, no evidence to indicate that chromosomal DNA distribution is a major determinant of Q-banding patterns, although it may have some influence in a few cases.

7.4.1 *Binding of quinacrine mustard and quinacrine to DNA and chromosomes*

There are two separate considerations in the understanding of the mechanism of Q-banding: the mechanism by which the dyes bind to the chromosomes, and the factors affecting their fluorescence. Obviously the amount of dye bound must be relevant to the degree of fluorescence which results, but as we shall see, the fluorescence can vary independently of the amount of dye bound to the chromosome. Because the dyes used for Q-banding have very low absorption, it is only rarely that direct measurements of dye binding have been made, and observations of fluorescence alone cannot distinguish between effects due to variation in dye binding, and differential excitation of fluorescence.

It seems certain that QM and quinacrine bind to DNA and chromosomes by somewhat different mechanisms, in spite of the fact that the resulting banding patterns are apparently indistinguishable. As already mentioned, the basis of QM selective fluorescence was intitially believed to be its binding to guanine residues in DNA, typical of alkylating agents (Caspersson *et al.* 1968). The occurrence of this reaction between QM and guanine or DNA was confirmed experimentally (Adkisson *et al.* 1971, Sengupta *et al.* 1971). However, Modest & Sengupta (1973) showed that the primary mode of binding of QM to DNA was by intercalation, with alkylation of guanine a secondary binding mode. Nevertheless, binding of QM is much stronger than that of quinacrine, and is confined to double-stranded DNA that contains guanine. Selander & de la Chapelle (1973a) showed that QM could also react to different extents with different histones, and speculated that this could be a factor in causing Q-bands.

There is no information on the distribution of the dye on QM-stained chromosomes. However, Adkisson *et al.* (1971) attributed the 'general moderate fluorescence' of chromosomes to alkylation, as it was prevented by depurination with acid. Other interpretations are possible, since depurination would effectively denature the DNA and inhibit intercalation; alternatively the acid might extract other chromosomal components, such as proteins, which might be essential for QM fluorescence. The intriguing finding by Adkisson et al. (1971), foreshadowed by Caspersson et al. (1969a), was that the fluorescence of the brightly fluorescent bands could be removed by mild acid treatment, whereas the weaker general fluorescence was stable to acid. The latter was therefore interpreted as a covalent bonding, through alkylation, whereas the bright fluorescence, which could be restored by restaining, was non-covalent, either ionic or intercalative.

The situation with quinacrine is somewhat clearer than that for QM. It has been known for many years that quinacrine binds to DNA by intercalation (Section 6.4). (Bontemps & Fredericq 1974, Davidson et al. 1978, Jones et al. 1979) and that this binding shows little base specificity (O'Brien et al. 1966, Baldini et al. 1981). The quinacrine side chain (see Fig. 7.1) is also bound to the DNA at low ionic strength, but is dissociated at high ionic strength (Davidson et al. 1978). Intercalative binding apparently also occurs with the DNA in chromosomes (Sumner 1986b); at high concentrations of quinacrine, the dye also binds ionically to the outside of the DNA molecule, but such binding is apparently not relevant to Q-banding, which can be obtained readily at low dye concentrations.

The distribution of quinacrine along chromosomes has been determined in several ways. Pachmann & Rigler (1972) measured the absorbance of chromosomes of *Vicia faba* stained with proflavine, and found that it was uniform. Proflavine and quinacrine produce similar banding patterns in *Vicia faba* (Caspersson et al. 1969b) although it does not follow that the mechanism of banding is the same. Hatfield et al. (1975) used tritiated quinacrine to show that the distal part of the long arm of the human Y chromosome bound no more dye than other, less brightly fluorescent chromosomes. Both these studies refer, to course, to heterochromatic regions of the chromosomes. A rather different result was obtained by Sumner (1985) in a study of the centromeric heterochromatin of mouse chromosomes, using X-ray microanalysis. Although the mouse centromeres bound more quinacrine relative to the amount of DNA, when compared with the chromosome arms, the absolute amount of quinacrine in the centromeres was less, because the DNA concentration was less in the centromeres than in the arms.

Studies of the quinacrine distribution on the euchromatic arms of mammalian chromosomes indicate that here also the dye distribution is essentially uniform. Latt & Gerald (1973) stained chromosomes with a conjugate of quinacrine and polylysine, and obtained typical Q-banding. Since the polylysine had been shown to bind uniformly to the chromosomes, it was deduced that the banding pattern was the result of differential excitation of uniformly bound dye. Sumner (1977b) showed that Q-banded chromosomes became uniformly fluorescent when re-mounted in organic media. The effect was reversible, and the most probable explanation was uniform binding of the quinacrine. Both of these methods were rather indirect, but Sumner *et al.* (1981) showed, using X-ray microanalysis, that quinacrine distribution on CHO chromosomes was essentially uniform, paralleling that of the DNA, and was clearly not related to the Q-banding pattern of these chromosomes.

In conclusion, to produce Q-banding quinacrine is bound to DNA in the chromosomes, as it is to free DNA, by intercalation, with the side chain of the molecule bound ionically. At high dye concentrations, external ionic binding of quinacrine occurs, but this appears to be irrelevant for Q-banding. Quinacrine mustard, on the other hand, binds not only by intercalation, but also covalently by alkylation. This involves binding of the side chain to guanine residues in the DNA. However, binding of QM in brightly fluorescent heterochromatic regions is less strong, probably by intercalation alone, or by simple ionic binding. The latter seems less likely since Sumner (1986b) found that external, ionic binding of quinacrine had no part in Q-banding. The absence of covalent binding of QM in such regions could be due to lack of guanine residues in A + T-rich heterochromatin. The binding of quinacrine is clearly proportional to DNA concentration in euchromatic parts of the chromosomes, and thus Q-bands in euchromatin must result from differential excitation of fluorescence (Section 7.4.2). The situation in heterochromatin is not so clear. Although there are apparently no large differences in quinacrine binding between euchromatic and heterochromatic segments of chromosomes, it seems that small differences can occur in some cases (Sumner 1985), although it is not certain how important these are in promoting differential quinacrine fluorescence. The distribution of QM on Q-banded chromosomes is not known; by analogy with quinacrine, which produces the same banding patterns, it might be expected to be uniform; since, however, QM shows guanine specificity in its binding, a non-uniform distribution along chromosomes, related to variations in base composition (Section 5.3) might be expected.

Proteins bound to chromosomal DNA are known to inhibit the binding of dyes, including quinacrine, to DNA (Darzynkiewicz 1979, Darzynkiewicz & Traganos 1988). Although there is evidence for chromatin conformation (by implication, proteins) affecting quinacrine fluorescence (Section 7.4.2) there is nothing in the data cited earlier in this section to suggest that inhibition of binding of quinacrine by chromosomal proteins occurs to any significant extent on fixed chromosomes.

7.4.2 Differential excitation of quinacrine fluorescence

Since all the available evidence indicates that the binding of quinacrine to chromosomes is essentially uniform, it follows that Q-banding must be the result of more efficient excitation of fluorescence in some chromosomal regions than in others (differential excitation of fluorescence). (Before proceeding further, however, it is necessary to make the caveat that the quinacrine distribution on chromosomes has only been studied in very few species. The possibility that in some cases Q-banding could be due to non-uniform distribution of quinacrine, especially perhaps in heterochromatin, must be considered seriously.) A very substantial body of evidence indicates that the main cause of differential excitation of fluorescence is the base composition of the chromosomal DNA, and, more specifically, that the fluorescence of quinacrine is brighter when it is bound to A + T-rich DNA than when it is bound to G + C-rich DNA. The evidence comes from two sources: observations on the quinacrine fluorescence of chromosomal regions of known base composition, and measurements of quinacrine fluorescence in solution in the presence of pure DNA of known average base composition.

Ellison & Barr (1971) found that in the Drosophilid fly *Samoaia leonensis* certain chromosomal regions showed brilliant quinacrine fluorescence, and that these regions failed to incorporate any substantial amount of deoxycytidine during DNA replication. It was, therefore, concluded that these regions were very A + T-rich, a conclusion later supported by analysis of the DNA of this species (Mayfield & Ellison 1975). Similar information is now available for a wide variety of organisms, both plants and animals, invertebrates and vertebrates, and this is summarized in Table 7.1. With few exceptions, the correlation between DNA base composition and brightness of quinacrine fluorescence is very good. It must, nevertheless, be remembered that the resolution of the techniques used is not very high (particularly *in situ* hybridization using autoradiography), and that in several cases there is evidence for satellite DNAs being interspersed with other DNA fractions (Section 4.2.3). The base composition deduced from that of the satellite is therefore not necessarily that of the whole block of heterochromatin. Species of *Drosophila* also show complex patterns of satellite DNA distribution, as well as complex banding patterns. Since it would be impracticable to describe here all the complexities of the distribution of satellite DNAs in *Drosophila* spp., and their relationships to Q-banding patterns, the interested reader is referred to the various references given in Table 7.1. Only a brief summary will be attempted here. The satellite DNAs of *Drosophila* spp. are usually (but not always) very A + T-rich, and several such satellites may occur in any one species (e.g. Cordeiro-Stone & Lee 1976, Peacock *et al.* 1977). These satellites are located in heterochromatic segments of chromosomes, but each satellite has a somewhat different distribution. When the distributions of the different satellite DNAs are compared with the distributions of bright quinacrine fluorescence it is found that the former are not necessarily restricted to Q positive regions (Holmquist 1975a, Gatti *et al.* 1976, Wheeler *et al.* 1978). The situation becomes even more complicated when the Q-banding patterns of *Drosophila* chromosomes are compared with the patterns obtained using other fluorochromes reported to show specificity for A + T-rich or G + C-rich DNA (see Ch. 8). Although bright fluorescence with quinacrine or Hoechst 33258 is believed to be related to A + T-richness, the relationships between the brightness or dullness of quinacrine fluorescence and of Hoechst 33258 fluorescence are remarkably complicated (Holmquist 1975a, Gatti *et al.* 1976, Wheeler *et al.* 1978). The same is also true when other fluorochromes regarded as showing base specificity are used (Abraham *et al.* 1983).

There are various interpretations of these discordant results. Holmquist (1975a) concluded that while G + C-richness might be a sufficient requirement for the production of weak quinacrine fluorescence, A + T-richness was a necessary, but not sufficient requirement for bright quinacrine fluorescence. Gatti *et al.* (1976) and Wheeler *et al.* (1978) came to similar conclusions. However, Stocker *et al.* (1978), studying the chromosomes of *Rhynchosciara hollaenderi*, which has banding patterns as complicated as those of *Drosophila* spp. suggested that differences in chromatin condensation might well affect the fluorescence properties of a particular chromosome segment. In this connection, it is worth noting that Moser *et al.* (1981) have shown that the brightness of the quinacrine fluorescence of mammalian nuclei varies with the degree of chromatin condensation during the cell cycle. Although this is a very different system from fixed metaphase chromosomes, it does show that in certain circumstances quinacrine fluorescence can be influenced by factors other than DNA base composition. In spite of the generally good correlations shown in Table 7.1, it is nevertheless appropriate to quote the statement by Holmquist (1975a) that:

'A gloomy pall of doubt occludes the ability of *in vitro* DNA-fluorochrome data to predict the response of chromosomes to fluorochrome staining ... base sequence or base composition of AT-rich chromatin are definitely not related to that chromatin's Hoechst or quinacrine brightness in any simple way.'

The data and discussion above are concerned entirely with heterochromatic regions of chromosomes. Is there any evidence that the Q-banding pattern of the euchromatic parts of mammalian chromosomes is related to the base composition of the DNA? We have seen (Section 5.3, Table 5.2) that positive G-bands, which are equivalent to positive Q-bands, are relatively A + T-rich, while the negative bands are relatively G + C-rich. Korenberg & Engels (1978) have investigated this point specifically in relation to the quinacrine brightness of human chromosomes, comparing the Q-banding pattern with the pattern of uptake of labelled DNA precursors. Their results support the correlation between A + T-richness and bright quinacrine fluorescence. It must be emphasized that demonstration of a correlation is not the same as proving cause and effect. Because the positive Q-bands are late replicating, and probably contain DNA which is less active genetically, and may condense earlier (Section 5.3), it seems possible that there are factors present such as degree of chromatin condensation, and perhaps the presence of different proteins, which themselves might influence the brightness of quinacrine fluorescence. The existence of isochores (Bernardi *et al.* 1985), regions which are large on the molecular scale and consist of DNA of distinctive base composition, and their possible connection with G-bands, has already been discussed in Section 5.3. Since isochores appear to occur only in higher vertebrates, and not in other animals (Bernardi *et al.* 1985), their taxonomic distribution would apparently coincide with that of euchromatic Q-bands (Section 16.3.3, Table 16.1). However, the almost complete absence of euchromatic Q-bands in lower vertebrates, invertebrates and plants could, as with G-bands, be merely a technical artefact. For euchromatin, therefore, the cytological evidence only indicates a correlation between DNA composition and quinacrine fluorescence, and not a causal relationship.

The other line of evidence indicating a role for DNA base composition in Q-banding comes from studies on the effect of DNAs of different base composition on the fluorescence of quinacrine in solution. Weisblum & de Haseth (1972) were the first authors to show that, in solution, the fluorescence of quinacrine is enhanced by $A+T$-rich DNA, and quenched by $G+C$-rich DNA. All natural DNA samples tested quenched quinacrine fluorescence, to a degree related to their $G+C$ content, and quinacrine mustard gave similar results. These initial findings have been confirmed subsequently by many authors (Michelson *et al.* 1972, Pachmann & Rigler 1972, Selander 1973, Latt *et al.* 1974a, Comings *et al.* 1975). The phenomenon is thus very soundly established, and the question must therefore be asked whether these results can easily be applied to chromosomes, which differ in many ways from DNA solutions. In chromosomes, the DNA is at a much higher concentration, and is complexed with proteins, and as we have seen, there is not always a good correlation between base composition and quinacrine fluorescence in chromosomes. Comings & Drets (1976) calculated that, given known variations of quinacrine fluorescence with base composition, the variations in base composition along chromosomes were sufficient to account for Q-banding. Nevertheless, difficulties have been encountered in explaining certain phenomena. The relatively weak fluorescence of the centromeric heterochromatin of the mouse, which contains $A+T$-rich satellite DNA, has long been a problem, and is still not satisfactorily explained. Weisblum (1973) explained this weak fluorescence on the basis that fluorescence is not simply a function of DNA base composition, but that the degree of interspersion of $G+C$ base pairs in the DNA sequence was also important. Selander & de la Chapelle (1973b) also obtained evidence suggesting that the anomalous behaviour of mouse centromeric heterochromatin could be related to the base sequence of the satellite DNA. Curiously, neither of these papers seems to have investigated the effects of mouse main band and satellite DNA on quinacrine fluorescence. It should be noted that mouse satellite DNA contains stretches of several adjacent $A+T$ base pairs (Harbers & Spencer 1974, Biro *et al.* 1975) which might be thought favourable for quinacrine fluorescence. Simola *et al.* (1975) and Comings *et al.* (1975) found, in fact, no difference in the effect of mouse main band or satellite DNA on the fluorescence of quinacrine in solution. Bostock & Christie (1974), however, found less fluorescence with satellite than with main band DNA, but did not regard the difference as sufficient to account for the cytological fluorescence properties. Some further complications should be mentioned. Natarajan & Gropp (1972) claimed that the centromeric heterochromatin fluoresced strongly in interphase nuclei of the mouse, but that this fluorescence pattern could be destroyed by trypsin treatment. It is not clear, however, whether this is simply a consequence of the heterochromatin being more condensed in interphase, and that the heterochromatin is decondensed by trypsin. Dev et al. (1973) pointed out that in a related species, *Mus cervicolor*, the centromeric heterochromatin was brightly fluorescent with quinacrine, although the satellite DNA of this species is less A + T-rich than that of *Mus musculus*. In support of a role for guanine interspersion in reducing fluorescence, it has been reported that after photochemical destruction of guanine residues, the mouse centromeric heterochromatin showed bright quinacrine fluorescence (Ferrucci & Mezzanotte 1982). Finally, Sumner (1985) showed that mouse centromeric heterochromatin bound relatively more quinacrine than the euchromatic arms, but that the excitation of the fluorescence was less efficient in the heterochromatin. This could have been due to concentration-dependent quenching (Latt 1974a, Sumner 1985) rather than to guanine-induced quenching. In addition, there was a lower concentration of DNA in the centromeres, which alone would result in reduced fluorescence even if all other parameters were the same in euchromatin and heterochromatin.

The situation in mouse centromeric heterochromatin has been described at some length not only because it has been extensively studied, but also because it illustrates the problems in trying to establish the basis of quinacrine, or indeed any other, banding. Studies of the quinacrine fluorescence of different DNA fractions have been done in few other species, but the results are of great interest. Bostock & Christie (1974) found that there was little difference between the fluorescence of different DNA fractions from kangaroo rat (*Dipodomys ordii*) in the presence of quinacrine. This was despite G + C contents ranging from 38.5% (for main band) up to 66% (for one of the satellites). Nevertheless, the chromosomal regions containing the satellites showed substantially reduced quinacrine fluorescence. Gottesfeld et al. (1974) compared the quinacrine fluorescence of condensed and dispersed chromatin fractions from *Drosophila* embryos and rat liver. In both cases, the quenching of fluorescence by dispersed chromatin was much greater than that by condensed chromatin, although the DNA from each fraction showed similar fluorescence with quinacrine. This appears to be a clear case of an effect of proteins on quinacrine fluorescence, but it must be pointed out that the experiments were performed with unfixed chromatin, rather than with fixed chromosomes. Nevertheless, since Q-banding is less dependent on the effects of fixation than most banding techniques, these results may still be relevant.

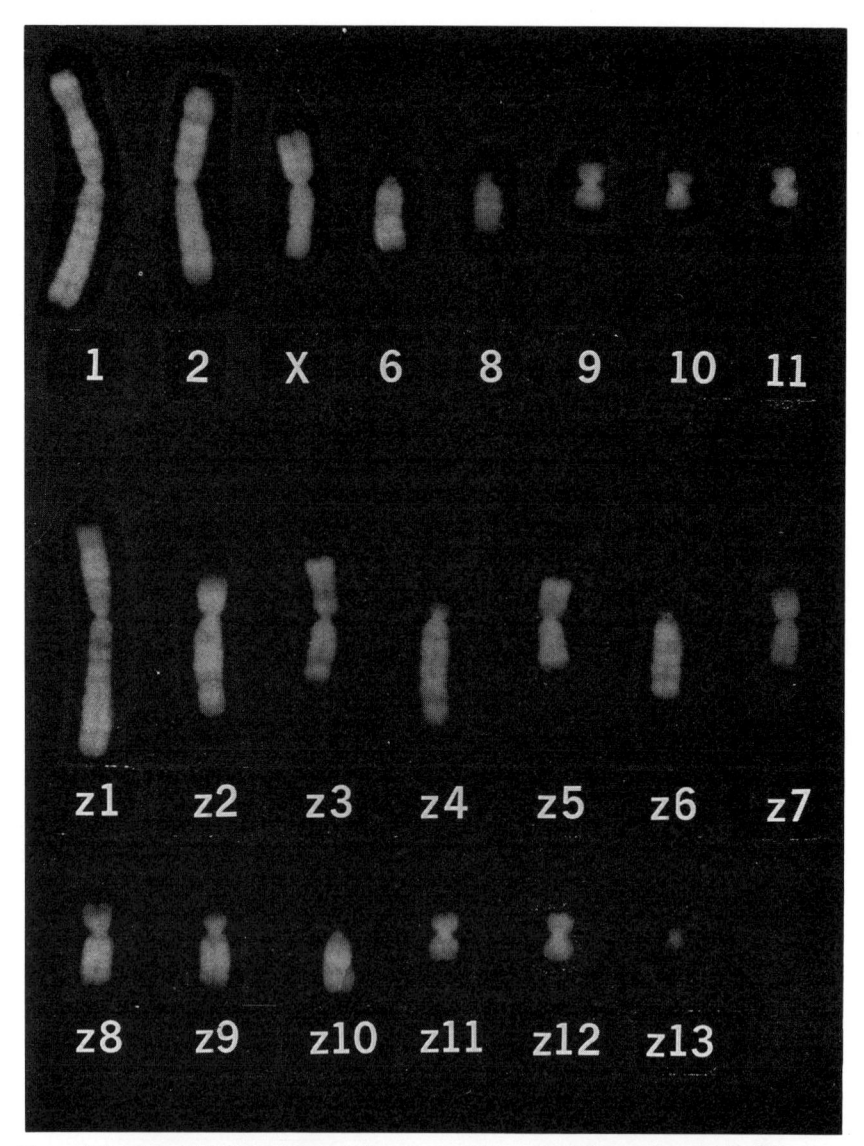

Figure 7.6 Q-banded karyotype of chromosomes from the Chinese hamster cell line CHO. Numbers without a prefix indicate unmodified chromosomes, identical to those in the intact animal; the z prefix indicates a chromosome modified from the corresponding normal one. Reproduced from Sumner *et al.* (1981) by permission of the authors and Springer-Verlag.

Sumner (1986b) approached the question of the base specificity of quinacrine directly, by measurements of fluorescence of nuclei containing DNAs of different base composition. The nuclei compared were those of human lymphocytes (60% A + T) and *Allium cepa* (67% A + T). Although *A. cepa* nuclei contain 5.5 times as much DNA as human ones, their quinacrine fluorescence was between 7.5 and 10.7 times greater, an effect most readily attributable to the difference in base composition.

An alternative approach to the investigation of the base specificity of quinacrine fluorescence has involved the use of other base-specific substances to inhibit quinacrine binding. It must be borne in mind that the base specificity of the binding of any such substances to chromosomes is probably no better established than the specificity of quinacrine itself. Nevertheless, the results obtained are in agreement with expectation, which gives one some confidence in the results. Schlammadinger *et al.* (1977) showed that the A + T-specific antibiotics netropsin and distamycin suppressed Q-banding of human chromosomes. Löber *et al.* (1978) also showed that netropsin binding caused the disappearance of Q-bands, and at the same time demonstrated that the G + C-specific antibiotic actinomycin D had no effect on Q-banding of human chromosomes. Thus these results again emphasize that quinacrine fluorescence is an indicator of A + T-richness of the DNA to which it binds. We must nevertheless conclude that there is not an absolute correlation between base composition and quinacrine fluorescence, and that, as stated earlier, A + T-richness of DNA may well be a necessary precondition for

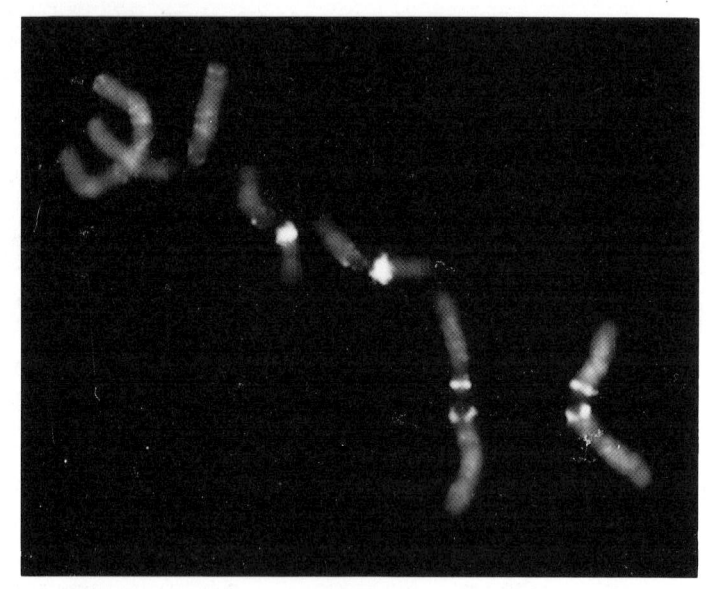

Figure 7.7 Q-banded metaphase chromosomes from an insect, *Culiseta longiareolata* (female). Note that differential fluorescence is confined to heterochromatin. Reproduced from Mezzanotte *et al.* (1979) *Genetica* **50**, 135–9 by permission of the authors and Kluwer Academic Publishers.

bright quinacrine fluorescence, but is only one of a number of requirements that need to be satisfied. The degree of interspersion of guanine residues, and the nature of associated proteins, are apparently other factors that might affect quinacrine fluorescence of chromosomes, although definitive evidence of this is not yet available.

Finally, it has been shown (Löber 1975) that the fluorescence of quinacrine is enhanced in the presence of organic solvents less polar than water. (Compare with the similar effect found by Sumner (1977b) on Q-banded chromosomes.) The enhanced fluorescence obtained with A+T-rich DNA could then be explained by the intercalation of the dye molecules into the apolar environment between A+T base pairs (see also Curtis & Horobin 1982). There is an interesting parallel here between the effect of hydrophobic regions of chromosomes promoting Giemsa binding in G-banding (Section 5.4), and in enhancing quinacrine fluorescence to produce Q-bands.

7.5 Applications of Q-banding

The applications of Q-banding are probably broader in scope than those of any other banding technique. Quinacrine produces distinctive patterns of fluorescence both in euchromatic (Fig. 7.6) and heterochromatic (Fig. 7.7) parts of chromosomes; the latter are commonly heteromorphic, and certain heterochromatic chromosome regions, particularly the human Y chromosome, can be reliably identified in interphase nuclei. The simplicity and reliability of Q-banding, even on difficult material, have helped to ensure its usefulness. However, although many cytogenetical discoveries were made using Q-banding (Section 7.1), it is probably much less used than formerly for routine chromosome identification in man and other mammals. This is almost certainly because of two great disadvantages generally associated with any fluorescence technique: impermanence of the preparations, and the tendency to fade during examination.

7.5.1 Euchromatic Q-bands

The applications of quinacrine to the identification of the chromosomes of higher vertebrates, through the production of distinctive patterns in the euchromatic parts of the chromosomes, are essentially similar to those of G-banding, which have been described in Section 5.5, and will not be repeated here. Many of the pioneering applications of Q-banding have also been described in the introduction to this chapter (Section 7.1). It should be noted that the standard human karyotype is based on Q-banding (ISCN 1985). Nevertheless, G-banding appears to have superseded Q-banding for many routine applications, and so far as I am aware, quinacrine has not been used for high resolution banding.

If Q-banding is less used than formerly is purely routine applications, it still has important uses in rather more specialized situations. First, there are situations where good chromosome banding is difficult to achieve, and material is not easily available, so that the vagaries of G-banding methods are not acceptable. Examples of this are studies of the chromosomes of cancer cells (Ohyashiki *et al.* 1985, Oshimura & Barrett 1985), and particularly studies of sperm chromosomes. In the latter case, the chromosomes from human sperm heads are made visible by the procedure of *in vitro* fertilization of hamster eggs. The method is complex and the yield small, so that it is essential to have a reliable method of identifying the sperm chromosomes. Q-banding has been favoured by the major laboratories engaged in this field (Martin *et al.* 1983, Brandriff *et al.* 1985).

Another difficult situation to which Q-banding has been applied is the identification of human pachytene bivalents (Caspersson *et al.* 1971b, de Torres & Abrisqueta 1978), although the results have scarcely been good enough to be adopted. Q-banding has also been used for isolated chromosomes (Wray & Stefos 1976), including chromosomes separated by a flow cytometer (Fantes *et al.* 1983). In the latter case, identification of the chromosomes is an important check on the sorting efficiency of the cell sorter.

In leukaemias (and indeed in other malignancies) it would be valuable to know from what type of cell abnormal chromosomes had been obtained. In a conventional chromosome preparation, fixed in methanol-acetic acid, this is impossible, as the cytoplasm is largely destroyed. Using other methods of preparation, however, morphological and immunological identification of cells has been achieved in combination with Q-banding (Knuutila & Teerenhovi 1989).

Second, Q-banding has been used as a method of identifying chromosomes in combination with some other procedure. For example, with *in situ* hybridization, prior G-banding is impracticable for identification of chromosomes when the sites of hybridization are to be detected by autoradiography, since the Giemsa dye mixture causes chemographic fogging of the autoradiographic emulsion, and is particularly difficult to remove completely. On the other hand, Q-banding can be carried out successfully before (Gosden *et al.* 1975a) or after (Lawrie & Gosden 1980) the *in situ* hybridization procedure without any interference with autoradiography. Q-banding has also been used successfully in combination with immunocytochemical methods for detecting *in situ* hybrids (Heyting *et al.* 1985). Quinacrine may also be used in combination with other fluorochromes (Section 7.2.4). In the methods described by Yoshida *et al.* (1975) and by Carlin & Rao (1982) to distinguish mouse chromosomes from those of other species in hybrid cells, Hoechst 33258 is used to pick out the mouse chromosomes by the brilliant fluorescence of their centromeres with this dye, while Q-banding is used to identify individual chromosomes. Finally, Q-banding may be used in combination with itself. When studying Q-band heteromorphisms of human and other mammalian chromosomes it is convenient and natural to make use of the Q-banding of the euchromatic parts of the chromosomes for purposes of identification.

7.5.2 Heterochromatic Q-bands

The use of quinacrine to label heterochromatic chromosome bands can be divided into four categories: identification of chromosomes, the characterization of heterochromatin, the study of heteromorphisms (which will be dealt with in Ch. 14), and studies of the human Y chromosome (to be discussed in Section 7.5.3).

In all organisms except the higher vertebrates, the production of euchromatic bands, whether G-, R- or Q-bands is rare, difficult and unreliable, and chromosome identification depends on demonstrating distinctive patterns of heterochromatin. Quinacrine has been widely used for this purpose, and a few examples of Q-banded karyotypes will be listed: in plants, *Allium carinatum* (Vosa 1971), *Brimeura* (Vosa 1979) and *Vicia faba* (Vosa & Marchi 1972b); in *Drosophila* spp. (Holmquist 1975a, Gatti *et al.* 1976); in amphibia (Schmid 1980b). Numerous other examples could, of course, be quoted.

The characterization of heterochromatin is closely related to the use of specific staining methods for heterochromatin for the purpose of identification. By characterization is meant the study of the reaction of a particular segment of heterochromatin to a number of different banding techniques. It has already been noted that heterochromatin is highly heterogeneous in its staining reactions (Section 4.2.7), and we shall return to this topic in Chapter 14; the heterogeneity of heterochromatin is closely associated with heteromorphism. The essential point is that segments that are heterochromatic by other criteria may show greater or lesser fluorescence with quinacrine and other fluorochromes than the remainder of the chromosome, or show no distinctive fluorescence at all. This had already been observed by Caspersson *et al.* (1969b) in the earliest studies on Q-banding of plant chromosomes. Vosa & Marchi (1972a) studied the chromosomes of seven species of plants, in which the heterochromatic segments, as defined by C-banding, show brighter or reduced quinacrine fluorescence, or no differentiation. The complications found in the chromosomes of *Drosophila* spp. have already been referred to. In human chromosomes, the major blocks of heterochromatin on the autosomes (1, 9 and 16) are all negative with quinacrine fluorescence, whereas the heterochromatic part of the long arm of the Y chromosome, as well as several small blocks of heterochromatin on various autosomes, show brilliant quinacrine fluorescence (Evans *et al.* 1971). Further examples are given in Table 7.1 and in Chapter 14.

Like all heterochromatic bands, heterochromatic Q-bands are heteromorphic. Again this matter will be dealt with in detail in Chapter 14, but a few examples of applications of the heteromorphism of human heterochromatic Q-bands will be given here, as an illustration of the value of

Q-banding. As previously noted, such bands vary not only in size, but also in intensity of fluorescence, and are therefore valuable markers. These heteromorphisms are, in general, inherited in a normal Mendelian fashion, although there is weak evidence for preferential transmission of bright Q-band heteromorphisms (Geraedts & Pearson 1974, Robinson et al. 1976), which has neither been explained nor confirmed. Nevertheless, results are sufficiently consistent to allow a high degree of confidence in paternity testing (Schnedl 1974, Olson et al. 1986). Similarly, Q-band heteromorphisms have been used to exclude the possibility of contamination with maternal cells in cultures of chorionic villus cells (Miny et al. 1985), and to distinguish donor and recipient cells in bone marrow transplants (Khokhar et al. 1987). Q-band heteromorphisms have also been used extensively to determine the origin of the extra chromosome in cases of trisomy of the acrocentric chromosomes, especially trisomy 21 (Magenis et al. 1977, Meulenbroek & Geraedts 1982, Hassold et al. 1985), and the origin of the extra sets of chromosomes in cases of triploidy (Jacobs et al. 1978, Meulenbroek & Geraedts 1982).

7.5.3 Human Y chromosome in metaphase and interphase

The human Y chromosome is a spectacular example of a heterochromatic Q-band that shows marked heteromorphism (Fig. 7.8). Even before the discovery of Q-banding, it was well known that the human Y chromosome could vary substantially in size, even between normal individuals; it was shown by the use of Q-banding that most, but not all, of this variation was in the brilliantly fluorescent heterochromatic segment (Bobrow et al. 1971, Schnedl 1971b, see also Section 14.4.7). The brilliantly fluorescent segment (Yq12) is a valuable marker for the Y chromosome, and has been used in studies of Y chromosome rearrangements and translocations (Robinson & Buckton 1971). However, its most interesting and distinctive use has been in the identification of the Y chromosome in interphase nuclei (Fig. 7.9) where is appears as a brilliantly fluorescent spot (Pearson et al. 1970). Such spots are commonly known as Y-bodies, or, to use a non-committal term, F-bodies (indicating fluorescent bodies) and have been used as an indicator of an XY chromosome constitution. In a study of buccal mucosa cells, Robinson (1971) showed that some 67% of nuclei from 46XY men contained a Y-body, whereas similar bodies were only seen in 5% of 46XX women. In 47XYY men, 83% of nuclei contained Y-bodies, and a high proportion of these contained two spots. It is clear from this and similar studies that the interphase Y-body is a good marker for the Y chromosome so long as it can be detected in a high proportion of cells, but technical factors can affect its visibility. The small proportion of fluorescent bodies from individuals with a 46XX chromosome constitution may be due partly to contaminating bacteria which fluoresce strongly, but an important source of fluorescent bodies in interphase nuclei may be brilliantly fluorescent

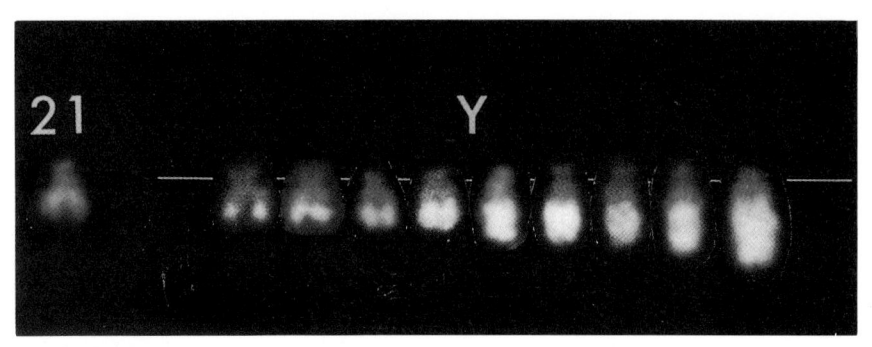

Figure 7.8 Variation in size of the brilliantly quinacrine fluorescent heterochromatin of the human Y chromosome, with chromosome 21 shown for comparison of total size. Reproduced from Yamada & Hasegawa (1978) *Human Genetics* **44**, 89–98 by permission of the authors and Springer-Verlag.

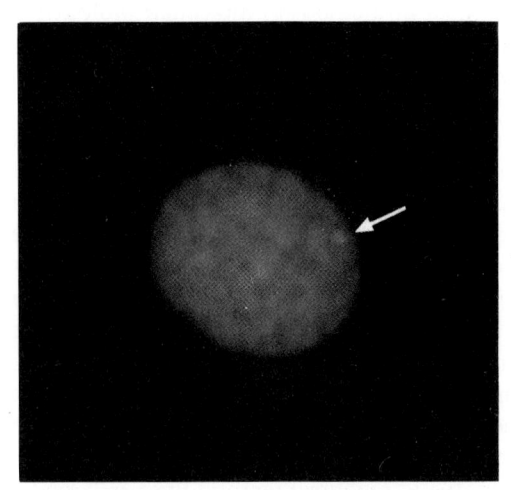

Figure 7.9 Fluorescent Y body (arrow) in an interphase nucleus from a buccal epithelial cell from a man.

regions of other chromosomes. Caspersson *et al.* (1970b) have shown that the brilliantly fluorescent segments of chromosome 3 and of the acrocentric chromosomes (13–15, 21, 22) may be distinguishable in some cases in interphase nuclei. Cases where such bodies could be confused with the Y chromosome are rare, but in doubtful cases it would be necessary to identify the Barr body (inactive X chromosome) or perform complete metaphase chromosome analysis.

Quinacrine fluorescence of Y-bodies has been used in determining chromosomal sex in nuclei from a variety of cell types: buccal mucosa (George 1971), lymphocytes (Polani & Mutton 1971), fibroblasts (Cramer & Hansen 1972), and hair root cells (Schwinger *et al.* 1971). For a recent review of this field, see Tishler (1985). Y-body staining has been used extensively for prenatal sex determination, using fetal cells obtained in a variety of ways. Shettles (1971) detected the presence or absence of Y-bodies in chorionic cells from cervical mucus, and was able to correlate these observations with the delivery of boys or girls. Schwinger *et al.* (1978) studied amniotic fluid cells, and Newburger & Latt (1979) studied fetal phagocytes. Fetal cells have also been detected in maternal blood by the presence (if the fetus is male) of Y-bodies in a small proportion of the nuclei (Schröder 1975, Iverson *et al.* 1981).

For most of the purposes described above, full chromosome analysis is now generally preferred to observation of the interphase Y-body, but there are, nevertheless, some situations where observations of the chromosomes are not appropriate. In what has been termed 'forensic genetics', observations of Y-bodies have been used to determine the sex of the 'donors' of bloodstains in cases of violent crime (Tishler 1980). Obviously it would be impossible to perform chromosome analysis on dried bloodstains, but sufficient cells of adequate quality can be obtained for sex determination. Here again, however, the observation of interphase Y-bodies has been rendered obsolete, this time by the development of methods of molecular biology for analysing the DNA in such samples. Such methods, indeed, provide a much more detailed and precise analysis in many situations than can be obtained using cytological procedures.

Perhaps one of the most valuable, yet controversial, applications of Y-body fluorescence has been in the study of the sex chromosome status of human spermatozoa. Once it had been shown that the Y chromosome could be detected in interphase nuclei by its fluorescence, it was natural to try quinacrine staining of sperm (Barlow & Vosa 1970). As expected, approximately half of the sperm heads contained a Y-body (Fig. 7.10). Similar observations were soon made by several other workers, who agreed that rather less than 50% of the sperm heads contained a Y-body (Sumner *et al.* 1971b, Moscetti *et al.* 1972, Schwinger *et al.* 1976, Klasen & Schmid 1981). Percentages of Y-bodies may be under 40%, but are often in the region of 45–47% or even more in good preparations. The deficiency from the expected 50% is no doubt largely due to technical factors. Even in somatic interphase nuclei from males, less than 100% contain Y-bodies, and in the

Figure 7.10 Fluorescent Y bodies (arrows) in human spermatozoa.

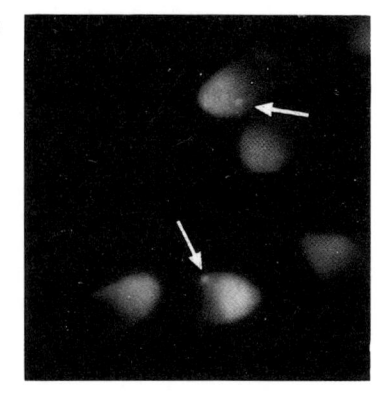

more densely compacted sperm heads it is to be expected that the Y-bodies would be more difficult to identify. Confirmation that spermatozoa with a Y-body do indeed contain a Y chromosome, while those without a Y-body contain an X chromosome, comes from DNA measurements. The difference in DNA content of sperm with and without a Y-body is not different statistically from that which would be expected from the difference in size of the X and Y chromosomes (Sumner et al. 1971b, Pearson et al. 1974, Sumner and Robinson 1976). The presence or absence of a Y-body in human sperm heads therefore appears, given good technical quality, to be a reliable indicator of the sex chromosome constitution of these spermatozoa. As such, this method has been used extensively for monitoring experiments aimed at separating X- and Y-bearing spermatozoa. A brief review of such experiments is provided by Ericsson & Glass (1982).

A small proportion of human spermatozoa contain two fluorescent bodies, and it was immediately assumed that such sperm contained two Y chromosomes. The proportion of sperm with two fluorescent bodies is usually in the range of one or two percent, but may be somewhat lower or as high as 5% (Pearson & Bobrow 1970b, Sumner et al. 1971a, Pawlowitzki & Pearson 1972, Moscetti et al. 1972, Schwinger et al. 1976, Klasen & Schmid 1981). This was a surprising result, for two reasons. First, the frequency of XYY males in the human newborn population is much lower, about 1 per 1000 males (Hook & Hamerton 1977). Since they are fully viable, and not selected against before birth, this would imply heavy selection against YY-bearing spermatozoa. Selection against sperm on the basis of their genetic content is believed not to occur. Second, if the frequency of spermatozoa with two fluorescent bodies reflects the rate of non-disjunction of the Y chromosome at meiosis, and this rate is typical of all chromosomes, the total rate of aneuploidy in human sperm would be almost unbelievably high (Pearson et al. 1974). In fact, measurements of spermatozoa with two fluorescent bodies suggest that they do not contain the extra DNA that a second Y chromosome would provide (Sumner & Robinson 1976). Although observations on such sperm, using a Y chromosome-specific DNA probe, seemed to confirm the existence of YY-bearing spermatozoa (Joseph et al. 1984), direct observations of chromosomes from decondensed spermatozoa provide no evidence for such a high level of Y chromosome non-disjunction (Martin et al. 1983, Brandriff et al. 1985). It must, therefore, be stated that for practical purposes spermatozoa with two fluorescent bodies do not contain two Y chromosomes, although there must be sufficient YY-bearing spermatozoa to account for the very small proportion of XYY males in the population. Although the induction by various chemicals of higher levels of spermatozoa with two fluorescent bodies has been taken as induction of Y chromosome non-disjunction (Kapp & Jacobson 1980), such an interpretation cannot be accepted (Allen et al. 1986).

The finding that the percentage of human spermatozoa with a Y-body is sometimes rather low (even less than 40%), the occurrence of spermatozoa with two fluorescent bodies that clearly do not represent two Y chromosomes, and the rare occurrence of three or more Y-bodies in a sperm head (Beatty 1977) have tended to bring into disrepute the use of Y-body fluorescence with quinacrine as a means of determining the sex chromosome constitution of human spermatozoa (Beatty 1977, Gledhill 1983). In my opinion this is an unjustified and far too pessimistic view, although there can be no doubt that skill and experience are required for good results. With good technique, and an understanding of the problems of technique and interpretation, the method can be regarded as reliable for 'sexing' human sperm in bulk. Individual sperm cannot, of course, be 'sexed' with 100% certainty, and as already explained, spermatozoa with two fluorescent bodies must not be interpreted as having a YY chromosome constitution.

Y-bodies cannot be used to determine the sex chromosome constitution of spermatozoa in any other species studied so far. One of man's closest relatives, the gorilla, has a fluorescent segment on his Y chromosome, but also has large fluorescent segments on autosomes, so that the Y chromosome is not, in fact, distinguishable in interphase nuclei and sperm heads (Seuanez *et al.* 1976b). Although F-bodies have been reported in a substantial proportion of sperm heads in various other mammals, these species do not have brightly fluorescent segments on their Y chromosomes, and no DNA measurements have been done to support a correlation between the presence of F-bodies and the sex chromosome constitution.

8 Banding with fluorochromes other than quinacrine

8.1 Introduction

Several fluorochromes other than quinacrine have been used successfully to produce chromosome banding. Although none of them is as important as quinacrine for practical purposes, some of them have valuable, often specialized applications. It should be emphasized that in general the fluorochromes described in this chapter produce banding without any pretreatment, unlike R-banding with acridine orange (Section 6.2.2). Nevertheless, it has been found useful in several cases to use fluorochromes in combination with another stain, often non-fluorescent. Such a procedure may either permit double staining of the same chromosome with two different patterns, which can be revealed by excitation with light of appropriate wavelengths; or a completely different pattern of banding may be produced. Finally in this chapter we shall consider light-induced banding (Section 8.6), in which there is a genuine pretreatment of the chromosomes before staining, as opposed to counterstaining with another substance.

Apart from convenience, it is appropriate to deal with light-induced banding in this chapter, since, like nearly all the other methods to be described here, it appears to depend largely on base-specific interactions with the chromosomal DNA. Most of the fluorochromes mentioned in this chapter have been shown to have preferences, in solution, for either A + T-rich or G + C-rich DNA, and although the interactions of some of these compounds with chromosomes are rather complex, especially when used with counterstains, the banding patterns they produce on chromosomes are most readily understood in terms of base specificity. Table 8.1 lists the fluorochromes to be covered in this chapter, together with their DNA base specificity in solution and data on their excitation and emission wavelengths.

8.2 Ethidium: a fluorochrome that shows no base specificity

Ethidium (Fig. 8.1) is a fluorochrome whose fluorescence is greatly enhanced when it intercalates in double-stranded DNA (LePecq & Paoletti 1967). There is only slight evidence that its binding or fluorescence is affected by the

Table 8.1 Properties of the fluorochromes described in this chapter.

Fluorochrome	Base specificity	λ_{max} excitation	λ_{max} emission	Reference
Ethidium	None	520nm	590nm	LePecq & Paoletti 1967
Hoechst 33258	A + T	356nm	465nm	Latt & Wohlleb 1975
DAPI	A + T	355nm	450nm	M.S.Lin et al. 1977
DIPI	A + T			
Daunomycin	A + T	505nm	~565nm	Comings & Drets 1976
			600nm	Johnston et al. 1978, Latt 1977
Adriamycin	A + T			
Dibutyl-proflavine	A + T	472nm	505nm	Müller et al. 1973
D287/170	?A + T			
D288/45	?A + T	400nm	460nm	Schweizer et al. 1987
D288/48	?A + T	400nm	460nm	
Chromomycin A$_3$	G + C	435nm	560nm	Crissman et al. 1978
Olivomycin	G + C	430nm	545nm	
Mithramycin	G + C	440nm	570nm	
7-amino-actinomycin D	G + C	551nm	655nm	Gill et al. 1975

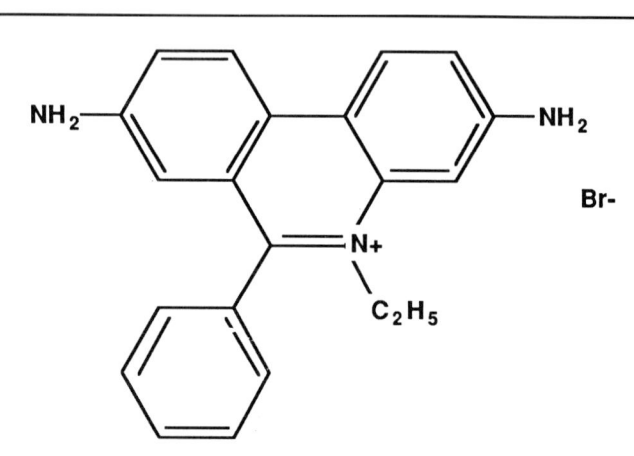

Figure 8.1 Structural formula of ethidium bromide.

base composition of the DNA and it is generally regarded as showing no significant base specificity (Distèche et al. 1980, Langlois et al. 1980). In most cases, chromosomes show uniform fluorescence after ethidium staining (e.g. Sumner et al. 1973, Mezzanotte et al. 1985), but there are two papers describing banding with ethidium. Caspersson et al. (1969b) found that the ethidium fluorescence pattern of the chromosomes of the plants *Trillium erectum* and *Scilla sibirica* was the opposite of that obtained with quinacrine mustard. A similar situation was found in the chromosomes of Bennett's wallaby by Pearson *et al.* (1971). In this species there were dull areas around the centromeres with quinacrine fluorescence which were bright with ethidium. In neither paper did the authors offer any explanation of the differential fluorescence pattern.

8.3 Banding with fluorochromes showing specificity for A + T-rich DNA

Several fluorochromes which have been reported to show stronger fluorescence with A + T-rich DNA have been used to produce banding patterns on chromosomes (Table 8.1). Although it might be supposed that such fluorochromes would simply be alternatives to quinacrine, which also shows fluorescence enhancement with A + T-rich DNA (Ch. 7), the fluorochromes to be considered here do not, in general, produce banding patterns identical to Q-bands, and have their own distinctive properties and applications. There are various reasons for this. The fluorochromes to be described in this section are structurally different from quinacrine (Fig. 8.2) and as a result, they apparently do not intercalate in DNA molecules. Probably because they do not intercalate, their fluorescence is not quenched by guanine, so that in this respect also they differ from quinacrine. As a general rule, they show less distinct euchromatic bands than quinacrine, but tend to show distinctive patterns of heterochromatin fluorescence.

8.3.1 Hoechst 33258

After quinacrine, Hoechst 33258 (2,2-(4-hydroxyphenyl)-6-benzimidazolyl-6-(1-methyl-4-piperazyl)-benzimidazol-trihydrochloride, Fig. 8.2) is probably the most important fluorochrome for chromosome banding, and was the next to come into general use. Its use for banding was pioneered by Hilwig & Gropp (1972), who applied it to chromosomes of mouse species (*Mus musculus* and *Mus poschiavinus*) and of the Algerian hedgehog (*Erinaceus algirus*). In the mouse chromosomes they observed that the centromeric heterochromatin fluoresced strongly, while the arms showed a faint Q-banding pattern. The hedgehog chromosomes, however, were generally bright, with weak fluorescence in the heterochromatic segments. Hilwig & Gropp therefore concluded that Hoechst 33258 was specific for only certain types of heterochromatin.

Since its introduction into cytogenetics, Hoechst 33258 has been used to stain the chromosomes of a wide variety of mammals (e.g. Seth & Gropp 1973, Seth *et al.* 1973, Raposa & Natarajan 1974), insects, especially *Drosophila* spp. (Holmquist 1975a, Gatti *et al.* 1976, Pimpinelli *et al.* 1976a) and plants (e.g. Vosa 1974, Schweizer & Nagl 1976, Mogford 1977) (Fig. 8.3). As

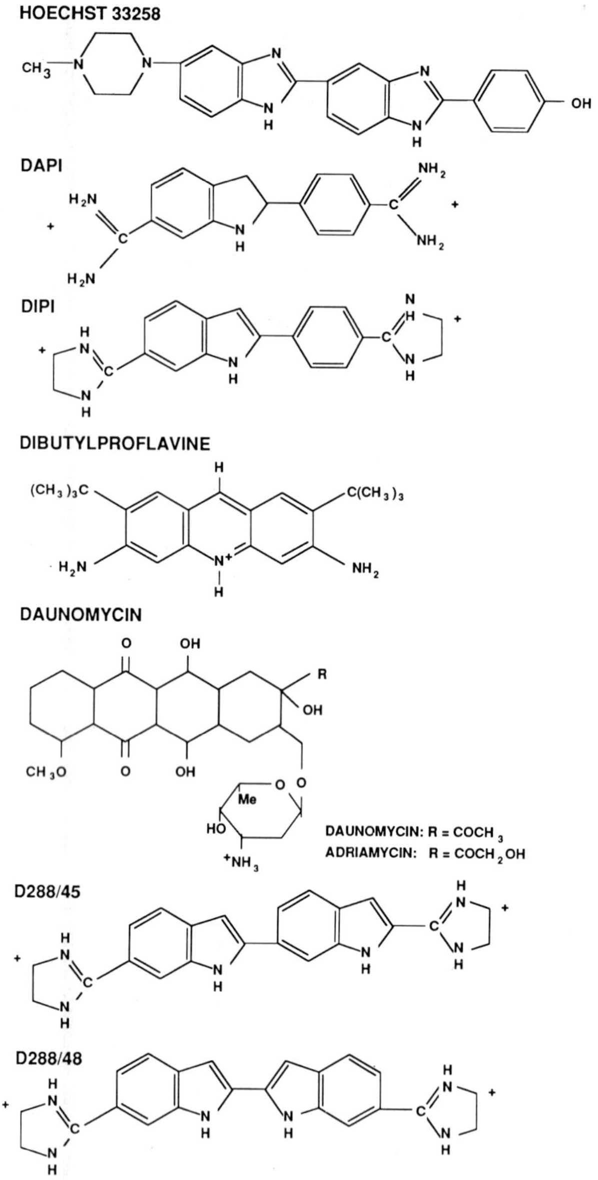

Figure 8.2 Structural formulae of fluorochromes showing preference for A +T-rich DNA.

(a)

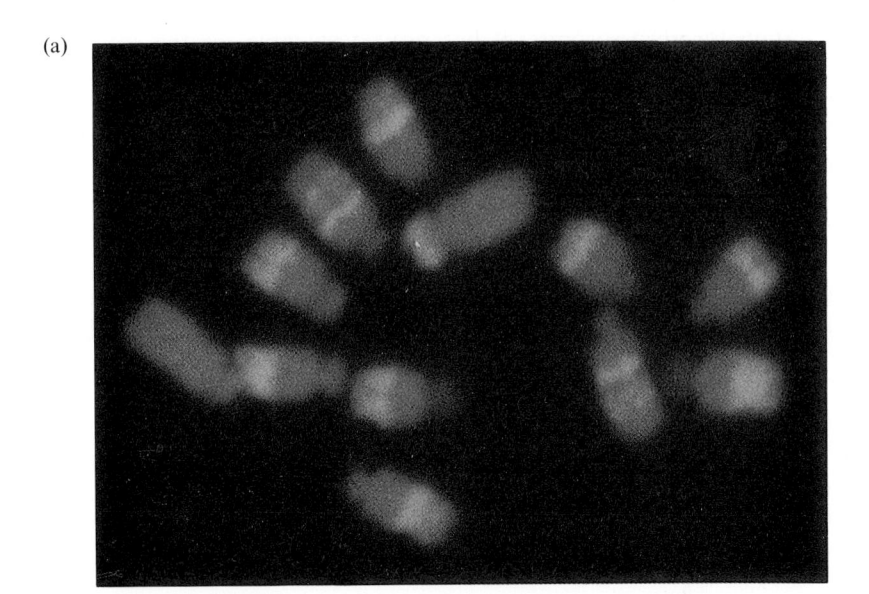

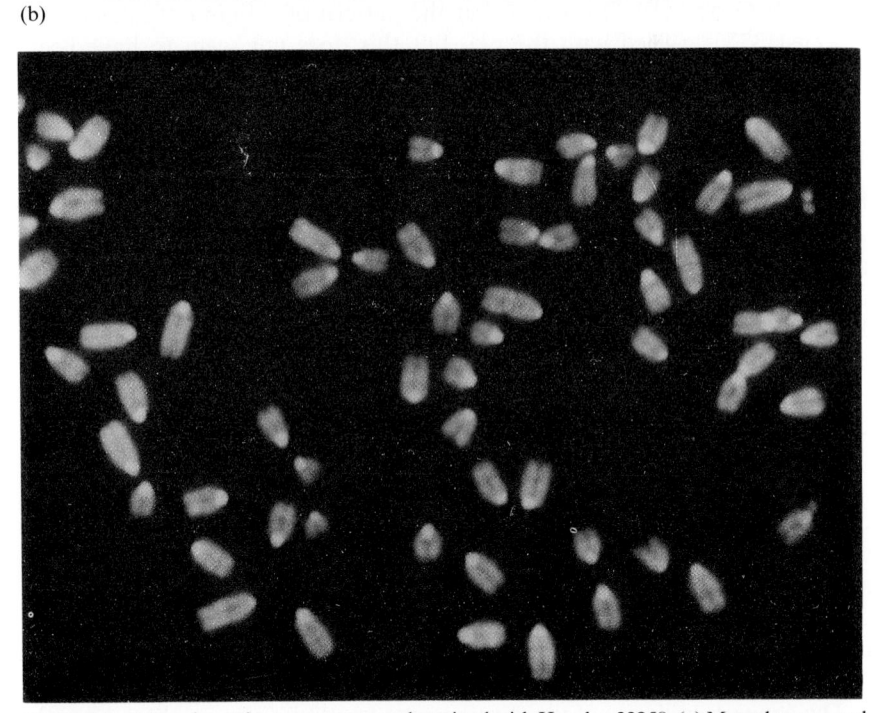

(b)

Figure 8.3 Metaphase chromosome spreads stained with Hoechst 33258. (a) Metaphase spread from *Ornithogalum zeyheri* Micrograph kindly provided by Dr C. G. Vosa. (b) Chromosomes from a mouse cell line; note the strong fluorescence of the centromeric heterochromatin.

a direct stain for chromosomes, it has proved valuable for the study of heterochromatin, although the differentiation of euchromatic bands with Hoechst 33258, being weaker than with quinacrine, for example, has not been found so useful. Staining is generally carried out as follows (Hilwig & Gropp 1972, Seth & Gropp 1973, Jalal et al. 1975):

1. Stain for 10 min in a solution containing approximately $0.05\mu g$ of Hoechst 33258/1 ml of saline (e.g. Hanks' balanced salt solution) or phosphate-citrate buffer. It is convenient to make up a more concentrated (50 $\mu g/ml$) stock solution which is refrigerated, and diluted as required.
2. After staining, the chromosome preparations are rinsed with distilled water, and mounted. Initially, glycerol was recommended as a mountant, but subsequently aqueous buffers (Jalal et al. 1975, Gatti et al. 1976) or distilled water (Misawa et al. 1977) have been used. Addition of 1% sodium dithionite to the mountant inhibits fading under illumination.
3. Preparations should be examined using ultra-violet illumination.

Hilwig & Gropp (1975) showed that the pattern of staining of nuclei with Hoechst 33258 varied with the pH, but this does not seem to have been investigated systematically for chromosomal staining. However, the pattern of chromosomal staining can be altered, at least in some cases, by altering the concentration of the staining solution (Gatti et al. 1976). It seems, therefore, to be essential when describing staining patterns produced by Hoechst 33258, to specify precisely the staining conditions.

Hoechst 33258 has a number of applications in cytogenetics apart from its use as a straightforward fluorochrome. It has been used in combination with quinacrine, to identify chromosomes in interspecific hybrid cells (Carlin & Rao 1982). Mouse chromosomes can be distinguished by the brilliant fluorescence of their centromeric heterochromatin, while quinacrine is used to identify individual chromosomes of each species, since it produces more distinct euchromatic bands. Hoechst 33258 has also been found very valuable in the study of chromosome replication (Section 11.3), and when applied to cells in culture, it can inhibit the condensation of certain specific regions of chromosomes (Section 11.2.2). In flow cytometry, this fluorochrome has been used, generally in combination with another, G + C-specific fluorochrome such as chromomycin, to resolve the chromosome complements of certain mammals (Langlois et al. 1980). It has also had applications in measurement of nuclear DNA (Cowell & Franks 1980, Cowden & Curtis 1981).

The general similarity of the chromosome banding patterns obtained with Hoechst 33258 and with quinacrine, as well as the degree of Hoechst fluorescence found in chromosomal regions of known base composition (Table 8.2), indicated that bright fluorescence with Hoechst 33258 might indicate regions of chromosomes whose DNA was rich in A + T base pairs. This preference of Hoechst 33258 for A + T-rich DNA was confirmed by

Table 8.2 Fluorescence of chromosomal regions of known base composition, as identified by *in situ* hybridization, with Hoechst 33258, DAPI, dibutyl proflavine and daunomycin.

Species	Satellite DNA composition	Fluorescence with: Hoechst 33258	DAPI	Dibutyl-proflavine	Daunomycin
Plants					
Scilla sibirica	G + C-rich (Timmis *et al.* 1975, Deumling 1981)		− ve (Deumling & Greilhuber 1982)		
Secale cereale	Slightly G + C-rich (Appels *et al.* 1978)	+ ve (Sarma & Natarajan 1973, Vosa 1974)			
Vicia faba	A + T-rich (Cionini *et al.* 1985)				+ ve (Gill & Nadjar 1977)
Insecta					
Drosophila melanogaster	A + T-rich (Peacock *et al.* 1977)	+ ve (Holmquist 1975a, Gatti *et al.* 1976)			
Drosophila nasuta	A + T-rich (Ranganath *et al.* 1982)	+ ve (Lakhotia & Kumar 1978)			
Drosophila virilis	A + T-rich (Gall *et al.* 1971)	+ ve (Gatti *et al.* 1976)			
Rhynchosciara hollaenderi	A + T-rich (Eckhardt & Gall 1971)	+ ve (Stocker *et al.* 1978)			
Birds					
Gallus domesticus	G + C-rich (Stefos & Arrighi 1974)		− ve (Fritschi & Stranzinger 1985)		
Mammals					
Saimiri sciureus	A + T-rich (Lau *et al.* 1977)	+ ve (Jalal *et al.* 1974)			
Mus musculus	A + T-rich (Jones & Robertson 1970)	+ ve (Jalal *et al.* 1975)	+ ve (Schnedl *et al.* 1977c)	+ ve (Distèche & Bontemps 1974)	
Peromyscus eremicus	G + C-rich (Hazen *et al.* 1977)	− ve (Jalal *et al.* 1974)			
Bos taurus	G + C-rich (Kurnit *et al.* 1973)	− ve (Jalal *et al.* 1975)	− ve (Schnedl *et al.* 1977c)	− ve (Distèche & Bontemps 1974)	

studies of the fluorescence enhancement of this dye in solution when mixed with DNAs of different base composition (Weisblum & Haenssler 1974, Comings 1975b, Latt & Wohlleb 1975). These authors found that, in fact, the binding of Hoechst 33258 to any sort of DNA enhanced its fluorescence, but that A + T-rich DNA caused greater enhancement than G + C-rich DNA. The absence of quenching by G + C-rich DNA might help to explain the differences in fluorescence patterns produced by quinacrine and by Hoechst 33258. If, for example, it is correct that mouse centromeres show relatively weak fluorescence with quinacrine because there are sufficient guanines in the A + T-rich satellite to quench the quinacrine fluorescence effectively (Weisblum 1973), a different result would be expected with Hoechst 33258 whose fluorescence is not quenched by guanine.

Subsequent studies have revealed more details about the interactions of Hoechst 33258 with DNA, and have shown that these differ in several ways from those of quinacrine with DNA. The A + T specificity of Hoechst 33258 appears to result from binding at low dye:DNA ratios, producing strong fluorescence (Latt & Wohlleb 1975, Jorgenson *et al.* 1988). At higher dye:DNA ratios, a different type of binding occurs, which is apparently largely ionic and results in low fluorescence. Since staining of chromosomes is normally done from very dilute dye solutions, this latter type of binding is probably irrelevant to the interpretation of banding with Hoechst 33258. Unlike quinacrine, Hoechst 33258 does not intercalate in DNA, but at low dye:DNA ratios binds in the minor groove of the DNA molecule (Jorgenson *et al.* 1988, Teng *et al.* 1988) with a preference for a sequence of at least four A + T base pairs. In such a position very precise interactions are possible between the dye molecule and DNA bases, and specific sequences of four or five base pairs have been described to which Hoechst 33258 binds (Martin & Holmes 1983, Harshman & Dervan 1985, Dervan 1986).

It is not clear that knowledge of the precise base sequences favoured by Hoechst 33258 is of much value in understanding the fluorescent banding patterns on chromosomes. For one thing, since there is a variety of these sequences, the dye will only indicate the aggregate of sites of such sequences on the much larger scale of the chromosome. Perhaps even more important are observations which indicate that the pattern of fluorescence obtained with Hoechst 33258 is affected by other factors as well as the DNA base composition. Both Latt & Wohlleb (1975) and Stokke & Steen (1986) showed that the number of binding sites for the dye was greatly reduced in chromatin compared with DNA or with dehistonized chromatin. Since proteins are lost during fixation with acetic acid (Section 3.5) such effects may be less marked or more localized in fixed chromosomes. The observations of Holmquist (1975a) and of Gatti *et al.* (1976) on chromosomes of *Drosophila* spp. that the presence of A + T-rich DNA was a necessary, but not sufficient, condition for bright Hoechst 33258 fluorescence support the possibility that in some cases chromosomal proteins may block staining. Another question is whether the differences in base composition between different DNA fractions are sufficient to account for the patterns seen in chromosomes. Simola *et al.* (1975) found no differences in the fluorescence of Hoechst 33258 with mouse satellite and main band DNA, although the sites of these DNA fractions fluoresce very differently in chromosomes. In spite of these reservations, however, it still seems safe to say that DNA base composition is the predominant factor in determining Hoechst 33258 banding patterns on chromosomes.

8.3.2 DAPI and DIPI

DAPI (4'-6-diamidino-2-phenylindole, Fig. 8.2) was developed as a trypanocide, and was introduced as a fluorochrome for chromosome banding by Schweizer & Nagl (1976). These authors showed that the pattern of fluorescence obtained with DAPI on the chromosomes of the orchid *Cymbidium* was very much the same as that found after staining with quinacrine or Hoechst 33258. In human chromosomes, DAPI banding was found to be essentially the same as that produced by Hoechst 33258; that is, the euchromatic parts of the chromosomes are Q-banded, while the heterochromatic segments are generally bright (as with Hoechst 33258 but unlike quinacrine) (Schweizer 1976, Schnedl *et al.* 1977a). In *Drosophila virilis*, however, certain chromosomal regions seem to show different patterns of fluorescence with DAPI and Hoechst 33258 (Abraham *et al.* 1983). The closely related compound DIPI [4'-6-bis(2'-imidazolinyl-4'-5'-H)-2-phenylindole, Fig. 8.2] was shown to produce patterns on chromosomes indistinguishable from those produced by DAPI (Schnedl *et al.* 1977a, b).

DIPI is difficult to obtain commercially, and most work has been done with DAPI. Evidently, as regards the patterns produced on chromosomes, DAPI has little advantage compared with Hoechst 33258, and the two compounds can apparently be used almost interchangeably. DAPI is, however, particularly stable to illumination, and DIPI even more so (Schnedl *et al.* 1977a, c), an important advantage when fading of fluorescence is a serious problem in studying chromosome banding. Indeed, DAPI can be made completely resistant to fading by adding suitable substances to the mountant (Hamada & Fujita 1983), although it is not known whether such additives are compatible with the demonstration of banding. DAPI has been used in combination with the G + C-specific fluorochromes chromomycin A_3 and mithramycin to produce complementary banding patterns merely by changing the wavelength of the exciting light (Section 8.5.1); with this combination there is no significant interaction of the fluorochromes. DAPI has been found useful in studies of chromosome replication (Section 11.3), for inhibiting chromosome condensation (Section 11.2.2), and for measuring DNA. In the last category, DAPI has been used not only to measure total DNA in nuclei (Cowden & Curtis 1981), mycoplasma and viruses (Russell *et al.* 1975), kinetoplasts (Hajduk 1976), and isolated on gels (Kapuściński & Yanagi 1979), but also to measure DNA base composition (Leeman & Ruch 1982). For such applications not only is its resistance to fading valuable, but also its great sensitivity. However, the most distinctive and probably the most important application of DAPI in cytogenetics came when Schweizer et al. (1978) counterstained chromosomes with the antibiotic distamycin, thereby highlighting certain heterochromatic regions. The Distamycin/DAPI and related techniques are described in Section 8.5.2.

DAPI or DIPI staining of chromosomes is easily accomplished by using a very dilute solution of the fluorochrome for a few minutes, followed by brief rinsing and mounting in buffer. Schweizer & Nagl (1976) stained for 5–10 min in a solution containing 0.2 μg of DAPI/ml of McIlvaine's phosphate buffer, pH 7, and subsequent workers have made only trivial alterations to the method. Schnedl et al. (1977b) used 0.3–0.5 μg of DAPI or DIPI in an aqueous solution, while M.S.Lin et al. (1977) used 1 μg/ml and Leeman & Ruch (1982) 2 μg/ml. Schnedl et al. (1977b) mounted their preparations in McIlvaine's buffer, pH 5.5, M.S.Lin et al. (1977) used phosphate buffer, pH 7.5, and Leeman & Ruch (1978, 1982) favoured glycerol, or buffered glycerol. Clearly this is a robust technique, the results of which are not significantly affected by moderate changes in the method.

The identity of the banding patterns produced by DAPI and by Hoechst 33258 strongly suggests that these two fluorochromes react with chromosomes in the same way. In other words, DAPI as well as Hoechst 33258 (Section 8.3.1) shows a preference for A + T-rich DNA. This is supported by the staining reactions observed in chromosomal regions of known base composition (Table 8.2); blocks of A + T-rich heterochromatin fluoresce brightly, while blocks of G + C-rich heterochromatin show weak fluorescence.

The evidence for the A + T specificity of DAPI for DNA in chromosomes is supported by observations on the interactions between DAPI and DNA in solution, although some of these reports are contradictory. Kapuściński & Szer (1979) reported that DAPI forms fluorescent complexes with DNA containing A + T base pairs but not G + C base pairs. Manzini et al. (1983) reported stronger binding to A + T-rich DNA, although Cavatorta et al. (1985) stated that there was no base specificity in the binding of DAPI to DNA, but that A + T-rich DNA enhanced DAPI fluorescence much more than G + C-rich DNA did. Whatever the precise details, there can be no doubt that DAPI shows much greater fluorescence with A + T-rich DNA than with G + C-rich DNA. The actual mechanism of binding is also uncertain. It is clear that the binding of DAPI to DNA is very strong (Kapuściński & Skoczylas 1978), and that the dye only binds to double-stranded DNA, and not to single-stranded DNA or RNA (Hajduk 1976, Kapuściński & Szer 1979). In fact DAPI binding seems to require that the DNA helix be in the B conformation (Kapuściński & Szer 1979). Earlier reports suggested that DAPI intercalated into the DNA molecule (Kapuściński & Skoczylas 1978), but more recently this has been ruled out, and it seems more probable that DAPI, like Hoechst 33258, binds to clusters of A + T pairs in the minor groove of the DNA (Manzini et al. 1983, Kubista et al. 1987).

An observation particularly relevant to the interpretation of DAPI fluorescence patterns on chromosomes is that the binding of DAPI to DNA is scarcely affected by chromosomal proteins. Kapuściński & Skoczylas (1978) and Darzynkiewicz & Traganos (1988) showed that histones had only a small effect on the number of DAPI-binding sites on DNA. Coleman *et al.* (1981) and Cowden & Curtis (1981) reported no discernible effect of degree of chromatin condensation or stage of the cell cycle on DAPI fluorescence of nuclei, while Monaco & Rasch (1982) found that DAPI fluorescence of nuclei remained brilliant during spermiogenesis in a wide variety of mammals, although staining with other dyes declined markedly. All these observations lead one to the conclusion that patterns of DAPI fluorescence on chromosomes are determined predominantly by DNA base composition, and are probably not influenced significantly by chromosomal proteins, although it should be pointed out that Darzynkiewicz *et al.* (1984) found that DAPI fluorescence of unfixed nuclei could be increased by extraction with acid. Since chromosome banding is almost invariably carried out on material treated with an acid fixative, the observations of Darzynkiewicz *et al.* (1984) are probably not relevant here.

8.3.3 *Dibutyl-proflavine*

2,7-di-*t*-butyl proflavine (Fig. 8.2) is a proflavine derivative that binds specifically to A + T base pairs of DNA without intercalation (Müller *et al.* 1973). It was introduced into cytogenetics by Distèche & Bontemps (1974), although it has not been generally used. As would have been expected, it produces bright fluorescence with the centromeric heterochromatin of mouse chromosomes, and dim fluorescence with that of bovine chromosomes. On human chromosomes, dibutyl proflavine produces weak banding on the chromosome arms, and bright fluorescence of the major blocks of constitutive heterochromatin on chromosomes 1, 9, 16 and Y, in general agreement with the patterns produced by Hoechst 33258 and DAPI (Sections 8.3.1 and 8.3.2).

Dibutyl proflavine is believed to bind externally to the DNA molecule, possibly in the major groove (Bontemps *et al.* 1974). The dye molecule apparently requires two adjacent binding sites, at least one of which must be an A + T pair. In the presence of histones, the strength of binding of the dye to DNA is apparently lowered, but without affecting the number of binding sites. Thus the pattern of dibutyl-proflavine fluorescence on chromosomes is likely to be determined primarily by DNA base composition, and to be largely uninfluenced by chromosomal proteins.

8.3.4 *Daunomycin and adriamycin*

Daunomycin and adriamycin (Fig. 8.2) are two closely related anthracycline antibiotics with a side chain consisting of the amino-sugar daunosamine.

They were introduced into cytogenetics by C. C. Lin & van de Sande (1975) who showed that when used to stain human chromosomes these antibiotics produced a banding pattern similar in all detail to that produced by quinacrine. That is, clear bands were produced along the chromosome arms, with especially bright fluorescence at the centromere of chromosome 3, on the short arm of chromosome 13, and on the distal part of the long arm of the Y. Other heterochromatic regions, which are strongly fluorescent with Hoechst 33258 and DAPI, did not fluoresce brightly. Daunomycin fluorescence is orange-red, and reported to be much more stable than that of quinacrine, although less bright. C.C.Lin & van de Sande (1975) unfortunately named the bands produced by daunomycin, D-bands, in spite of the apparent identity of the patterns with Q-bands. This identity with Q-bands was confirmed for the chromosomes of *Vicia faba* by Gill & Nadjar (1977), but otherwise neither daunomycin nor adriamycin appears to have been used for chromosome banding.

Although these two antibiotics have not been of any practical value in cytogenetics, they are important because of the information they provide on banding mechanisms, and particularly on the mechanism of quinacrine banding. The resemblance of the patterns produced by daunomycin to those produced by other fluorochromes discussed in the previous chapter and earlier in this chapter would suggest that daunomycin fluorescence is an indicator of A + T-rich DNA. At the same time, the existence of small but significant differences in the banding patterns produced by these fluorochromes must be taken to indicate either a subtle difference in specificity, or a difference in susceptibility to possible factors such as blocking of binding sites by chromosomal proteins. In fact, Schnedl (1978) has classified 'A + T-specific' fluorochromes into two groups, based on their fluorescence with certain regions of human heterochromatin (Table 8.3), showing clear differences in behaviour between the quinacrine group of fluorochromes, and the Hoechst 33258 group. It is to the former group that daunomycin clearly belongs, and knowledge of its behaviour may help to understand that of quinacrine.

It has been known for many years that daunomycin intercalates in DNA (Ward *et al.* 1965, Waring 1970). Ward *et al.* (1965) indicated that there was little base specificity in this intercalation; Plumbridge & Brown (1979) reported that both daunomycin and adriamycin intercalate in the B form of DNA, but not in the A form. These observations suggest that the binding preference of daunomycin is not important for banding, and that base-specific enhancement or quenching of fluorescence is the necessary factor. This has been investigated with rather surprising results. When the fluorescence of daunomycin was studied in the presence of DNAs with G + C contents between 31% and 100%, the quenching of fluorescence was found to be very large and similar throughout this range (Comings & Drets 1976, Johnston *et al.* 1978). Only with synthetic A + T polymers was bright fluorescence obtained. This implies either that the daunomycin fluorescence pattern is determined by the accessibility of its binding sites on chromosomes,

Table 8.3 Staining properties of human heterochromatic segments with quinacrine-type and Hoechst 33258-type fluorochromes

	Quinacrine-type fluorescence	Hoechst 33258-type fluorescence
Fluorochromes	Quinacrine Daunomycin	Hoechst 33258 DAPI DIPI Dibutyl-proflavine
Heterochromatin of chromosome:		
1	−	+
3	+	−
4	+	−
9	−	+
13	+	−
14		−
15		−
16	−	+
Y	+	+

or that it is indicating sites of extremely A + T-rich DNA on the chromosomes. It is not at present possible to distinguish between these alternatives. Nevertheless, as Johnston et al. (1978) have pointed out, stretches of poly dA.poly dT have been reported from a wide variety of eukaryotes, so it is not inconceivable that they may contribute to the banding patterns produced by daunomycin, and presumably by quinacrine.

The great sensitivity of daunomycin and quinacrine fluorescence to quenching by DNAs with quite moderate G + C contents is probably sufficient to explain the existence of blocks of heterochromatin that fluoresce with the Hoechst 33258 group of fluorochromes, but not with the quinacrine group. Such regions would be more A + T-rich than the remainder of the chromosomes, but still contain an adequate level of guanine, possibly interspersed at appropriate intervals (Weisblum 1973) to quench the fluorescence of susceptible fluorochromes. The existence of heterochromatic regions fluorescing brightly with the quinacrine group is probably explicable on the same principles. Such regions would be extremely A + T-rich, possibly completely lacking guanine, so that quenching of quinacrine and daunomycin fluorescence would be very small; such DNA would not, however, give such marked enhancement of the fluorescence of the Hoechst 33258 group of fluorochromes, so that with such fluorochromes these regions would not appear distinctive.

8.3.5 D287/170 and some related compounds

D287/170 is one of a number of modifications of DAPI synthesised by Professor O. Dann of Erlangen. It was introduced into cytogenetics by Schnedl *et al.* (1981a), who showed that when it was applied to human chromosomes they show largely uniform fluorescence, but that certain regions were highlighted. These regions which show especially strong fluorescence are the heterochromatin of chromosomes 9 and Y, and a region on the short arm of chromosome 15. Certain centromeric C-bands and regions on the short arms of the acrocentric chromosomes also showed somewhat greater fluorescence than the rest of the chromosomes. The general pattern resembles, in part, that produced by the distamycin/DAPI sequence (Section 8.5.2), but unlike distamycin/DAPI, does not stain the C-bands of chromosomes 1 and 16.

D287/170 is evidently a useful fluorochrome for distinguishing certain types of heterochromatin. Additional studies on human heterochromatin are those by Babu & Verma (1986a), who used it to distinguish at least seven types of heterochromatin on the basis of their staining properties, and by Tommerup (1982) who applied it to interphase nuclei. In interphase, D287/170 apparently produces bright fluorescence only with the heterochromatin of chromosome 9; it was suggested, therefore, that the ploidy of a nucleus could be deduced by counting the number of fluorescent bodies within it. This fluorochrome has also been used to distinguish certain classes of heterochromatin on the domestic pig (*Sus scrofa*) (Schnedl *et al.* 1981b), and in members of the Canidae (Mayr *et al.* 1986).

Although no formal studies have been made of the specificity of D287/170, certain suggestions have been made (Schnedl *et al.* 1981a). The close structural resemblance of D287/170 to DAPI might suggest an affinity for A + T-rich DNA, as would the Q-banding pattern produced throughout the chromosome arms when used at very low concentrations, and the bright fluorescence obtained with mouse centromeric DNA. However, the restricted distribution of D287/170-positive sites suggests that this fluorochrome does not simply show specificity for A + T-rich DNA, like the dyes previously considered in this section, but may also show sequence specificity. In the pig, D287/170 shows strong fluorescence at sites which are distamycin/ DAPI positive, quinacrine-negative, and show moderate chromomycin fluorescence (Schnedl *et al.* 1981b). The precise specificity of D287/170 is intriguing, but clearly it has already proved valuable as an empirical tool for identifying certain classes of heterochromatin.

The banding properties of another two dyes related to DAPI and to D287/170 have been described briefly (Schweizer *et al.* 1987). These fluorochromes, D288/45 and D288/48 (Fig. 8.2) are isomers, yet D288/48 binds to DNA with much higher affinity and greater A + T specificity, and produces patterns in euchromatin and heterochromatin that resemble Q-banding. D288/45, on the other hand, produces different and weaker fluorescence of heterochromatin, but can be substituted for DAPI in the distamycin/DAPI procedure (Section 8.5.2) much more successfully than D288/48. Evidently minor structural differences in the dye molecule can have important effects on their staining patterns.

8.4 Banding with fluorochromes showing specificity for G + C-rich DNA

Fluorochromes which show specificity for G + C-rich DNA are fewer and less diverse than those showing A + T specificity. Nevertheless, such fluorochromes have both practical and theoretical value. If two fluorochromes had opposite specificities for DNA base composition in solution, and produced complementary banding patterns on chromosomes, then it would be a reasonable deduction that the base composition of chromosomal DNA was the major factor in determining the banding patterns. If the banding patterns with the different fluorochromes were not complementary, then it might be necessary to invoke other factors to explain them, although, as we have already seen (Section 8.3) two fluorochromes having the same general base specificity may in fact produce banding patterns that are different in detail. Such differences can be explained by different responses to variations in base composition without having to invoke factors such as blocking of binding sites by chromosomal proteins.

The G + C-specific fluorochromes that have been used for staining chromosomes fall into two classes: the chromomycinone antibiotics, and 7-amino-actinomycin, a fluorescent derivative of actinomycin D.

8.4.1 *Chromomycinone antibiotics: chromomycin, mithramycin and olivomycin.*

These three antibiotics are very similar structurally (Fig. 8.4) and have been used interchangeably for chromosome banding and for staining DNA in cytological preparations. Chromomycin was first used by Schweizer (1976) who found that in plant chromosomes heterochromatic regions that showed bright chromomycin fluorescence had weak DAPI fluorescence and *vice versa*. On human chromosomes, chromomycin produced what appeared to be an R-banding pattern (Ch. 6), although the contrast was low and the patterns indistinct. Schweizer noted that mithramycin produced the same results as chromomycin. Van de Sande *et al.* (1977) shortly afterwards reported on the use of olivomycin for chromosome banding, and obtained clear fluorescent R-banding patterns on human chromosomes, and bright fluorescence of the G + C-rich centromeric heterochromatin of bovine chromosomes. Schnedl *et al.* (1977c) used mithramycin on various mammalian chromosomes, and confirmed the complementarity of patterns produced by this fluorochrome and by the A + T-specific fluorochrome DIPI, both in

CHROMOMYCINONE ANTIBIOTICS

R₁ and R₂ are different sugar substituents in Chromomycin, Mithramycin and Olivomycin.

7-AMINO-ACTINOMYCIN D

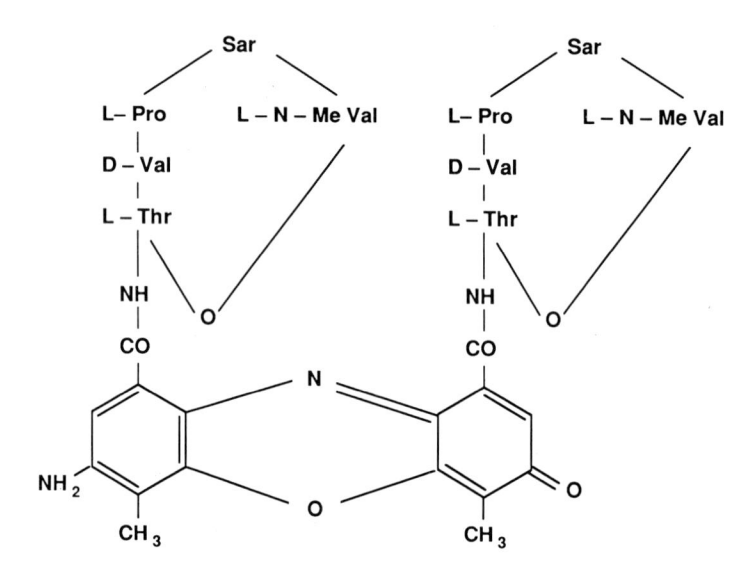

Figure 8.4 Structural formulae of fluorochromes showing preference for G + C-rich DNA.

euchromatic and heterochromatic segments of chromosomes. Since these pioneering studies, chromomycin and its congeners have been used as a simple means of R-banding (e.g. Jorgenson et al. 1978, C.C.Lin et al. 1980a) (Fig. 8.5), often in combination with an A + T-specific fluorochrome to obtain Q- and R-banding on the same chromosome, or with a counterstain to increase contrast (Section 8.5). Brothman (1987) found R-banding with chromomycin to be a useful means of identifying chromosomes after *in situ* hybridization. Chromomycin fluorescence is a useful method for characterizing heterochromatin (e.g. Leeman & Ruch 1983). In addition, chromomycin and mithramycin have been used for determination of base

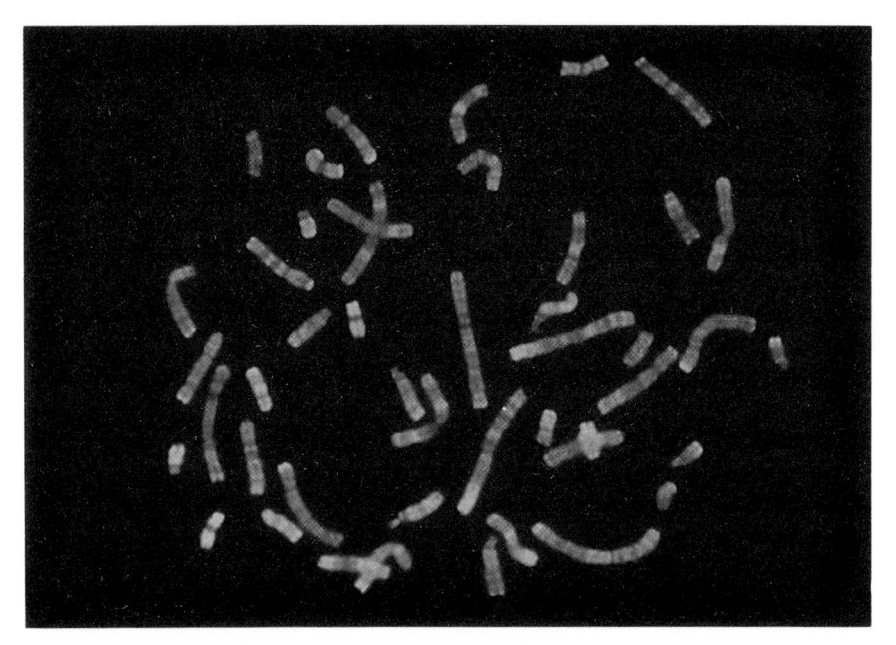

Figure 8.5 Human metaphase chromosome spread stained with chromomycin A$_3$. Micrograph kindly provided by Dr P. F. Ambros.

composition in nuclei and chromosomes (Hauser-Urfer *et al.* 1982, Leeman & Ruch 1982, 1984), and mithramycin fluorescence has been extensively used to measure DNA in flow cytometry (Crissman *et al.* 1978, Larsen *et al.* 1986) and in slide-based microfluorimetry (Coleman *et al.* 1981).

A curious phenomenon was reported by Prantera *et al.* (1979) and confirmed by Mostacci *et al.* (1980). While previous authors have all agreed that chromomycin produces an R-banding pattern on mammalian chromosomes (see above), these two groups reported that the pattern of fluorescence could be alternated between Q-banding and R-banding simply by altering the combination of excitation and barrier filters in the fluorescence microscope. Ultraviolet excitation produced Q-banding, while blue light excitation produced R-banding. In fact, the 'Q-banding' pattern was more akin to that produced by Hoechst 33258 or DAPI, in that the heterochromatin of human chromosomes 1 and 16, and that of mouse chromosomes fluoresced brightly. These interesting results have neither been repeated nor explained, but evidently need to be taken into account in explaining the mode of action of these fluorochromes. It may be significant that Mostacci *et al.* (1980) reported that best results were obtained by first observing the R-banding pattern, and then switching the filter combination to observe Q-bands after the R-bands had faded.

Staining of chromosomes with chromomycin, mithramycin or olivomycin is accomplished simply by staining in a dilute solution of the fluorochrome buffered near pH 7. Details vary according to the experimenter. Most workers have used a fluorochrome concentration between 0.1 mg and 1 mg/ml, and have stained for between 10 and 30 min (Schweizer 1976, van de Sande et al. 1977, Schnedl et al. 1977c, Leeman & Ruch 1982). Old chromomycin solutions may give better results than freshly prepared solutions (P. F. Ambros, personal communication), and the addition of 4% paraformaldehyde to the staining solution reduces swelling of the chromosomes (Ambros & Schweizer 1987). Magnesium ions are required for the binding of these antibiotics to DNA, so magnesium chloride (10mM) is generally added to the staining solution; however, van de Sande et al. (1977) and Schnedl et al. (1977c) obtained good results without adding any magnesium. It may possibly be significant that Prantera et al. (1979) and Mostacci et al. (1980) used much lower concentrations of chromomycin (5 µg/ml) and of magnesium chloride (2.5mM) than other workers. Slides stained with chromomycin or one of the other fluorochromes are normally mounted in the same buffer used for staining, although Leeman & Ruch (1982) used a mixture of buffer and glycerol. A serious problem with these fluorochromes is that they fade very rapidly when illuminated. Schweizer (1976) recommended finding the chromosomes by phase contrast microscopy before exposing them to the exciting radiation. In addition, it has been found that the fluorescence of slides that have been stored in the dark for 24 hr after staining is more stable (Schweizer 1976, Schnedl et al. 1977c, Prantera et al. 1979). P. F. Ambros (personal communication) recommends incubating the slides, after staining and mounting, for 3 days at 37°C in the dark to stabilize the fluorescence. Amemiya & Gold (1987) advocated mounting chromosomes stained with chromomycin in glycerol containing sodium hydroxide to improve differentiation of the bands.

The preference of chromomycinone antibiotics for binding to G + C-rich DNA in solution is well established (Ward et al. 1965, Behr et al. 1969, van de Sande et al. 1977, Berman et al. 1985). They also bind preferentially to double-stranded DNA, rather than to single-stranded DNA or to RNA (Ward et al. 1965). Chromomycin and mithramycin apparently do not intercalate in the DNA molecule (Waring 1970), and it is now thought that these antibiotics bind in the minor groove of DNA (Gao & Patel 1989, Sarker & Chen 1989). The Mg^{2+} ion (or certain others) is an essential part of the DNA-chromomycin complex (Ward et al. 1965, Berman et al. 1985). Evidence that chromomycinone antibiotics also show a preference for G + C-rich DNA when bound to chromosomes comes from several sources. First, there is a good correlation between the base composition of certain heterochromatic regions of chromosomes and their fluorescence with chromomycin (Table 8.4). Regions containing G + C-rich DNA show bright fluorescence, and *vice versa*. Second, these antibiotics produce an R-banding pattern on the euchromatic parts of mammalian chromosomes, the opposite of the pattern produced by fluorochromes of known A + T specificity. It is, moreover, known that the positive R-bands are relatively G + C-rich (Sections 5.3 and 6.3, Table 5.2). Third, photolysis of nuclei in the presence of

Table 8.4 Fluorescence of chromosomal regions of known base composition, as identified by *in situ* hybridization, with chromomycin A_3, mithramycin and 7-amino-actinomycin D.

Species	Satellite DNA composition	Fluorescence with:		
		Chromomycin	Mithramycin	7-amino-actinomycin D
Plants				
Scilla sibirica	G + C-rich (Timmis et al. 1975, Deumling 1981)	+ ve (Deumling & Greilhuber 1982)		
Birds				
Gallus domesticus	G + C-rich (Stefos & Arrighi 1974)	+ ve (Fritschi & Stranzinger 1985)		
Mammals				
Mus musculus	A + T-rich (Jones & Robertson 1970)		− ve (Schnedl et al. 1977c)	
Bos taurus	G + C-rich (Kurnit et al. 1973)		+ ve (Schnedl et al. 1977c)	+ ve (Latt 1977)

methylene blue, a process believed to destroy guanine residues preferentially, very drastically reduces the chromomycin fluorescence of such nuclei (Ferrucci & Sumner 1984).

In spite of the abundant evidence that the DNA base composition is an important, probably the fundamental, factor in determining the banding patterns produced by chromomycin, mithramycin and olivomycin, it is clear that other factors can affect the fluorescence of these substances when bound to nuclei or chromosomes. Mithramycin fluorescence of nuclei has been shown to be affected by the method of fixation (Larsen et al. 1986), stage of the cell cycle (Rabinovitch et al. 1981), degree of chromatin condensation (Monaco & Rasch 1982), and by extraction of protein from cells (Darzynkiewicz et al. 1984, Darzynkiewicz & Traganos 1988). Although none of these results is directly relevant to chromosome banding, they do imply that the fluorescence (and probably the binding) of these antibiotics can be influenced by accessibility of the DNA. In practice, it seems likely that chromosomal proteins have only minor effects on banding with these fluorochromes. For example, Schweizer (1976) obtained relatively poor euchromatic bands with chromomycin, while van de Sande et al. (1977) found very clear banding patterns with olivomycin. This sort of difference could well be due to subtle differences in the quality of the chromosome preparations used.

8.4.2 7-amino-actinomycin D

It has long been known that the antibiotic actinomycin D binds preferentially to G + C-rich DNA (Sobell 1973). It is not itself fluorescent, but

7-amino-actinomycin D (Fig. 8.4) has been synthesized (Modest & Sengupta 1974) and has been shown to be fluorescent. Its properties have been investigated by Modest & Sengupta (1974) and by Gill et al. (1975). The 7-amino derivative retains the G + C specificity of the parent compound, and its fluorescence is progressively enhanced by DNAs of increasing G + C content. Its fluorescence, which is red, is however relatively weak, and perhaps because of this it has not been used extensively in cytogenetics. Latt (1977) failed to find banding with 7-amino-actinomycin D in the euchromatic parts of human and bovine chromosomes, but as expected the G + C-rich centromeric heterochromatin of bovine chromosomes showed brighter fluorescence (Table 8.4). Although the fluorescence of 7-amino-actinomycin D bound to nuclei is apparently virtually unaffected by the degree of condensation of chromatin (Cowden & Curtis 1981), extraction of protein from unfixed nuclei resulted in a 13-fold increase in the fluorescence of this compound (Darzynkiewicz et al. 1984). The banding patterns produced on chromosomes by 7-amino-actinomycin D could, therefore, be influenced by accessibility of the DNA as much as by DNA base composition.

8.5 Banding with fluorochromes in combination

Following the successful application of several different fluorochromes singly for chromosome banding, many reports have been published of fluorochromes used in combination. The combinations used have been diverse, and the results of such combinations also highly varied. Fluorescent substances have been combined with other fluorescent substances, or with non-fluorescent substances; substances showing one type of DNA base specificity have been combined with substances showing the opposite or the same base specificity, or a base-specific substance has been combined with one showing no base-specificity. Sometimes the different substances appear to show no obvious interaction with each other, or they may compete for binding sites on the chromosomal DNA, or interact by energy transfer. Substances may be combined to produce complementary banding patterns on the same chromosomes, merely by altering the wavelength of the exciting light; or the contrast of an existing single banding pattern may be enhanced; or a completely new pattern may emerge. Clearly the whole subject of the use of fluorochromes in combination is very broad, but fortunately it is possible to reduce it to a relatively simple classification (Table 8.5). Moreover, relatively few of the combinations which have been tried are of significant practical importance, and only two such combinations require more detailed consideration (Sections 8.5.1 & 8.5.2).

Although the number of combinations of fluorochromes with other substances which have been tested is quite large (Table 8.5), the variety of effects produced by such combinations is rather limited. Several fluorochromes showing A + T specificity have been combined with other fluorochromes

Table 8.5 Combinations of fluorochromes used for chromosome banding.

Combination	Effect/application	Reference
Two dyes: Non-fluorescent/fluorescent		
A + T-specific/G + C-specific*		
Methyl green/7-amino-actinomycin D*	Enhanced R-bands	Latt et al. 1980a,b
Crystal violet/chromomycin*	Enhanced R-bands	Schweizer 1981
Distamycin/chromomycin*	Enhanced R-bands and G + C-rich heterochromatin	Schweizer 1981
Methyl green/chromomycin*	Enhanced R-bands and G + C-rich heterochromatin	Schweizer 1976, Sahar & Latt 1978, 1980, Latt et al. 1980a,b
Netropsin/chromomycin*	Enhanced R-bands and G + C-rich heterochromatin	Latt et al. 1980b, Sahar & Latt 1980, Schweizer 1981
Methyl green/coriphosphine*	Enhanced R-bands, plus some C-bands	Schweizer 1981
Distamycin/mithramycin*	NORs[†] enhanced (Amphibia)	Schmid 1980a
Malachite green/mithramycin*	Enhanced R-bands	Schweizer 1981
Netropsin/olivomycin*	Enhanced R-bands, plus some C-bands	Jorgenson et al. 1978
G + C-specific/A + T-specific*		
Actinomycin/DAPI	Enhanced Q-bands	Schweizer 1976
Echinomycin/DAPI*	Enhanced Q-bands, quenching or enhancement of heterochromatin	Schnedl et al. 1980
Actinomycin/Hoechst 33258*	Enhanced Q-bands	Jorgenson et al. 1978, Latt et al. 1980a,b
7-amino-actinomycin D/Hoechst 33258*	Enhanced Q-bands	Latt et al. 1980a
Chromomycin/Hoechst 33258*	Enhanced Q-bands	Latt et al. 1980a,b
Echinomycin/Hoechst 33258*	Enhanced Q-bands, fluorescence of heterochromatin suppressed	Latt et al. 1980b, Schweizer 1981
Actinomycin/quinacrine*	Enhanced Q-bands including heterochromatin. Enhancement of Y-body	Latt et al. 1980a,b; Newburger & Latt 1979
7-amino-actinomycin D/quinacrine*	Enhancement of Q +ve heterochromatin	Latt et al. 1980a
A + T-specific/A + T-/specific*		
Distamycin/DAPI*	Specific fluorescence of certain regions of heterochromatin	See Section 8.5.2
Methyl green/DAPI*		
Netropsin/DAPI*		
Netropsin/Hoechst 33258*		
Distamycin/quinacrine*	Enhancement of interphase Y-body	Klasen & Schmid 1981
Non-specific/A + T-specific*		
Ethidium/quinacrine*	Enhancement of interphase Y-body	Hollander et al. 1976

Table 8.5 (*cont.*)

Combination	Effect/application	Reference
A + T-specific/Non-specific*		
Distamycin/acridine orange*	R-bands	C.C. Lin *et al.* 1980
G + C-specific/non-specific*		
Actinomycin/acridine orange*	Q-bands	C.C. Lin *et al.* 1980
	Bright Q +ve and Hoechst 33258 +ve heterochromatin	Mezzanotte & Ferrucci 1981, 1983
Two fluorescent dyes		
A + T-specific*/A + T-specific*		
Hoechst 33258*/quinacrine*	Distinction and identification of human and mouse chromosomes	Carlin & Rao 1982 (Section 8.3.1)
A + T-specific*/G + C-specific*		
DAPI*/chromomycin*	Complementary R- and Q-banding	Schweizer 1976, 1980
	Base composition determination	Leeman & Ruch 1982, 1983, 1984
Hoechst 33258*/chromomycin*	Complementary R- and Q-banding	Sahar & Latt 1978
	Flow cytometry of chromosomes	Gray *et al.* 1979, Langlois *et al.* 1980
DAPI*/mithramycin*	Complementary R- and Q-banding	Leeman & Ruch 1978
	Base composition determination	Hauser-Urfer *et al.* 1982
DIPI*/mithramycin*	Complementary R- and Q-banding	Schnedl *et al.* 1977c
Hoechst 33258*/mithramycin*	Base composition determination	Hauser-Urfer *et al.* 1982
Quinacrine*/mithramycin*	Base composition determination	Hauser-Urfer *et al.* 1982
Combination of three substances		
Chromomycin*/distamycin/ DAPI*	R-bands plus certain regions of heterochromatin (Section 8.5.2)	Schweizer 1980
Combination of four substances		
Actinomycin/DAPI*/ distamycin/ DAPI*	Sequential Q-banding and heterochromatic bands	Taniwaki 1985
Chromomycin*/distamycin/ DAPI*/actinomycin	Sequential R-, heterochromatic and Q-banding	Taniwaki 1985

*Indicates the fluorescent component of this combination
† Nucleolar organizers

showing G + C specificity, with the result that the different patterns obtained with the individual fluorochromes can be demonstrated on the same chromosomes. Thus a Q-banding or an R-banding pattern can be produced simply by altering the wavelength of excitation of fluorescence. This has proved to be a particularly valuable type of combination, with several different fields of application, and will be described in more detail in Section 8.5.1. The use of a base-specific non-fluorescent substance followed by a fluorochrome showing no base specificity is simply understood also. Actinomycin, a G + C-specific substance, inhibits binding of acridine orange, which can therefore only bind to A + T-rich DNA; a pattern similar to Q-banding therefore results (C.C. Lin et al. 1980). Conversely, if the A + T-specific substance distamycin is used instead of actinomycin, an R-banding pattern is produced.

The commonest situation listed in Table 8.5 is that in which a non-fluorescent substance having a certain base specificity is combined with a fluorescent substance with the opposite specificity. This generally results in the enhancement of the banding pattern produced by the fluorochrome in the euchromatic parts of the chromosome. Thus chromomycin alone produces rather an ill-defined R-banding pattern on human chromosomes, but when the non-fluorescent dye methyl green is used as a counterstain, for example, R-banding of much greater contrast is obtained (Latt et al. 1980a,b, Schweizer 1981). The effects of such combinations on the fluorescence of heterochromatic segments of chromosomes are less consistent, enhancement of fluorescence being seen in some cases, and suppression of fluorescence in others. In many situations the fluorescence of heterochromatic regions, like that of euchromatic bands, is significantly enhanced, but the combination of the bis-intercalating, G + C-specific antibiotic echinomycin with an A + T-specific fluorochrome can suppress the fluorescence of certain heterochromatic segments (Latt et al. 1980b, Schnedl et al. 1980, Schweizer 1981).

It may be wondered how counterstaining with a substance having opposite DNA base specificity can enhance banding patterns. Two distinct mechanisms are involved: binding competition and energy transfer. If two substances were chosen, one of which was bound only to A + T base pairs, and the other to G + C base pairs, little or no competition for binding sites would be expected. In fact, this is not the situation we find when staining chromosomes. The fluorochromes used show preferences for DNA of a particular base composition, and generally have binding sites which occupy at least two and commonly more base pairs, which may contain a mixture of A + T and G + C base pairs. Obviously some binding competition may occur in such situations. In the case of quinacrine, we have already seen that it shows little binding specificity (Section 7.4.1), and its A + T specificity is due to base-specific enhancement of fluorescence. Thus if a G + C-specific counterstain is used with quinacrine, the quinacrine will be displaced from G + C-rich parts of the chromosomes, and remain in the A + T-rich segments, where its fluorescence will be enhanced most efficiently.

Binding competition cannot adequately explain the effect of counterstaining in all cases, and there is both theoretical and observational evidence that the second mechanism, energy transfer, is important in understanding what Schweizer (1981) has called 'counterstain-enhanced chromosome banding.' The effects of energy transfer in chromosome banding have been studied especially by Latt and his colleagues (Latt et al. 1979, 1980a,b, Sahar & Latt 1978, 1980; see also Langlois & Jensen 1979, Langlois et al. 1980). Energy transfer can occur when there is overlap between the fluorescence emission spectrum of one dye (the donor) and the absorption spectrum of another (the acceptor). Energy transfer is highly sensitive to the distance between donor and acceptor, and the spatial separation at which energy transfer is 50% efficient is in the region of 2–5 nm, equivalent to the distance occupied by between 5 and 15 base pairs of DNA. The effect of energy transfer is quenching of donor fluorescence, and, if the acceptor is also a fluorochrome, enhancement of acceptor fluorescence. These phenomena have been demonstrated experimentally and quantified for mammalian chromosomes using various combinations of dyes (Latt et al. 1979, Langlois & Jensen 1979). Combinations which act by energy transfer are listed in Table 8.6. It must be noted that even where energy transfer is the predominant interaction between two dyes, binding competition may also occur, and will tend to reinforce the effects of energy transfer.

The combination of two substances showing the same base specificity might be regarded as an unprofitable stratagem. Indeed, the suppression of quinacrine banding by netropsin (Schlammadinger et al. 1977, Löber et al. 1978) has already been cited as evidence for the A + T specificity of Q-banding (Section 7.4.2). However, we have also seen that not all fluorochromes that show an A + T preference produce identical banding patterns

Table 8.6 Pairs of dyes used for chromosome banding which interact by energy transfer.

Donor	Acceptor	Effect	Reference
Hoechst 33258	7-amino-actinomycin D	Enhanced Q-banding	Sahar & Latt 1978; Latt et al. 1979
Hoechst 33258	Actinomycin	Enhanced Q-banding	Latt et al. 1980a,b
Hoechst 33258	Chromomycin A_3	Enhanced Q-banding	Latt et al. 1980a,b
Hoechst 33258	Ethidium	No change in contrast	Latt et al. 1979
7-amino-actinomycin D	Methyl green	Enhanced R-banding	Latt et al. 1980a,b
Chromomycin A_3	Methyl green	Enhanced R-banding	Sahar & Latt 1978; Latt et al. 1980a, b
Quinacrine	Actinomycin	Enhanced Q + ve heterochromatin	Sahar & Latt 1978; Latt et al. 1980a,b
Quinacrine	7-amino-actinomycin D	Enhanced Q + ve heterochromatin	Latt et al. 1979

(Section 8.3.4). A particularly successful combination of two substances, both of which show specificity for A + T-rich DNA, is the counterstaining of DAPI fluorescence with distamycin (Schweizer et al. 1978). When applied to human chromosomes, the distamycin suppressed most of the DAPI fluorescence, but certain specific heterochromatic regions continued to fluoresce brilliantly. The distamycin/DAPI technique has thus become a valuable method for labelling certain chromosomal regions, not only in man but also in certain other organisms, and is considered fully in Section 8.5.2.

It is possible to use combinations of more than two substances to produce multiple banding patterns on the same chromosomes. Such procedures are useful when close comparison of different patterns is needed, as for example in the precise localization of a chromosome breakpoint, and particularly if material is in short supply. The combination of chromomycin with distamycin /DAPI (Schweizer 1980) permits identification of chromosomes by their R-banding pattern, as well as demonstrating the specific bands that show distamycin/DAPI fluorescence; this combination will be described in Section 8.5.2. Taniwaki (1985) has used combinations of no less than four substances (two fluorescent, two non-fluorescent) to produce a variety of different banding patterns both in euchromatin and heterochromatin (Table 8.5).

8.5.1 Chromomycin/DAPI and similar combinations

The combination of chromomycin, olivomycin or mithramycin (antibiotics showing a preference for G + C-rich DNA) with DAPI or DIPI (which show an A + T preference) has proved particularly useful and versatile. There have been three fields of application of such combinations: demonstrations of Q- and R-bands on the same chromosomes, simply by changing the wavelength of exciting light; measurement of DNA base composition in nuclei and chromosomes; and identification of chromosomes in flow cytometry.

The use of two fluorochromes for demonstrating Q- and R-bands on the same chromosomes was introduced independently by several authors. Schweizer (1976) combined chromomycin with DAPI; Schnedl et al. (1977c) combined mithramycin with DIPI; Leeman & Ruch (1978) used mithramycin with DAPI; and Jorgenson et al. (1978) combined olivomycin and Hoechst 33258. Whatever the combination chosen, the effects on mammalian chromosomes are similar. Excitation with ultraviolet (UV) light excites the A + T-specific fluorochrome, producing a Q-banding pattern, while blue light excitation produces an R-banding pattern due to the chromomycinone antibiotic (Fig. 8.6). Leeman & Ruch (1978) point out that narrow band excitation filters are to be preferred, to ensure that only one fluorochrome is excited at a time. They also point out that staining conditions are more critical when two dyes are used together than when either is used alone, so that a proper balance is achieved. The reader is referred to the papers cited for technical details.

An interesting application of the chromomycin/DAPI combination was

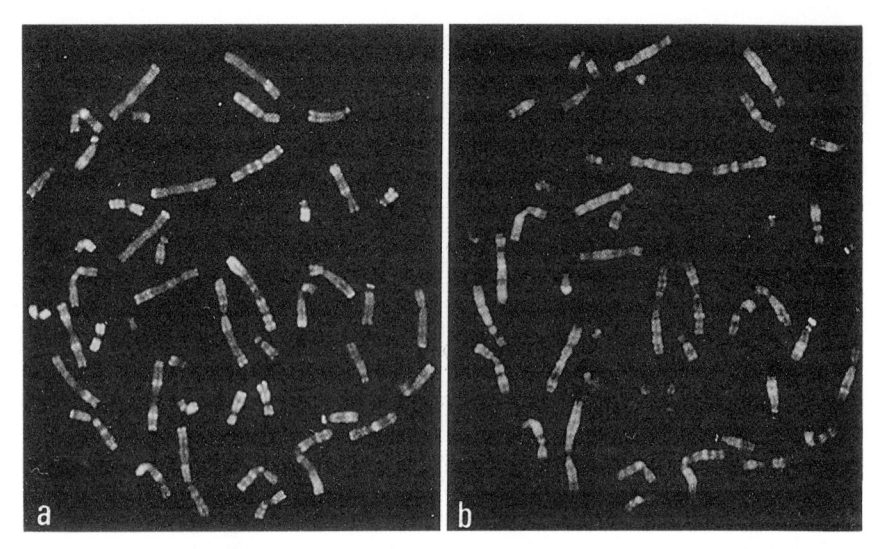

Figure 8.6 Double staining of the same human metaphase chromosome spread showing (left) chromomycin A$_3$ fluorescence, and (right) DAPI fluorescence after counterstaining with actinomycin D. Note that chromosome segments showing bright fluorescence with one fluorochrome are very weakly fluorescent with the other. Micrographs kindly provided by Professor D. Schweizer.

made by Ambros & Sumner (1987), who used this combination to show the equivalence of pachytene chromomeres and Q-bands in human chromosomes. Arguing that the banding patterns produced by these fluorochromes were due to DNA base composition, and would thus not be affected by differences in chromosome structure between meiotic prophase and mitotic metaphase, they showed that it was indeed the chromomeres that generally showed strong fluorescence with DAPI, but not with chromomycin. This provided an objective confirmation of what had previously only been deduced by subjective morphological comparisons. It should be noted, however, that certain chromomeres, mainly terminal ones, showed stronger fluorescence with chromomycin than with DAPI. These terminal chromomeres thus appear to be G + C-rich, and are not equivalent to Q-bands.

In mammalian chromosomes, the heterochromatin generally shows complementary patterns with fluorochromes showing A + T and G + C preference, depending on the base composition of the heterochromatin. This is also largely true of the heterochromatin of plant chromosomes (e.g. Schweizer 1976). For the plant *Anemone blanda*, however, Leeman & Ruch (1978) found a large number of brightly fluorescent intercalary bands with DAPI, and with chromomycin only one or two brightly fluorescent bands; in this case, therefore, the patterns were not complementary.

In a series of papers, Leeman & Ruch and their colleagues (Leeman & Ruch 1982, 1983, 1984, Hauser-Urfer *et al.* 1982) have described the use of chromomycin/DAPI and similar combinations for measuring the DNA base composition of nuclei, chromosomes, and even individual chromosome bands in

COMBINATIONS 181

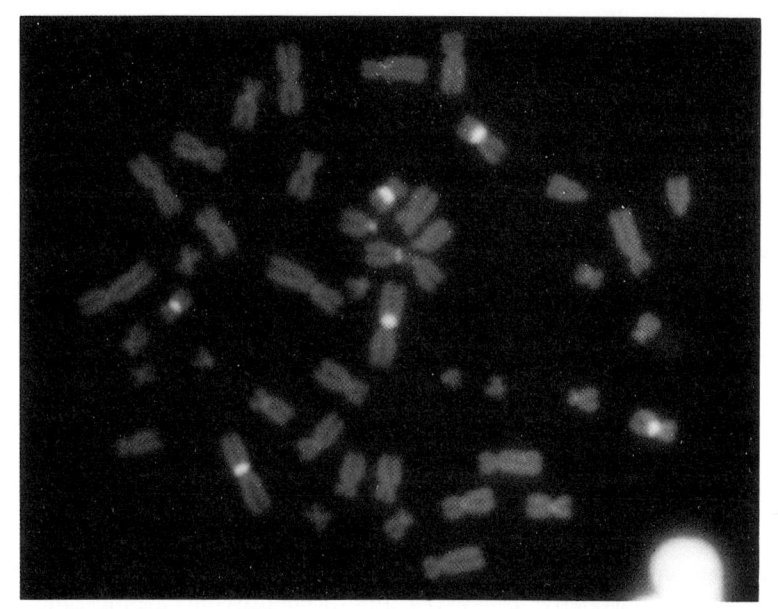

Figure 8.7 Distamycin/DAPI fluorescence of a human metaphase chromosome spread. Note that bright fluorescence is restricted to heterochromatin of chromosomes 1, 9, 15, 16 and Y. Reproduced from Sumner (1983) *Science Progress* **68**, 543–64 published by Blackwell Scientific Publications.

various plant and animal species. The method depends on measuring the ratio of fluorescence of the two fluorochromes in double-stained preparations.

The use of two fluorochromes having different base specificities has become a standard method in the identification of chromosomes by flow cytometry. Certain human chromosomes cannot be distinguished satisfactorily simply on the basis of DNA content, but the added parameter of base composition measurement permits the separation of many of these chromosomes that have similar DNA contents (Gray *et al.* 1979, Langlois *et al.* 1980).

8.5.2 *Distamycin/DAPI (DA/DAPI) staining*

The combination of the non-fluorescent antibiotic distamycin with the fluorochrome DAPI (Schweizer *et al.* 1978) has, as already noted, proved to be one of the most valuable banding methods using a fluorochrome with a counterstain. With this method, certain regions of heterochromatin show strong fluorescence, whereas all other chromosomal regions are faintly fluorescent. In human chromosomes (Fig. 8.7), the distamycin/DAPI-positive regions are the centromeric heterochromatin of chromosomes 1, 9 and 16, the short arm of chromosome 15, and the heterochromatin of the long arm of the Y chromosome (Schweizer *et al.* 1978), although in a few cases DA/DAPI fluorescence has been reported on the short arms of other acrocentric chromosomes (e.g. Bühler & Malik, 1988). The technique has

Table 8.7 Species reported to have distamycin/DAPI-positive segments on their chromosomes.

Species	Reference
Man (*Homo sapiens*)	Schweizer et al. 1978
Gorilla (*Gorilla gorilla*)	Schmid et al. 1986
Chimpanzee (*Pan troglodytes*)	Schweizer 1983
Orangutan (*Pongo pygmaeus*)	Schweizer et al. 1979
Dog (*Canis familiaris*)	Mayr et al. 1983
Wolf (*Canis lupus*)	Mayr et al. 1986
Blue fox (*Alopex lagopus*)	Mayr et al. 1986
Pig (*Sus scrofa*)	Schnedl et al. 1981b
Harvest mouse (*Micromys minutus*)	Schmid et al. 1984a
Duck (*Anas platyrhynchos*)	Mayr et al. 1989
Fish (*Poecilia sphenops*)	Haaf & Schmid 1984
Grasshopper (*Warramaba virgo*)	Schweizer et al. 1983
Grasshopper, 7 species	John et al. 1985

proved particularly useful for investigations of pericentric inversions of chromosome 9 (Gosden *et al.* 1981b), and for identifying chromosome 15 in rearrangements (Chamberlin & Magenis 1980, Ledbetter *et al.* 1980, Buckton *et al.* 1985). Other applications of DA/DAPI fluorescence in humans have been for the identification of specific chromosomes in male meiosis (Haaf *et al.* 1986a), and in the identification of triploidy from the number of fluorescent spots in interphase nuclei (Tommerup & Vejerslev 1985). Distamycin/DAPI-positive bands have also been found in certain chromosomes from other mammals (Table 8.7). Other species of mammals completely lack distamycin/-DAPI bands, however, including the mouse (Schnedl *et al.* 1980, but see Buhariwalla & Blecher 1983), and the red fox (*Vulpes vulpes*) (Mayr *et al.* 1986). The only non-mammalian species from which DA/DAPI bands have been reported are a fish, a bird and certain species of grasshoppers (Table 8.7). A study of chromosomes from a wide variety of plant species failed to discover any DA/DAPI positive bands (Schweizer 1983).

Distamycin/DAPI banding is performed as follows (Schweizer *et al.* 1978):

1 Chromosome preparations are first treated for 5–15 min in a solution of 0.1–0.2 mg of distamycin A/ml of McIlvaine's citrate-phosphate buffer, pH 7.
2 The slides are then rinsed briefly with the same buffer, and stained for 5–15 min with DAPI (0.2–0.4 μg/ml of McIlvaine's buffer, pH 7).
3 Slides are best mounted, after rinsing in buffer, in a 1:1 mixture of glycerol and McIlvaine's buffer (Schweizer 1983).

As with most banding techniques, poor results are obtained if the slides are too fresh or too old.

Since the original technique was published, Schweizer (1980) has added chromomycin staining to the distamycin/DAPI procedure. This permits the identification of the chromosomes by their R-banding patterns, as well as producing DA/DAPI bands on the same chromosomes, merely by altering the wavelength of illumination. Other variations of the DA/DAPI technique have been few. Schweizer (1983) pointed out that the order of using distamycin and DAPI was unimportant, although if distamycin were used second, a lower concentration of this antibiotic was required. Donlon & Magenis (1983), erroneously believing distamycin to be no longer available commercially, used methyl green instead, and obtained essentially similar results. Methyl green is, no doubt, more widely available than distamycin, and also considerably less expensive. Meyne *et al.* (1984) used a netropsin/DAPI procedure for flow cytometry; as with distamycin counterstaining, heterochromatin on human chromosomes 1, 9, 15, 16 and Y showed bright fluorescence.

Neither the precise mechanism of distamycin/DAPI staining, nor the exact nature of the substances stained, are yet understood. It was pointed out at the start that the distribution of DA/DAPI bands in human chromosomes corresponded to those sites known to be rich in 5-methylcytosine (Schweizer *et al.* 1978), but it soon became clear that this could not be a general correlation. For example, mouse satellite DNA is highly methylated, but not DA/DAPI positive (Schnedl *et al.* 1980). The pattern of DA/DAPI fluorescence in human chromosomes also resembles that of segments resistant to digestion by the restriction endonuclease *Hae* III (Babu & Verma 1986a). This enzyme cuts DNA at the sequence GGCC, and resistance to this enzyme is presumably an indication of A + T-richness; this however, has in any case been deduced from the staining properties of DA/DAPI-positive segments, and from the known specificity of DAPI (Schweizer 1983). A correlation between the presence of certain repetitive DNA sequences and DA/DAPI fluorescence in human chromosomes has also been reported (Schwarzacher-Robinson *et al.* 1988).

Both distamycin and DAPI bind preferentially to A + T-rich DNA, and the real question is why some chromosomal regions apparently show a higher affinity for DAPI than for distamycin, and *vice versa*. One approach has been to test different combinations of substances which show preferences for A + T-rich DNA (Schnedl *et al.* 1980, Schweizer 1981, 1983). Some of the successful combinations have been referred to earlier in this section, and in Table 8.5. Schweizer (1983) reported on the affinity for DNA, the A + T binding preference, and the ability to produce DA/DAPI banding of a number of derivatives of DAPI; these observations indicated that in general a high specificity for A + T-rich DNA is necessary if DA/DAPI banding is to be produced with distamycin counterstaining. This alone cannot explain differential staining with distamycin/DAPI. If it is assumed that binding competition is the sole interaction between distamycin and DAPI, then these two substances must differ in some way in their affinity for A + T-rich DNA, and the most plausible explanation is that base sequence is involved. Although extensive information is available on the sequence specificities of a number of substances, including distamycin (e.g. Harshman & Dervan 1985, Klevit *et al.* 1986), no such information appears to be available for DAPI. It is interesting to note that Hoechst 33258, netropsin and distamycin all have similar, but not necessarily identical binding sites. The latter two substances can both be used as counterstains with Hoechst 33258 to produce DA/DAPI banding. It therefore seems likely that if distamycin/DAPI banding does depend solely on DNA base sequence, the method is detecting some quite subtle differences.

8.6 Light-induced banding

Light-induced banding is the induction of banding patterns on chromosomes by photolysis in the presence of a suitable dye, followed by acridine orange or coriphosphine staining (Mezzanotte 1978, Mezzanotte *et al.* 1979, 1981, 1982). Photolysis of DNA in the presence of methylene blue or certain other dyes is known to cause destruction of guanine residues (Simon & van Vunakis 1962); such regions of DNA are effectively denatured, and would fluoresce red with acridine orange or coriphosphine, dyes which fluoresce green with double-stranded DNA but red with single-stranded DNA. This type of procedure has been applied to the chromosomes of a variety of species: *Drosophila* spp. (both polytene and mitotic chromosomes) (Mezzanotte 1978, Mezzanotte *et al.* 1981, 1982), *Culiseta longiareolata* (Diptera) (Mezzanotte *et al.* 1979, 1982), and the laboratory mouse (Mezzanotte *et al.* 1982). Results were consistent with the supposed mechanism of light-induced banding, in that chromosomal regions containing A + T-rich DNA showed green acridine orange fluorescence (that is, their DNA had shown negligible denaturation as it contained little guanine to be destroyed by photo-oxidation), whereas other parts of the chromosomes fluoresced red. In the mouse, where the centromeric heterochromatin is less A + T-rich than much of *Drosophila* heterochromatin, the centromeres fluoresce orange, while the chromosomes are brownish-red. Further evidence that the results of light-induced banding reflect DNA base composition was provided by the experiments of Ferrucci & Sumner (1984). Using nuclei of different DNA base composition, and fluorochromes of different base specificity, they found that the effects of photo-oxidation could be explained largely in terms of destruction of guanine. There was, however, some evidence that photolysis might also be producing DNA-protein crosslinks which might have affected some staining reactions.

Photo-oxidation of chromosomes using the techniques just described has been used to make an elegant test of the hypothesis that the weak quinacrine fluorescence of certain chromosomal regions containing A + T-rich DNA is due to interspersion of guanine residues (Weisblum 1973). Certain regions of heterochromatin in *Drosophila* spp. and the centromeric heterochromatin of mouse chromosomes all show weak quinacrine fluorescence in untreated chromosomes, despite having A + T contents ranging from 65% to 86%. After photo-oxidation, all these regions show bright quinacrine fluorescence (Ferrucci & Mezzanotte 1982).

Although light-induced banding has had only limited use, it is a valuable addition to the armoury of banding techniques, particularly as it produces its base specificity by a mechanism unrelated to that of quinacrine (Ch. 7), and the fluorochromes described earlier in this chapter. Its main disadvantage is lack of sensitivity, since only the most highly A + T-rich chromosomal segments are really distinctively demonstrated.

8.7 Concluding remarks

This chapter has attempted to give a broad view of the use of fluorochromes other than quinacrine for chromosome banding. Many such fluorochromes have valuable practical applications, albeit more restricted than those of quinacrine, or such methods as G- and C-banding, but perhaps the use of fluorochromes for chromosome banding has been most useful for the information it provides about chromosome organization.

Almost without exception, the banding patterns produced by the fluorochromes described in this chapter have been interpreted in terms of DNA base composition. There is overwhelming evidence from studies in solution that the fluorochromes can interact with DNA in a base-specific manner, some showing preferences for A + T-rich DNA and others for G + C-rich DNA. The evidence that chromosome banding is also a result of DNA base specificity also seems very strong. This evidence comes from two sources. First, there is a good correlation between the staining properties of specific chromosomal regions with particular fluorochromes, and the known base composition of these regions, whether heterochromatic or euchromatic. Second, fluorochromes with preferences for A + T-rich and G + C-rich DNA generally give complementary patterns on the same chromosomes. Nevertheless, reasonable doubts have arisen as to whether the patterns produced by fluorochromes, including quinacrine as well as those described in this chapter, can be attributed exclusively, or even predominantly, to DNA base composition. The arguments and evidence have been considered in detail already, and will only be summarized here. The arguments are of many kinds, and it should be recognized that there could be at least a grain of truth in them. It was questioned many years ago (Comings & Drets 1976) whether variations in base composition in the euchromatic parts of chromosomes were adequate to explain banding with certain fluorochromes. Certainly it appears in some cases that the expected differences in average base composition would, at best, produce quite small changes in fluorescence. In other cases, a particular chromosomal region of known base composition has failed to show the expected degree of fluorescence with a particular fluorochrome, of which the archetype is the mouse centromeric heterochromatin, consisting of A + T-rich satellite DNA which nevertheless shows weak quinacrine fluorescence. The existence of distamycin/DAPI fluorescence patterns shows that two fluorochromes with the same general DNA base preference need not have identical patterns of binding to chromosomes. To some extent these anomalies might be explained by the observations that, in some circumstances, proteins can affect the binding of fluorochromes (and non-fluorescent dyes) to chromosomes. The presence of chromosomal proteins that would affect the quantum efficiency of fluorochromes bound to chromosomal DNA is also likely, at least in principle. To what extent proteins actually help to determine chromosome banding patterns is not known. The effect could be small, but significant in specific regions; the possibility of such effects has certainly not been eliminated.

In spite of all these possibilities, there is good reason to suppose that many of the apparently anomalous patterns obtained with fluorochromes are not anomalies at all, because the binding of many fluorochromes to DNA shows a preference for particular base sequences (Dervan 1986, Low *et al.* 1986), and even closely related molecules can show subtle differences in their sequence specificities (Harshman & Dervan 1985). There is good reason to suppose that base sequence preference is important in determining chromosome banding patterns. This was foreshadowed many years ago when Weisblum (1973) suggested that the weak fluorescence of mouse centromeric heterochromatin was due to interspersion with guanine residues. Although the known sequence of mouse satellite DNA did not appear to support this contention, more recent evidence shows that destruction of guanine residues can indeed enhance the quinacrine fluorescence of mouse centromeric heterochromatin (Ferrucci & Mezzanotte 1982). As a more recent example, distamycin/DAPI banding can be most plausibly explained in terms of different base sequence preferences of the two competing substances.

Although the practising cytogeneticist may not be particularly concerned about the precise details of how fluorochromes bind to chromosomes, the matter is not completely academic. If the precise sequence specificities of fluorochromes were known, they could become even more valuable tools in studies of chromosomes than they are at present, and provide a useful complement to restriction endonucleases (Section 12.3). Satellite DNA fractions of similar DNA base composition might, nevertheless, have very different base sequences, and be easily distinguishable by the use of appropriate fluorochromes. This could be of great practical importance in cytogenetics, and the distamycin/DAPI method may possibly be only the first of many possible combinations of this type. Perhaps of even more importance is the possibility of recognizing the specific sequences associated with different aspects of chromosomal organization, such as the TATA boxes and the G + C-rich islands which are associated with different classes of genes (Bird 1986, Nussinov 1987). It does not seem too fanciful to contemplate the future use of better characterized and newly-developed fluorochromes as specific probes for a variety of chromosomal functions.

9 Nucleolar organizers (NORs)

9.1 Nature and organization of NORs and nucleoli

In recent years there has been extensive study of the structure and function of nucleolar organizers (nucleolus-organizing regions, NORs) and nucleoli, and the whole subject has been covered in a number of reviews. The purpose of this briefer account is to provide sufficient background for an understanding of the banding techniques which specifically stain the nucleolar organizer regions of chromosomes, and of the behaviour of these chromosomal segments.

The nucleolar organizers are the chromosomal sites of the genes for ribosomal RNA (rRNA), which is synthesized and processed into pre-ribosomes in the nucleoli, and which ultimately becomes part of the mature ribosomes in the cytoplasm. This account will deal in turn with the organization of the genes for ribosomal RNA, with their sites on the chromosomes, the nucleolar organizers, and with the structure and function of the nucleoli. Historical aspects have been amply covered in several of the reviews on the subject (e.g. Smetana & Busch 1974, Ghosh 1976, Schwarzacher & Wachtler 1983) and will not be reiterated here.

Ribosomes contain a large number of proteins, and four kinds of RNA, which are designated by their sedimentation constants: 5S, 5.8S, 18S and 28S (Wilson 1982). 5S RNA is synthesized by separate genes which do not occupy a structurally differentiated site on chromosomes, and which have not been shown to have any distinctive staining properties. The 5.8S, 18S and 28S RNAs, on the other hand, are all derived from a high molecular weight precursor, 45S RNA. It should be noted that in many lower organisms, the largest ribosomal RNAs are slightly smaller and have rather lower sedimentation constants (Attardi & Amaldi 1970); however, the terms 18S and 28S RNA will be used here, not only for convenience, but also because the main emphasis will be on mammals and other organisms which have rRNA species of this size.

No doubt because ribosomal RNA is by far the most common class of RNA in the cell, and is therefore needed in huge quantities, the rRNA genes are present in multiple copies in all eukaryotes that have been studied in detail (Birnstiel et al. 1971, Long & Dawid 1980). In general, the gene number is roughly correlated with the size of the genome, although large differences can occur between closely related species with genomes of similar size, and there is also good evidence for heteromorphism within a species. A few species have less than a hundred copies of ribosomal RNA genes, but most have a few hundred, and certain plants and amphibia with large genomes have thousands of ribosomal genes (Long & Dawid 1980).

The ribosomal genes consist of a number of blocks of rDNA which alternate with non-transcribed spacers (Long & Dawid 1980, O. L. Miller 1981). As well as the sequences coding for 5.8S, 18S and 28S rRNA, the transcribed region also contains spacers, which are described as external or internal, depending on their position. The typical arrangement of these segments is illustrated in Figure 9.1. The overall size of the transcribed region varies between species, essentially due to variation in the size of the transcribed spacers (Long & Dawid 1980). The non-transcribed spacers also vary considerably in size between different organisms, and even within the same animal (O. L. Miller 1981). Transcription of rDNA is mediated by RNA polymerase I, whereas most genes use RNA polymerase II for transcription, and 5S RNA is transcribed by RNA polymerase III (Sommerville 1986). Another feature of rDNA is that it is often G+C-rich, sometimes very markedly so (Attardi & Amaldi 1970, Birnstiel et al. 1971, O. L. Miller 1981, Wilson 1982). This G+C-richness is largely due to the 18S and 28S sequences themselves, the spacer regions being more variable.

Evidence for an association between nucleoli and specific segments of chromosomes was obtained over 50 years ago. It was found that at telophase new nucleoli were formed at specific sites on chromosomes that appeared as constrictions (Ghosh 1976, Stahl 1982, Schwarzacher & Wachtler 1983, Goessens 1984). Although nucleoli were always associated with the same chromosomal regions, it was not until much more recently that conclusive proof was obtained that these secondary constrictions were the actual sites of the ribosomal genes. This proof was obtained using *in situ* hybridization,

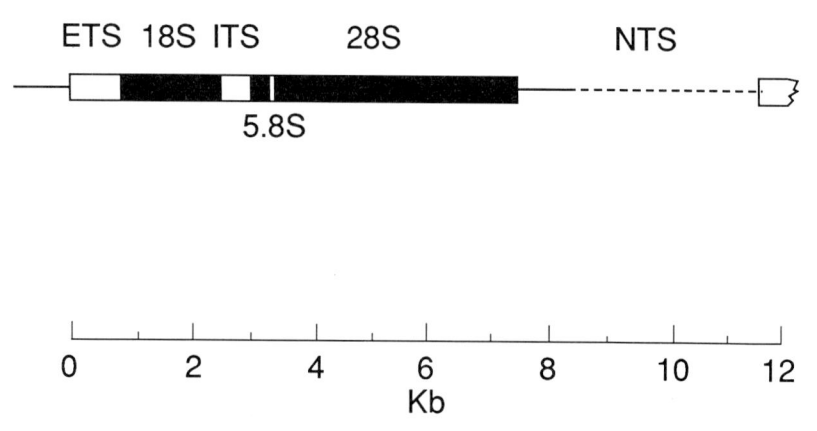

Figure 9.1 Arrangement of the genes for 18S and 28S ribosomal RNA (from *Xenopus laevis*). ETS = external transcribed spacer; ITS = internal transcribed spacer; NTS = non-transcribed spacer. Reproduced from Long & Dawid (1980).

by which means the presence of rDNA in those secondary constrictions that organize nucleoli was established (Henderson *et al.* 1972, Hsu *et al.* 1975). Thus, as stated by Schwarzacher & Wachtler (1983), the definition of a NOR as: 'a specific region of a chromosome which causes the formation of a nucleolus in interphase,' while still perfectly true, can be reworded in the light of modern research to state that: 'The NOR is that region of a chromosome which contains the main rRNA genes (18S and 28S rDNA).' It must be stressed that not every secondary constriction on chromosomes (that is, constrictions other than the centromere, or primary constriction) is a NOR. Other secondary constrictions, for example, consist of inactive heterochromatin. There is also some evidence that not all sites of rDNA appear as secondary constrictions, and that in some cases the NOR contains more condensed chromatin (Goessens 1984). Long & Dawid (1980) published extensive data on the chromosomal location of NORs in a wide variety of organisms, mostly animals, determined by *in situ* hybridization or by genetic means.

The structure and function of the nucleolus itself have been described in numerous recent reviews (Smetana & Busch 1974, Ghosh 1976, Goessens & Lepoint 1979, Stahl 1982, Schwarzacher & Wachtler 1983, Goessens 1984, Fakan & Hernandez-Verdun 1986, Sommerville 1986, Jordan 1987). It is clear that the nucleolus is quite a complicated and variable organelle, consisting of several components, the agreed nomenclature of which has been summarized by Jordan (1984). Nevertheless, its basic plan and the functional relations of its components have now been worked out and are fairly straightforward. A diagrammatic representation of the nucleolus is shown in Figure 9.2. Fibrillar centres are areas of low electron density in nucleoli that contain rDNA and RNA polymerase I. There is, however, little evidence for transcription or the presence of RNA in fibrillar centres. In fact, transcription of ribosomal genes apparently occurs in the dense fibrillar component which lies around the fibrillar centres. The outer part of the nucleolus consists of pre-ribosomal particles, 15 nm in diameter, which are the initial stage of packing of the ribosomal RNA. The outer surface of nucleoli often has clumps of condensed chromatin on it, which appear to be connected with the DNA of the fibrillar centre.

During mitosis, RNA synthesis, including synthesis of rRNA, ceases, and the nucleolus normally disappears. At prophase, the dense fibrillar component disappears, and the granular component disperses into the nuclear sap; at metaphase, however, some of the granular material may become attached to the chromosomes. In some cases this nucleolar material forms a sheath round the chromosomes (Paweletz & Risueño 1982). The material of the fibrillar centres can be recognized in some cell types throughout mitosis, always attached to the chromosomes. At telophase, new nucleoli are formed at specific loci on chromosomes. First, the dense fibrillar component appears adjacent to the fibrillar centres, and the granular component appears shortly afterwards.

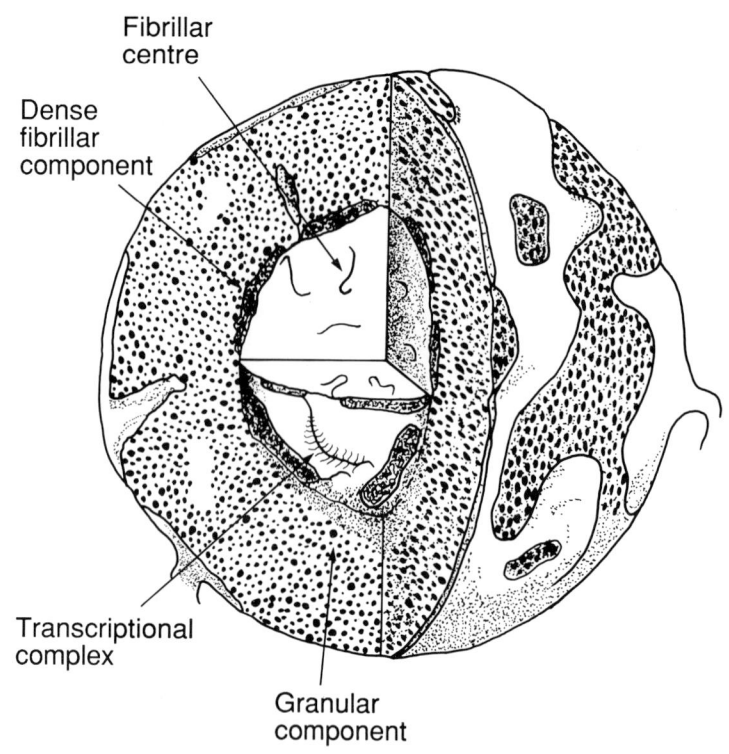

Figure 9.2 Diagram of a nucleolus. Redrawn from Goessens & Lepoint (1979) by permission of the authors and Société francaise de Microscopie éléctronique.

Table 9.1 Properties of some nucleolar proteins (after Sommerville 1986).

Protein	Molecular weight (KDa)	Function	Location Nucleolus	Mitosis
RNA polymerase I	190, 120 65, 42, 25 (5 subunits)	Transcription of rDNA	Fibrillar centres	NOR
C23/nucleolin	110/100	?Regulation of chromatin conformation & transcription (Jordan 1987)	Dense fibrillar component	NOR
Structural	180	Structural	Dense fibrillar component	Dispersed
B23	37	?rRNP processing	Granular component	Dispersed
Structural	145	?Storage & transport	Cortex	Dispersed
Ribocharin	40	Nucleocytoplasmic transport	Granular component	Chromosome surfaces
S1	45	Ribosome structure	Granular component	Chromosome surfaces

As well as nucleic acids, the nucleoli contain a large number of protein components (Busch *et al.* 1982), only a few of which have been characterized in any detail. These include structural proteins, and proteins which appear to be involved in nucleo-cytoplasmic transport, as well as proteins more directly involved in the synthesis of rRNA and production of ribosomes. The location of these proteins in interphase and during mitosis, as well as some other of their properties, are listed in Table 9.1. See Sommerville (1986) for a recent summary of information on nucleolar proteins.

9.2 Methods for selective staining of NORs

We have seen (Section 9.1) that in mitosis the RNA of the nucleolus becomes dispersed, and although some of it may become attached to the chromosomes, it is no longer associated specifically with the NORs. Specific staining of the NORs must therefore depend either upon DNA or protein. In fact, although NORs are ultimately defined by their DNA sequences, no practical NOR staining technique is based on properties of DNA. Schweizer (1976) reported that NORs of certain plant chromosomes could be selectively demonstrated with chromomycin or mithramycin fluorescence, and since then there have been several reports of the staining of NORs with G+C-specific stains in a wide variety of organisms (Deumling & Greilhuber 1982, Schmid 1980a, Schweizer *et al.* 1983, Fox & Santos 1985, Cau *et al.* 1988). However, such methods would not distinguish between NORs and other G+C-rich chromosomal segments, such as certain blocks of heterochromatin (Tables 4.1 & 8.4). Gagné (1981) showed that the NORs of human chromosomes were particularly resistant to digestion with DNase I, but again this was not a practical technique as the rest of the chromosome structure was extensively damaged.

Practical methods for NORs depend upon protein staining. The first such method was N-banding (Matsui & Sasaki 1973). Giemsa staining after extraction of histones and nucleic acids resulted in strong staining only at known sites of rDNA. This procedure will be considered in detail in Section 9.3. Two years later it was discovered (or rather rediscovered, see Section 1.2) that NORs could be stained specifically with silver (Howell *et al.* 1975, Goodpasture & Bloom 1975). Silver methods have now become the preferred procedures for NOR staining, and their use, mechanism and applications are described in detail in Section 9.4. Wang & Juurlink (1979) described a method for staining NORs with an acidified methanolic solution of the protein stain Coomassie brilliant blue. Although the procedure was simple and used a stable staining solution, the contrast of the staining was much lower than that obtainable with silver, and the method has only been used occasionally (e.g. Czaker & Mayr 1982). Haaf *et al.* (1988b) have used immunofluorescence with antibodies to RNA polymerase I to label NORs.

9.3 N-banding (Giemsa staining of NORs)

The original N-banding technique of Matsui & Sasaki (1973) involved extraction of nucleic acids and histones from chromosomes, followed by Giemsa staining. In a variety of species, N-bands were reported at presumed sites of NORs, although in human chromosomes they believed the satellites, rather than the secondary constrictions, were the sites of the staining. Faust & Vogel (1974) doubted whether the N-bands corresponded to the NORs themselves, but instead believed them to represent a specific kind of heterochromatin located adjacent to the NORs. However, using a modified technique, the N-bands were shown to correspond to the known sites of NORs in several species, including the satellite stalks (secondary constrictions) in human chromosomes (Matsui 1974, Funaki *et al.* 1975).

With the introduction of silver staining for NORs (Section 9.4), comparisons were made. In human chromosomes the same sites were stained by both methods (Archidiacono *et al.* 1977) although Taylor & Martin-de Leon (1980) found that N-banding was less sensitive, and could not be obtained with old slides. In the chromosomes of the plant *Vicia faba*, Schubert *et al.* (1979) showed that N-banding was less specific for NORs than silver staining, and indeed it has been shown in many plants and some insects that N-banding procedures stain distinctive types of heterochromatin not associated with NORs (Section 4.5). This inevitably reduces the value of N-banding as a method for demonstrating NORs, in which role it has, in any case, been largely superseded by silver staining.

9.3.1 Methods for N-banding

Matsui & Sasaki (1973) used successive extractions with trichloracetic and hydrochloric acids to extract nucleic acids and histones to leave only N-bands stainable with Giemsa. Chromosome preparations were treated as follows:

1. 5% trichloracetic acid at 85–90°C for 30 min, followed by a brief rinse in tap water.
2. 0.1N hydrochloric acid at 60°C for 35–40 min, followed by thorough rinsing with tap water.
3. Giemsa, diluted 1:10 with phosphate buffer, pH 7.0, for 60 min.

As an alternative to the trichloracetic acid treatment, successive digestions with DNase and RNase could be used (see also Gagné 1981).

Funaki *et al.* (1975) described an 'improved' method for N-banding which consisted of incubation in 1M sodium dihydrogen phosphate, pH 4.2, for 15 min at 96°C, followed by staining for 20 min in Giemsa, diluted 1:25 with 1/15M phosphate buffer, pH 7.0. Similar procedures have been used not only for NOR staining, but also for heterochromatin staining by N-banding methods (Section 4.4).

9.3.2 The mechanism of N-banding

The original method of N-banding, based on procedures known to extract nucleic acids and histones, indicated that the Giemsa-stainable substance was non-histone chromosomal protein, a conclusion supported by cytochemical tests (Matsui & Sasaki 1973, Matsui 1974, Funaki *et al.* 1975). Matsui (1974) showed that after the complete N-banding procedure, all the DNA was extracted from nuclei, and all but 6% of the proteins. Subsequent work has attempted to define these resistant non-histones in more detail.

Buys & Osinga (1984) prepared extracts of nucleolar proteins, subjected them to gel electrophoresis, and stained the gels with Giemsa. Four bands, corresponding to molecular weights of 29, 37, 78 and 104 KDa, were found to stain with Giemsa, and also with silver (Section 9.4). These authors presented evidence that these proteins stained with Giemsa because they were phosphorylated, and not because they had DNA and RNA bound to them. Using a similar approach Matsui *et al.* (1986) obtained rather different results. These authors identified three non-histones which stained with Giemsa: component A of 88 KDa, B of 55 KDa, and C of 41 KDa. Component A was found to be acid soluble, and was therefore eliminated as a likely N-band protein. Both component B and component C were enriched in nucleoli and firmly bound, and showed strong and specific binding to rDNA. An antibody to component C, when used for immunofluorescence, labelled NORs, and the same antibody suppressed transcription of rDNA in vitro by RNA polymerase I. Component C was found to be a very abundant protein in relation to the amount of rDNA.

Obviously there are substantial discrepancies between the results of these two groups, in spite of their similar approaches: each reports a completely different set of Giemsa-stainable proteins. It should be emphasized that Matsui *et al.* (1986) characterized their component C in much more detail, and found it to have many properties appropriate to an N-band protein: Giemsa staining, specific binding to rDNA, localization at NORs, and the ability of an antibody to the protein to suppress rDNA transcription. Some of these properties would not necessarily be required of a NOR protein, but would hardly be expected of a protein not associated with NORs. In spite of this, the identity of N-banding proteins cannot yet be regarded as having been established with certainty.

9.4 Silver staining of NORs (Ag-NOR staining)

The staining of nucleoli with silver has been known for about 90 years (Section 1.2), and the silver staining of NORs on chromosomes was observed 70 years ago (Carleton 1920), although it was not until 1975 that the regular application of silver staining to demonstrate NORs on chromosomes began. It is not clear to what extent the new techniques for staining metaphase chromosomes were derived from the older ones, but there can be no doubt that a new field has been opened up by the application of silver staining to chromosomes. This field includes not only mitotic and meiotic chromosomes, but also the nucleoli of interphase nuclei at the level of both the light and electron microscopes. Silver staining for NORs has thus become an essential addition to cytogenetic methods.

Silver staining has been used to demonstrate numerous different structures in cells and tissues. In chromosomes, besides NORs, the following structures can be stained: heterochromatin (Section 4.6), decondensed heterochromatin (Haaf *et al.* 1984), histones (Black & Ansley 1964, 1966), chromosome cores (Howell & Hsu 1979, Kaiserman & Burkholder 1980), the synaptonemal complexes of pachytene chromosomes (Pathak & Hsu 1979, Fletcher 1979) and kinetochores (Section 10.2.2). One might wonder whether silver staining could be relied upon to give specific staining of NORs, or whether there would always be a degree of uncertainty about the nature of the structure being stained. While it is true that the only certain method for identifying sites of NORs is *in situ* hybridization to ribosomal genes, silver staining, given proper methodology and interpretation, is nevertheless remarkably specific for NORs.

9.4.1 Silver staining methods

Silver staining methods for NORs (Ag-NOR staining) have evolved from complexity to simplicity, at the same time improving in reliability and cleanliness. The original ammoniacal silver–satellite (AS-SAT) technique of Howell *et al.* (1975) involved several stages: fixation of the chromosome preparations with formalin, treatment with an ammoniacal silver prestaining solution, a treatment with dilute sodium hydroxide, and then finally staining with a mixture of formalin and ammoniacal silver. The Ag-ammoniacal silver (Ag-AS) technique of Goodpasture & Bloom (1975), published in the same year, involved pipetting 50% silver nitrate on to the chromosome preparation, covering it with a coverslip, and exposing the slide to a photoflood lamp. Subsequently the chromosomes were exposed to a mixture of ammoniacal silver and formalin. Development of the staining had to be monitored under the microscope. The same authors later published a variant of this Ag-AS technique in which the chromosomes, covered with silver nitrate solution, were exposed to warmth rather than light (Bloom & Goodpasture 1976). More importantly, they found that NOR staining could be obtained by incubation in 50% silver nitrate without any subsequent treatment (the Ag-I technique). The method of Denton *et al.* (1976) is very similar to the original Ag-AS method of Goodpasture & Bloom (1975), although they do remark that the slide is heated during exposure to the photoflood lamp. Thus it may be, that in spite of the well known sensitivity to light of silver salts, the heating effect of lamps may be their main mode of action.

Although the original Ag-I method of Bloom & Goodpasture (1976) was not particularly reliable, similar methods which only involve incubation of the slides in silver nitrate have been published (Ved Brat *et al.* 1979, Brown & Loughman 1980). Hizume *et al.* (1980) and Sato (1985) used the method successfully on plant chromosomes, the latter paper including ultrastructural observations. Kodama *et al.* (1980) used a piece of nylon cloth to cover the specimen after 50% silver nitrate had been added. This was to produce cleaner preparations without the deposition of silver particles over them. Sozansky *et al.* (1984), instead of using heat, irradiated the slides, covered with silver nitrate solution, with a UV lamp, while Münke & Schmiady (1979) used both heat and light simultaneously.

An alternative approach to silver staining of NORs was pioneered by Olert (1979), who mixed 50% silver nitrate solution with dilute formic acid, and incubated the slides in this solution. This procedure was modified by Howell & Black (1980), who made a colloidal developer by adding formic acid to a dilute gelatin solution. This colloidal developer was mixed with the silver nitrate solution immediately before use, and the slides incubated on a hotplate. A detailed protocol for this very successful method is given below. A similar procedure, but using a lower temperature, was employed by Gold & Ellison (1983), who also treated their slides with sodium thiosulphate to stabilize the silver staining. Methods of this type have been used for plant chromosomes by Mehra *et al.* (1985) and for ultrastructural studies by Ploton *et al.* (1984).

In spite of the introduction of simplified methods, many authors continue to use variants of the Ag-AS method, in which an initial incubation in silver nitrate is followed by treatment with an ammoniacal silver-formalin mixture. Such procedures have commonly been used for ultrastructural work (Hernandez-Verdun *et al.* 1980, Angelier *et al.* 1982, Medina *et al.* 1986), but also for light microscopy of animal chromosomes (Fernandez-Gomez *et al.* 1983). In the last paper, interesting modifications were the use of ammoniacal silver carbonate instead of ammoniacal silver formalin, and subsequent stabilization with thiosulphate.

Various authors have studied the effects of pretreatments on Ag-NOR staining, usually with the intention of improving the specificity of staining, and eliminating non-specific deposition of silver. Ferraro *et al.* (1982) found potassium cyanide useful for this purpose, while Thiebaut *et al.* (1984) used Schiff's reagent. Gosálvez *et al.* (1986) used a pretreatment to obtain a different effect. They incubated preparations of grasshopper chromosomes in 2 × SSC at 60°C for 15 min, and found that this not only improved the silver staining of the 'main' nucleoli, but also revealed the presence of other silver staining material that the authors describe as micronucleoli. This is an interesting variant of Ag-NOR staining, but it has not yet been shown that the 'micronucleoli' contain ribosomal nucleic acid sequences.

The method of Howell & Black (1980) for silver staining of NORs is performed as follows. Standard chromosome preparations are used, preferably within two or three days after they were made. Older preparations accumulate dust and other contamination which stain with silver. Two solutions are used:

(1) Colloidal developer: 2 g of gelatin is dissolved in 100 ml of deionized water, with stirring and gentle heating. When the gelatin is dissolved, 1 ml of formic acid is added. This solution is stable for about two weeks.
(2) Silver nitrate solution: 4 g of silver nitrate is dissolved in 8 ml of deionized water. This solution is stable indefinitely in the dark.

To stain NORs, 2 drops of the colloidal developer and 4 drops of the silver nitrate solution are mixed, and pipetted on to the chromosome preparation. The mixture is covered with a coverslip, and the slide placed on a hotplate pre-heated to 70°C. The solution turns yellow and then golden brown. At this stage (1–2 min) the coverslip should be washed off and the slide rinsed thoroughly with a generous stream of deionized water. The slide can then be blotted dry and mounted.

Chromosomes stained in this way appear light yellow, with the NORs appearing as dense black dots. Occasionally other structures are more darkly stained than the chromosome arms, such as centromeric heterochromatin (for example, in mouse chromosomes), and chromosome cores, but these can be distinguished easily from NORs by their form, and by their colour, which is generally yellowish-brown rather than black.

Ag-NOR stained chromosomes may be viewed by transmitted light, but the yellow colour of the arms can be difficult to see, and the outline not clear. Visibility can be improved by using phase-contrast microscopy, in which case the NORs may appear as bright bodies. Ploton *et al.* (1986) have discussed the use of various optical methods for viewing Ag-NOR stained preparations. It is probably simpler to use a counterstain, such as Giemsa, so that the chromosomes can be seen easily by transmitted light; a few minutes staining with a dilute Giemsa solution should be adequate. Excessive staining makes the NORs difficult to see. A counterstained Ag-NOR preparation is illustrated in Figure 9.3.

It is often essential to be able to identify the chromosomes that bear NORs, and to this end Ag-NOR staining has been combined with G-, Q-, R- and C-banding. The procedure for identifying the chromosomes may be carried out before or after the Ag-NOR staining, and the results of the two procedures may be examined simultaneously or successively. In general it is more convenient if the Ag-NOR staining can be seen at the same time as the other banding method, otherwise it is necessary to photograph and restain the same chromosome preparations. Tantravahi *et al.* (1977) studied the possibility of silver staining after Q-, C-, G- and R-banding, and were successful with all combinations, including a Q-banding/C-banding/Ag-NOR staining sequence. However, silver staining was reduced after G-banding treatments, and completely abolished by prolonged trypsin digestion. Verma *et al.* (1981) also obtained good Ag-NOR staining after using a fluorescent R-banding technique. Some workers have destained the chromosomes before silver staining (Bloom & Goodpasture 1976, Tantravahi *et al.* 1977, Romagnano *et al.* 1987), but this does not seem to be necessary. With these methods the banded chromosomes must be photographed before silver staining, which is a disadvantage. If the chromosomes are banded for identification after the silver staining, both the banding and the Ag-NOR staining are visible simultaneously. The banding technique generally has to be modified to take account of the effects of silver staining on the chromosomes. 'Post-banding' has been done successfully with G-banding (Zankl & Bernhardt 1977, Howell & Black 1978, Czaker & Mayr 1980), R-banding (Dipierri & Fraisse 1983a), Q-banding (Lau *et al.* 1978, Markovic *et al.* 1978) and various other fluorescence techniques (Buys *et al.* 1978, Wachtler & Musil 1980). Mikelsaar & Schwarzacher (1978) removed the silver from their preparations before restaining with quinacrine mustard,

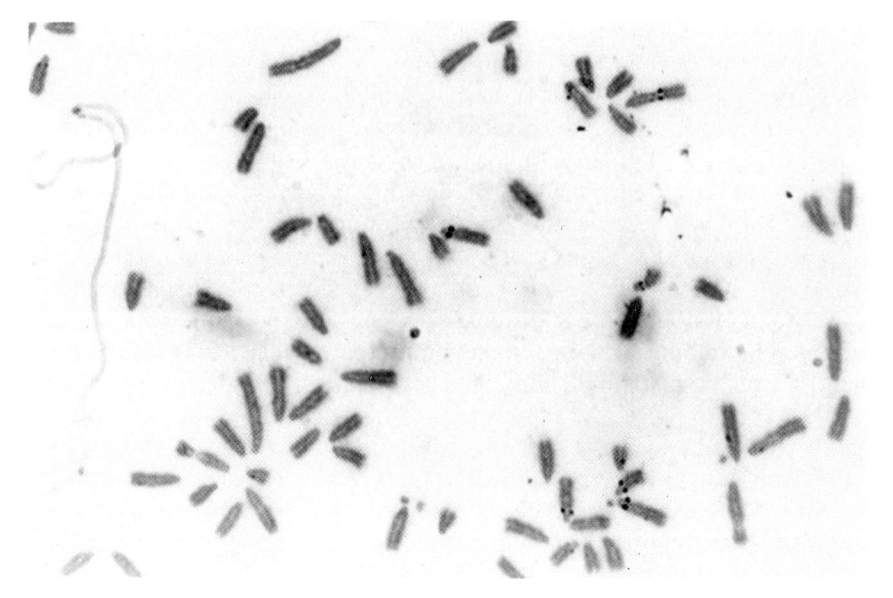

Figure 9.3 Ag-NOR-stained metaphase chromosome spread from a mouse cell line, counterstained with Giemsa.

but in fact it is possible to observe both the fluorescent bands and the silver-stained NORs simultaneously using a combination of epi-illumination and transmitted light. Simultaneous demonstration of NORs and C-bands has been obtained by first carrying out the C-banding pretreatments, then silver staining to reveal the NORs, and finally staining with Giemsa to show up the C-bands (Rufas *et al.* 1983b, Tuck-Muller *et al.* 1984).

9.4.2 *The nature of the silver-stained material of NORs*

Four aspects of the nature of the silver stained material of NORs need to be considered: the functional, and the morphological aspects of the staining; the identity of the macromolecule or macromolecules which are stained by silver; and the precise chemical grouping responsible for the reduction of silver salts to metallic silver in the NORs. All these aspects have been subject to a good deal of investigation, but as so often in the field of chromosome banding, we have at present only partial or conflicting answers.

It has become clear that silver staining does not reveal all sites on chromosomes that contain rDNA genes. Studies on a variety of species showed that sites of Ag-NOR staining on chromosomes corresponded to sites of ribosomal genes identified by *in situ* hybridization (Goodpasture & Bloom 1975, Tantravahi *et al.* 1976, Warburton & Henderson 1979, Schubert & Wobus 1985). However, in humans for example, it is very rare for all ten sites of rDNA to be labelled with silver, although this could be because some sites contain rather few rDNA genes, below the threshold of sensitivity of silver staining. Clear evidence for the absence of silver staining although ribosomal genes were present came from studies of human-mouse somatic hybrid cells, in which transcription of the human ribosomal genes was suppressed, and Ag-NOR staining lacking on the human chromosomes (D. A. Miller *et al.* 1976). It was shown that the human acrocentric chromosomes, which were not silver stained in the hybrid cells, showed Ag-NOR staining in the parental human cells. The same group also obtained evidence that in hybrid cells, Ag-NOR staining was suppressed on chromosomes of the species whose chromosomes tended to be selectively eliminated from the hybrid cell (O. J. Miller *et al.* 1976). Other evidence for an association between Ag-NOR staining and ribosomal gene activity comes from various sources. The amount of silver staining of interphase nucleoli is much greater than that of the metaphase chromosomes in the same preparation, which may be taken to reflect the suppression of transcription at mitosis. During vertebrate spermatogenesis, there are substantial changes in Ag-NOR stainability which are well correlated with changes in patterns of ribosomal RNA synthesis (Hofgärtner *et al.* 1979, Schmid *et al.* 1982, 1983b); similar results have also been reported for human oogenesis (Hartung *et al.* 1979). De Capoa *et al.* (1983) showed that in the embryonic development of the amphibian *Xenopus laevis*, Ag-NOR staining appeared at the same time as the start of embryonic rRNA synthesis, while Morton *et al.* (1983) showed a direct correlation between the amounts of Ag-NOR staining of human chromosomes and synthesis of 18S and 28S rRNA.

All the above observations indicate that Ag-NOR staining is an indicator of ribosomal gene activity (Hubbell 1985), and indeed it has often been used on this understanding (Section 9.4.3). Results from experiments to test this hypothesis are, however, somewhat ambiguous. Various substances known to inhibit transcription of ribosomal genes (actinomycin D, mercuric chloride, ethidium bromide and cordycepin) were found to have no significant effect on Ag-NOR staining (Raman & Sperling 1981, Verschaeve et al. 1981, Sanchez-Pina et al. 1984) although MPB, an inhibitor of RNA synthesis, did reduce silver staining (Hubbell 1985). Such treatments are scarcely natural, and Raman & Sperling (1981) found reduction of Ag-NOR staining in contact-inhibited cells, and complete loss of staining in long-quiescent lymphocytes. Of course, it is still possible to obtain Ag-NOR staining on metaphase chromosomes, in which transcription is suppressed, so it is clear that silver staining does not disappear immediately when transcription stops. The experiments using inhibitors of transcription may merely indicate that complete loss of Ag-NOR staining is a slow process, although large variations apparently occur during a single cell cycle. Baldini et al. (1988) obtained an increase in rRNA synthesis in cultured cells by treating them with growth hormone, an increase which was correlated with increased Ag-NOR staining. Ferraro & Lavia (1983, 1985) treated cells with 5-azacytidine and obtained an increase in Ag-NOR staining. The rationale of these experiments was that 5-azacytidine is incorporated into DNA at replication, suppresses methylation of DNA, and thereby permits a greater level of transcription. These authors' results were, therefore, in accordance with expectation, although they did not actually demonstrate an increase in rRNA synthesis. Moreover, 5-azacytidine substitution can have substantial effects on chromosome morphology, including decondensation (Section 11.2.3). In this connection it is worth noting the observation of Haaf et al. (1984) that decondensed non-nucleolar chromosome segments can become stainable with silver. In spite of all these uncertainties, which really centre around details of the mechanism of Ag-NOR staining, there seems to be little doubt that, in normal circumstances, the amount of silver staining is correlated with the degree of transcription of ribosomal genes. The presence of silver may, however, indicate the potential for transcription rather than the actual occurrence of transcription (Medina et al. 1986).

Schwarzacher et al. (1978) have clearly shown that the sites of Ag-NOR staining associated with chromosomes are not actually within the chromatin itself, but lie at the sides of the chromatids (Fig. 9.4). In interphase, the silver staining occurs largely in the fibrillar centres (Hernandez-Verdun et al. 1980, Fernandez-Gomez 1983), although staining of the dense fibrillar component can apparently also occur (Ploton et al. 1984, Fakan & Hernandez-Verdun 1986). Thus the silver-stainable material appears to be largely in a transcriptionally inactive part of the nucleolus (the fibrillar centre, see Section 9.1),

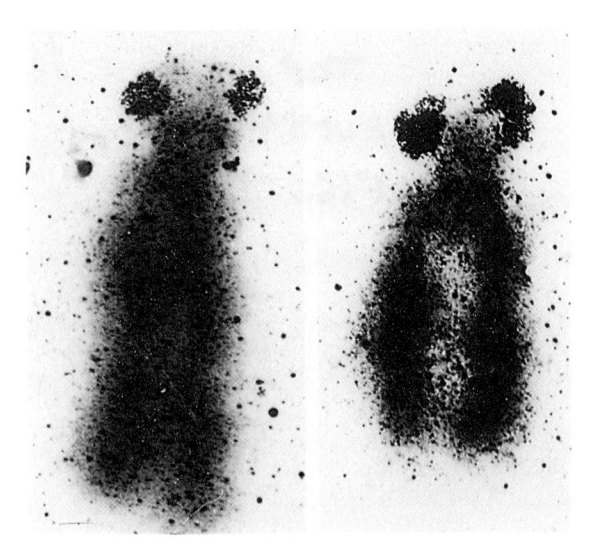

Figure 9.4 Electron micrographs of Ag-NOR-stained human metaphase chromosomes, showing the silver stained material lying at the sides of the chromatids, and not within them. Reproduced from Schwarzacher *et al.* (1978) by permission of the authors and S. Karger AG.

and less regularly in the dense fibrillar component, which appears to be the site of transcription. The site of silver staining has been investigated at very high resolution by Angelier *et al.* (1982), who applied Ag-NOR staining to spreads of transcriptional units from nucleoli. In such preparations, the silver was deposited exclusively on the transcribed part of the rDNA and not on the untranscribed spacer regions. Within the transcribed regions the silver was located on the DNP axis rather than on the newly synthesized RNP fibrils. Labelling with silver did not occur on non-nucleolar transcriptional units. These results clearly indicate an association between Ag-NOR proteins and transcription of rDNA, a point to which we shall return below.

Early work on Ag-NOR staining indicated that the substance to which the silver bound was non-histone protein. This is still true, in spite of dissenting views (Clavaguera *et al.* 1983), and the knowledge that silver can be used to stain DNA as well as proteins on electrophoresis gels (Sommerville & Wang 1981). Treatments which extract nucleic acids from chromosomes have no effect on Ag-NOR staining, while treatment of chromosome preparations with proteases quickly prevents Ag-NOR staining (Howell *et al.* 1975, Goodpasture & Bloom 1975, Schwarzacher *et al.* 1978, Olert *et al.* 1979, Buys & Osinga 1980a). Attempts have been made to identify and characterize these proteins by separating nucleolar proteins on gels, which are then stained with silver as for Ag-NOR staining. Busch and his colleagues have studied two nucleolar proteins which they have named B23 and C23. Both stain with silver after an Ag-NOR technique, and they have molecular weights of 37 KDa and 100 KDa respectively (Lischwe *et al.* 1979). Antibodies to these proteins were prepared, and used to determine the sites of the proteins in cytological preparations. C23 was found in the fibrillar centres of interphase nucleoli, and in the metaphase NORs, whereas B23 was located peripherally in the nucleolus (Lischwe et al. 1981, Ochs et al. 1983, Smetana et al. 1984, Spector et al. 1984). Ochs & Busch (1984) showed that antibody against C23 would block silver staining of NORs and nucleoli, whereas antibody against B23 had no effect on silver staining. This group therefore believed that C23 was the Ag-NOR protein. Olson et al. (1983) reported that C23 appeared to have a preference for binding DNA in the non-transcribed spacer regions of rDNA (an observation not consistent with the results of Angelier et al. 1982, see above), while Herrera & Olson (1986) found that C23 appeared to be associated with nascent preribosomal RNA.

Hubbell et al. (1979) also found the silver-binding proteins B23 and C23 in nucleoli. In their experiments, however, the nucleoli could still be stained with silver after extraction of these two proteins, and they attributed the genuine Ag-NOR staining to a third protein which, however, they did not characterize further. Pfeifle et al. (1986) also extracted a nucleolar protein similar to C23, but which they called pp105. Although pp105 bound both Giemsa and silver, its distribution in chromosomes and nucleoli did not correspond to that expected for the Ag-NOR protein; another protein, pp135, seemed to have more appropriate characteristics. Williams et al. (1982) did not find any silver-stainable proteins similar to C23 in their nucleolar extracts; instead, there was a pair of proteins (MW = 190 and 195 KDa) which bound silver. The authors speculate that this pair of proteins could represent the large subunit of RNA polymerase I, which is expected to be present in active nucleoli. Unfortunately no further work on this point appears to have been published. Finally, Buys & Osinga (1984) reported the presence in nucleolar extracts of four proteins, of 104, 78, 37 and 29 KDa, that stained with both Giemsa and silver. The proteins of 104 KDa and 37 KDa could well be equivalent to C23 and B23 respectively, and the authors speculate that their 29 KDa protein might be equivalent to the Ag-NOR protein of Hubbell et al. (1979). The identity of the 78 KDa protein could not be established, and no silver-staining protein equivalent to RNA polymerase I was present.

It must be clear that there is at present no unanimity on the identity of the silver-stainable proteins in NORs. C23 and similar proteins are strong candidates, but at the same time there are several other possible proteins, which so far have not been adequately characterized. C23 has most of the right properties, except that it seems to be associated with the non-transcribed spacers, whereas Angelier et al. (1982) found silver deposits associated with transcriptional units. Here again there seems to be an anomaly, since silver staining is associated above all with the transcriptionally inactive fibrillar centres of the nucleoli, regions which nevertheless contain RNA polymerase I (Scheer & Rose 1984).

The chemical groups responsible for Ag-NOR staining have not yet been identified with certainty. Most of the nucleolar proteins regarded as substrates for silver staining are phosphoproteins (Lischwe et al. 1979, Buys & Osinga 1984, Pfeifle et al. 1986). This idea is supported by observations that

Ag-NOR staining can be decreased or abolished by digestion with phosphatases (Satoh & Busch 1981, Hubbell 1985). Inevitably there are dissenting views. Olert *et al.* (1979) proposed that the silver bound to carboxyl groups, not to phosphates. This hypothesis was supported by Buys & Osinga (1984), who nevertheless believed that the binding of Giemsa to the same nucleolar proteins was due to phosphate groups. Another possibility is that silver staining is due to protein sulphydryl groups. NORs certainly contain high concentrations of protein sulphydryls and disulphides (Buys & Osinga 1980a) and sulphydryl groups are able to bind silver ions and reduce them to metallic silver. De Capoa *et al.* (1982) claimed to have evidence that silver stainability of NORs was due directly to protein sulphydryl groups. However, the results of Olert *et al.* (1979), Buys & Osinga (1984) and Hubbell (1985) did not support the involvement of SH groups in Ag-NOR staining. Clearly it is not possible yet to state exactly which groups on which proteins are responsible for Ag-NOR staining. No doubt some of the confusion is due to different workers using different materials (i.e. different sets of proteins extracted from nucleoli from different species using different procedures), and some to the use of different methods of investigation. Moreover, the formation of visible silver deposits is itself a complex process, requiring successive stages of binding of silver ions, reduction to metallic silver, followed by an autocatalytic accumulation of silver to produce a visible deposit (Gallyas 1982). It seems quite possible that different chemical groups on the Ag-NOR proteins might influence different stages of the process of silver staining, although the consistency with which Ag-NOR staining can be obtained in, apparently, all eukaryotes tested indicates that a common mechanism is involved.

9.4.3 *Applications of silver staining of NORs*

Whatever the mechanism of Ag-NOR staining, silver staining for NORs has become one of the standard banding methods, and Ag-NOR staining is an essential part of the characterization of a species' karyotype. Using these methods, NORs have now been localized in a large number of species. Table I in Howell (1982) lists a large number of reports of NOR localization, mainly in vertebrates. Since then, many additional studies of NOR location have been made in various species, and no attempt will be made to give a complete listing. It is, however, worth noting observations on additional species of plants (Medina *et al.* 1983, von Kalm & Smyth 1984, Mehra *et al.* 1985), on molluscs (Kawano *et al.* 1987) and on insects (Gosálvez *et al.* 1986, Wallace & Newton 1987). Particularly interesting is the observation of Ag-NOR staining in the ciliate protozoan *Tetrahymena pyriformis* (Sabaneyeva *et al.* 1984). Taken together, all these observations show that silver staining can probably be used universally to demonstrate NORs in eukaryotes.

We have seen in the previous section (Section 9.4.2) that Ag-NOR staining is a reflection of nucleolar activity rather than a straightforward indication of the presence of ribosomal genes. Ag-NOR staining has, therefore, been applied in a number of situations to determine the transcriptional activity of rDNA. Some of these have been referred to in the previous section (9.4.2), for example, the suppression of Ag-NOR staining in the chromosomes of one species in cell hybrids (D. A. Miller *et al.* 1976, O. J. Miller *et al.* 1976). The same phenomenon can also occur in organismal hybrids: for example, in hybrids between *Aegilops* and rye, the activity of the rye NORs was suppressed (Cermeno & Lacadena 1985), while in trout hybrids, brook trout NORs were inactive and only rainbow trout NORs were stained with silver (Ueda *et al.* 1988), and in mules the donkey NORs were active and the horse NORs inactive (Kopp *et al.* 1986). However, the NORs of one genome are not invariably suppressed in interspecific crosses (Weide *et al.* 1979). In the grasshopper *Eyprepocnemis plorans*, addition of a B chromosome to the karyotype affects the activity of NORs on other chromosomes (Cabrero *et al.* 1987). Dhaliwal *et al.* (1988) described dosage compensation in Bennett's wallaby, in which the NORs are on the X chromosome. Nevertheless, the Ag-NOR staining due to the single male X was similar to that due to the two female X chromosomes.

Ag-NOR staining has proved to be particularly useful in studying nucleolar activity during gametogenesis. In the spermatogenesis of mammals (Hofgärtner *et al.* 1979, Schmid *et al.* 1983a) and other vertebrates (Schmid *et al.* 1982), silver staining of NORs occurs throughout meiotic prophase up to the end of pachytene. Staining is then absent from diakinesis, through metaphase I to metaphase II, but then reappears in early spermatids, indicating a post-meiotic reactivation of rDNA transcription. The silver staining disappears again as the spermatids develop and elongate. Hartung *et al.* (1979) have also reported changes in Ag-NOR staining during human oogenesis; here the amount of staining increases to a maximum at diplotene. Evidence was also obtained for the presence of numerous micronucleoli, which were interpreted as the products of amplified ribosomal genes. Observations have also been made of NOR activity during the meiotic cycle in grasshoppers (Satya-Prakash & Pathak 1984, Santos *et al.* 1987).

At a later stage of the life cycle, various authors have looked for changes in Ag-NOR staining with ageing (Buys *et al.* 1979b, Denton *et al.* 1981, Das *et al.* 1986). A consistent finding was a reduction in Ag-NOR staining with increasing age. Evidence has also been sought for changes in activity of NORs in neoplastic cells, although results seem to be rather inconsistent. Hubbell & Hsu (1977) and Brasch & Smyth (1987) found no consistent differences between the silver staining of chromosomes of normal and neoplastic cells, whereas in cells from patients with chronic myelocytic leukaemia both Mamaev *et al.* (1985) and Sato *et al.* (1986) found that silver staining tended to be reduced. Nevertheless, Ag-NOR staining has now been recommended as a diagnostic tool in human (Anon 1987) and canine (Bostock *et al.* 1989) histopathology, diagnosis depending mainly on patterns of silver staining in interphase nuclei rather than on staining of metaphase chromosomes.

It has already been mentioned that in human chromosomes, it is rare for all the sites of rDNA to be stained with silver. In addition, there are often clear differences in size between those NORs that are stained. These patterns of silver staining are consistent within an individual (D. A. Miller *et al.* 1977, Mikelsaar & Schwarzacher 1978), but differ between individuals (Varley 1977) and are inherited in a Mendelian fashion (Mikelsaar *et al.* 1977b, Markovic *et al.* 1978, Taylor & Martin-de Leon 1981). Ag-NOR staining in humans thus forms a heteromorphic system comparable with C-banding, and will be considered further in Chapter 14. Heritability of Ag-NOR staining patterns has also been reported for other species of mammals (Henderson & Bruère 1980, Arruga & Monteaguido 1989). There are, nevertheless, some minor variations in this system. In a study of human twins, Weltens *et al.* (1985) found significant non-genetic variation of Ag-NOR patterns, although this was not sufficient to obscure the genetically defined pattern. Ferraro *et al.* (1981) found several different patterns of Ag-NOR staining in human fibroblasts, and concluded that there were several different clones in the population, each with a specific heritable pattern. Similar results have not, however, been reported by other workers, and in general Ag-NOR staining patterns can be regarded as essentially constant within an individual. Some Ag-NOR stained human chromosomes are particularly distinctive, with double sites of silver deposition on the short arm of the chromosome (Sozansky *et al.* 1984, Verma & Rodriguez 1985). On the other hand, use of high resolution banding methods does not increase the detecta-

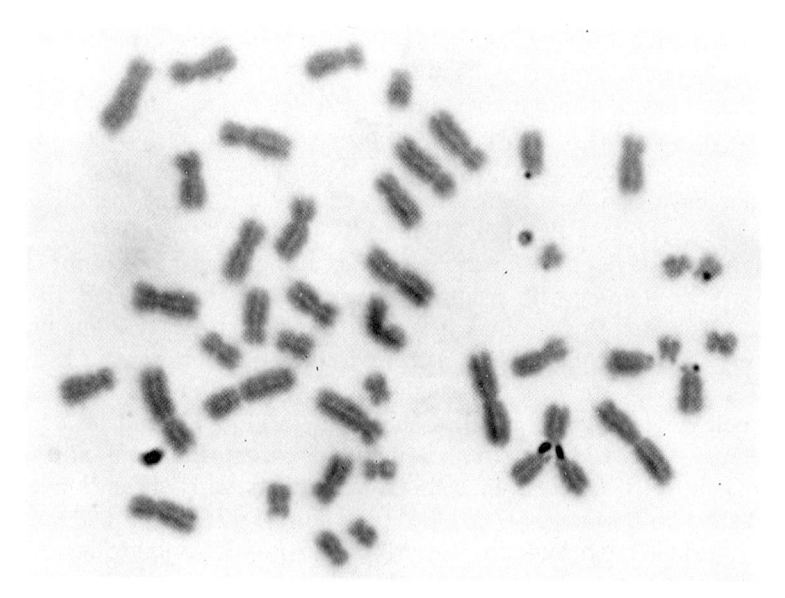

Figure 9.5 Satellite association of human chromosomes, stained with silver.

bility of Ag-NORs, nor segment them into smaller subunits (Verma *et al.* 1984).

The human NOR-bearing chromosomes are frequently associated by their short, NOR-bearing arms (for a review, see Houghton & Houghton 1978). In conventional preparations, no physical connection is visible between associated chromosomes, and the definition of an association is somewhat arbitrary. With Ag-NOR staining, the associated chromosomes are clearly linked by a deposit of silver (Denton *et al.* 1976) (Fig. 9.5). D. A. Miller *et al.* (1977) showed that the frequency with which a particular chromosome was involved in associations was related to the size of its Ag-NOR region. Satellite association between the human acrocentric chromosomes is clinically important because of the possibility that Robertsonian translocations between these chromosomes might be facilitated, and because they provide an obvious mechanism for the formation of trisomic genomes. Evidently the degree of Ag-NOR staining, which in turn may be regarded as an indication of the involvement of these chromosomes in nucleolus formation, is likely to be an important parameter affecting these phenomena.

10 Kinetochores

10.1 Introduction

10.1.1 Definitions

The names 'centromere' and 'kinetochore' were both coined by light microscopists to indicate that region of the chromosomes, usually the primary constriction, which is attached to the spindle at cell division. In recent years the two words have come to have rather different meanings, and 'kinetochore' is generally used to mean the structure, which can be observed by electron microscopy, to which the spindle microtubules become attached (Stack 1974, Ris & Witt 1981, Rieder 1982, Godward 1985). 'Centromere' is now generally used in a somewhat broader sense, indicating the primary constriction, within which the kinetochore proper is located, and often also the associated pericentromeric heterochromatin. In this chapter the term kinetochore will be used, not because the objects being discussed have necessarily been investigated by electron microscopy, nor because it has been shown that the spindle microtubules are attached to them, but because it is nevertheless believed that they do correspond to the objects so defined. In addition, it is important to distinguish the objects described in this chapter from centromeric heterochromatin revealed by methods such as those described in Chapter 4.

10.1.2 Structure and chemistry of kinetochores

The structure of kinetochores has been studied in considerable detail, and it is clear that various types exist, especially in many protozoa and fungi. Among higher organisms, however, only two types of kinetochore are found: the trilaminar type and the ball-and-cup type. The trilaminar type consists essentially of an inner dense layer of material in contact with, and apparently an integral part of, the centromeric chromatin, which is separated from the outer dense layer by an electron lucent layer (Fig. 10.1). Such kinetochores are found in vertebrates, some insects, and various protozoa, etc (Bostock & Sumner 1978, Rieder 1982, Godward 1985). The ball-and-cup type of kinetochore consists of a 'ball' of less dense material sitting within a pocket of chromatin, and is characteristic of higher plants, but has also been found in certain insects (Bostock & Sumner 1978, Rieder 1982). The distinction between the two types of kinetochore may, perhaps, not be completely rigid, since the trilaminar kinetochore found at metaphase apparently passes through a stage resembling the ball-and-cup structure at prophase (Heneen 1975).

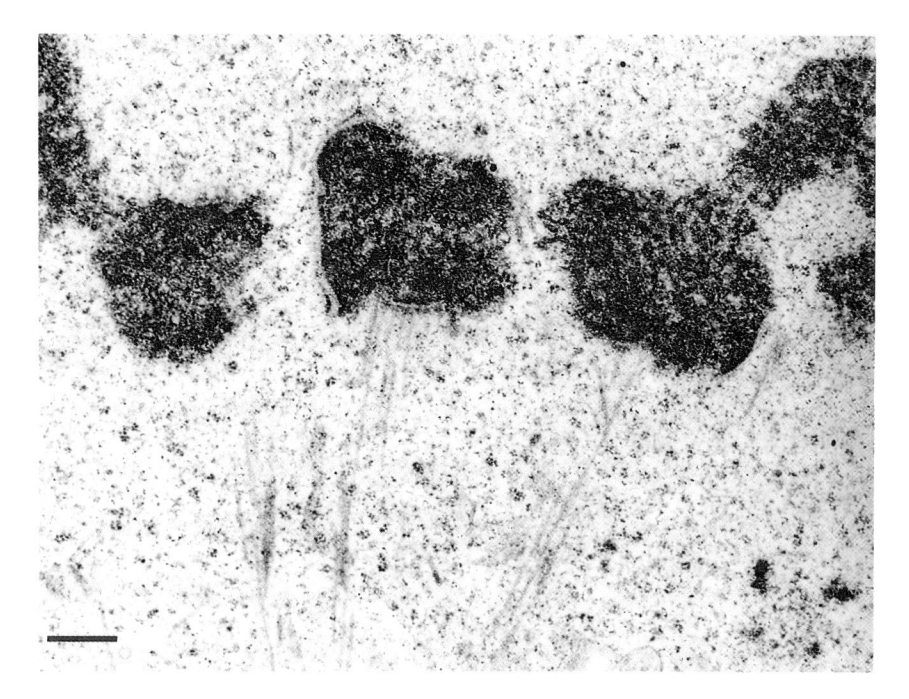

Figure 10.1 Electron micrograph showing trilaminar kinetochores of Chinese hamster chromosomes. The bar represents 0.5 μm.

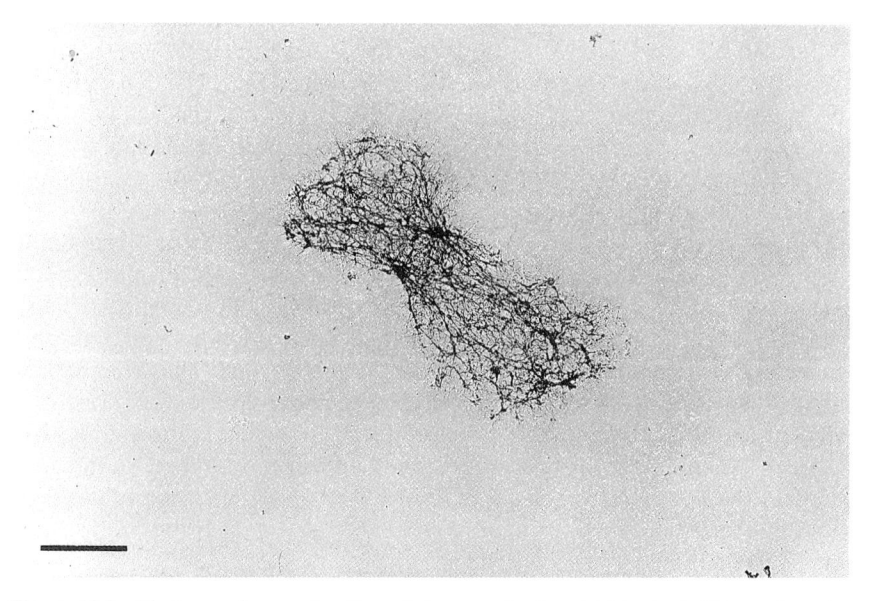

Figure 10.2 Electron micrograph of a whole mount of a dehistonized Chinese hamster chromosome. The residual chromosome scaffold shows a pair of dense bodies, believed to be the kinetochores. The bar represents 1.5 μm. Reproduced from Hadlaczky *et al.* (1981a) by permission of the authors and Springer-Verlag.

Details of kinetochore structure are not visible in whole mount preparations of chromosomes, but the kinetochores appear as paired dense regions (Fig. 10.2) which are the most resistant parts of the chromosome to procedures that disperse chromatin, or to enzymes that digest chromosomes (Moses & Counce 1974, Rattner et al. 1978, Hadlaczky et al. 1981a,b). Evidence that these objects are indeed the kinetochores is provided by the observation that they may have microtubules attached to them (Rattner 1986). The kinetochores appear to be an integral part of the chromosome scaffold (Earnshaw et al. 1984). With appropriate pretreatments, small paired dots can also be seen at the centromeres by light microscopy without staining (Evans & Ross 1974, Clapham & Östergren 1978), although the identity of these dots with kinetochores has been disputed (Roos 1975).

Kinetochores apparently consist of DNA, RNA and proteins. The evidence comes mainly from the results of ultrastructural cytochemical staining and digestion with specific enzymes. Several authors have reported the presence of RNA in the kinetochores of both plants and animals (e.g. Braselton 1975, Esponda 1978, Rieder 1979). DNA can also be detected in the inner and outer layers of the trilaminar kinetochore, using a specific EM staining method (Ris & Witt 1981). It has been proposed that the DNA in the outer layer of the kinetochore is arranged in chromatin fibres (Ris & Witt 1981, Rattner 1986). The nature of the kinetochore DNA has not yet been established. In yeast it is known that a special DNA sequence is contained in the centromeric region, and is necessary for proper chromosome segregation (Carbon 1984), but this is an organism with very small chromosomes and no differentiated kinetochore. In humans, specific repetitive DNA sequences (the alphoid satellites) are associated with the centromeres (rather than with the centromeric heterochromatin) (Jabs et al. 1984, Aleixandre et al. 1987) but the relationship of these to the kinetochore itself, although spatially close (Wong & Rattner 1988, Masumoto et al. 1989), has not yet been established.

Kinetochores undoubtedly contain several proteins, although much more has still to be discovered about them. Much has been learnt from the use of autoimmune CREST sera, a subject which will be considered in detail in Section 10.2.3. Tubulin is apparently an important component (Pepper & Brinkley 1977, Mitchison & Kirschner 1985). In fact, the kinetochores appear to act as capture sites for the spindle microtubules; for a review of current ideas on the mode of functioning of kinetochores, see Mitchison (1988). The spindle microtubules generally appear to terminate in the outer layer of the trilaminar kinetochore (Ris & Witt 1981). The microtubules do not bind directly to the kinetochore DNA, but appear to be bound via some other protein (Hancock & Burns 1987).

Certain plants and animals do not have a localized kinetochore, but have holocentric chromosomes in which there is a diffuse kinetochore over all or most of the length of the chromosomes (reviewed by Godward 1985). The attachments of such chromosomes to the spindle are varied in structure, and will not be described in detail here. In fact, no staining methods strictly specific for kinetochores appear to have been applied to such chromosomes, although it would be very interesting to do so.

10.2 Staining methods for kinetochores

Staining methods for kinetochores appear to be among the oldest of banding methods, although many of the early observations were rather casual, and there was some uncertainty about the nature of the objects stained. Levan (1946) tabulated several of these early observations, beginning with those made by Metzner (1894) on the chromosomes of *Salamandra*. In general, this early work involved tissue fixed with a mixture containing osmium tetroxide, followed by embedding and sectioning, and staining with haematoxylin. Levan's own work (Levan 1946) employed a pre-fixation treatment with mercuric nitrate, and crystal violet staining. Östergren (1947) also stained plant kinetochores with crystal violet, after fixation with an aqueous mixture of ethylene glycol and acetic acid, and post-fixation with a chromic fixative (see also Clapham & Östergren 1978). These latter methods appear to work by the swelling of the chromosomes, only the kinetochores retaining their compactness and stainability.

Interesting though these older methods are, they have not been adapted to use on chromosome spreads, and are now of historical interest only. Modern methods for kinetochore staining belong to three classes: methods using Giemsa staining (C_d-banding and related methods), silver staining, and immunocytochemical labelling with autoimmune CREST serum. Each of these classes is considered in more detail in the following sections.

10.2.1 C_d staining and other Giemsa methods for kinetochores

Eiberg (1974) described 'a new banding technique which reveals two identical dots at the place of the centromere, one on each chromatid' (Fig. 10.3). The dots were the same size on all (human) chromosomes, and it was suggested that they might 'represent organelles associated with the spindle fibres.' The method was named C_d staining, to indicate centromeric dots.

The method involves the following stages:

1   Chromosome preparations are incubated in Earle's BSS medium (pH 8.5–9.0) for 45 min at 85°C.
2   Slides are stained with Giemsa.

In the original method, the cells were fixed in a series of methanol-acetic acid mixtures of decreasing ratios, although subsequent users of the method have used standard 3:1 methanol-acetic acid fixation. Eiberg (1974) did, however,

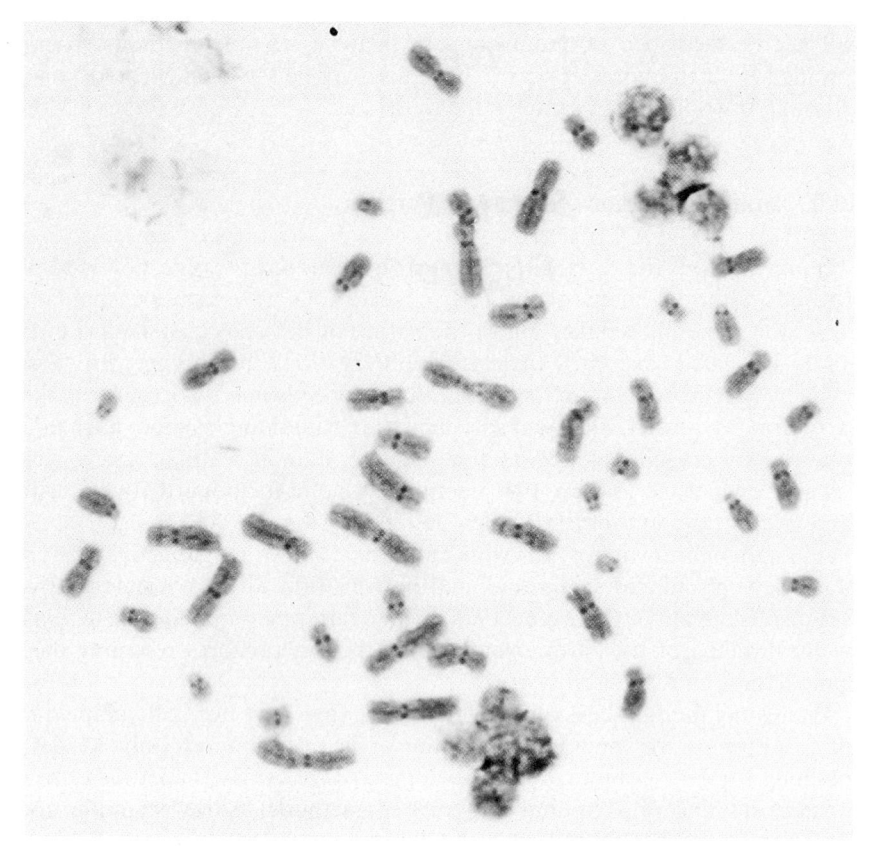

Figure 10.3 C_d-banding of a human metaphase chromosome spread. Reproduced from Eiberg (1974) by permission of the author. Copyright © 1971 Macmillan Magazines Ltd.

emphasize the importance of standardizing the method of preparation of the chromosomes if C_d-bands were to be produced routinely.

C_d-banding has been used to study the behaviour of kinetochores in stable dicentric chromosomes. It is generally believed that for a dicentric chromosome to be stable, one of the centromeres must be inactivated, unless the two centromeres are so close together that they cannot orient independently at cell division. Such inactivated centromeres do not appear as a chromosomal constriction. In these stable dicentric chromosomes, it has been found that only one centromeric site, the one that still appears as a constriction, is stained by the C_d method (Nakagome et al. 1976, Daniel 1979, Maraschio et al. 1980, Lambiase et al. 1984). Although the inactive centromere lacks a C_d-band, a C-band is normally retained, although in one case it had been lost (Nakagome et al. 1986). These observations emphasize the distinction between C-bands and C_d-bands, and show that the former cannot be used reliably to mark the sites of active centromeres.

Lau & Hsu (1977) investigated mouse chromosomes formed by Robertsonian fusion, using C_d staining. When two acrocentric chromosomes fuse to form a metacentric one, their kinetochores would be very close together, and could function as one without any inactivation. Lau & Hsu inhibited the condensation of the mouse heterochromatin by culturing in the presence of Hoechst 33258 (Section 11.2.2), thereby permitting the separate resolution of very close kinetochores. They found that in many cases two (or occasionally more) sets of kinetochores (as demonstrated by C_d staining) were retained in the biarmed chromosomes, although in others only a single set remained.

Older men and women sometimes have chromosomes that lack a primary constriction. Nakagome et al. (1984) showed that such chromosomes also lack C_d-bands. If C_d-bands truly represent active kinetochores, then their absence in these chromosomes would no doubt contribute to the aneuploidy that develops in the cells of older individuals (Galloway & Buckton 1978).

Objects which are presumed to be kinetochores have been demonstrated in the chromosomes of certain plants, using what are essentially C-banding methods (Stack 1974, G. E. Marks 1975). Indeed, it is not uncommon to obtain paired dots at the centromeres after excessive C-banding treatment, and it is probable that such 'residual C-bands' are in fact kinetochores. Only very limited use has been made of this sort of technique to demonstrate kinetochores, and indeed such use cannot be recommended because of the possibility of confusion between genuine kinetochores and small centromeric C-bands.

A final method which involves staining kinetochores differentially with Giemsa should be noted. Sato & Sato (1982) exposed preparations of *Allium* chromosomes to UV light in the presence of Hoechst 33258, a procedure which caused photolysis leaving only the outline of the chromosomes, and especially paired centromeric dots. Similar results have been reported for animal chromosomes (Buys & Osinga 1982), but these procedures are not of any practical value.

The mechanism of C_d-banding has not been investigated, nor the mechanism of the other Giemsa staining methods for kinetochores just mentioned. However, it seems very likely that they all depend on the greater resistance of kinetochores to dispersion and destruction compared with other chromosomal components (Section 10.1.2).

10.2.2 Kinetochore staining with silver

The tendency of Ag-NOR staining methods (Section 9.4.1) to stain centromeric regions of chromosomes in addition was noted early on (Goodpasture & Bloom 1975). Since then a number of procedures have been published in which kinetochores as well as NORs can be stained distinctively with silver (Fig. 10.4).

Silver staining methods for kinetochores are not numerous, nor do they differ substantially from those used for NOR staining (Section 9.4.1).

Denton *et al.* (1977) employed the method they had previously used for Ag-NOR staining, with a pretreatment in very dilute sodium hydroxide (pH 8.5). Ved Brat *et al.* (1979) merely incubated the chromosome preparations with a 50% silver nitrate solution for 16–18 h at 37°C. The Kt staining method of Brown & Loughman (1980) differed from the previous one only in a pretreatment with hot phosphate buffer. Rufas *et al.* (1982) did not find any pretreatment necessary, and simply used a strong silver nitrate solution, adjusted to pH 3 with formic acid, for 5–15 min at 60°C. None of these methods is specific for kinetochores; as well as staining NORs, as already mentioned, such methods also stain chromatid cores in meiotic chromosomes of grasshoppers (Rufas *et al.* 1982, 1983a).

The useful practical applications of silver staining of kinetochores have been limited, although it must be emphasized that an unambiguous marker for the centromeres of chromosomes is always valuable. This is particularly true in meiotic studies (Rufas *et al.* 1982, 1983b, Shi *et al.* 1982, Sudman & Greenbaum 1989). Rocchi *et al.* (1982) made the interesting observation that the kinetochores of all mouse autosomes plus the X could be stained with silver, but that the Y chromosome was consistently unstained. They speculated that the staining method they used may not, in fact, be staining the kinetochores (which the Y chromosome must possess) but some other structure associated with the centromeric region. The precise nature of the silver-stained material at the centromeres is unknown, although the results

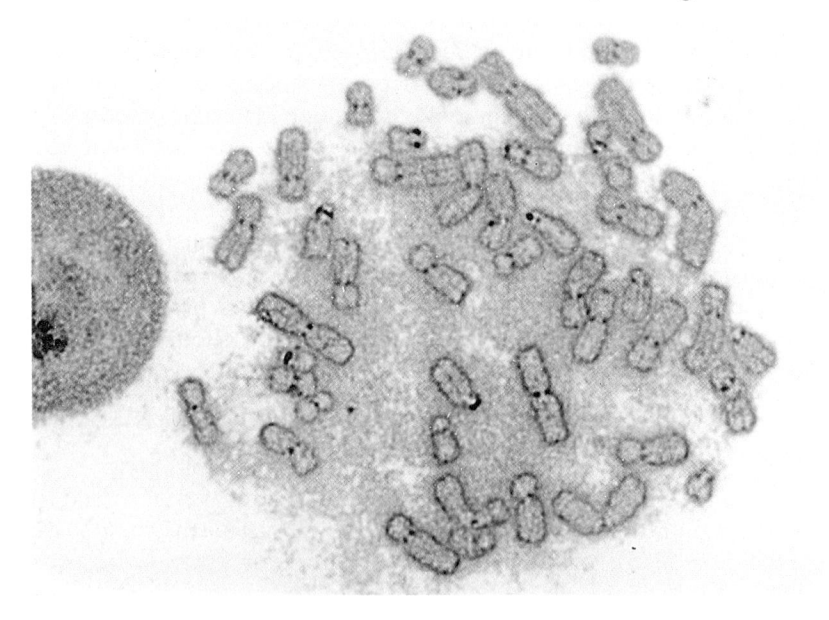

Figure 10.4 Silver staining of kinetochores of human chromosomes. Note that the NORs are also stained. Micrograph kindly provided by Mr G. Spowart. Reproduced from Sumner (1983) *Science Progress* **68**, 543–64, published by Blackwell Scientific Publications.

of Buys & Osinga (1980a) suggest that the deposition of silver could be due to sulphydryl groups.

10.2.3 Immunocytochemical staining of kinetochores with CREST serum

Undoubtedly the most important method for staining kinetochores is the immunocytochemical procedure using CREST serum. CREST serum is obtained from patients with a relatively mild form of the autoimmune disease scleroderma, the name being an acronym of the main symptoms: calcinosis, Raynaud's phenomenon, esophageal dysmotility, sclerodactyly and telangiectasia. For a review of this and other autoantibodies to nuclear antigens, and their clinical significance, see Tan (1989). It was shown by Moroi et al. (1980) that CREST sera could be used to produce a speckled pattern of immunofluorescence in nuclei, and to label specifically two small centromeric dots on each chromosome. Proof that the CREST serum was actually labelling the trilaminar kinetochore was obtained by Brenner et al. (1981), using immunoelectron microscopy. Since then, CREST serum has become an invaluable tool, not only for labelling the kinetochores throughout the cell cycle in normal and abnormal situations, but also for analysing kinetochore composition and function.

Immunolabelling of kinetochores using CREST serum differs from virtually all other banding techniques described in this book in that it cannot normally be carried out on conventional, methanol-acetic acid-fixed chromosome preparations (but see Oaks et al. 1987 and Earnshaw et al. 1989). The antigens recognized by CREST serum are very sensitive to fixation and are destroyed when chromosomes are prepared in the standard way. It is usual, therefore, to swell the mitotic cells with a hypotonic solution, and then spread them on slides by centrifugation using a Cytospin (Shandon Southern Instruments Ltd.) or similar centrifuge. The preparations are then subjected to a gentle fixation procedure, such as methanol or acetone at $-20°C$ for 30 min, or periodate-lysine-paraformaldehyde (PLP) (McLean & Nakane 1974) for 10 min at 4°C, or a brief formaldehyde fixation. After fixation, slides may be stored in the freezer for a limited period, but it is best to use them as fresh as possible. The staining procedure is straightforward:

1 A drop of CREST serum, diluted 1:200 with PBS (phosphate-buffered saline) is placed over the chromosomes for 30–60 min at room temperature.
2 The preparations are washed for a total of 15 min in three lots of PBS.
3 The sites of CREST binding are labelled with FITC-labelled antibody to human IgG, diluted 1:5 or 1:10 with PBS, for 30–60 min.
4 The slides are again washed with three lots of PBS (total time 15 min).
5 Finally the chromosomes are counterstained with a fluorochrome for DNA.

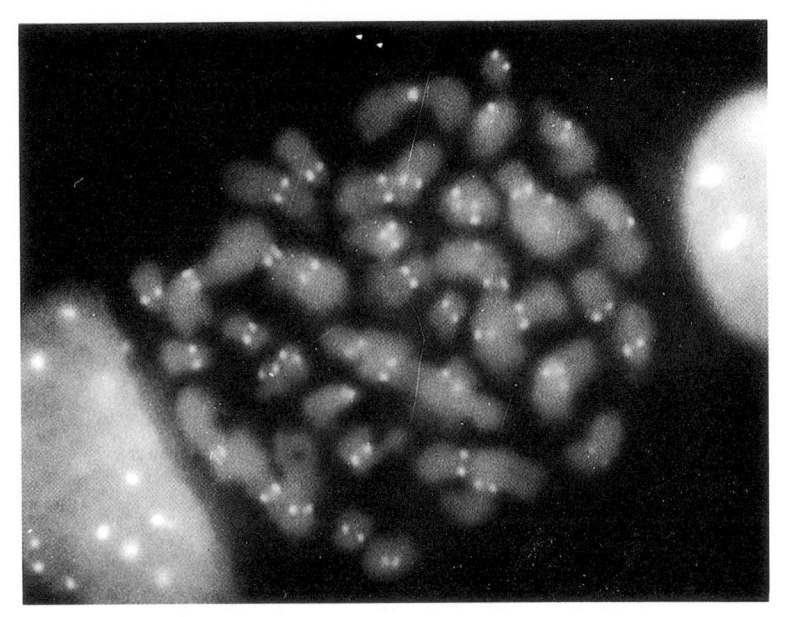

Figure 10.5 Human metaphase chromosome spread labelled with CREST serum followed by FITC-labelled anti-human IgG, to show kinetochores. Counterstained with ethidium bromide. Micrograph kindly provided by Mr G. Spowart. Reproduced from Sumner (1989).

A typical result of CREST/FITC staining is shown in Fig. 10.5. On the chromosomes, paired dots are stained at all the centromeres. The counterstain used may be ethidium, in which case it can be seen simultaneously with the FITC fluorescence, or it may be Hoechst 33258 or DAPI, which are not excited at the wavelength of light used to excite FITC fluorescence. There are advantages and disadvantages with both procedures. It is obviously useful to be able to see the whole chromosome at the same time as the specific kinetochore fluorescence, but there is a danger of overstaining, so that the ethidium fluorescence can swamp the weak kinetochore fluorescence. Moroi *et al.* (1980) stained for only 5 min with an ethidium bromide solution containing only 10 μg/ml of dye. With Hoechst 33258 or DAPI there is no problem with swamping the FITC fluorescence, but separate observations have to be made at the different wavelengths, and two photographs taken for comparison.

The dilutions specified above for the immunocytochemical reagents are for guidance only, since the optimal dilutions, particularly of the CREST serum, may vary substantially according to the samples used. It will, therefore, probably be necessary to experiment to achieve the best results. Obviously, for the greatest economy with these valuable reagents, the highest dilution that gives an adequate signal should be used. CREST sera may be obtained through rheumatology departments, or from certain suppliers of immunological reagents.

Slides should be mounted in PBS, with or without addition of glycerol.

Use of an additive to stabilize fluorescence is highly desirable, and such mountants are now available commercially (e.g. Citifluor, Citifluor Ltd., London). A fast film is recommended for photography.

As with any immunocytochemical method, a variety of labels can be used instead of fluorescein. Other fluorochromes that have been used are rhodamine (del Mazo et al. 1986) and Texas red (Earnshaw & Migeon 1985), which have advantages of different colour, greater brightness and greater stability compared with fluorescein. Horseradish peroxidase has the advantage that it can be seen by transmitted light, and has been used to label kinetochores both for light microscopy (Moroi et al. 1981) and for electron microscopy (Moroi et al. 1981, Brenner et al. 1981). Colloidal gold has also been used successfully to label kinetochores for electron microscopy (Valdivia et al. 1986, Sumner & Speed 1987).

CREST sera have been used to study the kinetochores in normal and abnormal mitotic cells of mammals, in meiotic cells and during subsequent stages of spermatogenesis, and in interphase. No reports have yet appeared of the use of CREST sera on the chromosomes of lower vertebrates, invertebrates or plants, although Cox & Olmsted (1984) reported the presence of antigens that react with CREST serum in frog, *Drosophila*, and the protozoan *Tetrahymena*. In human and other mammalian chromosomes, Cherry & Johnston (1987) and Cherry et al. (1989) reported that the kinetochores of different chromosomes were of different sizes, those of the human Y chromosome being particularly small. They postulated that this might lead to non-disjunction of the Y, with consequent aneuploidy. Peretti et al. (1986) and Earnshaw et al. (1987b) obtained similar results. On the other hand, Fantes et al. (1989) found no difference between the fluorescence of the Y kinetochore and that of the kinetochores of other human chromosomes.

As with C_d staining, CREST immunostaining has been used to study inactive centromeres in stable dicentric chromosomes. Merry et al. (1985) found reduced kinetochore fluorescence at the inactive centromere of a stable human dicentric (9;11) chromosome, but in a stable dicentric X chromosome, Earnshaw & Migeon (1985) found that the inactive centromere had no kinetochore fluorescence. Peretti et al. (1986) examined several stable human dicentrics, and found that the inactive centromeres could be either labelled or unlabelled, although labelled inactive centromeres were less intensely fluorescent than active centromeres. In stable mouse dicentric chromosomes also, only one centromere shows kinetochore fluorescence with CREST sera (Vig & Zinkowski 1986, Zinkowski et al. 1986). However, newly formed mouse dicentrics, and rat dicentric and multicentric chromosomes, showed as many pairs of labelled kinetochores as there were centromeres (Zinkowski et al. 1986). Wandall (1989) was able to correlate the presence or absence of CREST labelling in dicentrics with the presence or absence of a trilaminar kinetochore seen by electron microscopy.

The ability of CREST sera to label the pre-kinetochores of interphase nuclei was recognized very early on during their application to kinetochore labelling (Moroi *et al.* 1980). In general, the number of CREST-labelled dots in interphase corresponds to the number of chromosomes (Moroi *et al.* 1981), and the dots become double bodies during the G2 phase of the cell cycle (Brenner *et al.* 1981), which is presumably the time of replication of kinetochores. Several workers have looked at the distribution of kinetochores in interphase, and found non-random patterns. In CHO cells, pre-kinetochores were found associated with the nuclear envelope and with the surface of the nucleoli (Moroi *et al.* 1981). Cox *et al.* (1983) and Earnshaw *et al.* (1984) found that the kinetochores were associated with the nuclear matrix and the chromosome scaffold. In rat kangaroo and Indian muntjac (species chosen because their low chromosome numbers makes interpretation much easier) various non-random arrangements of interphase kinetochores were found (Hadlaczky *et al.* 1986): arrangement of pre-kinetochores in pairs, suggesting association of homologues; clustering of kinetochores at one pole of the nucleus (Rabl configuration, representing the persistence of the telophase distribution); and an arrangement in which two groups of kinetochores form mirror images of each other. In the Chinese muntjac (*M. reevesi*, 2N = 46), the pre-kinetochores clustered into aggregates in interphase, similar to the arrangement produced by the large 'compound kinetochores' of the Indian muntjac (2N = 6 ♀, 7 ♂) (Brinkley *et al.* 1984); the authors suggest that the aggregation of kinetochores could well have led to

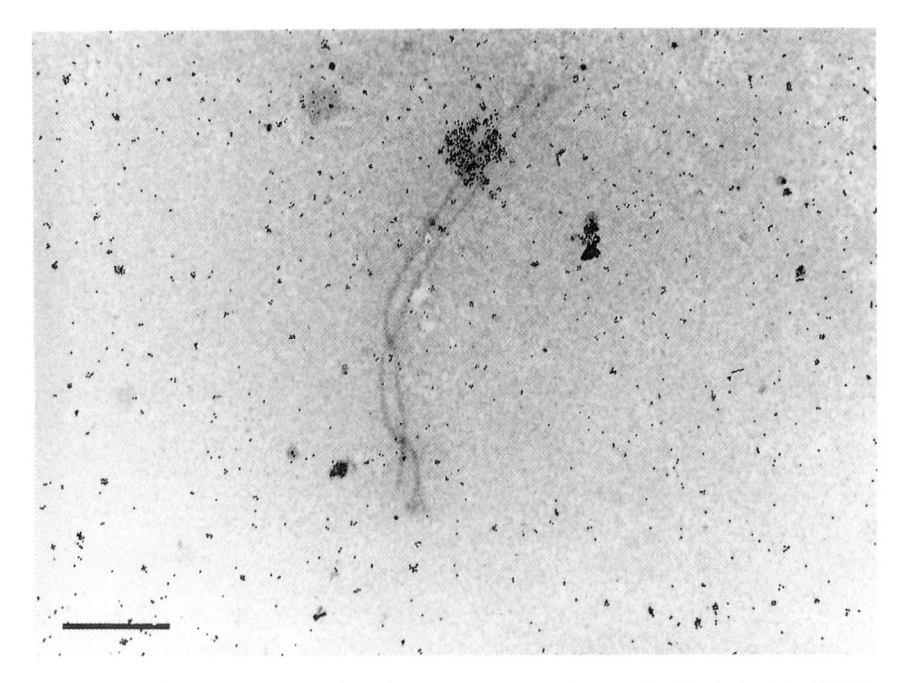

Figure 10.6 Electron micrograph of human synaptonemal complex labelled with CREST serum and colloidial gold-anti-human IgG to show position of kinetochore. The bar represents 1 μm.

chromosome fusion and the greatly reduced chromosome number found in the latter species. Aggregation of pre-kinetochores has also been reported during spermatogenesis, particularly in spermatids (Brinkley et al. 1986, del Mazo et al. 1986). As the sperm head condenses, immunofluorescence with CREST sera is eventually lost, but this is simply because the immunological reagents cannot penetrate the highly condensed chromatin; if the mature sperm heads are decondensed artificially, the pre-kinetochores again become stainable (Sumner 1987). In mouse Sertoli cells, the pre-kinetochores apparently become even more highly aggregated, being restricted to only two or three spots (Brinkley et al. 1986, del Mazo et al. 1986). It is not clear whether these spots represent aggregations of the total number of pre-kinetochores in these cells, or whether for some reason, the majority of chromosomes no longer have stainable pre-kinetochores. A report on CREST staining of kinetochores throughout mouse meiosis has been published by Brinkley et al. (1986). They describe clustering of kinetochores during pachytene, and there are also indications that centromere replication may be delayed in the first meiotic division, which, it has been suggested, may be a necessary requirement for proper disjunction of homologues. Immunolabelling of the kinetochores on synaptonemal complexes at pachytene has also been reported, using colloidal gold as a label (Sumner & Speed 1987) (Fig. 10.6); this method was used to assess the extent of pairing of the human X and Y chromosomes.

CREST sera have become very valuable reagents, not only for studying

Table 10.1 Molecular weights of kinetochore proteins recognized by CREST sera.

Reference	Molecular weights (KDa)	Species
Cox et al. 1983	14, 20, 23 34	Human, rat kangaroo, mouse, hamster, Indian muntjac
Earnshaw et al. 1984	77, 110	Human
Ayer & Fritzler 1984	33	Human
Guldner et al. 1984	19.5	Human
Nishikai et al. 1984	70	Rabbit, calf
Earnshaw & Rothfield 1985	17, 80, 140 50	Human Rat
Spowart et al. 1985	19.5	Human
Valdivia & Brinkley 1985	18, 80	Human, Chinese hamster
McNeilage et al. 1986	19.5, 72	Human
Hadlaczky et al. 1986	18	
Palmer et al. 1987	17	Human (not rat or chicken)
Kingwell & Rattner 1987	19.5, 50	Human, Indian muntjac
Hadlaczky et al. 1989	59	Mouse, human (mitosis only, not interphase)

Table 10.2 Classification of kinetochore proteins.

Class*	CENP designation†	Molecular weight range (KDa)
1	CENP-A	14–19.5
2	CENP-D	20–25
3	CENP-E	33–34
4	CENP-B	70–80
5	CENP-C	110–140
6	CENP-E	50

* Classification of Kingwell & Rattner (1987)
† Classification of Earnshaw & Rothfield (1985) (CENP-A, B & C), extended by Kingwell & Rattner (1987). CENP = CENtromere Protein.
For full references see Table 10.1, and Kingwell & Rattner (1987).

kinetochores cytologically during mitosis, meiosis and interphase, but also for the molecular analysis of kinetochore components, as a means of establishing their function. Different CREST sera recognize a number of different kinetochore proteins (Table 10.1), and indeed some will also react with non-kinetochore proteins (Jeppesen & Nicol 1986). Although at first sight there is an almost bewildering variety of such proteins, in fact only two or three make up the majority of reports. An attempt to classify these kinetochore proteins has been made by Earnshaw & Rothfield (1985) and supplemented by Kingwell & Rattner (1987) (Table 10.2). To complicate matters further, different CREST sera may recognize different epitopes on the same kinetochore protein, and similar epitopes can be present on different kinetochore proteins (Earnshaw & Rothfield 1985, Earnshaw et al. 1987a). Most of the kinetochore proteins identified by CREST sera are apparently present in all species of mammals so far tested, with only odd exceptions (Table 10.1). However, it does seem that whereas some sera will produce equal labelling of the kinetochores in all chromosomes in a particular species (Fantes et al. 1989), other sera will produce consistently different levels of labelling on the kinetochores of specific chromosomes (Peretti et al. 1986, Cherry & Johnston 1987, Earnshaw et al. 1987b). Earnshaw et al. (1989) have shown that the amount of CENP-B varies from one chromosome to another, and that this antigen is present at both active and inactive centromeres in stable dicentric chromosomes. CENP-C, on the other hand, is present in roughly similar amounts on all chromosomes, but is undetectable at inactive centromeres. These findings help to explain the different results obtained on normal and dicentric human chromosomes by earlier workers, and serve to emphasize the heterogeneity of CREST sera. In critical work, it is desirable to define the antigens with which the sera react, in order to obtain a full understanding of the observations made.

Preliminary attempts have been made to characterize in more detail the function of kinetochore proteins recognized by CREST sera. Ayer & Fritzler (1984) believed that the 33 KDa protein recognized by their serum was a histone variant associated with condensed chromatin. Similarly, Palmer *et al.* (1987) claimed that the 17 KDa protein, CENP-A, co-purified with core histones and with nucleosomes, and suggested that it was a centromere-specific core histone. Balczon & Brinkley (1987) showed that an 18 KDa kinetochore antigen (again probably CENP-A) was not necessary for the kinetochore to organize the assembly of spindle microtubules; instead, the 80 KDa antigen (CENP-B) appeared to be intimately involved in the binding of tubulin.

Altogether CREST sera have become immensely valuable not only for the detailed analysis of kinetochore organization, but also as a cytological tool. For the latter purpose it has two disadvantages, both consequences of the lability of the kinetochore antigens. The first disadvantage is the necessity to use special preparative and fixation procedures, so that CREST immunostaining cannot be applied to conventional chromosome preparations if they should appear to show interesting centromeric features. Nevertheless, two papers have been published in which the authors have succeeded in applying CREST staining to methanol-acetic acid fixed chromosomes (Oaks *et al.* 1987, Earnshaw *et al.* 1989). The other disadvantage is that chromosome preparations suited for CREST staining do not readily lend themselves to identification of specific chromosomes. Again it appears that this disadvantage can be overcome, as reports have been published of Q-banding being combined with CREST staining (Earnshaw & Migeon 1985, Cherry & Shah 1987). Even if these limitations cannot be wholly overcome, however, CREST sera have proved themselves to be the reagents of choice for labelling kinetochores in both cell division and interphase.

11 Banding produced by treatment of living cells: condensation and replication banding

11.1 Introduction

The treatment of living cells with various substances, either in the intact organism or in cell cultures, often has significant effects on the appearance of the chromosomes after the cells have been fixed. Various types of effects can be distinguished, although it must be emphasized first, that there appears to be a continuum of effects rather than discrete classes, and second, that the same substance can be used to produce different effects on chromosomes according to the circumstances in which it is used. Some substances, when applied to living cells, show an overall effect on the condensation of chromosomes, either inhibiting contraction during prophase (e.g. Ikeuchi & Sasaki 1979, Matsubara & Nakagome 1983, Buys *et al.* 1984c), or causing excessive contraction of the chromosomes (e.g. Andersen *et al.* 1983, Andersen & Rønne 1983). Other treatments induce morphological differentiation along the length of the chromosomes, by inducing secondary constrictions (Sasaki & Makino 1963), fragile sites (Sutherland & Hecht 1985), or by inhibiting contraction of heterochromatin (Sections 11.1.1 and 11.2), or of other segments of chromosomes (Stubblefield 1964). Finally, some substances affect chromosomes in living cells in such a way that the fixed chromosomes can be banded simply by Giemsa staining, or by a special treatment which, in the absence of the *in vivo* treatment, does not produce bands (e.g. replication banding, Section 11.3). By a strict application of the definition of chromosome banding as variation in staining properties along the chromosomes in the absence of obvious structural differentiation (Section 1.1), the subject matter of this chapter should probably be restricted to the final category, but this seems unrealistic for two reasons. First, most of those substances that appear to produce banding patterns *sensu stricto* probably in fact also produce some morphological differentiation of the chromosomes, however slight; and second, many of the morphological differentiations produced by treatment of the living cells are closely related to, or identical with, different sorts of bands in the strict sense, so that a consideration of them is desirable for a full understanding of banding. The boundaries of the subject matter of this chapter will, therefore, be extended to include certain types of induced morphological differentiation, although others, such as secondary constrictions and fragile sites, will only be mentioned in passing. Certain secondary constrictions have, of course, been mentioned already as the sites of nucleolar organizers (Section 9.1).

11.1.1 Classification of the effects on chromosomes of treatment of living cells

Some agents, when applied to living cells, clearly produce patterns on chromosomes which resemble G- or R-bands; that is, their effects are primarily in the euchromatic parts of the chromosomes. Other substances obviously affect primarily the heterochromatic parts of chromosomes. While it is important to be able to classify these effects in terms of known types of banding patterns, it is in practice even more useful to classify them by the mechanisms by which the effects are produced. Two principal effects can be recognized: inhibition of chromosome condensation, and labelling of newly replicated DNA. In the former case, inhibition of condensation can be recognized as the elongation and thinning of specific segments of chromosomes. In the latter, patterns of DNA replication can be detected by incorporation of DNA base analogues which can later be identified by suitable staining procedures, thus providing an analogue of the long-standing practice of identifying newly replicated DNA by incorporation of tritiated thymidine, followed by autoradiography. We can, therefore, recognize two more classes of banding: condensation banding and replication banding. These are considered in detail below, in Sections 11.2 and 11.3 respectively.

11.2 Inhibition of condensation

As chromosomes pass through mitosis, they change from long thin threads at early prophase to shorter and thicker structures at metaphase. Thus it seems most appropriate to describe the effect of treatments which leave some chromosomal segments extended and narrow as inhibition of condensation; other terms which have been used, such as despiralization and decondensation, imply that a condensed chromosome segment has subsequently been unravelled, an interpretation that seems unlikely in most cases.

A variety of agents have been found to inhibit condensation of chromosome segments, and these are listed in Table 11.1. They comprise many classes of compounds, plus a purely physical agency, but three groups are of particular importance. These are cold treatment (Section 11.2.1), Hoechst 33258 and related substances that bind preferentially to A+T-rich DNA (Section 11.2.2), and 5-azacytidine, a DNA base analogue (Section 11.2.3), which will all be considered in detail below. There is little that can usefully be said about most of the other substances listed in Table 11.1; they have

Table 11.1 Agents causing differential inhibition of condensation of chromosomes.

Agent	Cells	Effect	Reference
Cold treatment	Many species	Heterochromatin uncondensed	Section 11.2.1
Spindle poisons			
Colchicine	Ladybirds	Heterochromatin contracts more than euchromatin	Smith 1965
Colcemid	Chinese hamster	Late replicating regions uncondensed	Stubblefield 1964
Oxyquinoline	Various plants	Stretching of centromeric regions	Tjio & Levan 1950
Nucleic acid base analogues			
5-azacytidine	Many species	Heterochromatin uncondensed	Section 11.2.3
Bromodeoxyuridine	Many species	Mild decondensation of regions where BrdU incorporated	Section 11.3
Fluorouridine	Human lymphocytes	G-banding	Rønne & Andersen 1978a
Cytosine arabinoside	Human lymphocytes	Decondensation to form chromomeres	Bastide *et al.* 1981
DNA-binding ligands			
Hoechst 33258			Section 11.2.2
Distamycin A	Many species	Heterochromatin uncondensed	
DAPI			
Berenil	Human lymphocytes	Heterochromatin of Y uncondensed	Haaf *et al.* 1989
Actinomycin D	Chinese hamster	Differential inhibition of condensation	Arrighi & Hsu 1965
Actinomycin D	Human lymphocytes	Decondensation to form chromomeres	Bastide *et al.* 1981
Actinomycin D	Human lymphocytes	G-banding	Rønne & Andersen 1978b
Actinomycin D	Termite spermatogonia	'G-band type'	Fontana & Goldoni 1986
Actinomycin D + ethidium bromide	Chinese hamster	Chromosomes extended, unidentified banding pattern	Hsu *et al.* 1973
Daunomycin	Human lymphocytes	G-banding	Rønne 1977a
Ethidium bromide	Human lymphocytes	G-banding	Rønne & Andersen 1978b
Mitomycin C	Human cells	Heterochromatin uncondensed	Cohen & Shaw 1964
		Chromomere pattern	Nowell 1964

Table 11.1 (cont.)

Agent	Cells	Effect	Reference
Agents affecting proteins			
Cycloheximide	Human lymphocytes	G-banding	Rønne 1977b
Histidinol } Parafluoro phenylalanine }	Human lymphocytes	G-banding	Andersen & Rønne 1978
2,2'-dithio-dipyridine	Human lymphocytes	Random pattern of inhibition of condensation	Kosztolanyi & Bühler 1978
Dithioerythritol	Human lymphocytes	G-banding	Rønne 1979
Fucidin	Human lymphocytes	G-banding	Sandermann & Rønne 1977
Puromycin	Human lymphocytes	G-banding	Rønne & Andersen 1978b
Miscellaneous			
α-amanitin } Methotrexate }	Human lymphocytes	G-banding	Rønne & Andersen 1978b
Rifampicin + amphotericin		G-banding	Shafer 1974
Tuberculostatic drugs	Human lymphocytes	? G-banding	Caratzali *et al.* 1972
Hydroxylamine	Chinese hamster cells	'Banded pattern'	Somers & Hsu 1962
Streptonigrin	Human lymphocytes	Centromeres and other regions attenuated	Cohen *et al.* 1963
Beryllium salts	*Allium* root tips	Negative staining segments	Fiskesjö 1988

neither become useful practical tools, nor has any clear mechanism for their action been suggested. A few substances, however, deserve a less abrupt dismissal. The ability of the spindle poisons colchicine and Colcemid to induce banding patterns by inhibition of condensation is an important point to note, if only because these substances are so widely used to arrest dividing cells at metaphase for chromosome studies. Bromodeoxyuridine (BrdU) is used essentially for the study of replication patterns, and will be described fully in Section 11.3; nevertheless, it should be mentioned here that incorporation of BrdU into chromosomes can inhibit their condensation. A third substance to be noted is actinomycin D, which like Hoechst 33258 and related compounds binds to chromosomal DNA, but unlike them appears to produce a G-banding pattern (Hsu *et al.* 1973).

11.2.1 Cold treatment

The effect of cold treatment of plants in producing 'differential reactivity' of their chromosomes was reported by Darlington & La Cour (1940), who described the phenomenon thus:

> 'Under ordinary conditions their chromatids have a uniform staining capacity, and they also have a uniform diameter, ... Under special

conditions this uniformity breaks down. Certain segments in each chromosome are reduced to half the standard diameter of the chromatid and appear under-stained from late prophase until anaphase. They remain their usual length.' (Fig. 11.1).

This description of the effect of exposure to cold on the heterochromatic segments of chromosomes of the plants *Paris*, *Trillium* and *Fritillaria* can scarcely be bettered, although the explanation in terms of 'nucleic acid charge' is meaningless in the light of modern knowledge of chromosome organization.

Although *Trillium* chromosomes have remained the most popular subject for the study of the effects of cold treatment on heterochromatic segments (Wilson & Boothroyd 1944, Boothroyd 1953, Schweizer 1973, Fukuda & Grant 1980), similar effects have been described in a wide variety of plant species (Wilson & Boothroyd 1944, Yamasaki 1956, Dyer 1963, Baumann 1971). Cold-induced differential heterochromatic segments have also been reported from animal chromosomes: those of the amphibia *Triturus* spp. (Callan 1942, Rudak & Callan 1976) and *Ambystoma* (Signoret 1965), and

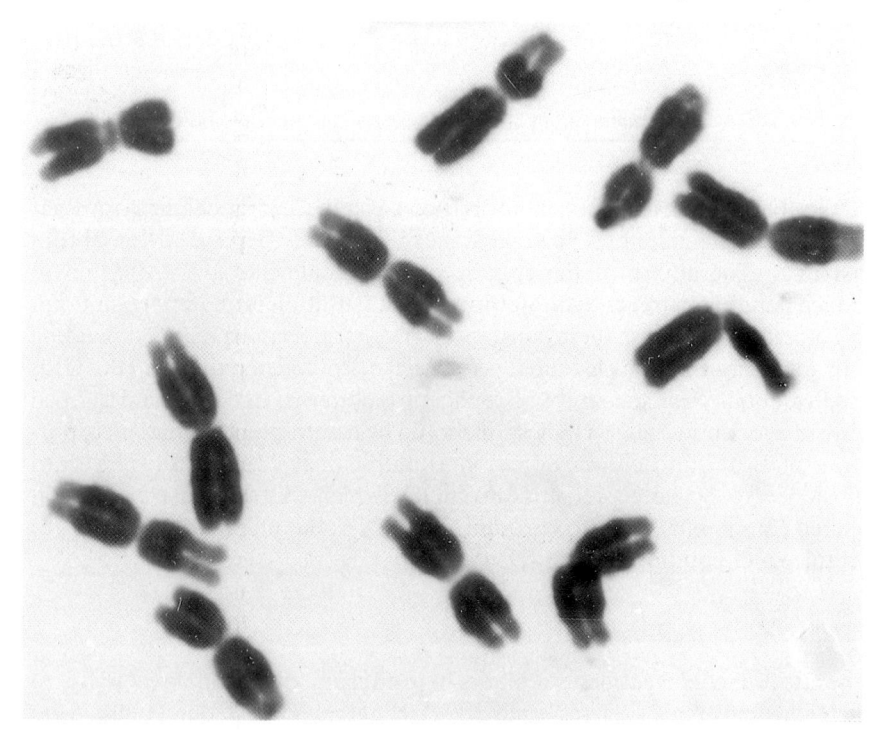

Figure 11.1 Effect of cold treatment on chromosomes of the plant *Tulbaghia cernua*; note the reduced width of certain, mainly terminal, segments of the chromosomes. Reproduced from Vosa (1985) in *Advances in chromosome and cell genetics*, A. K. Sharma & A. Sharma (eds), 79–104 by permission of the author and IBH Publishing House, India.

also in human chromosomes (Hampel & Levan 1964, Shiraishi 1970). In general, the weakly stained, thinner regions of the chromosomes seen after cold treatment correspond to regions known, by other criteria, to be heterochromatic, although not all heterochromatin shows differential reactivity after cold treatment. Indeed, it may be that the majority of blocks of heterochromatin in the majority of species, both animal and plant, do not show cold-induced differential reactivity. Exceptions to this rule that cold-induced segments correspond to heterochromatin occur in human chromosomes, where the pale, cold-induced segments are late replicating (Shiraishi 1972), and the patterns on the chromosomes show a general resemblance to R-banding (Shiraishi 1970) and in *Ambystoma macrodactylum*, where the cold-induced segments are not C-banded (Kezer *et al.* 1980).

The mechanism by which cold-induced differential reactivity is produced is not properly understood. Early explanations were made essentially in terms of DNA content, even though the concepts of the role of DNA in chromosome organization 50 years ago seem curious to a present-day reader. Nevertheless, the idea that these segments were 'undercharged with nucleic acid' (Darlington & La Cour 1940) was not an unreasonable suggestion, which took many years to refute. However, Woodard & Swift (1964) showed that no loss of DNA was induced by cold treatment, and that disappearance of cold-induced segments at room temperature did not involve DNA synthesis. Boothroyd & Lima-de-Faria (1964) also showed that the differentiation could be induced by cold after DNA synthesis had been completed.

An alternative explanation for the appearance of cold-induced segments is that they represent regions of differential chromatin packing. Although Darlington & La Cour (1940) stated that such segments 'remain their usual length', the measurements of Wilson & Boothroyd (1944) and of Boothroyd (1953) showed that differential contraction did, in fact, occur. This conclusion was supported by the results of electron microscopy; however, while Braselton (1973) and La Cour & Wells (1974) found that cold-induced segments in plants were less condensed than the remainder of the mitotic chromatin, Rudak & Callan (1976) found that such segments in amphibian chromosomes were more tightly compacted, so that different sorts of changes appear to occur in different organisms. The mechanism by which cold treatment causes these changes does not seem to have been studied.

Since the development of post-fixation banding procedures, there has been little incentive to use cold treatment to produce longitudinal differentiation of chromosomes, and the procedures are now largely of historical interest. Nevertheless, cold treatment may still be regarded as a useful technique, in certain plants and amphibia, for differentiation of a particular class of heterochromatin.

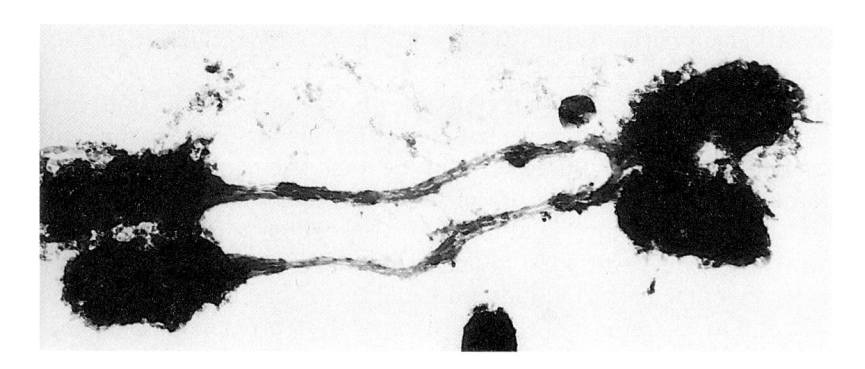

Figure 11.2 Electron micrograph of a mouse chromosome showing inhibition of condensation of the heterochromatin after culture in the presence of Hoechst 33258. Micrograph kindly provided by Professor J. Gosalvez.

11.2.2 Effects of Hoechst 33258 and similar compounds

Hilwig & Gropp (1973) showed that when mouse cells were grown in the presence of Hoechst 33258, the constitutive heterochromatin of the mouse chromosomes was uncondensed, appearing as a long thin thread (Fig. 11.2). Since then similar effects have been reported in a wide variety of animal species, and it has also been found that they can be induced by a variety of ligands that show preference for A + T-rich DNA. As well as Hoechst 33258, ligands that inhibit condensation of heterochromatin are distamycin and DAPI; netropsin has also been used occasionally (Thust & Rønne 1980).

The different ligands do not necessarily produce the same effects in a particular species, and not all heterochromatic segments in any one species are affected in the same way. Even in the mouse, subtle differences in the effect of Hoechst 33258 on the heterochromatin of different chromosomes can be detected (Marcus *et al.* 1979b). Originally it was supposed that Hoechst 33258 was without effect on human chromosomes in living cells (Kim & Grzeschik 1974), but this was found to depend on cell type. In mouse-human hybrid cells, and in fibroblasts and lymphoblasts, but not in peripheral blood lymphocytes, the heterochromatin of the human Y is greatly extended under the influence of the dye (Marcus *et al.* 1979a, c). Other blocks of heterochromatin, on the human autosomes, are affected only to a limited extent. The differences in reaction to Hoechst 33258 in different types of cells are probably due to differences in the permeability of the membranes of these cells (Marcus *et al.* 1979a), and should not be attributed to cell-specific differences in the organization of heterochromatin. Pimpinelli *et al.* (1976b) showed, in fact, that even in human peripheral blood lymphocytes, inhibition of condensation could be obtained if the concentration of Hoechst 33258 was high enough.

Hoechst 33258 also inhibits heterochromatin condensation in *Drosophila* spp. (Pimpinelli *et al.* 1975, Gatti *et al.* 1976), and in several species of marsupials (Hayman & Sharp 1981). In these cases, not all heterochromatic regions were affected, and it is clear that this procedure can be used to demonstrate a subset of heterochromatic bands. In the Chinese hamster, however, it is not the limited regions of heterochromatin whose condensation is inhibited, but a larger number of segments throughout the euchromatic parts of the chromosomes (Fig. 11.3) (Rocchi et al. 1976, Thust & Rønne 1981); no attempt seems to have been made to compare the resultant pattern with, for example, the G-banding pattern of the same chromosomes.

DAPI and distamycin, when applied to living cells, produce generally the same effects on chromosomes as Hoechst 33258. Although they inhibit the condensation of heterochromatin in most of the species studies, the detailed patterns observed after treatment of human chromosomes differ somewhat between these two compounds, and from the pattern induced by Hoechst 33258 (Table 11.2). As with Hoechst 33258, the heterochromatin of the human Y chromosome is particularly susceptible to the effect of distamycin (Jotterand-Bellomo 1983, Schmid et al. 1984c), and these authors have suggested that this could prove a useful diagnostic tool. In the great apes and man as a whole, it is generally the quinacrine-bright heterochromatin that is preferentially uncondensed (Schmid et al. 1981).

Like Hoechst 33258, DAPI and distamycin produce a general lengthening

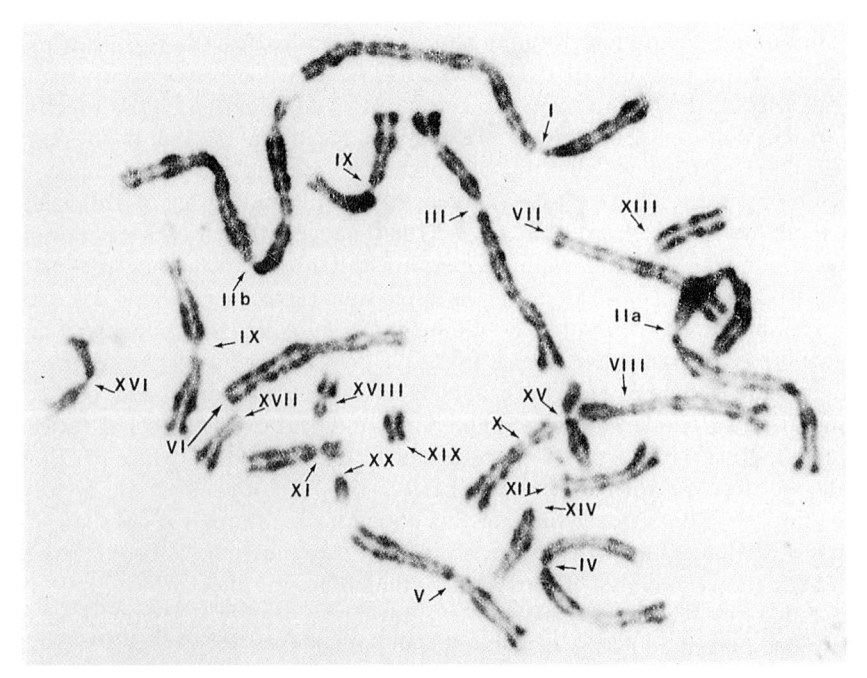

Figure 11.3 Inhibition of condensation of Chinese hamster chromosomes by Hoechst 33258, showing regions of undercondensation throughout the length of the chromosomes. Reproduced from Thust & Rønne (1981) by permission of the authors and the publishers of Hereditas.

Table 11.2 Patterns of human heterochromatin decondensation induced by A + T-specific ligands (after Rocchi et al. 1979).

Substance	Affected chromosome segments*						
	1c	1q	2q	3c	9c	16c	Y
DAPI	+	+	+	−	+	+	+
Hoechst 33258	+	+	+	+	+	−	−
Distamycin A	+	−	−	+	+	−	+

*c = centromeric region
9 = heterochromatic segment on long arm adjacent to centromere
Y = heterochromatic segment on long arm of Y chromosome

of Chinese hamster chromosomes, and all three compounds produce essentially the same, reproducible pattern of segmentation (Prantera et al. 1977, Rocchi et al. 1980). In chromosomes of cattle, a general lengthening under the influence of distamycin is also observed, but the G + C-rich centromeric heterochromatin condenses normally (Rønne et al. 1982). Distamycin also inhibits the condensation of quinacrine-bright heterochromatin in the chromosomes of *Drosophila melanogaster* (Faccio Dolfini & Bonifazio Razzini 1983). In other species of *Drosophila*, however, the correlation between quinacrine brightness and inhibition of condensation is much less good (Faccio Dolfini 1987).

In human, mouse and Chinese hamster chromosomes, the regions whose condensation is inhibited by Hoechst 33258, DAPI or distamycin can be readily stained with silver, in the manner used for staining NORs (Section 9.4) (Haaf et al. 1984, Prantera et al. 1986, Rocchi et al. 1986). The same effect is observed whether the uncondensed regions are euchromatic or heterochromatic (Fig. 11.4). The reason for the induction of this silver stainability is not clear: Haaf et al. (1984) suggest that the silver staining material might be similar to that found in NORs, while Prantera et al. (1986) and Rocchi et al. (1986) attribute the silver staining to the chromosome core.

The mechanism of inhibition of condensation by ligands showing preference for A + T-rich DNA seems fairly clear in outline, even though many details are still lacking. The general correlation, already noted, between inhibition of condensation and the A + T-richness of the affected regions indicates that the effect is due to base specific binding of the ligands to DNA. The fact that the correlation is not perfect, as in *Drosophila* species (Faccio Dolfini 1987) does not invalidate this conclusion, although it does suggest that other factors may also be involved. Nor is it known how the binding of these ligands inhibits chromosome condensation, although it may be assumed that the binding to DNA of certain proteins necessary for condensation is prevented by the presence of the ligand.

Careful studies show that both Hoechst 33258 and DAPI are effective in the G2 phase, after DNA synthesis is complete (Marcus et al. 1979a, Prantera et al. 1981). Evidence that Hoechst 33258 inhibits condensation

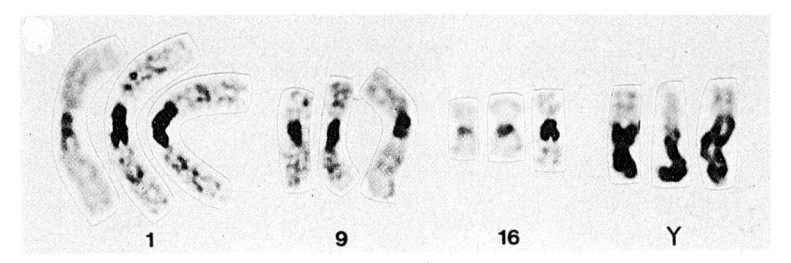

Figure 11.4 Silver staining of undercondensed regions of human chromosomes induced by culture in the presence of DAPI. Reproduced from Haaf *et al.* (1984) by permission of the authors and Springer-Verlag.

rather than decondensing segments of chromosomes that were already condensed was obtained by Marcus *et al.* (1979b). First, there is no effect on chromosomes that are already condensed, and second, prolonged incubation of mitotic cells in the presence of both Colcemid and Hoechst 33258 does not result in the latter decondensing the chromosomes.

It should be noted that although actinomycin D can inhibit condensation of euchromatic bands (Section 11.2), ligands that show a preference for G + C-rich DNA have not, so far, been found to inhibit the condensation of G + C-rich heterochromatin (Thust & Rønne 1981). This may, perhaps, be related to some differences in their mode of binding to DNA when compared to that of ligands which bind to A + T-rich DNA.

11.2.3 5-azacytidine

Like cold treatment, and treatment with Hoechst 33258 and similar compounds, treatment of living cells with 5-azacytidine (5-azaC) inhibits the condensation of chromosomes or parts thereof. However, it differs from these other agents in that it must be incorporated into the chromosomal DNA, and must therefore be present during the S phase to produce its effect. It is an analogue of the naturally-occurring base cytidine, and will thus be preferentially incorporated into G + C-rich DNA.

The inhibition of chromosome condensation by 5-azaC seems to have been reported first by Fucik *et al.* (1970), who studied the chromosomes of root tip meristems of *Vicia faba*, but most work on the effects of 5-azaC has been done on the chromosomes of humans and other mammals. Viegas-Péquignot & Dutrillaux (1976) recognized four effects of this compound on human chromosomes. The first effect, which they described as segmentation, was due to the inhibition of condensation of the G-band regions of the chromosomes, producing a pattern almost indistinguishable from R-banding (Fig. 11.5a). In more extreme cases, the segmentation of the chromosomes is much more extensive, producing a pulverized appearance, when it becomes impossible to recognize many of the chromosomes. A third effect is the extension of the heterochromatin (Fig. 11.5b). Schmid *et al.* (1983a) pointed out that the regions of heterochromatin involved, on chromosomes 1, 9, 15, 16, and Y, were all characterized by bright distamycin/DAPI staining, and contained highly methylated DNA. Undercondensation of these blocks of heterochromatin is favoured if the 5-azaC is applied at low doses during the last hours of culture, whereas higher doses tend to produce the R-banded and pulverized appearances (Schmid et al. 1984a). The fourth effect observed by Viegas-Péquignot & Dutrillaux (1976), and also by Schmid et al. (1983a), is the association of chromosomes (Fig. 11.5c). A fifth effect was reported by Haaf et al. (1986b), who showed that when cells were allowed to go through two cycles of replication in the presence of 5-azaC, the chromatid that was bifilarly substituted with 5-azaC was greatly elongated compared with the unifilarly substituted chromatid, producing some grotesque figures (Fig. 11.5d). The condensation of the inactive X chromosome is inhibited by 5-azadeoxycytidine (5-aza-dC) in a small proportion of human female cells (Haaf et al. 1988a). All these effects may be regarded as manifestations of the ability of 5-azaC and 5-aza-dC to inhibit chromosome condensation, the differences being due solely to the time and concentration in which this DNA base analogue is presented, thereby affecting which chromosomal regions are substituted.

Viegas-Péquignot & Dutrillaux (1981) found that 5-azaC, when applied to cultures during late S-phase, inhibited condensation of the R-band positive heterochromatin (presumed to be G+C-rich) in the chromosomes of *Bos taurus*, and in certain species of primates. However, the effect of 5-azaC is not confined to G+C-rich chromatin. In the chromosomes of chimpanzee (Schmid & Haaf 1984) and gorilla (Schmid et al. 1986) numerous blocks of heterochromatin are affected by 5-azaC, some of which are also distamycin/DAPI positive and therefore presumably A+T-rich (Section 8.5.2). On the other hand, several of the distamycin/DAPI positive bands are unaffected by 5-azaC. Moreover, different blocks of heterochromatin are affected to different extents by 5-azaC. Clearly the action of 5-azaC on chromosomes is not simple, and is evidently not restricted to G+C-rich DNA. Indeed, as already noted, under the appropriate conditions the relatively A+T-rich G-bands are decondensed, giving an R-banding pattern. It should be noted that, contrary to the situation with Hoechst 33258 and related compounds (Section 11.2.2), chromosomal regions uncondensed after 5-azaC treatment do not stain with silver (Haaf et al. 1984).

When cells replicate their DNA in the presence of 5-azaC, this base analogue is incorporated into the newly synthesized DNA, and its effects on chromosome condensation are therefore attributed to its incorporation into chromosomal DNA. Thus, although 5-azaC inhibits chromosome condensation, it does not indicate the sequence of condensation, but rather the sequence of replication. The pattern produced by 5-azaC is determined essentially by when it is incorporated in the DNA during S phase, and it is not effective during G2. Note, however, that Haaf & Schmid (1989) claim that 5-aza-dC can still inhibit condensation when supplied after DNA replication is complete.

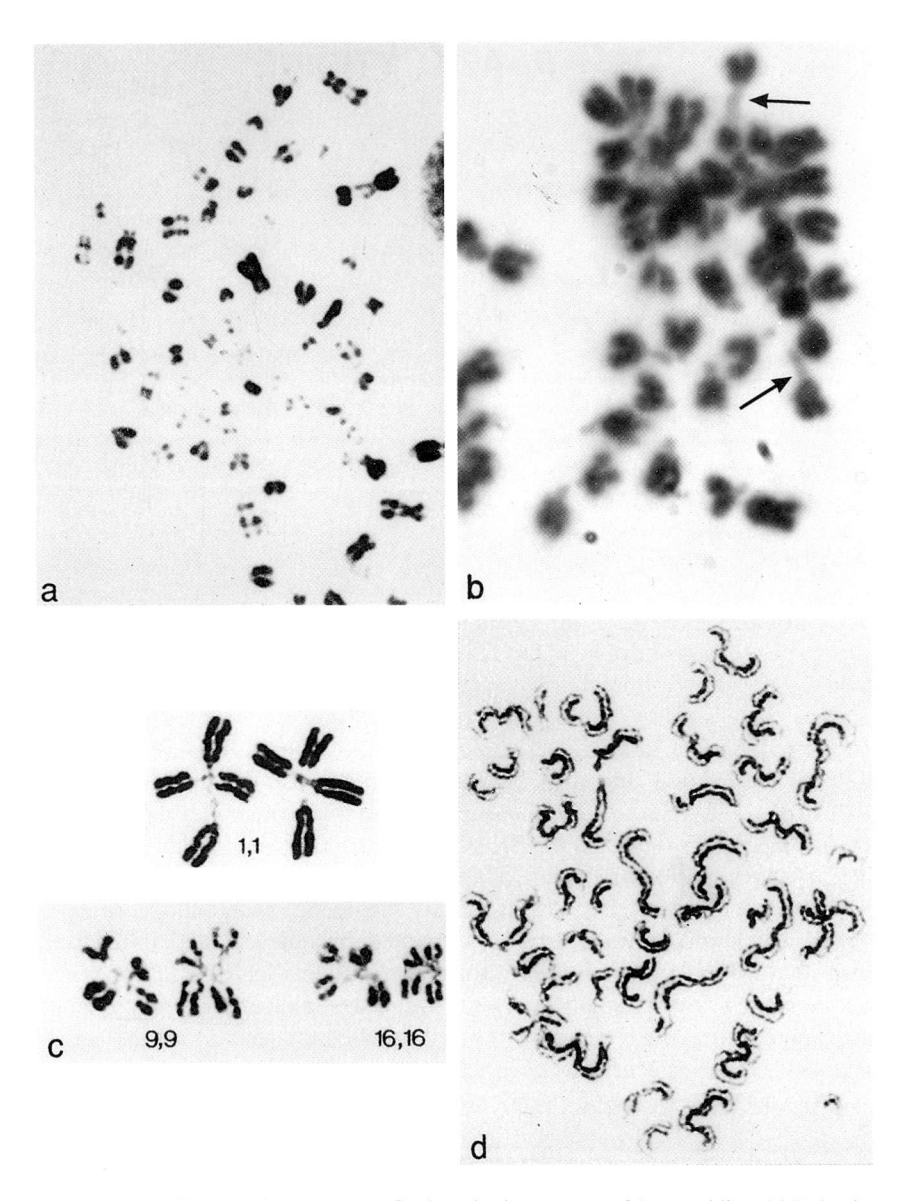

Figure 11.5 Effects on chromosomes of culture in the presence of 5-azacytidine. (a) Induction in human chromosomes of a pattern resembling R-banding. Reproduced from Schmid *et al.* (1984a) by permission of the authors and Springer-Verlag. (b) Inhibition of condensation of mouse heterochromatin; note the great extension of the centromeres (arrows). (c) Induction of associations between heterochromatin in human chromosomes. Reproduced from Schmid *et al.* (1983a) by permission of the authors and S. Karger. (d) Sister chromatid differentiation in human chromosomes, after culture through two cell cycles in the presence of 5-azaC. The chromatid that is bifilarly substituted with 5-azaC is elongated, producing some grotesque figures. Reproduced from Haaf *et al.* (1986b) by permission of the authors and Springer-Verlag.

5–AZACYTIDINE

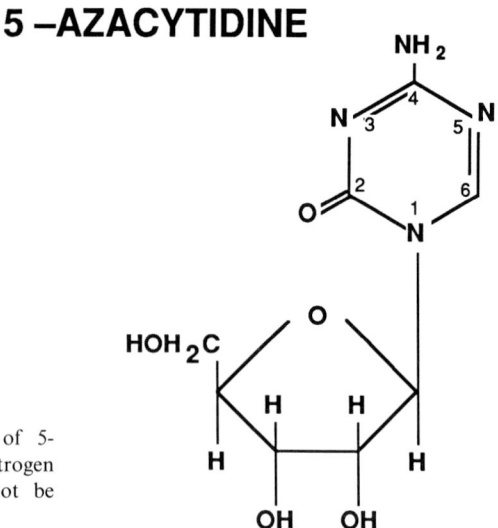

Figure 11.6 Structural formula of 5-azacytidine (5-azaC). Note the nitrogen atom at position 5, which cannot be methylated.

It is not known how 5-azaC inhibits chromosome condensation, although various suggestions have been put forward. The simplest is that the different structure of 5-azaC compared with cytidine (Fig. 11.6) could inhibit condensation by a steric effect (Schmid & Haaf 1984). However, 5-azaC also inhibits the methylation of newly synthesized DNA, not simply because it cannot itself be methylated, but also because it inhibits the methylase involved (Christman 1984, Santi et al. 1984). Why the under-methylation of DNA should inhibit chromosome condensation is not known, but there is nevertheless one situation in which a reduced level of DNA methylation is associated with extended chromosome segments. Mammalian satellite DNAs, which normally are heavily methylated, become less methylated than main band DNA during spermatogenesis (Adams et al. 1983). When chromosomes from human spermatozoa are examined, the regions of heterochromatin which contain the satellite DNA are found to be extended (Rudak et al. 1978, Martin et al. 1983).

11.3 Replication banding: BrdU substitution

If cells replicate their DNA in the presence of a suitably labelled precursor, the DNA becomes labelled in such a way as to indicate the time or sequence of replication. This has become a very powerful tool, not only to study patterns of replication within chromosomes (replication banding), but also as a means of distinguishing between sister chromatids (sister chromatid differentiation). Only the former is the subject of this section, although the technology and mechanism of sister chromatid differentiation are similar, the essential differences being in the period for which the label is administered, and the length of time for which the cells are allowed to grow after they have replicated their DNA in the presence of the label. A comparison of the two procedures is shown diagrammatically in Figure 11.7.

The original label used for the study of DNA replication patterns was tritiated thymidine, which was detected by autoradiography. Although the autoradiographic resolution was not as good as could have been wished, much valuable work was done using this label (O. J. Miller 1970). We have seen in the previous section (11.2.3) that the patterns produced in chromosomes by 5-azacytidine are essentially replication patterns, but with the added complication that 5-azaC substitution in the DNA is detected by inhibition of chromosome condensation. There is, however, no doubt that the label of choice for DNA replication studies in chromosomes is 5-bromo-deoxyuridine (BrdU) (Fig. 11.8). Once it was discovered how to detect DNA containing BrdU using relatively simple staining methods, there was an enormous explosion of research not only on replication patterns within chromosomes, but also in the field of sister chromatid exchange (for a review of the latter field, see Wolff 1982).

It is perhaps ironic that the original effects of BrdU on chromosomes that were reported involved the inhibition of chromosome condensation. Hsu & Somers (1961) showed that BrdU inhibited the condensation of heterochromatin in mouse and Chinese hamster chromosomes, and similar obser-

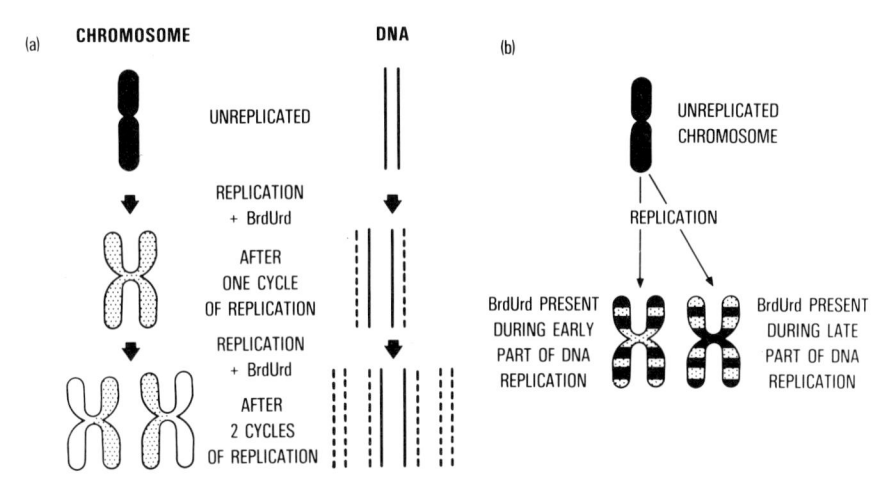

Figure 11.7 Diagrams illustrating the principle of detecting chromosome replication by BrdU substitution. (a) Sister chromatid differentiation. Initially each chromosome consists of a double strand of unsubstituted DNA. After one round of replication in the presence of BrdU each daughter DNA molecule contains one substituted (broken line) and one unsubstituted strand (continuous line); the chromatids therefore stain equally, but more lightly than the parental chromosome. After a second round of replication in the presence of BrdU, the daughter chromosomes each have one chromatid with DNA substituted in both strands, and the other chromatid substituted in only one strand. The former therefore stains more weakly. (b) Detection of replication patterns. BrdU is supplied to the chromosomes during only part of the replication (S) phase; those regions that incorporate BrdU show pale staining. Reproduced from Sumner (1983) *Science Progress* **68**, 543–64, published by Blackwell Scientific Publications.

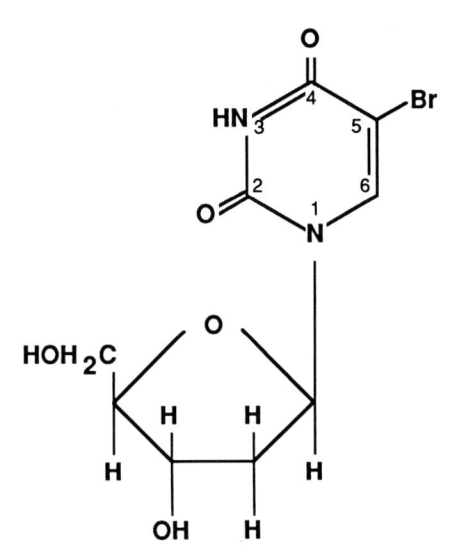

Figure 11.8 Structural formula of 5-bromodeoxyuridine (BrdU).

vations were made on human chromosomes by several authors (Kaback et al. 1964, Palmer 1970). More detailed studies showed that patterns of decondensation on human chromosomes could be obtained that corresponded very closely to R-banding (Zakharov et al. 1974) while in Chinese hamster chromosomes the late replicating regions were shown to be decondensed after BrdU treatment (Zakharov & Egolina 1972). Baranovskaya et al. (1972) also showed that the early and late replicating X chromosomes could be distinguished by this method. However, the remainder of this section will be concerned with methods in which BrdU substitution in the chromosomal DNA is detected by special staining methods, without any substantial decondensation of the chromosome. Such methods were originally developed for detecting sister chromatid exchanges (SCEs) (Latt 1973, Perry & Wolff 1974), but the technology was quickly adapted for studying replication patterns.

11.3.1 *Methods for producing replication banding*

Methods for producing replication bands on chromosomes consist of two parts: incorporation of BrdU into chromosomes, which must take place in the living cell, and the subsequent differential staining of the fixed chromosomes. Incorporation of BrdU is commonly done by culturing the cells in a medium that contains BrdU, a method which can be applied conveniently to any type of cell that can be cultured. This is indeed the best method, since the concentration of BrdU and the length of time for which it is supplied can be controlled accurately. When the cells of the organism cannot easily be cultured, the BrdU must be supplied in a different way. Plants are usually grown in a solution containing BrdU, so that the root tips have direct access to this base analogue. For certain animals, injection has been used (e.g. di Castro et al. 1979, Schempp & Schmid 1981). Different types of cells have different cell cycle times, so that the period of administration of BrdU must be adjusted to allow for this. In addition, it must be decided whether to label the DNA which replicates early in the S phase, or that which replicates later. The procedure given here is designed for human lymphocytes. Cultures are set up in the standard way, except that BrdU and certain other substances are added at an appropriate time. The concentrations of these substances are:

Bromodeoxyuridine	10^{-4}M
Fluorodeoxyuridine (FdU)	10^{-6}M
Deoxyuridine (dU)	6×10^{-6}M
Deoxycytidine (dC)	10^{-4}M

FdU and dU are not strictly necessary, but help to increase the incorporation of BrdU into the newly synthesized DNA. To obtain a late replication pattern, these substances are added at the beginning of culture, and then, 5–7 h before harvesting the culture, the cells are washed and transferred to standard culture medium containing 6×10^{-4}M deoxythymidine. This is known as a T pulse, as a result of which late replicating DNA is unsubstituted with BrdU.

To obtain an early replication pattern, the cells are grown in normal culture medium until 5–7 h before harvesting, and then BrdU and the other base analogues are simply added to the culture medium to give the prescribed concentrations. This is known as a B pulse. BrdU is light sensitive, and solutions containing it should be kept in the dark as far as possible, or handled in red light.

Differential staining methods for BrdU-substituted chromosomes fall into three main classes. Originally simple fluorescence methods were used, the only disadvantages of which are the impermanence of fluorescence, and the need for special equipment. The first fluorochrome to be used was acridine orange (Dutrillaux et al. 1973a), but Hoechst 33258 was introduced for the same purpose shortly afterwards (Latt 1973) and has been widely used. M. S. Lin et al. (1976) proposed the use of DAPI, and a few other fluorochromes have been used to demonstrate replication banding, although they have not been used routinely. In all cases, the BrdU-substituted chromatin shows reduced fluorescence. By no means all fluorochromes that bind to DNA can be used to demonstrate BrdU substitution, however.

The second class of methods for differential staining of BrdU-substituted chromatin are those which involve some post-fixation treatment followed by staining with a dye visible by transmitted light. The prototype of such methods is the Fluorescence-Plus-Giemsa (FPG) method of Perry & Wolff (1974). This popular method involves staining the chromosome preparations with Hoechst 33258, exposure to light, and staining with Giemsa, and is given in detail below. A number of minor variations of the same principle have been published (Buys et al. 1981b, Shiraishi et al. 1982, Ieshima et al.

1984), although other workers have used completely different pretreatments while retaining Giemsa staining. Korenberg & Freedlender (1974) omitted staining with Hoechst 33258 and exposure to light and obtained differentiation of substituted and unsubstituted DNA merely by treating with hot phosphate solution (pH 8.0 at 87–89°C for 10 min), followed by Giemsa staining. In all the methods described so far, the DNA substituted with BrdU is stained more weakly than unsubstituted DNA, but a reverse pattern of staining can be obtained using appropriate conditions. Thus Burkholder (1978) found that the unsubstituted DNA was darkly stained with Giemsa if pretreatment was with phosphate at high pH, but that if low pH treatment was used, the unsubstituted DNA stained weakly. Many similar observations have been made in the context of sister chromatid differentiation (e.g. Scheres *et al.* 1977, Takayama & Tachibana 1980, Jan *et al.* 1982) but these

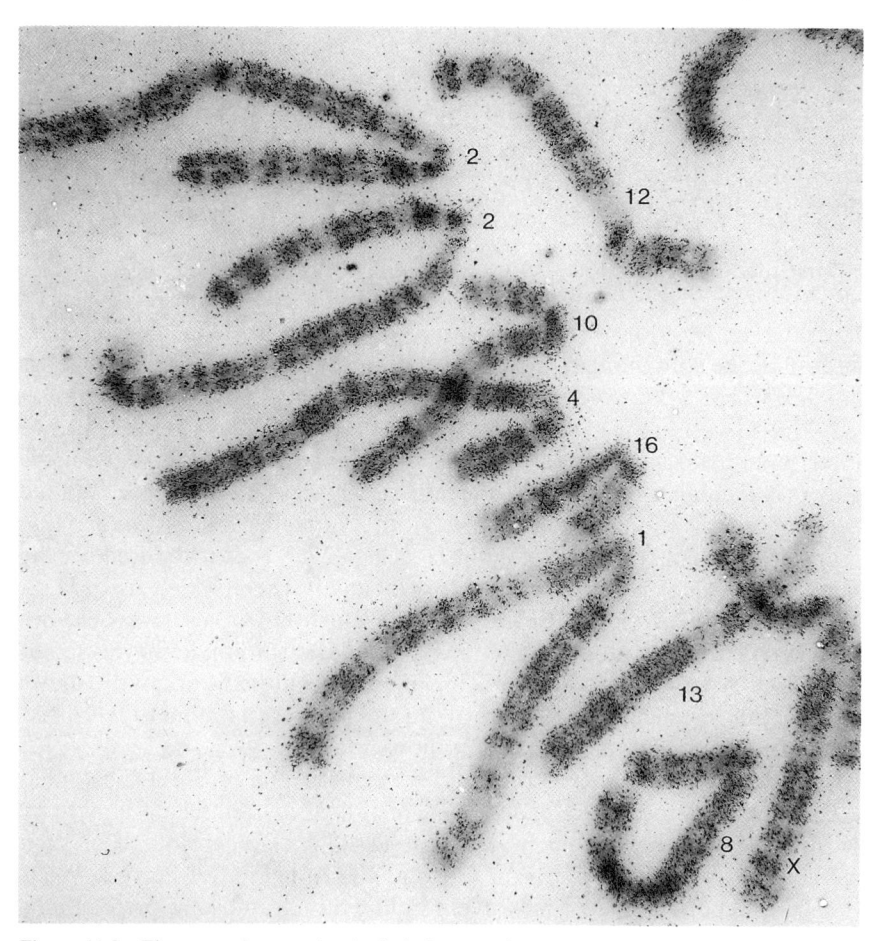

Figure 11.9 Electron micrograph of whole human chromosomes substituted with BrdU to show the late labelling pattern, detected by anti-BrdU and gold labelling, as described by Drouin *et al.* (1988b). Micrograph kindly provided by Dr P.-E. Messier.

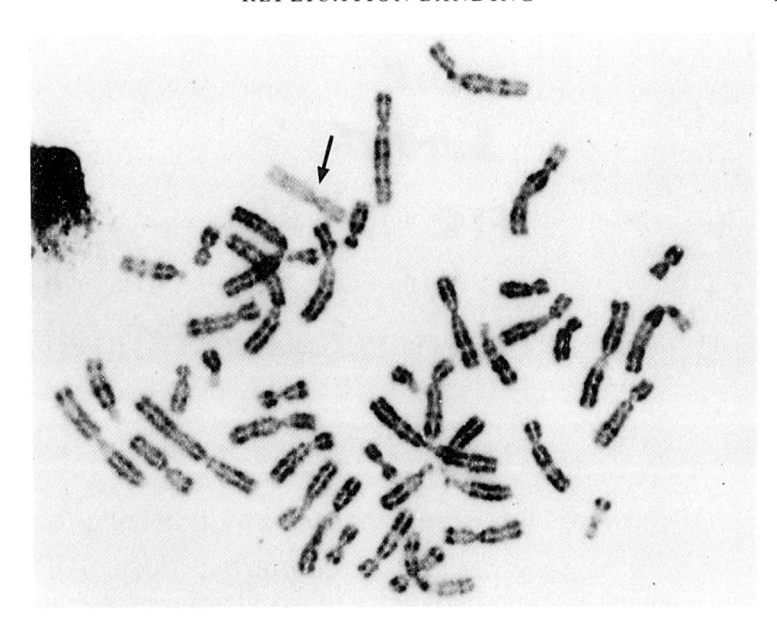

Figure 11.10 Human female metaphase chromosomes cultured in the presence of BrdU to show late replicating regions of the chromosomes (weakly stained). Note the late replicating, inactive, X chromosome (arrowed). Micrograph kindly provided by Miss K. E. Buckton. Reproduced from Sumner (1983) *Science Progress* **68**, 543–64, published by Blackwell Scientific Publications.

procedures have only recently been used for replication banding (Aghamohammadi & Savage 1990). Several staining methods other than Giemsa were used in these procedures, including silver staining, which was used for replication banding by Vogel *et al.* (1978); the BrdU-substituted chromatin failed to stain with ammoniacal silver nitrate, although the authors regarded the method as inferior to more conventional Giemsa staining methods.

The third class of method for detecting chromatin substituted with BrdU is those that use immunocytochemical procedures (Vogel *et al.* 1986, Latos-Bielenska *et al.* 1987). These methods are reported to show greater sensitivity than other differential staining methods, and will no doubt increase in importance now that antibodies to BrdU are available commercially. Denaturation of the chromosomal DNA is required before applying the antibody, which can then be detected using a suitably labelled second antibody. Vogel *et al.* (1986) used fluorescein, horseradish peroxidase and alkaline phosphatase as labels, and recently the method has been used for electron microscopy (Fig. 11.9) using colloidal gold as label (Drouin *et al.* 1988b).

The FPG staining method (Perry & Wolff 1974) is performed as follows:

1 After standard methanol-acetic acid fixation, the chromosome preparations are first stained for 15 min in a solution containing 0.5 μg Hoechst 33258/ml of deionized water.

2 After rinsing, the slides are mounted in deionized water or buffer, pH 6.8 (using buffer tablets), the coverslips sealed with rubber solution, and the slides exposed to light. If daylight is used, 12 h or more exposure is required (in Scotland), and a white fluorescent tube requires several hours. The most convenient source is a 20 w black light tube, which requires exposure for 5–15 min at a distance of 10–15 cm.

3 After exposure to light, the coverslip is removed, the slides are incubated in $2 \times$ SSC (0.3M NaCl + 0.03M tri-sodium citrate) for 30 min at 60°C, and rinsed.

4 Finally, the slides are stained in 3% Giemsa at pH 6.8 (Gurr's buffer tablets) for 2–3 min or as long as needed.

A result of this type of procedure is shown in Figure 11.10.

11.3.2 Mechanisms of differential staining after BrdU substitution

In the case of immunocytochemical labelling of BrdU-substituted DNA, differential staining is simply the result of specific attachment of antibody to antigen. In the other cases, more complex mechanisms are involved. Fluorescence methods are the simplest to understand, but even here there is apparently more than one mechanism. With Giemsa staining methods, and other methods involving dyes viewed by transmitted light, the requirement for pretreatments, and the reverse patterns that can be obtained in some situations, indicate that there are more intricate mechanisms. It should be noted that much of the work on mechanisms of differential staining after BrdU substitution has been done on differentially substituted sister chromatids, rather than on replication banding. Nevertheless, there is no good evidence that significantly different mechanisms are involved, although in the former case the distinction is between unifilarly and bifilary substituted DNA, while in replication banding the difference is between unsubstituted and unifilary substituted DNA.

Fluorescence of Hoechst 33258 is markedly quenched when it is bound to DNA containing BrdU (Latt 1973), and this is undoubtedly the primary mechanism of differential staining of BrdU-substituted chromatin (Latt *et al.* 1975). It should be noted, however, that differential fluorescence with Hoechst 33258 is not obtained in the region of pH 4, but only around pH 7–7.5. Irradiation of BrdU-substituted chromatin stained with Hoechst 33258 can also produce changes in chromatin that are important for subsequent differential staining with Giemsa (see below), but are only of secondary, if any, importance in producing differential fluorescence with Hoechst 33258. Differential fluorescence with other dyes does not appear to be a result of quenching by BrdU, but is apparently the result of preferential photolysis of DNA that contains BrdU, when it is irradiated in the presence of the fluorochrome; such an effect has been reported for acridine orange (Scheid 1976), DAPI (Scheid 1979), and mithramycin (Buys & Osinga 1980b).

Photolysis of DNA is apparently also involved in differential staining with Giemsa in those methods, such as FPG, which involve prior exposure to light, but not in all Giemsa staining methods. Numerous authors have shown that when chromosomes containing BrdU are irradiated either directly with UV light, or in the presence of a photosensitizing dye, DNA which has a higher level of BrdU substitution is preferentially destroyed (Goto et al. 1975, Scheid 1976, Burkholder 1979, Ockey 1980, Webber et al. 1981, Jan et al. 1984). After photolysis the preferential loss of DNA can be detected by staining with DNA-specific dyes or by autoradiography. Further evidence for the importance of photolysis is provided by the results of Scheid (1979) and of Jan et al. (1984), who showed that reagents that inhibit photolysis of DNA also inhibit differential staining of irradiated chromosomes substituted with BrdU. Loss of DNA due to photolysis would then be sufficient to account for the darker staining with Giemsa and other dyes of chromatin containing unsubstituted DNA. Subsequent treatment with salt solutions, acid, enzymes or urea would merely enable the photolyzed DNA to escape from the surrounding chromosomal proteins.

In those cases where reverse differential staining is produced (i.e. the chromatin with the higher level of BrdU substitution stains more darkly) photolysis of DNA is apparently not involved. Brief irradiation may indeed produce differential staining without loss of DNA (Webber et al. 1981, Takayama & Taniguchi 1986), and normal and reverse patterns of differentiation can be produced on the same chromosomes merely by altering staining conditions. A general explanation of these phenomena is not yet available, but there is considerable evidence that chromosomal proteins are involved. DNA containing BrdU binds histones significantly more strongly than unsubstituted DNA (M. S. Lin et al. 1976, Matthes et al. 1977). Both Burkholder (1982) and Buys et al. (1982) proposed that in some cases differential staining might result from tighter binding of BrdU-substituted DNA to chromosomal proteins, thereby permitting preferential loss of unsubstituted DNA. Tempelaar et al. (1982) obtained reverse differential staining after prolonged Feulgen hydrolysis, which they attributed to the stronger binding of BrdU-substituted DNA to chromosomal proteins. Irradiation of BrdU-substituted DNA also induces extensive DNA-protein crosslinking (Guo et al. 1986), and this might explain reverse differential staining after irradiation plus DNase treatment (Jan et al. 1985). Breakage of disulphide bonds in chromosomal proteins, either by chemicals or by irradiation, leading to alterations in chromosome structure without loss of DNA, has also been proposed as a mechanism involved in differential staining of BrdU-substituted DNA (Wolff & Bodycote 1977, Buys & Osinga 1981). Although a complete explanation of all aspects of differential staining after BrdU-substitution of DNA has not yet been obtained, it is reasonable to invoke interactions with chromosomal proteins as well as photolysis of DNA. In any particular situation, one mechanism or the other may predominate, or the final result may be due to interaction of the different factors.

11.3.3 Applications of replication banding

Substitution with BrdU as a means of studying DNA replication has become a standard procedure in a number of fields. For example, BrdU substitution has permitted the routine study of sister chromatid exchange, which has become a very important method for studying certain aspects of mutagenesis (Wolff 1982). It has also been applied in studies using flow cytometry, for analysis of the cell cycle (Rabinovitch *et al.* 1988). The present section is, however, limited to applications of replication banding, where BrdU is used to study variations in the time of DNA replication in different segments of chromosomes. Although it had long been known, from autoradiographic and biochemical studies, that not all segments of DNA in eukaryotic chromosomes were replicated simultaneously, it was not until the introduction of replication banding using BrdU that the precise patterns of replication could be determined. Using culture in the presence of BrdU, and staining by a variety of methods, several authors reported the patterns of early replicating and late replicating DNA in human chromosomes (Latt 1973, 1975, Dutrillaux 1975, Epplen *et al.* 1975). It was quite clear that in general early replicating segments of chromosomes corresponded to positive R-bands, whereas late replicating segments corresponded to positive G- or Q-bands. Such a correlation had indeed been found earlier using autoradiography of tritiated thymidine to detect replication patterns (Ganner & Evans 1971, Calderon & Schnedl 1973), but detailed patterns could not, of course, be obtained using autoradiography. Since the initial observations on human chromosomes, the correlation between early replication and R-bands, and between late replication and G- or Q-bands, has been confirmed for the chromosomes of many mammalian species: for example, Chinese hamster (Crossen *et al.* 1975), Syrian hamster (Cawood 1981), *Macaca mulatta* (Dzhemilev & Ataeva 1975), *Lagothrix lagotricha* (Dutrillaux *et al.* 1980a), laboratory rat (Shiraishi *et al.* 1982), rhesus monkey (Armada & Seuanez 1984) and the great apes (Weber *et al.* 1986).

Although there is a good correlation between replication patterns, on the one hand, and G-, Q- and R-banding patterns on the other, the correlation is not perfect. A small degree of asynchrony in the replication of homologous autosomes is quite usual (Latt 1974a, 1975, Cawood 1981), and small

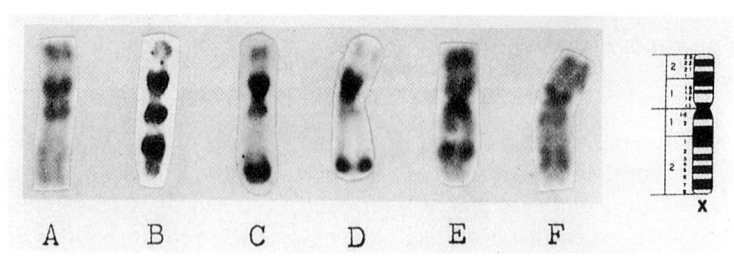

Figure 11.11 Different late replication patterns (A–F) of the human inactive X chromosome. Reproduced from Schmidt *et al.* (1982) by permission of the authors and Springer-Verlag.

differences between replication patterns in different tissues have also been found (Sheldon & Nichols 1981a, b). Detailed comparisons have shown that there are, in fact, many small differences between replication patterns, and G-, Q- or R-bands (Meer *et al.* 1981, Shafer *et al.* 1982, von Kiel *et al.* 1985). The most striking difference between replication patterns and conventional banding is, however, found in the inactive X chromosome in female mammals. Replication is not merely delayed, but several different patterns of replication have been described. Essentially all this work has been done in humans. Variations in the replication pattern of the late-replicating, inactive X chromosomes occur within the same tissue (Latt 1974b, Schmidt *et al.* 1982, Schmidt & Stolzmann 1984), but tissue-specific differences also occur (Willard 1977, Latt *et al.* 1981, Tsukahara & Kajii 1985). A number of replication patterns for the inactive X chromosome are illustrated in Figure 11.11. However, Schwemmle *et al.* (1989), using pulse labelling, found only a single replication pattern in inactive X chromosomes from both human lymphocytes and amniotic fluid cells.

In certain circumstances, the time of replication of particular bands can be altered. Karube & Watanabe (1988) found that in a case of acute myeloid leukaemia with a chromosome translocation, the chromosome segment at the site of translocation became early replicating instead of being late replicating as in the normal chromosome. Replication patterns in the human inactive X chromosome have been of particular interest in this context also. Inactive X chromosomes in hybrid cells can be reactivated by treatment with 5-azacytidine. Both Hors-Cayla *et al.* (1983) and Schmidt *et al.* (1985) found that several genes on the X chromosome could be activated while the chromosome remained late replicating with no change in replication pattern. However, Schmidt *et al.* (1985) also reported one case in which the order of replication was altered as a result of 5-azaC treatment. In the study by Hors-Cayla *et al.* (1983), the X chromosome could become early replicating although not all the genes studied were reactivated.

The results described so far show the close resemblances between replication banding and other types of banding, but it is important to realise that the replication patterns are not identical with, for example, G-banding. The different methods give different information, even if the patterns are largely similar. This is shown by evolutionary studies, from which it is clear that replication patterns can often provide the clearest evidence for homologies between the chromosomes of different species (von Kiel *et al.* 1985, Weber *et al.* 1986, see also Ch. 15). A rather similar situation occurs with the human sex chromosomes, in which there is a homologous replication pattern in the pairing segments of both X and Y chromosomes, although there is no obvious resemblance in their G-banding patterns (Müller & Schempp 1982, Schempp & Meer 1983).

The essential dissociation of replication banding from the other types of euchromatic banding becomes especially apparent when the chromosomes of organisms other than higher vertebrates are studied. As previously

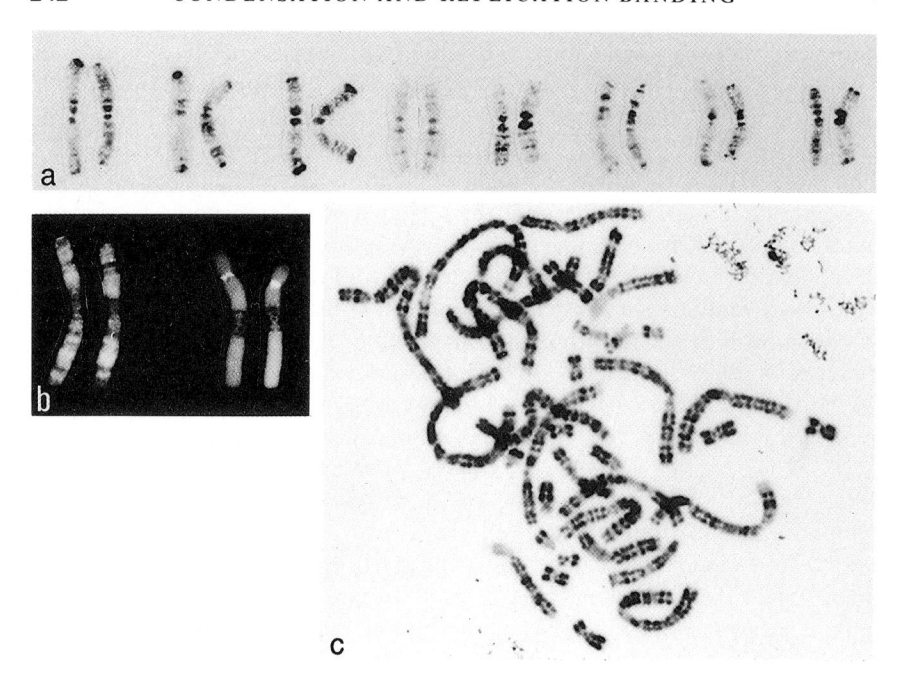

Figure 11.12 Replication bands in chromosomes of organisms other than higher vertebrates. (a) *Allium cepa*. Reproduced from Cortes & Escalza (1986) by permission of the authors and Kluwer Academic Publishers. (b) *Vicia faba*. Replication pattern of M chromosomes (left) and chromomycin fluorescence pattern (right). Micrograph kindly provided by Professor D. Schweizer. (c) *Hynobius retardatus* (Amphibia). Reproduced from Kuro-o *et al.* (1987) by permission of the authors and S. Karger AG.

remarked (Ch. 5), lower vertebrates, invertebrates and plants do not, in general, have euchromatic bands on their chromosomes, yet it seems quite clear that the chromosomes of these organisms do show replication bands (see also Table 16.1). In other words, their chromosomes, like those of higher vertebrates, are made up of small alternating early- and late-replicating segments. So far replication bands have been reported, from plants, in the chromosomes of species of *Allium* (Cortes & Escalza 1986, Cortes *et al.* 1980) (Fig. 11.12a) and of *Vicia faba* (Schweizer *et al.* 1990) (Fig. 11.12b), in several species of fish (Delany & Bloom 1984, Hong & Zhou 1985, Giles *et al.* 1988, Liu 1988), and among amphibia from frogs in the family Ranidae (Schempp & Schmid 1981) and from salamanders of the genus *Hynobius* (Kuro-o *et al.* 1986, 1987) (Fig. 11.12c). Although these reports are as yet few in number, they are important for the indication they provide that replication bands might be present in most, if not all, eukaryotes, whether or not they show conventional euchromatic bands. The quality of the replication banding that has been obtained is generally high, which suggests that it is only a matter of getting the conditions right to produce the bands. The production of replication banding in plants and lower vertebrates has, of course, the practical value of producing more detailed banding patterns as an aid to more accurate chromosome identification.

So far the process of replication of chromosomal DNA has been discussed as if it were clearly divided into early and late phases. Much of the literature also refers to early- and late-replicating DNA as if they were distinct fractions. Using biochemical methods, Klevecz et al. (1975) found that DNA synthesis in CHO cells occurred in three main stages, 10–20% of the DNA being synthesized in early S phase, 40% in mid S, and 40–50% in late S, whereas Holmquist et al. (1982) reported a sharp separation of DNA synthesis into early and late phases. Replication banding techniques have also been used to investigate this question. Earlier work indicated that intermediate, as well as early and late replication patterns could be obtained (Dutrillaux 1975, Epplen et al. 1975), which could be interpreted as meaning that DNA synthesis is continuous throughout the S phase. However, with the relatively long pulses of BrdU used, it was not possible to analyse the S phase in detail. Schmidt (1980) used much shorter BrdU pulses during the S phase, and found a clear distinction into an 'early replication period' and a 'late replication period', with no overlap, in cultured human lymphocytes. Vogel et al. (1985) obtained similar results in CHO and HeLa cells, and found an abrupt transition from the early S-phase pattern to the late S-phase pattern. Vogel & Speit (1986) found that the synthesis in mouse cell lines followed a similar pattern, but that heterochromatin starts replication during early S phase, and finishes replication before most G-bands. Cawood & Savage (1985) investigated the question of whether DNA synthesis actually stopped completely between early S-phase and late S phase. Using Syrian hamster cells, and a double labelling technique (^{3}H-thymidine for 5 min followed by BrdU) they concluded that there was no period of complete cessation of DNA synthesis, since replication banding was never found without tritium labelling. Although BrdU incorporation is an important method for studying the course of chromosomal DNA synthesis, it is clear that it has not yet given an unequivocal answer to this problem. Some of the differences observed may be due to use of different types of cells, and there is also the possibility that the methods used to synchronize the cells at the beginning of S phase in these experiments might perturb the normal sequence of synthesis.

11.3.4 High resolution banding

High resolution banding has already been considered in Section 5.2.4, but a further mention is necessary here because BrdU is frequently used for the production of high-resolution bands, and the bands produced in such cases are obviously replication bands rather than conventional G- or R-bands. The object of high-resolution banding is, of course, to obtain more detailed patterns on elongated chromosomes, and thereby locate more precisely the sites of chromosome breaks, or detect smaller deletions.

Methods involving BrdU for the production of high-resolution banding are, in principle, the same as other high resolution methods: the cells are synchronized at the beginning of S phase, and harvested at an appropriate time afterwards, before the chromosomes have condensed completely. In most cases, the cell cycle is blocked by the usual agents: excess thymidine (Viegas-Péquignot & Dutrillaux 1978, Schollmayer *et al.* 1981, Romagnano & Richer 1985), methotrexate (Rønne 1985), fluorouracil (Rønne 1984), or fluorodeoxyuridine (de Braekeleer *et al.* 1985). The block is then released using standard culture medium containing BrdU, which is incorporated into the chromosomes during the subsequent DNA synthesis. Banding is obtained using the FPG or a similar technique. Rather surprisingly, the pattern obtained is normally similar to R-banding; that is, the BrdU has been incorporated mainly into the late-replicating DNA. However, G-band patterns can be obtained using conventional G-banding procedures. Viegas-Péquignot & Dutrillaux (1978) deduced, from the observation that BrdU was incorporated mainly into late-replicating DNA, that the block to DNA replication must occur in the middle of S phase rather than at the beginning

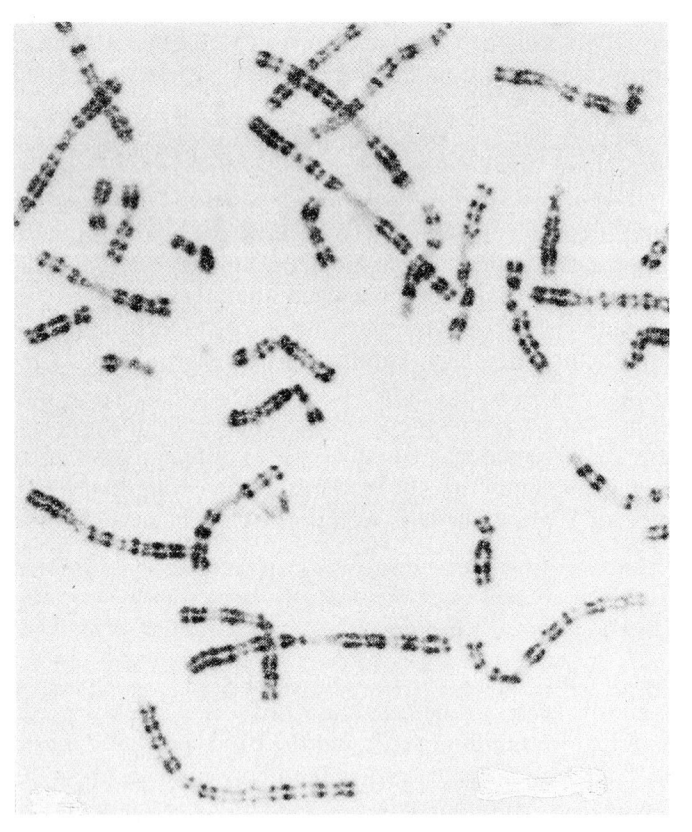

Figure 11.13 High-resolution replication banding of human chromosomes. Reproduced from Rønne (1984) by permission of the authors and the publishers of Hereditas.

of S. The BrdU not only provides a substrate for differential staining of the elongated chromosomes, but also tends to inhibit condensation (Schollmayer et al. 1981, Romagnano & Richer 1985, Drouin & Richer 1989) thus causing the chromosomes to be even longer. Other agents which inhibit condensation still more have been used in conjuction with BrdU, such as Hoechst 33258 (Rønne 1984, de Braekeleer et al. 1985) and ethidium bromide (Rønne 1985). Chromosomes treated by these procedures show very high quality banding and are aesthetically pleasing (Fig. 11.13).

Procedures for high resolution banding were simplified by Dutrillaux & Viegas-Péquignot (1981), who used BrdU to block DNA synthesis and produce cell synchronization. The BrdU is apparently incorporated into the early replicating DNA, up to a block in the middle of S phase. If the cells are subsequently grown in the absence of BrdU, a late-replication ('G-banding') pattern is produced, although a proportion of cells show early replication patterns. This suggests either that the BrdU block is not complete, or that synthesis can be blocked at several points.

It should be noted that these procedures can be used to produce not only G- and R-band patterns (which are in reality replication banding patterns), but can also be used to study C-bands (de Braekeleer et al. 1986) and NORs (Verma et al. 1984) at high resolution. In these cases the incorporation of BrdU into the chromosomes does not, as far as we know, alter the banding pattern, and so these are not examples of replication banding.

11.3.5 Lateral asymmetry

Replication banding is produced when BrdU is incorporated into DNA during only part of the S phase, so that each chromatid consists of segments which are substituted with BrdU, alternating with unsubstituted segments. Sister chromatid differentiation is produced when the cells go through two (or more) cycles of replication, so that one chromatid is unifilarly substituted (i.e. in only one strand of the DNA molecule) and the other is bifilarly substituted with BrdU (Fig. 11.7). If BrdU is incorporated throughout one complete S phase and no more, no differentiation, either longitudinal or lateral, would be expected, since each chromatid should contain a unifilarly substituted DNA molecule (Fig. 11.14a). In fact, in certain regions of heterochromatin, lateral asymmetry can be observed after cells have been grown in the presence of BrdU for an entire S phase (Fig. 11.15). The heterochromatin of one chromatid, therefore, has apparently incorporated more BrdU than that in the other chromatid. The explanation of this phenomenon is based on the original observations of lateral asymmetry in the heterochromatin of mouse chromosomes (M. S. Lin et al. 1974). In the mouse, of course, the heterochromatin is composed predominantly of a single type of satellite DNA, and this satellite shows marked asymmetry in the base composition of its two DNA strands (Flamm et al. 1967). One strand contains 22% thymine, whereas the other contains 45% thymine.

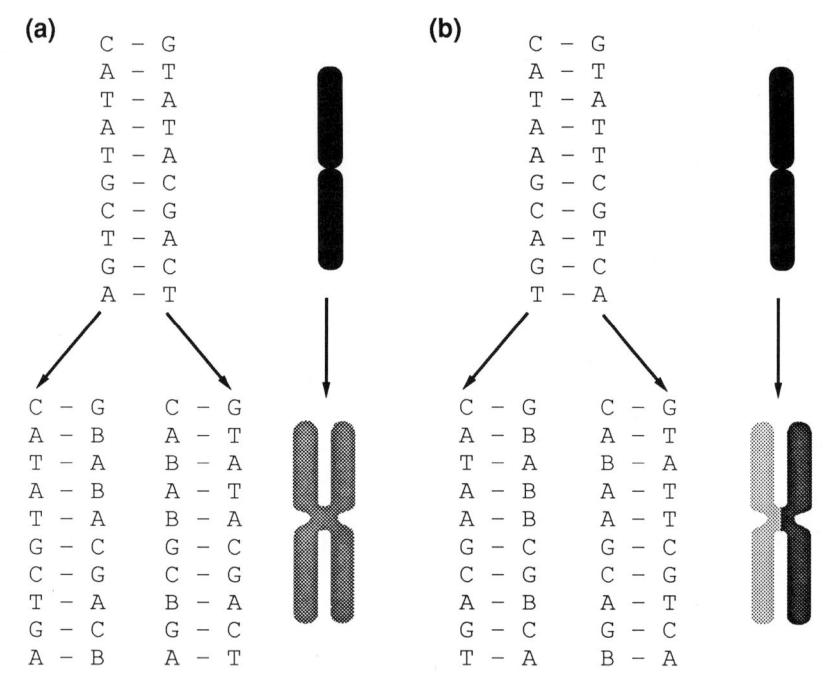

Figure 11.14 Diagram illustrating how lateral asymmetry can be induced by BrdU substitution in DNA with marked asymmetry in base composition between its two DNA strands. (a) No asymmetry, resulting in equal incorporation of BrdU (B) in both daughter molecules, producing an equal reduction of staining in both chromatids. (b) Pronounced asymmetry of base composition, so that one daughter molecule incorporates twice as much BrdU as the other, resulting in differential staining of the daughter chromatids.

Since BrdU is incorporated in place of thymine, lateral asymmetry would be expected to result (Fig. 11.14b). Although such inter-strand bias has not been independently established in most cases of lateral asymmetry, no other plausible explanation has been put forward. Indeed, the occurrence of lateral asymmetry in a chromosomal segment is now usually taken as evidence that the DNA in that segment shows interstrand bias.

Lateral asymmetry has been used to show that DNA polarity is maintained in metacentric chromosomes produced by Robertsonian fusion of normal mouse acrocentrics (M. S. Lin & Davidson 1974, Holmquist & Comings 1975). In such Robertsonian fusions, the lateral asymmetry shows a contralateral arrangement on either side of the centromere, an arrangement wholly consistent with the maintenance of DNA polarity across the centromere in the fused metacentric chromosome (Fig. 11.16).

Among human chromosomes, lateral asymmetry was first described in the heterochromatin of the long arms of the Y chromosome (Latt *et al.* 1974b, Limon & Gibas 1985), and has since been described for many, but not all, heterochromatic blocks in human chromosomes (Table 11.3). A number of points can be made about these observations. First, different authors have

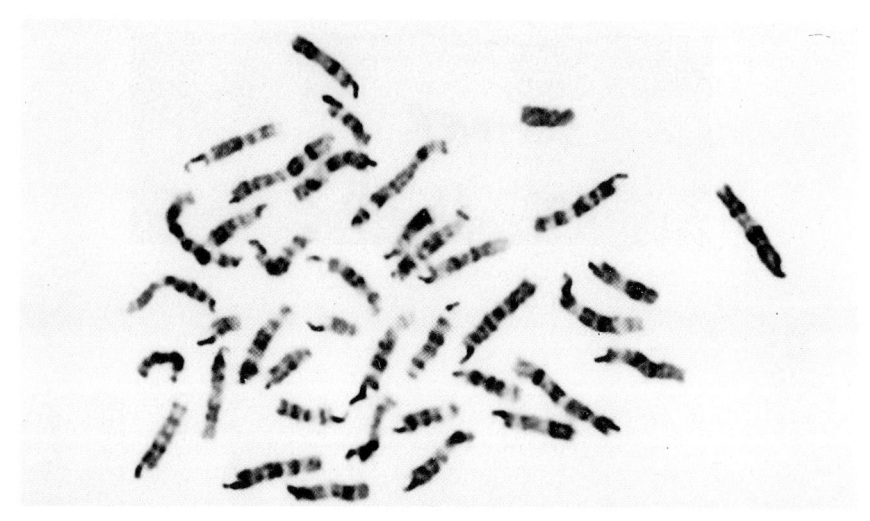

Figure 11.15 Lateral asymmetry in heterochromatin of mouse chromosomes. Reproduced from Takayama & Matsumoto (1982) *Chromosoma* **85**, 583–90 by permission of the authors and Springer-Verlag.

reported that different sets of chromosomes show lateral asymmetry. With some of the smaller blocks of heterochromatin, this may be due to nothing more than the difficulty of detecting lateral asymmetry in such small areas. Other technical factors may also play a part; for example, Brito-Babapulle (1981) reported that lateral asymmetry of the Y chromosome could not be observed after Giemsa staining at pH 10.4 (which could be used to demonstrate lateral asymmetry on several other chromosomes), although it was clearly demonstrable with Hoechst 33258 fluorescence (Latt *et al.* 1974b). Nevertheless, the report by the same author of lateral asymmetry in a particular chromosome in some individuals but not in others (Brito-Babapulle 1981) raises the possibility of heteromorphism for DNA composition in these regions (see also Ch. 14).

Second, the heterochromatin of certain chromosomes shows compound lateral asymmetry (Fig 11.17); that is, different parts of the block of heterochromatin show lateral asymmetry in opposite senses (Angell & Jacobs 1975, 1978). Kim (1975) attributed the appearance of compound lateral asymmetry to sister chromatid exchange within the heterochromatic region, but Angell & Jacobs (1978) showed that compound asymmetry is a genuine heteromorphism, and its expression is constant within an individual and is apparently inherited in a Mendelian fashion. Brito-Babapulle (1981) pointed out that, in chromosome 1 at least, the heterochromatin could be divided up into different blocks on the basis of other staining properties, and that these subdivisions of the heterochromatin might correspond to the regions showing lateral asymmetry of opposite sense.

Other species in which lateral asymmetry has been reported include, among mammals, the kangaroo rat (Bostock & Christie 1976) and the rhesus

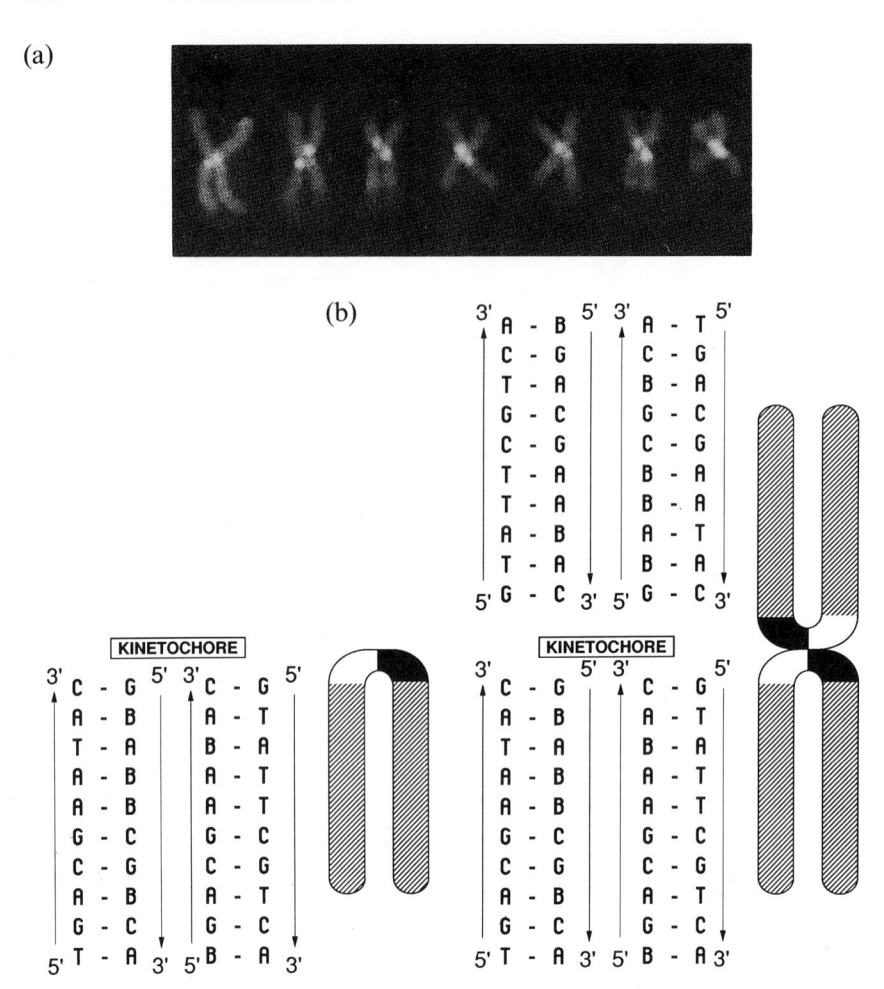

Figure 11.16 Lateral asymmetry in mouse metacentric chromosomes formed by Robertsonian fusion. (a) Contralateral asymmetry on either side of the centromere in mouse chromosomes. Reproduced from Lin & Davidson (1974) by permission of the authors. Copyright © 1974 by the AAAS. (b) Diagram illustrating how maintenance of 5'-3' polarity of DNA across the centromere must result in contralateral asymmetry of BrdU substitution and staining in the heterochromatin.

monkey (Ghosh *et al.* 1981). In the former species, strong interstrand bias of base composition of the satellite DNAs which comprise the heterochromatin is well established, and the occurrence of lateral asymmetry in the chromosomes is to be expected; the blocks of heterochromatin are also large, and compound lateral asymmetry is frequent. Lateral asymmetry has also been described for heterochromatic segments of the chromosomes of *Drosophila* spp. (Lakhotia *et al.* 1979); in *D. hydei* (Bonaccorsi *et al.* 1981) compound lateral asymmetry occurs, and in the Y chromosome it occurs in the absence of any highly repetitive DNA. Finally, Schubert & Rieger (1979) described

Table 11.3 Human chromosomes showing lateral asymmetry after replication in the presence of BrdU.

Chromosome	Site	Comments	Reference
1	Heterochromatin	Compound asymmetry	Angell & Jacobs 1975, Kim 1975
2–5	Centromere		Brito-Babapulle 1981
6	Band 6q12–6q14	Not heterochromatin; compound asymmetry	Emanuel 1978
7	Centromere		Brito-Babapulle 1981
9	Heterochromatin	Other authors report no asymmetry	Angell & Jacobs 1978
13, 14	Centromeres		Brito-Babapulle 1981
15	Heterochromatin		Angell & Jacobs 1975, 1978
16	Heterochromatin		Kim 1975, Latt *et al.* 1975
17	C-band		Galloway & Evans 1975
19	Centromere		Kim 1975
20, 21, 22	C-bands		Galloway & Evans 1975, Brito-Babapulle 1981
Y	Heterochromatin of long arm	Compound asymmetry. Depends on staining technique	Brito-Babapulle 1981

asymmetrical banding in the chromosomes of the plant *Vicia faba* after BrdU incorporation for one cell cycle. The bands were always interstitial and very small, and could not all be correlated in position with bands demonstrated by other techniques.

Lateral asymmetry has never shown any sign of becoming a widely used technique, but it is nevertheless an interesting phenomenon in its own right. It is, moreover, a good example of a molecular phenomenon that can be recognized at a cytological level. Its main value is in providing a further method of distinctively labelling specific chromosomal segments, usually blocks of heterochromatin. As heteromorphic markers, lateral asymmetries have some value in linkage analysis and determining origins of abnormal chromosomes, and in this respect have been found much more efficient than simple C-band heteromorphisms (Angell & Jacobs 1978).

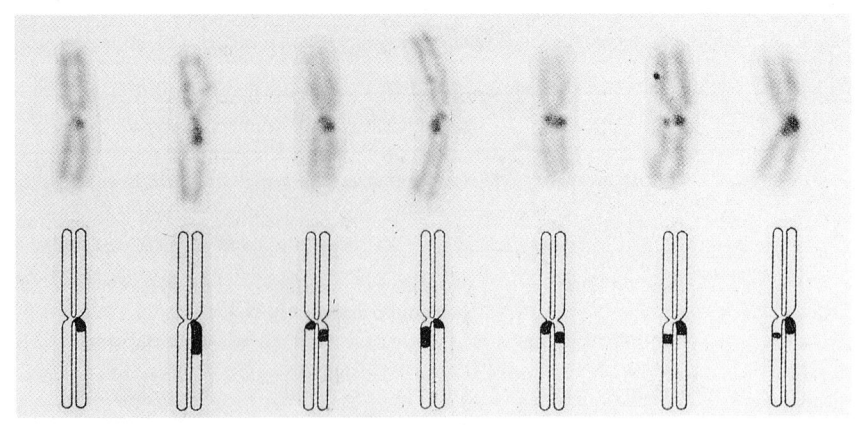

Figure 11.17 Compound lateral asymmetry in heterochromatin of human chromosome 1. Lower line shows a diagrammatic interpretation of the micrographs. Reproduced from Angell & Jacobs (1978) by permission of the authors. Copyright © 1978 by the American Society of Human Genetics. All rights reserved.

11.4 Unclassified effects of agents which induce banding in living cells

There are several reports of substances that induce bands in the chromosomes of living cells, without it being apparent that the type of banding is either condensation or replication banding. Such substances are listed, for the sake of completeness, in Table 11.4. Neither the text nor the illustrations of the papers quoted provide any indication that the substances mentioned cause any significant inhibition of condensation, although in some cases it is possible that further investigation might show that this does in fact occur. Nor is there any reason to suppose that the substances mentioned can be incorporated into newly synthesized DNA to produce replication banding.

Nevertheless, there is a good deal of similarity between the effects of the agents listed in Table 11.4, whose mode of action is as yet unexplained, and those listed in Table 11.1, which have been shown to inhibit chromosome condensation. First, the usual effect (on mammalian chromosomes) is to produce a G-banding pattern. Second, in both cases a number of the agents are known to interact with DNA, or (in the case of the cyclic compounds in Table 11.4) have been postulated to interact with DNA. It may then be that the majority of agents listed in Table 11.4 do in fact produce condensation banding, but that the degree of inhibition of condensation is relatively slight and therefore has not been recognized. Another possibility is that the agents in Table 11.4 interact with chromosomal DNA at the same sites as agents which inhibit condensation, but that they cause a change in chromosomal organization which affects staining but not condensation. There is no reason to suppose that blocking of Giemsa staining would be caused by any of the agents even under favourable conditions, let alone after fixation.

Table 11.4 Substances which induce banding by undetermined mechanisms when applied to living cells.

Agent	Cells	Effect	Reference
Heat treatment	Human lymphocytes	Stronger staining of heterochromatin	Kanda 1978
Cyclic compounds			
Tetracycline	Human lymphocytes	G-banding	Meisner *et al.* 1973a
Histamine, Thymidazol	Human lymphocytes	G-banding	Simeonova *et al.* 1979
Various dyes	Human lymphocytes	G-banding	Couturier & Lejeune 1976
Vitamins B_1, B_2, B_6 & PP	Human lymphocytes	G-banding	Simeonova & Raikov 1981
Imidazole, pyrimidine & its derivatives	Human lymphocytes	G-banding	Simeonova & Raikov 1982
DNA-binding ligands			
Chemical carcinogens	Human lymphocytes & Syrian hamster embryo cultures	G-banding	di Paolo & Popescu 1974
Alkylating agents	Mouse bone marrow	G-banding	Kitchin & Loudenslager 1976
Miscellaneous			
Hydroxyurea	Human lymphocytes & Syrian hamster embryo cultures	G-banding	Popescu & di Paolo 1974
Methyl p-benzo quinones	Human lymphocytes	G-banding	Folle & López de Griego 1987

The effect of heat treatment in producing stronger staining of heterochromatin (Kanda 1978) is in a different category, and should perhaps not be considered in this chapter at all. The heat treatment is applied during the hypotonic treatment of the cells immediately before fixation, and the temperatures are above physiological ones. It is improbable that the cells are still viable, therefore, and this method should perhaps be regarded simply as another type of C-banding method (Ch. 4) in which DNA is lost more readily from the less compact euchromatic chromosome arms.

11.5 General discussion

As shown earlier in this chapter, a wide variety of treatments applied to living cells can induce banding patterns when the chromosomes are subsequently fixed and stained. In spite of the lack of information about the mechanism by which many of these treatments produce their effects, it is clear that there are essentially two classes of effects: condensation banding and replication banding. In the former, the effect is inhibition of chromosome condensation, while in the latter it is labelling of newly replicated DNA. The essential difference between these and all other banding procedures is that they define processes, whereas all other methods define structures. Drouin *et al.* (1988a) have introduced the useful term dynamic banding for those methods in which there is 'interference with chromosome condensation by treatments during cell culture....' The definition can in fact usefully be made broader, for although incorporation of BrdU into chromosomes can indeed inhibit condensation, it does not necessarily do so to any marked extent, and its incorporation can be recognized in other ways. Dynamic banding can, therefore, be redefined as 'treatment of living cells with such agents that the site of processes occurring during the cell cycle (i.e. DNA replication or chromosome condensation) can subsequently be recognized in fixed chromosome preparations.' On the other hand, structural (or morphological) banding is 'the use of specific staining methods to localize fixed structural elements in chromosomes.'

So far as we can tell, replication banding gives a faithful view of patterns of replication within chromosomes, and as already remarked (Section 11.3.4) has become extremely valuable for this purpose. The situation with condensation banding is, however, rather different. No substance has yet been found which can be used to study patterns of early and late chromosome condensation in the absence of other confounding effects, in the way that BrdU can be used to study replication. In the case of Hoechst 33258 and similar compounds, inhibition of condensation is the result of binding of the dyes to A+T-rich chromatin, and they can therefore only provide information about the condensation of chromosomal segments of a particular base composition. To produce its effect, 5-azacytidine has to be incorporated into the chromosomal DNA, which is therefore essentially a replication pattern, although it is identified as inhibition of condensation. A substance which would inhibit chromatin condensation specifically, without any confusing side-effects, would be invaluable for the production of condensation patterns on chromosomes. Use of such a substance would permit comparison of patterns of condensation with replication patterns and with, for example, G-banding patterns. We have already noted the considerable, though not perfect, similarity of G-banding patterns and late-replication patterns, and the degree of resemblance of these to the pattern of condensation would also be of interest.

12 Banding with nucleases

12.1 Introduction

Induction of banding patterns in fixed chromosomes as a result of digestion with non-specific nucleases was first reported many years ago. Alfi *et al.* (1973) and Kato *et al.* (1974) used deoxyribonuclease (DNase) to produce a C-banding pattern on mammalian chromosomes. Since then various relatively non-specific nucleases have been used to produce banding patterns, but the most important applications of nucleases to chromosome banding have both aimed at greater specificity. DNase I has been used to detect hypersensitive sites on chromosomes, that are believed to be associated with active genes (Section 12.2.1), while restriction endonucleases, which cut specifically a variety of base sequences, have been used to produce distinctive patterns of banding on chromosomes (Section 12.3). Both these procedures are being actively developed and applied at present and are among the most important developments in contemporary chromosome banding.

12.2 Digestion of chromosomes with non-specific nucleases

As mentioned above, C-banding can be produced by digesting chromosomes with DNase under appropriate conditions (Alfi *et al.* 1973, Kato *et al.* 1974). Using gentler treatment, G-banding patterns can also be produced in mammalian chromosomes (Fig. 12.1) by digestion with DNase I (Burkholder & Weaver 1977), and with micrococcal nuclease or DNase II (Sahasrabuddhe *et al.* 1978). Gagné (1981) had difficulty in producing G- or C-bands with DNase I, since the digestion was too rapid, but did consistently observe the resistance of NORs to digestion. These methods were not practical procedures for demonstrating the different types of bands, but were nevertheless important for showing the differential sensitivity of different chromosome regions to nucleases, probably due to differences in DNA-protein interactions (Burkholder & Weaver 1975).

The specificity of DNase I digestion of chromosomes was improved by Schweizer (1977), who stained the chromosomes with chromomycin A_3 (which shows a preference for binding to $G+C$-rich DNA, Section 8.4.1) before digesting them with DNase I, after which they were stained with Giemsa. On human chromosomes, an R-banding pattern resulted, that is, regions that showed positive chromomycin fluorescence were darkly stained after the chromomycin/DNase I/Giemsa sequence. The banding pattern

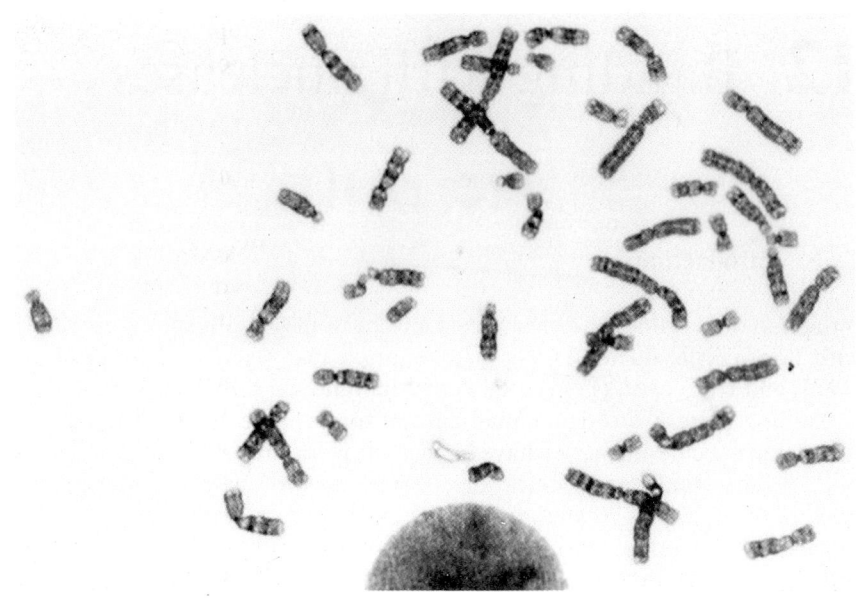

Figure 12.1 G-banding of human metaphase chromosomes induced by gentle digestion with DNase I.

produced by digestion was, however, sharper than that due to fluorescence alone. In the chromosomes of certain plants, C-bands that were chromomycin positive were also resistant to DNase digestion after chromomycin staining, whereas other C-bands were not resistant. Apart from the potential practical value of this procedure, these experiments demonstrated that patterns of chromomycin fluorescence on chromosomes are the result of differential binding of chromomycin, and not the result of differential excitation of fluorescence. Loidl (1985) applied similar procedures to the chromosomes of the plant *Allium flavum*. Before digestion, the chromosomes were treated with one of the following antibiotics or dyes, the binding preferences of which are indicated in brackets: actinomycin D (preference for G+C-rich DNA); chromomycin A_3 (G+C-rich DNA); distamycin A (A+T-rich DNA); and the diamidine derivative D288/27 (specificity not established). Actinomycin did not protect the chromosomes against DNase I digestion, but all the other treatments resulted in characteristic patterns of bands which were different subsets of C-bands. It should be noted that the D288/27-DNase I sequence produced some bands that were paler than the rest of the chromosome, as well as several darker bands.

Techniques such as these should help in distinguishing different classes of heterochromatin. To some extent they would be expected to give similar results to the direct application of base-specific fluorochromes, but, as Loidl (1985) pointed out, the present techniques do not require the use of a fluorescence microscope, nor in fact is it necessary to use a fluorescent base-specific compound. In practice, techniques of this sort have not come into general use. One reason for this is that it is simpler to distinguish different classes of heterochromatin using restriction endonucleases (Section 12.3).

12.2.1 DNase I hypersensitivity as an indicator of sites of potentially active genes

Whereas the methods described so far in this chapter have proved to be of rather academic interest and have not had any widespread practical applications, the study of DNase I hypersensitivity has become an important tool in studying chromosome organization.

DNase I hypersensitivity was first described by Weintraub & Groudine (1976) who showed that certain genes in tissues in which they were normally active were particularly susceptible to digestion by DNase I, while the same genes were not more susceptible to digestion than the remainder of the cellular DNA in tissues where they were not normally active. Since then, the subject of DNase I hypersensitivity has been studied intensively, and a number of reviews have been written (Yaniv & Cereghini 1986, Elgin 1988, Gross & Garrard 1988). The main conclusions to be drawn from this work were that DNase I hypersensitivity indicates a potential for transcription, rather than being an indication that transcription is actively occurring, and that the hypersensitivity is the result of an altered chromatin structure. This altered structure is apparently due to lack of nucleosomes, leaving the DNA open to attack by a variety of agents (Elgin 1988, Gross & Garrard 1988). HMG proteins are probably important in maintaining the hypersensitive structure (Weisbrod & Weintraub 1979).

The study of DNase I hypersensitive sites on chromosomes began when Gazit *et al.* (1982) showed that DNase sensitivity of known active genes was maintained in mitotic chromatin, and could be detected on fixed cytological preparations using nick translation. The procedure of nick translation is illustrated in Figure 12.2. Briefly, DNA polymerase can synthesize a new strand of DNA starting at a site that has been nicked by DNase I (or, indeed, any other agency); if the synthesis is carried out using suitably labelled nucleotides, the newly synthesized strand of DNA can be recognized, and, in cytological preparations, localized. Earlier work used as a label tritiated thymidine, the sites of incorporation of which were identified by autoradiography. Although valuable work has been done with this procedure, the resolution was necessarily inadequate for detailed studies of patterns on chromosomes, and latterly biotinylated nucleotides have been employed, which can be localized using suitably labelled antibodies to biotin, or using labelled streptavidin. Peroxidase (Kerem *et al.* 1984, Adolph & Hameister 1985) and alkaline phosphatase (Bullerdiek *et al.* 1986a) have been used as the labels.

Information on the distribution of DNase I hypersensitive sites on chromosomes, obtained using autoradiography, is necessarily limited in resolution, but must nevertheless be included here as it represents the pioneering work in this field. For example, Kerem et al. (1983) showed that the inactive X chromosome of the gerbil was unlabelled after nick translation, while in the Chinese hamster, Kuo & Plunkett (1985) found a lower level of labelling on the heterochromatic Y chromosome. Studies of male meiotic chromosomes have shown a limited number of specific sites of hypersensitivity in the XY pair; in the mouse these were believed to correspond to early replicating bands (Richler et al. 1987), while in men they were thought to represent telomeres and the pairing site (Chandley & McBeath 1987).

Use of biotin-labelled nucleotides, together with non-radioactive methods for detecting sites of incorporation, has produced more detailed patterns of DNase I hypersensitivity on chromosomes. Kerem et al. (1984) found that in human and Chinese hamster chromosomes the DNase I sensitive regions (which they called D-bands) usually corresponded to the negative G-bands (i.e. R-bands), although not all negative G-bands were DNase sensitive. Known regions of inactive constitutive heterochromatin were negative (Fig. 12.3). Such precise results have not always been obtained by other

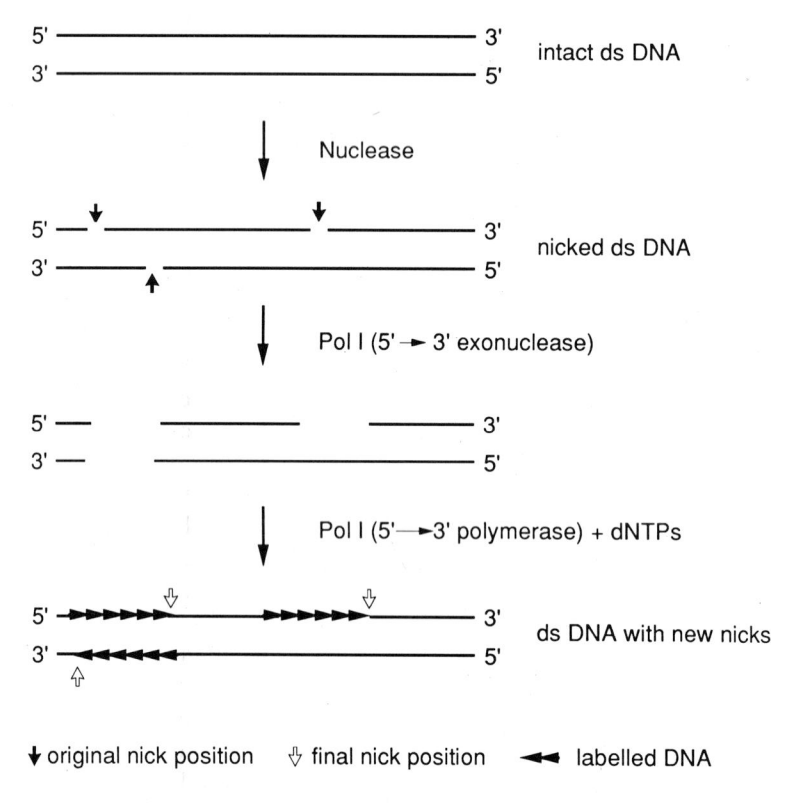

Figure 12.2 Diagram of nick translation of DNA. Redrawn from Arrand (1985) in *Nucleic acid hybridization*, B. D. Hames & S. J. Higgins (eds), 17–45. Oxford University Press.

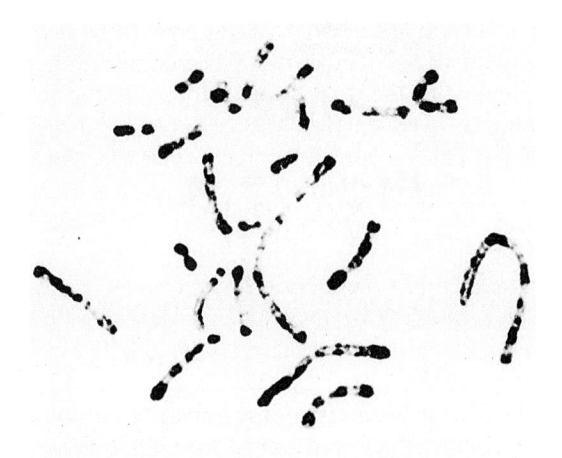

Figure 12.3 Pattern of DNase I-hypersensitivity of Chinese hamster chromosomes (D-banding). Reproduced from Kerem *et al.* (1984) by permission of the authors and Cell Press.

workers, and it is clear that this is at least partly the result of technical problems. Thus Adolph & Hameister (1985) obtained distinctive patterns of DNase I hypersensitivity along human and mouse chromosomes, but commented that 'No complete G- or R-type banding pattern was observed', although they indicated that the pattern resembled R-banding more closely. They also pointed out that too much DNase I digestion would lead to loss of DNA as a result of excessive nicking; in these circumstances the strongest staining occurs in the regions that have retained the most DNA, which are the blocks of centromeric heterochromatin. Murer-Orlando & Peterson (1985) also obtained labelling of DNase I hypersensitive regions in both R- and G-bands. The pattern of DNase I hypersensitivity found in human chromosomes by Bullerdiek *et al.* (1986a) appeared to correspond to R-banding, but they also found that the heterochromatin of chromosome 9, and particularly that of the Y chromosome, were heavily labelled. Indeed, Adolph & Hameister (1985) had already shown that the heterochromatin of the human Y showed as much DNase sensitivity as the euchromatin, but that the heaviest labelling was at the boundary of the euchromatin and heterochromatin.

The available evidence indicates that in general the sites of DNase I hypersensitivity that are revealed cytologically are indeed sites of potentially active genes, yet there are nevertheless some discrepancies. These discrepancies might either be the result of technical problems, or may genuinely show that regions of metaphase chromosomes can be hypersensitive for reasons other than the presence of potentially active genes. The evidence that cytologically demonstrable hypersensitive regions contain potentially active genes is that such genes are known to be hypersensitive, that the hypersensitivity is known to be maintained in metaphase chromosomes, and that the pattern of hypersensitivity on chromosomes corresponds, in general, with what is believed to be the distribution of genes on chromosomes (Section 5.3, Table 5.2). Genes are believed to be concentrated in R-bands (or negative

G-bands), and the patterns of DNase I hypersensitivity appear to be generally similar. There is also the evidence that inactive X chromosomes are not DNase I hypersensitive (Kerem *et al.* 1983), although Murer-Orlando & Peterson (1985) found no difference between the level of labelling of the inactive X, and of the rest of the chromosomes in the metaphase plate. Ferraro & Prantera (1988) also showed that human NORs, shown to be active by silver staining (Section 9.4.2) were also more sensitive to DNase I than inactive NORs.

On the other hand, it seems unlikely that the DNase I hypersensitive sites labelled by nick translation in meiotic XY bivalents are active genes, although the consistency of the distribution of labelled sites, and the low levels of labelling elsewhere, indicate that this is genuine hypersensitivity. The situation with labelling of the heterochromatin of the human Y chromosome is less clear. This is not a region which is believed to contain genes, whether potentially active or not, and it would not be expected to be labelled, as indeed Kerem *et al.* (1984) found. It is not easy to see why Adolph & Hameister (1985) should have obtained a different result using an essentially similar method, although it should be noted that when Bullerdiek *et al.* (1986a) obtained labelling of the Y heterochromatin, they used slides a week old. Most workers have used freshly prepared chromosome preparations for *in situ* nick translation, and although Murer-Orlando & Peterson (1985) obtained satisfactory results on preparations up to a month old, there is good evidence that fixed chromosomes undergo spontaneous nicking and other degradative processes with time (Nakamura *et al.* 1976, Mezzanotte *et al.* 1988). The possibility that the observed reaction is due to non-specific nicking should, therefore, always be tested using appropriate controls. Another variable factor is the concentration of DNase I used for detecting the hypersensitive sites. Different workers have used very different concentrations, varying from 0.2 ng/ml to 200 ng/ml in the papers quoted in this chapter. As noted earlier, different patterns are obtained with different amounts of enzyme, and in addition, chromosomes from different species appear to require different concentrations of enzyme for optimal results (Adolph & Hameister 1985). Evidently there are several technical variables which can influence the results of experiments to determine the pattern of DNase I hypersensitivity of chromosomes.

To conclude this section, it may be stated that the study of DNase I hypersensitivity is a potentially very powerful method for investigating the distribution of potentially active genes, and perhaps other specialized sites (Gross & Garrard 1988), on fixed chromosomes. Clearly there are technical problems still to be overcome, but already the technique has been used to demonstrate distinctive patterns on chromosomes, which are largely, if not yet entirely, consistent with what is known about the distribution of genes on chromosomes. Several questions still remain to be answered, however. The first concerns the consistency of the patterns; although they seem to be repeatable in a general sense, it is not yet known whether the patterns are repeatable in detail. Second, how closely do the patterns of DNase I hypersensitivity resemble R-banding, and are any differences consistent and significant? Third, it is well known that different sets of genes are active in different types of cells, and although such differences might not be visible at the relatively coarse level observed on metaphase chromosomes, where each block of stain must represent hundreds of genes, it would be logical to investigate patterns of DNase I hypersensitivity in different cell types. To carry out such a study successfully would obviously require a high degree of reproducibility in the technique. Fourth, it would clearly be of interest to extend the application of this technique to species other than mammals. In particular, it would be interesting to see whether those organisms such as insects and plants, whose chromosomes appear to lack G- and R-bands, show detailed patterns of DNase I hypersensitivity. It may well be that the distribution of potentially active genes, like replication patterns (Section 11.3.4), and patterns of pachytene chromomeres, forms a pattern that is generally similar to, but not necessarily identical with, G- and R-banding patterns, and that in some circumstances any one of these patterns can exist independently of some or all of the others. We shall return to this point in Chapter 16.

12.3 Digestion of chromosomes with restriction endonucleases

Type II restriction endonucleases are enzymes that cut DNA at specific nucleotide sequences, commonly of 4, 5 or 6 bases. The sequence is normally palindromic, and the enzymes may cut the DNA so as to leave blunt ends or staggered ends, depending on their specificity. Because of their strict sequence specificity, restriction endonucleases have become invaluable reagents in molecular biology. The first application of restriction endonucleases to metaphase chromosomes was by Lima-de-Faria *et al.* (1980). Since then restriction endonuclease banding has become widely used, and has in particular been found capable of differentiating between different classes of heterochromatin.

Extensive lists of restriction endonucleases, including their sequence specificity, have been published (Roberts 1990). Similar information is also available from the firms that supply these enzymes. A list of restriction endonucleases that have been used for studies on chromosomes is given in Table 12.1. The enzymes are named after the organisms, usually bacteria, from which they are obtained, the first (capital) letter being the first letter of the generic name, and the second and third letters being the first two letters of the specific name. These letters should be italicised. Other letters or numbers indicate a particular strain of the bacterium from which that enzyme was derived, and finally Roman numerals are used to distinguish between different enzymes from the same organism. Two enzymes which cut the same base sequence are known as isoschizomers.

Table 12.1 Restriction endonucleases used for chromosome banding in mammals.

Enzyme	Recognition sequence (a)	Iso-schizomers	Mean fragment length (bp)(e)	Banding patterns (Reference) (d)
Alu I	AG/CT		198	C (2–7,9,10, 13,15) C+NOR* (11)
Ava I	C/YCGRG		6399	Nil (3)
Ava II	G/GWCC			G (9) G,C−ve (mouse) (14)
Bam HI	G/GATCC			G (7)
Bst NI	CC/WGG (b)	Eco RII	506	C−ve (mouse) (8) G (9); Nil (10)
Cfo I	GCGC	{ Hha I, Hin PI }	2244	Nil (3)
Dde I	C/TNAG		212	C (3,6,10,15)
Dra I	TTT/AAA		>2000	C (17)
Eco RI	G/AATTC (f)		3013	C (4); G (5,10); G+C (14); Nil (1)
Eco RII	/CCWGG (c)	Bst NI	506	C (3,7,13); C+G (9)
Fnu DII	CG/CG		9670	Nil (10)
Hae III	GG/CC		405	C+G (3,5,6,9,10) C+G+NOR* (11) G (1,13)
Hha I	GCG/C	{ Cfo I, Hin PI }	2244	Nil (3,10,12)
Hind III	A/ACGTT		1873	G (5,7,10)
Hinf I	G/ANTC		292	C (9,13); C+ve/−ve (human) (3,6,10,11)
Hin PI	G/CGC	{ Cfo I, Hha I }	2244	Nil (12)
Hpa II	C/CGG (c)	Msp I	1747	Nil (3,12)
Mbo I	/GATC			C (3,6,9,10,13,14,15)
Msp I	C/CGG (b)	Hpa II	1747	G+C(+ve/−ve) (12); NOR* (3)
Rsa I	GT/AC (b)		493	C (6,10,13,15)
Sau 96I	G/GNCC			C−ve (mouse) (16)
Sma I	CCC/GGG		30389	Nil (17)
Ssp I	AAT/ATT			G (17)
Taq I	T/CGA (b)		1179	Nil (10); G (12)

* Nucleolar organizer
(a) The following abbreviations are used:
A=adenine, C=cytosine, G=guanine, T=thymidine, R=G or A, W=A or T, Y=C or T, N=A, C, G or T.
The oblique line indicates the site at which the enzyme cuts the DNA (where known).
All sequences are written in the 5'→3' direction.
(b) Not inhibited by methylation of cytosine (in the case of *Msp* I, this refers to the inner cytosine; methylation of the outer cytosine does inhibit the enzyme).
(c) Enzyme inhibited by methylation of cytosine (in the case of *Hpa* II, this refers to methylation of the inner cytosine).

Table 12.1 (*cont.*)

(d) C and G refer to the general type of banding, and do not necessarily imply a detailed correspondence with conventional C- or G-banding patterns. In particular, many enzymes demonstrate only a subset of C-bands. C-bands may be paler than the rest of the chromosome (C−ve) after treatment with certain enzymes.
(e) Mean fragment length is the average length that human DNA will be cut into by the enzyme (Bishop *et al.* 1983).
(f) Under certain conditions, the specificity of *Eco* RI is reduced, so that it will recognize any sequence RRATYY (Polisky *et al.* 1975), and under these conditions the mean fragment length is greatly reduced.

References
(1) Lima-de-Faria *et al.* (1980)
(2) Mezzanotte *et al.* (1983b)
(3) D. A. Miller *et al.* (1983)
(4) Lica & Hamkalo (1983)
(5) Mezzanotte *et al.* (1983a)
(6) N. O. Bianchi *et al.* (1984)
(7) Mezzanotte & Ferrucci (1984)
(8) D. A. Miller *et al.* (1984)
(9) Kaelbling *et al.* (1984)
(10) M. S. Bianchi *et al.* (1985)
(11) Babu & Verma (1986c)
(12) N. O. Bianchi *et al.* (1986)
(13) Ueda *et al.* (1987)
(14) Adolph (1988)
(15) Babu *et al.* (1988)
(16) Burkholder (1989)
(17) Sumner *et al.* (1990)

Many restriction enzymes fail to cut their recognition sequence if it is methylated. The methylation may be of adenine or cytosine, but of these only 5-methylcytosine occurs commonly in eukaryotic DNA. Tabulations are also available showing which enzymes are affected by methylation of the DNA, and which are not (Nelson & McClelland 1989). Several pairs of isoschizomers have now been isolated, one of which is inhibited by methylation, while the other is insensitive to methylation. Such pairs of enzymes are obviously of great value in studying the degree of methylation of DNA.

12.3.1 *Banding patterns produced by restriction endonucleases with conventional staining*

Restriction enzyme banding is simply carried out by treating the fixed chromosome preparations first with a solution of the restriction endonuclease, followed by staining, normally with Giemsa but sometimes with a DNA-specific fluorochrome. Because of the high cost of these enzymes, it is usual to make up small quantities of the enzyme solution, and typically 20 μl is applied to the horizontal slide and covered with a coverslip. Different enzymes are diluted in different buffers; appropriate buffers are recommended, and often supplied, by the firms that sell the enzymes. Each slide may be digested with between 1 and 30 units of enzyme, commonly for 1–2 h at 37°C, but sometimes for rather longer or even overnight. For such long incubations the coverslip is sealed with rubber solution to prevent evaporation.

Results of the application of restriction enzyme banding to mammalian

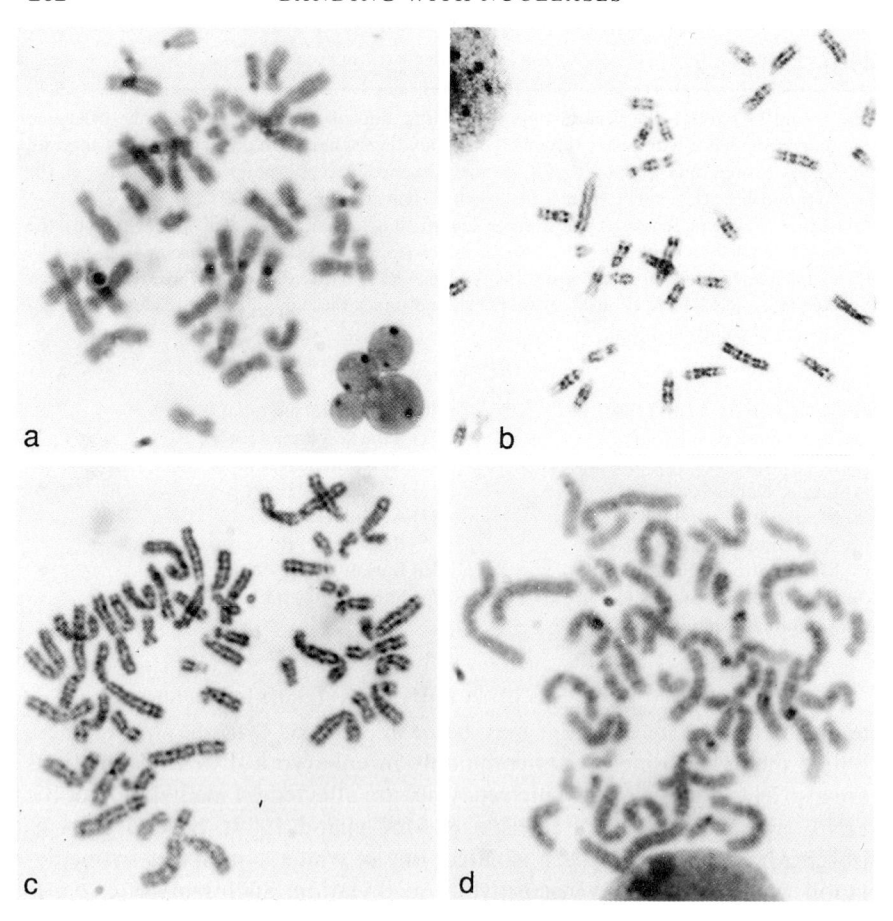

Figure 12.4 Banding patterns produced by digestion with restriction endonucleases and Giemsa staining. (a) C-banding pattern of human chromosomes digested with *Alu* I. (b) Negative staining of heterochromatin of mouse chromosomes after digestion with *Ava* II. Micrograph kindly provided by Dr S. Adolph. (c) G-banding pattern of human chromosomes digested with *Eco* RI. (d) Combination of C- and G-banding produced by digesting human chromosomes with *Hae* III.

chromosomes are listed in Table 12.1, and some examples illustrated in Figure 12.4. It will be seen that essentially three types of banding are produced. Several enzymes produce positively stained heterochromatic blocks, for example *Alu* I (Fig. 12.4a). With other enzymes, certain heterochromatic blocks are negatively stained; that is, they are paler than the chromosome arms. *Hinf* I in particular produces such an effect on certain human chromosomes, and a similar effect has also been reported for *Msp* I. It must be noted that a particular enzyme will usually not differentiate all the heterochromatic blocks, particularly in human chromosomes, and indeed restriction enzyme banding provides yet another means of distinguishing between different types of heterochromatin. Among human chromosomes, there is extensive heterogeneity in the digestibility of blocks of heterochromatin by different restriction endonucleases, and many blocks are also heteromorphic for size (D. A. Miller *et al.* 1983, N. O. Bianchi *et al.* 1984, M. S. Bianchi *et al.* 1985, Babu *et al.* 1988). These heteromorphisms will be described in more detail in a later chapter (Section 14.4). Similar variability has been found in the chromosomes of the great apes (N. O. Bianchi *et al.* 1985, Ferrucci *et al.* 1987); even homologous chromosomes in the different species (man, gorilla, chimpanzee and orangutan) do not necessarily have heterochromatin with similar digestibility by restriction enzymes. The heterochromatin of muntjac chromosomes is also quite variable in its resistance to digestion (Babu & Verma 1986c). The situation in mouse is rather different, since all the heterochromatic blocks (except that on the Y chromosome) behave similarly, although size heteromorphisms can be seen. *Ava* II and *Bst* NI, however, extract the whole of the C-band material except for a small region at the centromeric end (Fig. 12.4b) which, it is speculated, may represent a minor satellite (Kaelbling *et al.* 1984). In this case, treatment with restriction enzymes permits subdivision of the heterochromatin.

Several enzymes produce G-banding on mammalian chromosomes (Fig. 12.4c) and as a very rough rule, these tend to produce longer mean fragment lengths than those which produce C-bands. *Hae* III produces G- or C-banding, or a combination of the two (Fig. 12.4d), depending on the length of digestion (Mezzanotte *et al.* 1983a). This appears to be the only enzyme with which the digestion conditions consistently affect the banding pattern produced, although with a few other enzymes listed in Table 12.1 there are inconsistencies between reports by different authors.

Some restriction enzymes are reported to produce no banding at all. These all include the dinucleotide CG in their recognition sequence, and because of the rarity of this doublet (Russell *et al.* 1976) such enzymes cut too few sites in the DNA for it to be lost from the chromosomes. Table 12.1 shows that isoschizomers (many of which contain the CG doublet) tend to produce the same banding pattern, although an exception is found with the pair *Hpa* II and *Msp* I. Digestion by *Hpa* II is, however, inhibited if the internal cytosine is methylated, whereas digestion by *Msp* I is not inhibited by methylation of this cytosine. Since the cytosine of the CG doublet is commonly methylated, *Msp* I would be expected to digest more DNA than *Hpa* II, and the production of banding by the former but not by the latter is in accordance with expectation.

Restriction endonucleases have also been used to produce banding on the chromosomes of certain insects and lower vertebrates, and of one species of plant (Table 12.2). Since none of these organisms has G-bands on its chromosomes, it is not surprising that only heterochromatic bands are demonstrated. Although a few of the enzymes used fail to produce any banding pattern, the majority produce patterns which resemble C-banding, each enzyme often revealing a characteristic subset of bands. In a grasshopper, Gosálvez *et al.* (1987) noted that the distamycin-DAPI positive

Table 12.2 Restriction endonuclease banding in organisms other than mammals.

Species	Restriction enzymes used	References
Insects: Orthoptera		
Arcyptera tornosi	*Alu* I, *Hae* III, *Hin*d III, *Hin*f I	Gosálvez *et al.* (1987)
Baetica ustulata	*Alu* I, *Hpa* II, *Hap* II, *Msp* I, *Hae* III, *Hha* I	Sentis *et al.* (1989)
Oedipoda germanica	*Alu* I, *Dde* I, *Hae* III, *Hin*f I, *Hpa* II, *Mbo* I, *Msp* I, *Sau* 3A, *Taq* I	Lopez-Fernandez *et al.* (1989)
Pyrgomorpha conica	*Alu* I, *Mbo* I, *Hae* III	Lopez-Fernandez *et al.* (1988)
Insecta: Diptera		
Aedes albopictus	*Hha* I, *Hin* PI, *Hpa* II, *Msp* I, *Mbo* I, *Taq* I	N. O. Bianchi *et al.* (1986)
Culiseta longiareolata	*Alu* I, *Hha* I, *Hae* III, *Hpa* II, *Msp* I, *Dde* I, *Hin*f I, *Bam* HI	Marchi & Mezzanotte (1988)
Drosophila melanogaster	*Alu* I, *Hae* III	Mezzanotte (1986)
Drosophila virilis	*Alu* I, *Hae* III	Mezzanotte *et al.* (1987)
Sarcophaga bullata	*Alu* I, *Hae* III, *Hin*d III	Bultmann & Mezzanotte (1987)
Fish		
Salmo gairdneri	*Alu* I, *Hae* III, *Hin*f I, *Mbo* I, *Pvu* II	Lloyd & Thorgaard (1988)
Muraena helena	*Dde* I, *Hae* III, *Mbo* I	Cau *et al.* (1988)
Amphibia		
Anura, 4 spp. *Triturus alpestris*	*Eco* RI, *Hae* III, *Hin*d III, *Hpa* II, *Msp* I	Schmid & de Almeida (1988)
Odontophrynus americanus	*Bgl* II, *Dra* I, *Eco* RI, *Hpa* II, *Msp* I	Beçak *et al.* (1988)
Birds		
Gallus domesticus	*Eco* RI, *Hae* III, *Hin*d III, *Hpa* II, *Msp* I	Schmid & de Almeida (1988)
Plants		
Vicia faba	*Alu* I, *Bam* HI, *Mbo* I	Frediani *et al.* (1987)

heterochromatin, believed to be A+T-rich, was digested by *Hae* III, which cleaves the sequence GGCC. Obviously it is the occurrence of specific sequences that is important in determining whether particular regions of chromosomes will be digested by restriction endonucleases, and not the average base composition. In *Drosophila melanogaster* the specificity of digestion could be correlated with the known properties of the satellite DNAs in the different blocks of heterochromatin (Mezzanotte 1986). In both *Drosophila melanogaster* and *D. virilis*, restriction enzymes were used to digest polytene chromosomes, and specific bands were positively stained after digestion (Mezzanotte 1986, Mezzanotte *et al.* 1987).

Perhaps the most interesting of this work on Diptera is that of N. O. Bianchi *et al.* (1986) who compared results using enzymes that are inhibited by methylation of cytosine, with results using enzymes unaffected by methylation; their work included similar experiments on human cells. Enzymes that were inhibited by methylation extracted very little DNA from human cells, although if the DNA was demethylated by treating the cells with 5-azacytidine the same enzymes extracted as much DNA as enzymes that were unaffected by methylation. In the mosquito chromosomes, however, all enzymes extracted substantial amounts of DNA, indicating that there was little or no methylation of the DNA, in agreement with the belief that insect DNA is essentially unmethylated (Bird & Taggart 1980). It appeared, however, that the mosquito chromosomes were not wholly unmethylated, since digestion with *Hpa* II, which is inhibited by methylation, resulted in centromeric bands on most of the chromosomes, while digestion with *Msp* I, which is unaffected by methylation, or with *Hpa* II after treatment of the cells with 5-azacytidine, resulted in uniform digestion of the chromosomes, without C-bands. See also the work of Sentis *et al.* (1989) on grasshopper chromosomes. *Mbo* I also produced bands on mosquito chromosomes, which could be due to the presence of methyladenine, which is present in mosquito cells and which inhibits this enzyme, but it was not possible to test this hypothesis rigorously. Apart from their general utility in producing banding patterns on mosquito and other chromosomes, it is clear that the use of pairs of enzymes having different sensitivities to methylation can be a powerful tool to investigate the degree of methylation of different segments of chromosomes.

12.3.2 Restriction enzyme/nick translation

Not all workers using restriction enzymes on chromosomes have used the simple sequence of digestion followed by Giemsa staining described at the beginning of the previous section. The 'traditional' method does not, in fact, detect sites of enzyme attack directly, but only, it seems, by inducing DNA loss. Such a method is ineffective in detecting the rarer sites of cutting. Certain authors have, therefore, begun to experiment with nick translation as a means of detecting cuts in the chromosomal DNA produced by restriction endonucleases. As already described in Section 12.2.1, *in situ* nick translation involves the synthesis of labelled DNA at a nick in the existing DNA. The label can then be recognized by a suitable procedure, either autoradiographic or immunocytochemical (Fig. 12.5).

Bullerdiek *et al.* (1985) used *Eco* RI on human chromosomes, and in favourable cases obtained bands throughout the length of the chromosomes. Later, the same group used a wider variety of enzymes (*Eco* RI, *Dra* I, and *Sma* I) and obtained more consistent results on both human and Chinese

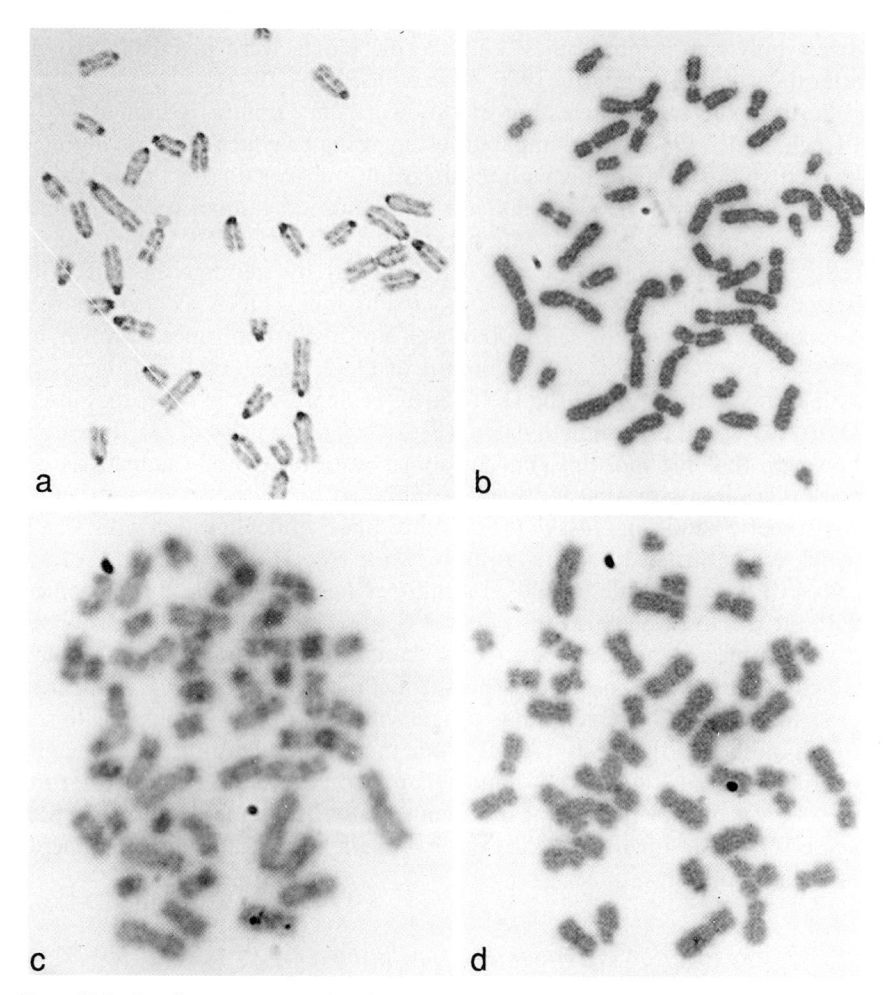

Figure 12.5 Banding patterns produced by nick translation of chromosomes after digestion with restriction endonucleases. (a) Strong labelling of mouse heterochromatin after digestion with *Ava* II. Reproduced from Adolph (1988) by permission of the author and Springer-Verlag. (b) Unlabelled centromeric regions of human chromosomes after digestion with *Eco* RI. (c) Strong labelling of centromeric and some terminal regions of human chromosomes after overnight digestion with *Dra* I. (d) Uniform labelling of human chromosomes digested with *Alu* I.

hamster chromosomes, although it did not seem possible to relate the banding pattern to either G- or R-banding (Bullerdiek *et al.* 1986b).

Adolph (1988) applied the restriction enzyme/nick translation procedure to mouse chromosomes, using the enzymes *Ava* II and *Eco* RI. As shown by Giemsa staining, *Ava* II digested the centromeric heterochromatin, while *Eco* RI did not. Using the nick translation procedure, the centromeric heterochromatin was unstained after *Eco* RI digestion, but strongly stained after *Ava* II (Fig. 12.5a), which is as expected. Staining of the chromosome arms was observed after both enzymes, but the patterns were not analysed.

Viegas-Péquignot et al. (1988) digested human chromosomes with Hha I, an enzyme whose activity is inhibited by methylation in the sequence GCGC, and found that nick translation produced an R-banding pattern on the autosomes. Surprisingly, the inactive X chromosome in female cells was particularly strongly labelled, and the authors concluded that this chromosome was undermethylated.

In a comprehensive study on human chromosomes, Sumner et al. (1990) tested a wide variety of restriction endonucleases having different sequence specificities, and compared the results obtained with nick translation and with Giemsa staining. A number of distinct nick translation patterns were obtained. Many enzymes produced patterns with negative C-bands (Fig. 12.5b) although in some cases these became strongly positive with prolonged digestion (Dra I, Hae III) (Fig. 12.5c). Alu I invariably produced a uniform nick translation pattern (Fig. 12.5d), although it gave a C-banding pattern with Giemsa, and extracted large amounts of DNA from the chromosome arms. Other enzymes produced weak R-banding patterns (Hpa II, Msp I, Ssp I). It is important to note that the nick translation and Giemsa staining patterns were not, in general, complementary. The authors concluded that, although the base sequence specificity of the restriction endonucleases must be the primary determinant of the nick translation patterns, they were also strongly influenced by the ability of the enzymes to penetrate to the chromosomal DNA, and of the digested DNA to diffuse out of the chromosome. Thus although nick translation methods are inherently more reliable than Giemsa staining for indicating the distribution on chromosomes of the recognition sites for restriction endonucleases, it is clear that so far these methods also have their limitations.

12.3.3 Studies on the mechanism of action of restriction endonucleases on chromosomes

The principle behind the digestion of chromosomes with restriction endonucleases is that, whether the chromosomes are stained with Giemsa or some other dye, or nick translated, any resulting pattern will be a consequence of the distribution of the enzyme recognition sites along the chromosomes. The available evidence indicates that things are not necessarily so simple.

The simplest case to consider is when a restriction enzyme preferentially digests a satellite DNA that is the predominant component of certain C-bands. Thus D. A. Miller et al. (1984) demonstrated an actual preferential loss of satellite DNA in mouse chromosomes after digestion with Bst NI, which produces negative C-bands in this species. Ava II, which produces the same effect, was shown by Adolph (1988) to produce a strong nick translation reaction in mouse centromeric heterochromatin. M. S. Bianchi et al. (1985) pointed out that Hinf I cleaves human satellites II, III and IV into small fragments, and that C-bands rich in these satellites appear negatively stained after digestion with this enzyme and Giemsa staining. However, I have been unable to reproduce this effect, nor obtain a strong nick translation reaction in C-bands with this enzyme; apparently, therefore, some factor must be preventing easy access of the enzyme to the DNA in my experiments (Sumner et al. 1990).

Evidence that resistance of certain C-bands to restriction enzymes is because their DNA lacks sites for cutting by the enzymes is less easy to come by. It must be remembered that the DNA in C-bands is, in any case, less easily extractable than euchromatic DNA, even by non-specific agents (Section 4.2.5). Evidence for a considerable degree of specificity in the action of restriction enzymes on C-bands comes from two sources. First, in human chromosomes, different enzymes produce different subsets of C-bands (D. A. Miller et al. 1983, M. S. Bianchi et al. 1985, Babu et al. 1988), and second, certain enzymes, as noted above, can extract all or most of the DNA from certain C-bands. Both sets of observations indicate that, under the conditions in which they are applied to chromosomes, restriction endonucleases have access to all parts of the chromosomes; as stated by D. A. Miller et al. (1983): 'Giemsa staining could be eliminated from every region of every chromosome.' C-bands can also be demonstrated using dyes which are stoichiometric for DNA, unlike Giemsa (Mezzanotte et al. 1983a,b, 1985, Mezzanotte & Ferrucci 1984, Ueda et al. 1987), and measurements of DNA have shown that there is loss of DNA from chromosomes when restriction enzymes are used to produce C-bands (M. S. Bianchi et al. 1985, Mezzanotte et al. 1985, Burkholder 1989). Cross-linking of DNA to chromosomal proteins with UV light prevents extraction of DNA and prevents banding, the whole chromosome remaining strongly stained with Giemsa (N. O. Bianchi et al. 1984). There is thus clear evidence that the C-bands are the consequence of the selective extraction of non-C-band DNA by the enzymes.

There is, however, good evidence that chromosomal proteins do affect the extraction of DNA. Lica & Hamkalo (1983) showed that the resistance of the centromeric heterochromatin of (unfixed) mouse chromosomes to *Alu* I or *Eco* RI was partly due to its highly condensed structure, since protein-free DNA from this condensed chromatin can be cleaved further by additional digestion with the same enzymes. In unfixed mouse chromosomes, *Bst* NI fails to digest satellite DNA, which must therefore be protected by proteins that are lost during fixation (Burkholder 1989). Evidence that access of the enzymes to heterochromatin is restricted also comes from studies using nick translation (Sumner et al. 1990). For example, *Dra* I fails to attack human C-band DNA during a short digestion, but attacks it strongly when digestion is prolonged. In spite of this, the chromosomes are C-banded with Giemsa, showing that the digested C-band DNA is not extracted. With *Alu* I the whole chromosome is evenly nick translated, although DNA is preferentially lost from the arms. Thus two factors, the accessibility of the DNA to the enzyme, and the ability of the digested DNA to diffuse out of the chromosome, can modify the pattern due to the restriction enzyme recognition sites.

D. A. Miller et al. (1983) estimated that DNA fragments produced by restriction enzymes that were as small as 100 bp would readily diffuse out of chromosomes, whereas fragments greater than 1000 bp would be retained. However, Burkholder (1989) showed experimentally that fragments greater than 4000 bp could be lost from unfixed chromosomes after digestion, while Sumner *et al.* (1990) found that *Dra* I digestion, which produced fragments predominantly greater than 2000 bp, caused extensive DNA loss from fixed chromosomes.

When we turn to the mechanism of induction of G-bands by restriction endonucleases, it is immediately striking that no enzyme induces R-banding with Giemsa staining, in spite of the wide variety of recognition sequences of the enzymes used. In addition, the G-bands are seen only after Giemsa staining, and not after staining with dyes that are stoichiometric for DNA (Mezzanotte *et al.* 1983a, 1985, Mezzanotte & Ferrucci 1984), and the amounts of DNA extracted by these enzymes are very small (Mezzanotte *et al.* 1985). The situation here thus seems to be very similar to that for conventional G-banding (Section 5.4), in which there is also no significant loss of DNA. Nick translation generally shows uniform nicking throughout the length of the chromosomes, and it may be surmised that within the positive G-bands the DNA is so constrained by chromosomal proteins that no conformational changes can occur that affect Giemsa staining. In the negative G-bands, however, the nicked DNA can undergo conformational changes much more readily, and in some way loses its ability to stain with Giemsa.

There is one exception to the rule that restriction enzymes that produce G-bands do not extract DNA. *Hae* III produces G-bands with Giemsa, produces a similar pattern of staining with ethidium (Mezzanotte *et al.* 1983) or propidium (Ueda *et al.* 1987), and extracts a considerable amount of DNA from chromosomes (M. S. Bianchi *et al.* 1985, Mezzanotte *et al.* 1985). Nick translation experiments show that the chromosome arms are uniformly nicked by this enzyme (Sumner *et al.* 1990) so that the banding pattern must result from preferential loss of DNA from positive R-bands.

12.4 General remarks on banding with nucleases

Any new banding method should be welcomed if it provides new information about chromosome organization. It is quite obvious that this is so for banding with many of the restriction endonucleases, which, particularly in human chromosomes, can demonstrate characteristic subsets of C-bands (D. A. Miller *et al.* 1983, N. O. Bianchi *et al.* 1984, M. S. Bianchi *et al.* 1985, Babu *et al.* 1988). Useful though this may be (and as yet these methods are too recent for a completely objective appraisal of their value), the potential of methods using restriction endonucleases must undoubtedly lie in their ability to identify specific DNA sequences. So far this has been attempted only to a limited extent, the best example being the use of the isoschizomers

Hpa II and *Msp* I to distinguish between methylated and non-methylated DNA sequences (N. O. Bianchi *et al.* 1986, Sentis *et al.* 1989).

In general the action of restriction enzymes on chromosomes has been assessed using Giemsa staining, the pattern of which was believed (in some cases wrongly) to indicate extraction of DNA. As we have seen, however, DNA extraction is not a very satisfactory method of assessing the action of enzymes on chromosomal DNA, and procedures using *in situ* nick translation will undoubtedly become more popular, in spite of their greater complexity. Indeed, such methods are the only way of labelling DNase I hypersensitive sites; the amount of DNA digestion in such cases is too small to be detected easily by any other method. In any case, it is more satisfying to have a positive label at the sites of enzyme attack, rather than simply a lack of staining.

Use of DNase I hypersensitivity to identify sites of potentially active genes is a powerful tool for analysing the functional organization of chromosomes. Use of suitable restriction chromosomes to identify specific nucleotide sequences which may be associated with genes will provide another approach to the same problem. At present, the patterns produced by such treatments are not clearly defined, although the patterns of DNase I hypersensitivity seem to show some resemblance to R-banding. More detailed analysis is required to establish the relationships between distribution of genes and other distinctive sequences (such as sites for initiation of replication, and perhaps of condensation) and 'conventional' banding patterns, and thereby establish a functional explanation of banding. Restriction endonucleases, with their ability to recognize specific sequences in DNA, should be very powerful tools for such studies.

12.5 Note on nomenclature

There is no generally agreed method of designating chromosome bands induced by nucleases, although various systems have been proposed. In this chapter the terms C-banding, G-banding and so on have been used only to indicate the general class of banding produced, and not to imply a detailed resemblance to conventional C- and G-band patterns. Various other terms have been coined in the literature. Kerem *et al.* (1984) used the term D-bands for the patterns of DNase I hypersensitivity on chromosomes; although the term D-bands had previously been used for fluorescent daunomycin bands (Section 8.3.4), the latter technique has not come into use, so that there is no problem of confusion. For bands produced by restriction endonucleases, several systems have been proposed. The simplest, used by various authors, is to use the name of the enzyme only (e.g. *Alu* I-bands), but this has the disadvantage of giving no indication of the pattern of banding. To overcome this, N. O. Bianchi *et al.* (1985) proposed the use of the name of the enzyme followed by the type of banding (C, G, or gap, the last referring to a negatively stained C-band). Thus C-banding induced by *Alu* I would be *Alu* I-C-banding. The same authors evidently felt that this system was not wholly satisfactory, since the next year they modified it using the symbol Re to indicate restriction enzyme banding, followed in brackets by the type of pattern, and again followed in brackets by the name of the enzyme [e.g. Re(C)(*Alu* I)] (N. O. Bianchi *et al.* 1986). These latter systems are undoubtedly better than just the use of the name of the enzyme, in that they do give an indication of the type of banding produced, yet they do suffer from the disadvantage that the pattern of banding is dependent on the method used to detect enzyme action. A band that was negative after restriction enzyme digestion and Giemsa staining might, for example, be positive using the same enzyme with nick translation.

With all other banding techniques, nomenclature is dependent on the method used, and not on the pattern produced, and while this itself is not the ideal system, it may well be the most practical when used in conjunction with a proper description of the method. No convenient shorthand can convey all the possible subtleties of technique or banding pattern, and it may well be most convenient simply to refer to DNase I-banding, *Hae* III-banding, etc. If it is necessary to indicate the method of detecting the action of the enzyme, then terms such as DNase I/nick translation banding, or *Hae* III/Giemsa banding could be used. (Note that '*Hae* III/G-banding' should not be used, as this would imply similarity to banding patterns produced by more conventional G-banding methods). The actual pattern can then be described by reference to that obtained using a more conventional method; for example, 'the C-band of chromosome 1 is negatively stained using the *Eco* RI/nick translation method' or '1qh is *Alu* I/Giemsa positive.'

13 Banding demonstrated by immunocytochemical methods

13.1 Scope of applications of immunocytochemical methods to chromosomes

Immunocytochemistry has become a very powerful tool for studying the organization of cells, and of chromosomes in particular (Bustin 1979). Such studies have included work at the molecular level, the arrangement of histones in nucleosomes, for example, and also extensive studies on polytene and lampbrush chromosomes, including observations on proteins connected with transcriptional activity. The scope of this chapter is more restricted, being confined to immunocytochemical procedures that have demonstrated longitudinal differentiation (i.e. banding) on mitotic metaphase chromosomes, a field that has been named 'immunocytogenetics' by Haaf *et al.* (1988b). Although interesting and valuable work is being done on meiotic chromosomes (e.g. Moens *et al.* 1987, Dresser *et al.* 1987) the antigens studied are not distributed in a banded pattern, and are therefore outside the scope of this chapter. Similarly, immunocytochemical studies of components of mitotic chromosomes that are evenly distributed along the length of the chromosomes, such as those in the chromosome core (Earnshaw & Heck 1985, Gasser *et al.* 1986) or peripheral components (McKeon *et al.* 1984) will not be considered here. Certain aspects of chromosome banding using immunocytochemical reagents have already been discussed in previous chapters, including studies on nucleolar organizers (Ch. 9), the use of CREST serum for labelling kinetochores (Section 10.2.3), and the use of antibodies to BrdU for the study of replication banding (Section 11.3). This chapter covers chromosome banding produced by two main classes of antibodies: those against DNA and its components, and those against chromosomal proteins, that is, histones and non-histones.

13.2 Antibodies to nucleic acids and nucleosides

Antibodies have been used to study the distribution of DNA bases along chromosomes, and also to investigate the conformation of chromosomal DNA. For a general review of antibodies to DNA, see Stoller (1986).

Practical details will not be given here, as they are described in the papers cited and also in textbooks on immunocytochemical methods: no distinctive methods are required for immunocytochemical studies of chromosomes. Essentially, the primary antibody binds to its specific site on the chromosome, and is in turn recognized by a secondary antibody, which is commonly labelled with fluorescein, although a peroxidase label has also been used, which is claimed to give much greater sensitivity as well as producing permanent preparations.

13.2.1 Antibodies to the four unsubstituted bases of DNA

In a series of papers, O. J. Miller and Erlanger and their colleagues studied the binding of anti-adenosine, anti-cytidine, anti-guanosine, and anti-thymidine (anti-A, anti-C, anti-G, and anti-T) to mammalian metaphase chromosomes. An important initial finding was that these antibodies would not bind to chromosomes unless their DNA was first denatured (Freeman et al. 1971). Denaturation is a critical step, and the banding pattern obtained with different antibodies depends to some extent on the method used for denaturation. Results obtained on the euchromatic segments of mammalian chromosomes are summarized in Table 13.1 and illustrated in Figures 13.1 and 13.2. Hot formamide is expected to produce general denaturation of chromosomal DNA. Ultraviolet irradiation has as its main effect the formation of pyrimidine dimers, particularly thymine dimers, and thus denaturation is likely to be more extensive in A + T-rich regions of the chromosomes, where the dimerized thymine can no longer pair with adenine. Photo-oxidation, on the other hand, is regarded as destroying guanine residues preferentially, so that cytidine residues will become unpaired and be available for reaction with the antibody. Thus both UV irradiation and photo-oxidation would be expected to reinforce the base specificity of the antibodies; the antibodies will react with a chromosomal segment not simply because it contains the appropriate nucleoside, but because it contains that nucleoside antigen it can also be specifically denatured, allowing access of the antibody. Bearing these points in mind, it is clear that anti-A and anti-T bind preferentially to positive Q- or G-bands, while anti-C binds predominantly to positive R-bands. Indeed, the patterns of anti-nucleoside antibody binding provided some of the first and clearest evidence for a relationship between chromosome banding patterns and DNA base composition. Further supporting evidence comes from the observations of Magaud et al. (1985) who obtained R-banding using an antibody to double-stranded DNA which was specific for poly dG-poly dC.

The binding of anti-nucleoside antibodies to heterochromatic segments of chromosomes is critically dependent on the method of denaturation. No binding could be obtained after formamide denaturation of the chromosomes (Dev et al. 1972c), and it was concluded that the highly repeated DNA in these regions annealed too quickly for antibody binding to occur (O. J.

Table 13.1 Banding patterns produced on mammalian chromosomes by different anti-nucleoside antibodies.

Method of denaturation (References)	Formamide, 65°C (1)	Ultraviolet irradiation (2, 3)	Photooxidation (3, 4)
Antibodies			
Anti-A	G/Q-bands	G/Q-bands	{ Uniform (4) { R-bands (3)
Anti-C	G/Q-bands	Uniform	R-bands (3, 4)
Anti-G	G/Q-bands	Uniform	
Anti-T	G/Q-bands	G/Q-bands	Uniform (4)
Anti-$\widehat{TT}$*		G/Q-bands	

*$\widehat{TT}$ = thymine dimer.
References
(1) Dev *et al*. 1972c
(2) Schreck *et al*. 1974
(3) Schreck *et al*. 1977a
(4) Schreck *et al*. 1973

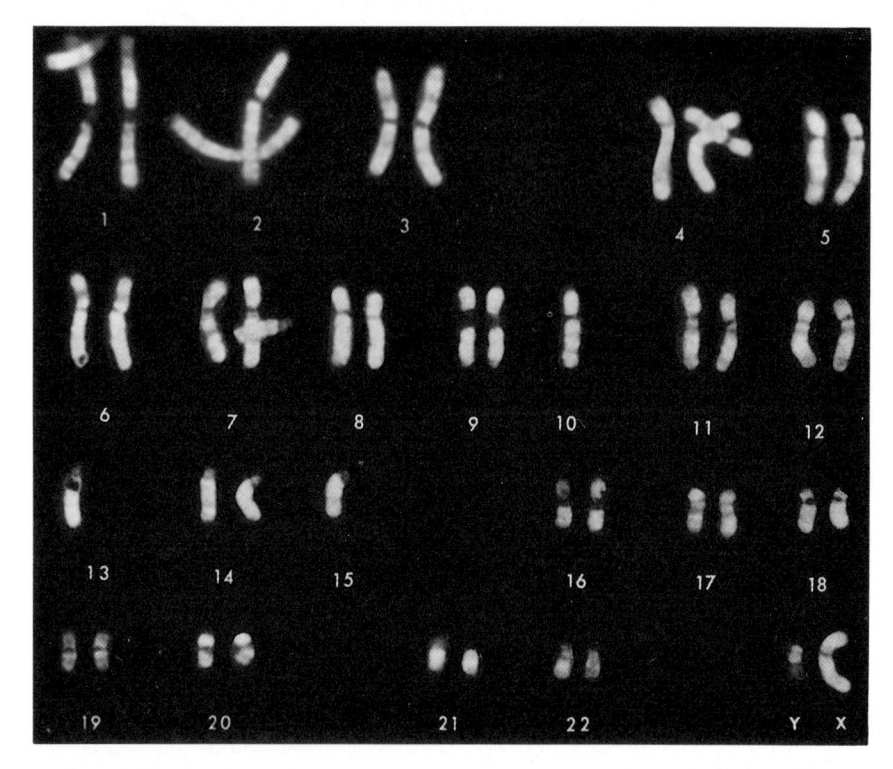

Figure 13.1 Pattern of anti-adenosine binding to human chromosomes after denaturation with formamide. Reproduced from Dev *et al*. (1972c) by permission of the authors and Academic Press.

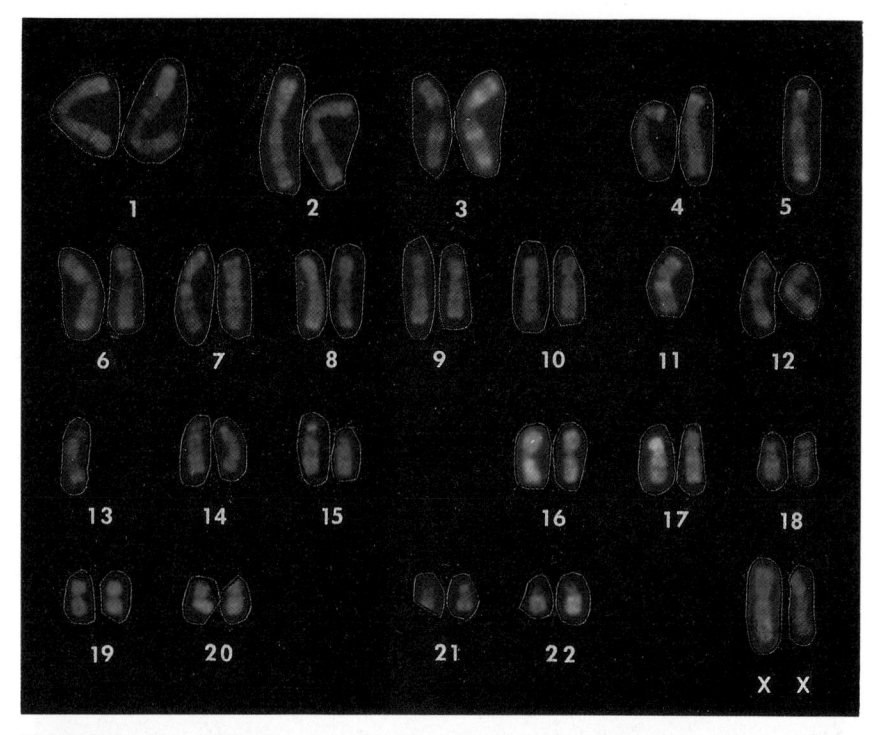

Figure 13.2 Pattern of anti-cytosine binding to human chromosomes after photo-oxidation. Reproduced from Schreck *et al.* (1973) by permission of the authors.

Miller & Erlanger 1975). The other methods of denaturation, UV irradiation and photo-oxidation, which produce irreversible denaturation, do permit binding of antibody, and the pattern of binding to heterochromatin is consistent with the base composition of the repetitive DNA contained within it (Table 13.2). It is, however, abundantly clear that the specificity is due to the combination of denaturation method and antibody used, and is not simply that of the antibody alone.

13.2.2 Anti-5-methylcytosine

Antibodies to 5-methylcytosine (5MeC) have been used to study the distribution of this modified base in a variety of organisms, mostly mammals. As with the antibodies already described, prior denaturation of the chromosomal DNA is required for antibody binding; heat treatment, UV irradiation, and photo-oxidation have all been used, and produce largely the same pattern of antibody binding, although some small differences have been reported. The major finding, in several species of mammals, is that 5-methylcytosine is concentrated in blocks of heterochromatin (Fig. 13.3). This is true whether the heterochromatin contains A + T-rich or G + C-rich DNA, and has been reported for the mouse (O.J. Miller *et al.* 1974, Schreck *et al.*

Table 13.2 Patterns of binding of anti-nucleoside antibodies to mammalian heterochromatin.

Antibody to: Method of denaturation		A UV	A Photo- oxidation	C UV	C Photo- oxidation	T UV
Region	Base composition					
Human (Schreck *et al.* 1973, 1974)						
1qh	A+T-rich	+ve				+ve
9qh	Complex	+ve		+ve		+ve
16qh	A+T-rich	+ve				+ve
Yqh	A+T-rich	+ve			−ve	+ve
Mouse (Schreck *et al.* 1974, 1977a)						
Centromeric heterochromatin	A+T-rich	+ve	−ve	−ve	−ve	+ve
Kangaroo rat (Schreck *et al.* 1977b)						
Centromeric heterochromatin	G+C-rich	−ve	+ve		+ve	
Non-centromeric heterchromatin	G+C-rich		+ve		+ve	

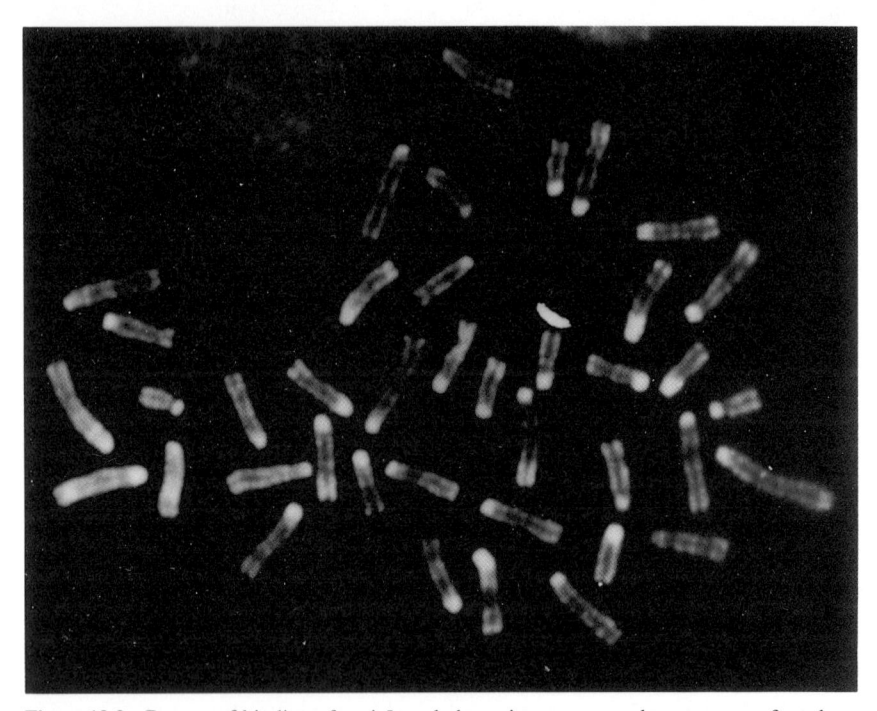

Figure 13.3 Pattern of binding of anti-5-methylcytosine to mouse chromosomes after photo-oxidation. Reproduced from Schreck *et al.* (1977a) by permission of the authors and Springer-Verlag.

1977a), humans (O. J. Miller *et al.* 1974, Schnedl *et al.* 1975, Lubit *et al.* 1976), the great apes (Schnedl *et al.* 1975), cattle, sheep, and goats (Schnedl *et al.* 1976), the kangaroo rat *Dipodomys ordii* (Schreck *et al.* 1977b) and the Indian muntjac (Vasilikaki-Baker & Nishioka 1983). Patterns of 5-MeC distribution show heteromorphisms just as other staining methods for heterochromatin do (O. J. Miller *et al.* 1974, Schnedl *et al.* 1975). However, anti-5-MeC does not bind equally to all blocks of heterochromatin. In humans, the major C-bands on chromosomes 1, 9, 16 and Y are well labelled, but the short arms of chromosome 15 also bind anti-5-MeC strongly, a feature that appears to be diagnostic for this chromosome amongst the human acrocentrics (O. J. Miller *et al.* 1974). In the kangaroo rat, strong binding of anti-5-MeC is confined to the centromeric heterochromatin, while the non-centromeric heterochromatin, which is composed of different satellites, binds the antibody no more strongly than euchromatin (Schreck *et al.* 1977b).

As mentioned above, minor variations in the pattern of binding of anti-5-MeC occur as a result of different denaturation conditions. In the mouse, UV irradiation results in completely uniform binding to the chromosome arms, whereas after photo-oxidation there is a distinct pattern on the arms, with the ends of the chromosomes in particular showing brighter fluorescence (Schreck *et al.* 1977a). In the Indian muntjac, photo-oxidation permitted binding of anti-5-MeC to many sites on the chromosomes, both euchromatic and heterochromatic, whereas only the centromere of chromosome 3 reacted after UV irradiation (Vasilikaki-Baker & Nishioka 1983). These observations only serve to emphasize the important influence of denaturation methods on the pattern of binding of anti-nucleoside antibodies.

Because of the evidence that methylation of cytidine is involved in control of gene transcription (Cooper 1983, Razin & Cedar 1984), it is not surprising that attempts have been made using anti-5-MeC to study differences in DNA methylation associated with repression of transcription. No difference in the degree of methylation could be detected when active and inactive X chromosomes of certain primates were compared (D. A. Miller *et al.* 1982). It is, however, clear that amplified ribosomal RNA genes that are transcriptionally inactive as shown by silver staining (Section 9.4.2) also bind anti-5-MeC strongly, indicating that methylation is an important factor in the inactivation of these regions (Tantravahi *et al.* 1981a,b).

13.2.3 Antibodies to single-stranded and double-stranded DNA

A limited number of studies have been carried out in which antibodies to single-stranded (ss) or double-stranded (ds) DNA have been applied to mammalian chromosomes. Results are few and inconsistent, and no doubt much of the inconsistency is because the antibodies have not been adequately characterized. This point will be discussed in more detail in the next section (13.2.4), which deals with antibodies to Z-DNA.

Mace et al. (1972) applied serum from a patient with systemic lupus erythematosis (SLE) to mouse chromosomes. This serum was stated to contain antibody to ssDNA, and failed to bind to conventionally fixed chromosomes. However, after the chromosomes were treated for C-banding (Ch. 4), the chromosome arms bound antibody strongly, whereas the centromeres did not. This was interpreted as demonstrating rapid annealing of the repeated DNA in the centromeric heterochromatin. Morin et al. (1977) also used an SLE serum, again apparently with specificity for ssDNA, since the chromosome preparations were first treated with UV light. These experiments produced patterns resembling R-banding, which is perhaps not what would be expected from comparison with studies using anti-nucleoside antibodies (Table 13.1).

Bosman & Nakane (1982) obtained uniform staining with anti-dsDNA derived from SLE serum, but Magaud et al. (1985) obtained R-banding with a monoclonal anti-dsDNA antibody; the latter, however, also showed specificity for dG-dC, which would be consistent with the pattern obtained. Mezzanotte et al. (1989) obtained yet another pattern of staining. Using two different monoclonal antibodies to dsDNA, they found a G-banding pattern on freshly made, methanol-acetic acid-fixed chromosome preparations. Staining became uniform both when the slides were aged for a few days before staining, and also if the fresh preparations were treated briefly with nucleases to nick the DNA before staining. It was concluded that the lack of staining in the negative G-bands was due to some conformational difference in the DNA, not recognized by the antibodies, which could be relieved by nicking the DNA. This point will be considered in more detail in the next section.

It cannot be claimed, on the basis of the results just described, that the use of antibodies against ss- or dsDNA has so far provided any conclusive insights into chromosome organization in general, or into chromosome banding in particular. From the variety of results obtained, it seems clear that it is not enough to describe antibodies as being specific for either single- or double-stranded DNA, since antibodies claimed to have the same specificity can produce different patterns on chromosomes. Evidently it will be necessary to characterize these antibodies much more precisely before they can be regarded as useful tools for studying chromosomes.

13.2.4 *Antibodies to Z-DNA*

In certain circumstances, DNA can form a left-handed, double-stranded helix known as Z-DNA (Jovin et al. 1983, Rich et al. 1984). In particular, Z-DNA can form under physiological conditions either if the DNA is methylated, or if it is supercoiled. Z-DNA is highly immunogenic, and many different antibodies have been prepared against it, some of which have been applied to chromosomes. Much of this work has been done on polytene chromosomes, but two papers describe work on mammalian metaphase chromosomes. In both a gerbil (*Gerbillus nigeriae*) and a monkey (*Cebus albifrons*), segments of G + C-rich, R-band positive heterochromatin bind anti-Z-DNA strongly (Viegas-Péquignot et al. 1982, 1983a). Blocks of heterochromatin in gerbils and humans, which were not G + C-rich and R-band positive, did not bind anti-Z-DNA. In addition to the staining of certain heterochromatic segments, the anti-Z-DNA also produced a weak but distinct R-banding pattern on the chromosomes of several species of mammals, with the ends of the chromosomes, the T-bands, showing a somewhat stronger reaction.

The question of whether chromosomes really contain Z-DNA, or whether the binding of anti-Z-DNA is some kind of artefact resulting from changes during preparation, has been the subject of intensive study. Studies on polytene chromosomes have shown that when they are isolated under physiological conditions, they bind very little anti-Z-DNA, but that a massive increase in antibody binding occurs after fixation with acetic acid (Hill & Stollar 1983, Robert-Nicoud et al. 1984). The pattern of binding to polytene chromosomes can also be altered by the amount of exposure to acid. Binding of antibodies to B-DNA is strong in both native and fixed polytene chromosomes, so it seems unlikely that the absence of binding of anti-Z-DNA in native chromosomes is due to masking by chromosomal proteins that are extracted by acetic acid. Hill & Stollar (1983) put forward an alternative view, that extraction of histones by fixation produces supercoiling in the DNA, the energy of which could drive the DNA from the B to the Z conformation. In support of this, treatment of fixed chromosomes with topoisomerase I, or limited nicking with DNase I, both treatments that would relax supercoiling, inhibit the binding of anti-Z-DNA to chromosomes. Thus it seems probable that native chromosomes only contain sequences that have the potential to form Z-DNA, but there is no incontrovertible evidence that the Z conformation is formed until the chromosomes are fixed.

A further complication is that not all antibodies to Z-DNA have the same specificity. Several classes of antibodies to Z-DNA can be recognized (Jovin et al. 1983), of which three have been applied to chromosomes (Arndt-Jovin et al. 1985), Nordheim et al. 1986). Class 1 antibodies recognize all Z-DNAs with equal affinity; class 2 antibodies have a preference for poly d(G-C) sequences, while class 3 antibodies have an additional preference for substitutions at the C5 position of the pyrimidines. While antibodies of all three classes produce generally similar patterns of binding on polytene chromosomes, preferential binding of one class of antibody can be seen in certain regions of chromosomes (Arndt-Jovin et al. 1985).

The various factors which can affect the binding of antibodies to Z-DNA have been described in some detail not only because of their intrinsic interest, but also because similar considerations undoubtedly apply to studies using antibodies to other classes of DNA. For example, we saw in the previous section (13.2.3) that three different patterns of binding to chromosomes could be obtained using three different antibodies all of which were described as being specific for dsDNA. Furthermore, the binding of one of these antibodies is affected by treatments that relax supercoiling, just as has been found with anti-Z-DNA. It is important when using immunocytochemical procedures to remember, first, that antibodies have exquisitely precise specificity, and second, that the antigens can be altered by the preparative procedures.

13.2.5 Antibodies to triple-stranded DNA

The sequence $(TC.AG)_n$ occurs relatively frequently in mammalian DNA, and at low pH or when exposed to negative supercoiling can form a structure called H-DNA (Htun & Dahlberg 1989). The best model of this structure consists of both single- and triple-stranded regions. Burkholder *et al.* (1988) labelled mammalian chromosomes with an antibody to triple-stranded DNA, and obtained a G and C-banding pattern. Although this pattern was clearest on fixed chromosomes, it could also be obtained on gently decondensed, unfixed chromosomes, which suggests that triplex DNA may be present in native chromosomes. However, fixation may enhance the formation of triplex DNA.

Possible functions of triple-stranded DNA in chromosomes are so far purely speculative, although Burkholder *et al.* (1988) suggest a possible role in chromosome condensation. Sequences capable of forming H-DNA are also present in regions involved in transcription, replication or recombination, however (Htun & Dahlberg 1989).

13.3 Antibodies to chromosomal proteins

13.3.1 Antibodies to histones

The histones are distributed essentially uniformly along the chromosomal DNA, the core histones forming the nucleosomes and histone H1 being located between the nucleosomes (Section 3.3). Any differential binding of antibodies to histones must therefore be due either to non-uniform distribution of post-synthetic modifications to the histones, such as acetylation or phosphorylation, or to changes caused by fixation. Fixation can certainly affect the binding of antibodies to histones (Mihalakis *et al.* 1976), and there is strong evidence for the extraction of histones during fixation with acetic acid (Section 3.5), so that the binding of anti-histone antibodies to fixed chromosomes should not necessarily be expected to reflect the histone distribution on the native chromosomes.

In fact, very few studies of the binding of anti-histone antibodies to fixed chromosomes have been carried out. Chromosomes that have been prepared with the use of acetic acid bind the antibodies essentially uniformly (Bustin *et al.* 1976, Turner 1982), although differences in intensity of staining, which do not amount to a banding pattern, do occur, apparently due to differences in accessibility (Turner 1982). The use of acetic acid, as in methanol-acetic acid fixation, reduces substantially the binding of antibodies to histones, and affinity for anti-H1 is lost completely (Pothier et al. 1975, Bustin et al. 1976), presumably as a result of extraction of histones. Bustin et al. (1976) obtained a spotty pattern of binding after acetic acid treatment, whereas Pothier et al. (1975) obtained clear banding patterns with anti-H4, which seemed to bear some resemblance to G-banding. Even prolonged acid treatment did not remove histones from the chromosomes completely, and the results of Pothier et al. (1975) provide a basis for the possible involvement of histones in G-banding (Section 5.4). Turner (1989) obtained a banding pattern on unfixed CHO chromosomes labelled with antisera to acetylated H4, but again it was not clear that there was any consistent pattern, or whether the banding represented differences in the level of acetylation, or merely of accessibility.

Results obtained with antibodies to histones are rather disappointing, and there are probably two reasons for this. One is the preparation of the chromosomes. Only chromosomes fixed with acetic acid have the good morphology required for banding, yet such fixation extracts or modifies histones. Chromosomes which are prepared so as to minimize loss of histones may well be too contracted to show clearly any differential binding of antibodies, and accessibility of the chromosomes may also be reduced, which complicates interpretation. Second, few of the antibodies to histones that have been applied to chromosomes have been adequately characterized, and unless the precise binding site of the antibody is known, the results will be of limited value. There can be little doubt that antibodies that recognize modified amino acids in histones, such as acetylated or phosphorylated residues which may be associated with transcription or condensation, will be of the greatest interest.

13.3.2 Antibodies to non-histone chromosomal proteins

Staining chromosomes with antibodies to non-histone chromosomal proteins has not been any more successful than the use of antibodies to histones. In addition there is the problem of obtaining sufficient quantities of non-histones, in sufficiently pure form, to raise antibodies. The situation is improving, however, and the ability to raise monoclonal antibodies, whose specificity can be clearly defined, will be a great help. In practice few of the published studies provide much of value, either the specificity of the antibody or the pattern of distribution on the chromosomes being inadequately defined.

Antibodies to the high mobility group proteins HMG 1 and HMG 2 bound to chromosomes of various species of mammals, and with mouse chromosomes anti-HMG 2 showed 'indications ... of some chromosome banding' (Smith et al. 1978). Disney et al. (1989) confirmed this for anti-HMG I, showing that the distribution of the protein was the same as that of G-bands on human chromosomes. Jeppesen & Nicol (1986) found that a certain autoimmune CREST serum bound to isolated unfixed chromosomes to produce differential staining, but were unable to identify the banding pattern, nor characterize the antigen in detail. In this case it seemed that the antigen was probably a tightly bound component of the chromosome core, that could not be purified without destroying its antigenicity.

D'Alisa et al. (1979) investigated the distribution of T antigen on the chromosomes of muntjac cells infected with SV40 virus. They found that the pattern of antibody binding corresponded clearly to G- or Q-bands, and pointed out that one of the T antigen binding sites in SV40 virus contains A + T-rich DNA. Since the G- or Q-bands contain A + T-rich DNA (Section 5.3), the binding of T antigen to such regions of chromosomes would not be unexpected. However, the chromosomes used in these studies were prepared by methanol-acetic acid fixation, and the possibility that the pattern of antigen binding was the result of selective extraction from certain regions of the chromosomes was not investigated.

Maul et al. (1986) used immunocytochemistry to localize topoisomerase I in the centromeres of mouse cells, and suggested that the enzyme may be involved in some aspect of mitosis, or in the maintenance of chromosome structure. Two other studies describe the location to well-defined regions of chromosomes of proteins of defined molecular weight. Will & Bautz (1980) raised an antiserum to a non-histone chromosomal protein which had a molecular weight of 38 KDa. This antiserum bound to the heterochromatin of the polytene chromosomes of several species of *Drosophila*, but results for mitotic chromosomes have not been described. Schonberg et al. (1987) described two monoclonal antibodies to non-histones. One, of 38 KDa, was preferentially located in positive R-bands, while the other, of 175 KDa, was located in C-bands. Both antigens were conserved, since immunostaining could be obtained in both animal and plant chromosomes. Further details of this work, only published in abstract form so far, would be very welcome; the association of specific proteins with specific types of chromosome bands is an important milestone in understanding chromosome banding, and indeed chromosome organization as a whole.

14 Chromosome banding polymorphisms, heteromorphisms and variants

14.1 Introduction

When the chromosomes of a particular species are examined carefully, a good deal of variability between homologous chromosomes, both within and between individuals, can often be seen. Some of this variability is morphological, and had been recognized long before the introduction of banding techniques (although banding methods can often help in describing the nature of the morphological change). Much of the variability involves differences in the size or staining properties of heterochromatic segments and NORs, and such variability is the subject of the greater part of this chapter. In the case of heterochromatic bands and NORs, it is the bands themselves that vary, whereas in the case of morphological variations such as translocations or inversions, banding patterns are essentially an aid to identifying the alteration, and the bands themselves do not vary.

Different terms have been used to describe the variations between homologous chromosomes which are the subject of this chapter. None is ideal, and in fact there is probably no one term that covers all situations. E. B. Ford (1960) defined polymorphism as: 'the occurrence together in the same habitat of two or more discontinuous forms of a species in such proportions that the rarest of them cannot be maintained merely by recurrent mutation.' It is clear that many examples of chromosome variation cannot be confidently described as polymorphisms according to this strict definition, since variations in size or staining properties can rarely be described as discontinuous with any certainty. The most unambiguous examples of chromosomal polymorphism are inversions and other structural rearrangements: only two forms, the normal and rearranged, are possible, and these are clearly discontinuous. The Supplement to the Paris Conference (ISCN 1985) introduced the term *heteromorphism*, which does not have associated with it any special genetical definition, and is perhaps a more neutral term, merely indicating variation between homologous chromosomes without implying that the different forms are necessarily discontinuous, or indeed that they need to occur together in the same habitat. The term *variant*, 'a specimen slightly differing from the type,' tends to imply a form that is rare or even unique compared with a normal type, and scarcely seems appropriate when different forms are present in the population at substantial frequencies. Nevertheless, all three terms are used in the literature, in many cases apparently with the same meaning. It would certainly be most satisfactory if the terms polymorphism, heteromorphism and variant could be used only in their strict senses, but in many cases it may not be clear which term is appropriate. Information on frequency, geographical distribution, and on whether the size distribution of the different forms is discontinuous or not, is usually inadequate. In the present chapter the term heteromorphism will be used in a neutral sense, to indicate the occurrence of different chromosomal forms, without the strict limits implicit in the use of the term polymorphism, which will only be used when the situation clearly corresponds to that defined by E. B. Ford. 'Variant' will be used to describe forms differing from the norm, without any implications as to whether or not they are new mutations.

By their nature, chromosomal polymorpohisms, heteromorphisms and variants are heritable. Suspicions that inheritance is not always according to strict Mendelian principles will be discussed in Section 14.4.8. In the case of Ag-NOR variation, it must be remembered that the amount of silver staining reflects not only the number of rRNA genes on the chromosome, but also their activity, which can vary with cell type and stage of development (Section 9.4.2). Technical factors are a further problem in these studies. In this connection, it should be noted that the sizes of C-bands on the chromosomes of monozygotic twins are not entirely concordant (Pedrosa et al. 1983), a finding which is presumably largely due to technical factors.

Heteromorphism of heterochromatin and NORs is probably universal throughout the animal and plant kingdoms. Obviously only a small proportion of species has been examined chromosomally, and even fewer have been studied adequately for heteromorphism. Nevertheless, the writer knows of no species in which a large outbred population has been investigated without some degree of heteromorphism being found. As we shall see below, heteromorphism is widespread in groups such as higher plants, insects and vertebrates, which have been studied intensively by cytogeneticists.

14.2 Mechanisms of formation of heteromorphisms

In those species in which there is heteromorphism for translocations or inversions of euchromatic parts of chromosomes, the mechanism of formation of the heteromorphism is the same as its description, even if we do not understand in detail how translocations or inversions come about. Translocations and inversions can also occur in heterochromatin, and indeed partial or complete inversion of heterochromatin is an important cause of heteromorphism in humans (Section 14.4). However, the main question to be posed in this section is how blocks of heterochromatin and NORs can vary in size.

Several authors have proposed unequal crossing-over as a mechanism for generating size heteromorphisms, and the available evidence was collated by Kurnit (1979). Emphasis has been laid on the content of highly repetitive DNA sequences in many blocks of heterochromatin, so that recombination could occur between homologous sequences that were separated (in molecular terms) by large distances. Thus unequal crossing-over between two sequences *abcabc* could lead to the shortened sequence *abc* and the lengthened sequence *abcabcabc*.

Unequal meiotic crossing-over is not likely to be a significant source of generation of heterochromatic variants, since meiotic exchange in heterochromatin is virtually unknown (John 1976, Kurnit 1979). Mitotic crossing-over is therefore favoured as the mechanism for generating size diversity in heterochromatin. Direct evidence for such a mechanism comes from various sources. A few workers have reported mosaicism for heterochromatin in human chromosomes, which can only have arisen by a somatic event (Craig-Holmes *et al.* 1975, Simi & Tursi 1982, Wahlström *et al.* 1985); however, it is not clear from the published data whether the sizes of the mosaic blocks of heterochromatin are in accordance with expectation from the hypothesis of somatic crossing-over. Craig-Holmes *et al.* (1975) remarked that they had never found unequal C-bands on the two sister chromatids of the same chromosome, and suggested that the mosaics had most probably arisen from somatic crossing-over between homologues rather than from faulty replication. There is, however, experimental evidence that changes in size of blocks of heterochromatin can occur by sister chromatid exchange. Mitomycin C, which is a potent inducer of sister chromatid exchanges, also produces alterations in heterochromatin (Hoehn & Martin 1972). Obviously heterochromatic variants must be passed through the germ line, and since they appear to be produced largely by mitotic rather than meiotic crossing-over, the exchanges presumably occur in the mitotically dividing oogonia and spermatogonia. Evidence for new mutations in the germ line is provided by the observation of heteromorphic C-bands which do not occur in the parents of the individual (Craig-Holmes *et al.* 1975, Simi & Tursi 1982).

NORs also vary in size, and in *Drosophila* it has been shown that rRNA genes are amplified by unequal mitotic sister chromatic exchanges (Tartof 1973). This finding supports the proposal that heterochromatin is amplified or diminished by such a mechanism, and indicates a common mechanism for heterochromatin and NORs, both of which, of course, contain repeated DNA sequences.

Nothing that has been said so far precludes changes in heterochromatin by processes such as deletion, or amplification by mechanisms similar to those that give rise to homogeneously staining regions (HSRs) (Section 5.5.5). Evidence for or against such mechanisms is simply not available.

14.3 Quantitative assessment of heteromorphic bands

For a satisfactory study of heteromorphic bands, it is necessary to have some sort of estimate of their relative, if not their absolute, sizes, or in the case of fluorescence heteromorphisms, their intensities. In many cases, blocks of heterochromatin are merely described as larger or smaller than normal, and little or no attempt at more sophisticated classification has been made except for human chromosomes. This discussion will, therefore, be confined to attempts to estimate the sizes of human heteromorphic bands, although most of the principles can be applied equally well to the chromosomes of other organisms.

The necessity of classifying heteromorphic bands was recognized early on, and the Supplement to the Paris Conference report (ISCN 1985) gave a method of classifying the size of both Q- and C-bands, and the intensity of Q-bands, into five classes. Many workers have, however, felt unable to distinguish as many classes, and have only recognized small, medium and large categories. Whatever the number of classes recognized, this approach has several difficulties. It is subjective, so that comparisons between different laboratories are difficult; it takes no account of the differential contraction of euchromatin and heterochromatin; and finally, it artificially places blocks of heterochromatin into a small number of fixed categories which may not correspond to those which actually occur. At present, it is not established whether the size distribution of blocks of heterochromatin is continuous, or whether it falls into discrete classes, and if so how many. Evidence has been presented supporting both points of view.

Patil & Lubs (1977) proposed using the short arm of chromosome 16 as a standard against which to compare the sizes of C-bands, and valuable work has been done using this principle. However, it still does not take account of the problem of differential contraction. Clearly, to deal with this problem satisfactorily, as well as the question of the true distribution of C-band size, actual measurements are required. Since C-bands are very small objects, measurement is difficult and tends to be subject to substantial errors. In addition, it is often difficult to decide where to draw the boundary of a C-band, and finally, technical factors in the C-banding procedure are clearly an important variable in assessing C-band size. Nevertheless, many workers have successfully made measurements of C-bands using a variety of procedures: simple length measurements, either by hand or using a digitizer (Beltran *et al.* 1979, Friedrich & Therkelsen 1982, Maes *et al.* 1983); simple area measurements made using an image-analyser (Lopetegui 1980); or using sophisticated microdensitometric procedures (Drets & Seuanez 1974, Mason *et al.* 1975, Erdtmann *et al.* 1981). A high resolution image analyser has also been used to measure areas of silver-stained NORs, apparently with a high degree of reproducibility (Schmid *et al.* 1982). Measurements are necessary if the effect of chromosome contraction on C-band size is to be assessed. The relationship between the length of a block of heterochromatin and the total length of the chromosome has been studied by several workers, and it has been found that although the euchromatin contracts at a greater rate than the heterochromatin, there is a more or less linear relationship between the two, so that it is possible to normalize the C-band size for a given degree of chromosome contraction (Schmiady & Sperling 1976, Baliček et al. 1977, Friedrich & Therkelsen 1982). In principle, these procedures solve most of the major problems in measuring C-bands, although they cannot circumvent the problem of variation in staining.

A number of radically different procedures for measuring C-bands have also been proposed. Harrison et al. (1985a) proposed that the size of heterochromatic segments could be measured more accurately on scanning electron micrographs of chromosomes. Although the greater magnification and resolution of this method is undoubtedly a great advantage for measurement, it has yet to see any actual application. Another approach is based on observations that C-bands often appear to consist of a number of discrete blocks; to quantify the C-band, it is only necessary to count the number of distinct blocks of which it is composed (Drets & Seuanez 1974, Wahedi & Pawlowitzki 1987). Pawlowitzki's group used reflection microscopy to make the sub-divisions of the C-bands, which they called C_e-bands (for centromeric elevation), more readily visible, even in C-bands that were apparently uniformly stained. The reliability of this technique has yet to be tested, and it should be added that there is little evidence that the subdivisions of the C-band are necessarily of the same size.

Probably the most accurate method of measuring the variability in size of heterochromatic segments is by measuring the DNA content of whole chromosomes, either on slides (Geraedts et al. 1975, Sumner 1977a, Wall & Butler 1989) or by flow cytometry (Harris et al. 1986). Cytochemcial methods for DNA which have excellent stoichiometry are available (Feulgen for microdensitometry on slides, fluorescence methods for flow cytometry), and measurements are unaffected by the degree of contraction of the chromosomes, nor is there any difficulty of defining the edge of the C-band. On the other hand, a direct measure of C-band size cannot be obtained, but only the difference in DNA content between chromosomes having different sized C-bands. Of course, the utility of the method depends on the heterochromatin being additional to a fixed amount of euchromatic DNA, which certainly appears to be true in humans (Geraedts et al. 1975, Sumner 1977a), and is probably the normal situation. Flow cytometry is probably the most precise method of making DNA measurements, and has the additional advantage that prior identification of the heteromorphic chromosomes is not required. If two homologous chromosomes differ in DNA content, even by as little as one thousandth of the total genome, they can appear as separate peaks on a flow karyotype (Harris et al. 1986). Some of these small heteromorphisms involve pairs of homologues which cannot be distinguished by conventional cytogenetical procedures.

Measurements of quinacrine fluorescence heteromorphisms in humans

(Section 7.5.2) have not generally been attempted, and assessment of such heteromorphisms is still essentially subjective, and is additionally bedevilled by problems of fading, and variation in fluorescence intensity from one part of a metaphase to another. Nevertheless, Schnedl *et al.* (1977a) described a photometric method for quantifying quinacrine heteromorphisms from photographs, and claimed greatly improved discrimination between homologous chromosomes. Overton *et al.* (1976) emphasized the importance of using a range of exposures in photographic printing to show up optimally both brightly and dimly fluorescent heteromorphic segments.

From the comments above it should be clear that assessment of the size of heteromorphic C-bands is a somewhat imprecise process, and that assessment of the brightness of heteromorphic Q-bands is highly subjective. These deficiencies in measurement must be borne in mind when considering the work to be described below. Although there is no doubt that large differences in blocks of heterochromatin can easily be recognized, more subtle differences must necessarily be harder to detect.

14.4 Human chromosome heteromorphisms

More is known about banding heteromorphisms in humans than in any other species. As well as being possibly the most complex system of heteromorphisms yet identified, human heteromorphisms have attracted considerable interest because of their potential and actual applications as markers, as for example in paternity testing and gene mapping, and because of their possible phenotypic effects. In this section the heteromorphisms will be described in detail, an account given of studies of the inheritance of heteromorphisms and their distribution in populations, their applications listed, and work on the possible phenotypic effects of heteromorphisms reviewed.

All human chromosomes have blocks of heterochromatin at and adjacent to their centromeres, and those on chromosomes 1, 9 and 16 are particularly large. In addition, the Y chromosome has a large block on the distal segment of the long arm, and the acrocentric chromosomes of the D group (13–15) and G group (21–22) have complex arrangements of heterochromatin involving the short arms and satellites. All, or nearly all, of these blocks of heterochromatin show variations in size, sometimes quite considerable, and except for the acrocentric chromosomes (including the Y chromosome) show variations in position. The C-band may be in either the long arm or the short arm adjacent to the centromere, or part may be in each arm (Phillips 1980); the distribution of the heterochromatin between the arms varies with the chromosome, but with chromosomes 1, 9 and 16 the heterochromatin is normally on the long arm side of the centromere.

In most, but probably not all, cases, the C-banded material comprises the whole of the heterochromatin on any one chromosome. When a variety of different techniques are used to characterize the heterochromatin, it is apparent, however, that different blocks of heterochromatin react in different ways, and that these other techniques usually stain only part of the total heterochromatin. Since the differently staining subbands that make up a block of heterochromatin can vary independently of each other both in size and position, it will be realized that human heterochromatin is remarkably complicated, and moreover that it will often be impossible to give a single description for a particular block.

Having said that virtually all human chromosomes show heteromorphism in size and position, and sometimes staining properties of their heterochromatin, it must be admitted that for many chromosomes the centromeric heterochromatin only forms a small block, whose variation and heterogeneity are not striking. In the following sections, therefore, omission of mention of a particular chromosome is an indication that its heterochromatin shows no remarkable variation beyond what has already been described.

14.4.1 Chromosome 1

Human chromosome 1 generally has one of the largest blocks of heterochromatin in the karyotype, normally situated on the long arm next to the centromere (1qh), although occasionally part or whole of the heterochromatin may be in the short arm. These alterations are generally referred to as partial or total inversions respectively, although direct evidence that they arise by such a mechanism seems to be lacking. Variation in size is also extensive; some of the range of C-band variants of chromosome 1 is illustrated in Figure 14.1.

The C-band of chromosome 1 is also G-band positive, but Q-band negative, and negative with the fluorochrome D287/170 (Babu & Verma 1986a). It is also entirely resistant to digestion by the restriction endonuclease *Alu* I (Babu & Verma 1986b). Other techniques, however, stain only part of the heterochromatin. Magenis *et al.* (1978) found that the C-band contained a G–11 positive sub-band whose position within the C-band was variable. Two individuals with an average-sized C-band had no G–11 band, and two others with a large C-band had two G–11 bands within it. Against

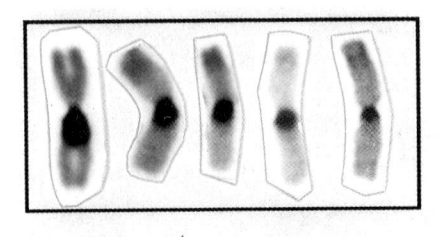

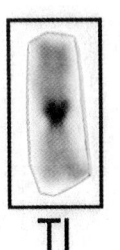

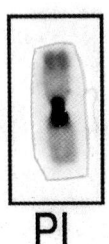

Figure 14.1 Variation of the heterochromatin of human chromosome 1. Left, size variations. Right, total (TI) and partial (PI) pericentric inversions.

this background, statements that a particular sub-band occupies a particular position within the C-band should be taken as applying only to the individuals described in a particular paper. Sigmund & Schwarz (1979) reported that the C-band consisted of a proximal mithramycin positive, G + C-rich segment, and a distal DAPI-positive, A + T-rich segment, the two segments being independently variable in size; in two individuals the C-band consisted of two mithramycin-positive and two DAPI-positive bands. Distamycin/DAPI fluorescence is reported to cover all except the proximal part of the C-band, as is staining after *Hae* III digestion (Babu & Verma 1986b). Other restriction endonucleases produce characteristic patterns within the C-band of chromosome 1, as they do in other chromosomes (Babu *et al.* 1988), although the relationships between these and other sub-bands have yet to be established. It seems probable that the C-band of chromosome 1 consists of only a small number of sub-bands, each of which has a characteristic response to each of a variety of banding methods. It may also be that each contains a characteristic satellite DNA, although this has yet to be established. Gosden *et al.* (1981a) showed that a sequence derived from human satellite III hybridized to chromosome 1 in proportion to the size of the C-band, but no attempt was made to relate the amount of hybridization to the size of individual components of the C-band. Such correlative studies would be valuable, but would require a considerable effort to characterize the C-bands from a large number of individuals by a wide variety of methods.

14.4.2 *Chromosomes 3 and 4*

Chromosomes 3 and 4 both have a small centromeric band which fluoresces brightly with quinacrine, varies in size and position, and may be completely absent (Fig. 14.2). Recent work (Babu & Verma 1986a) indicates that the quinacrine-positive bands are not equivalent to the C-bands on these chromosomes. On chromosome 3, the C-band at the primary constriction is digested by *Alu* I, whereas the brilliant Q-band is resistant to this enzyme; these two sub-bands can vary in size independently. Yurov *et al.* (1987) showed that the amount of alphoid satellite DNA on chromosome 3 was correlated with the size of the brilliant Q-band. On chromosome 4, there is also sometimes an *Alu* I resistant band that corresponds to the brilliant Q-band; this material corresponds to only part of the C-band (Babu & Verma 1986d).

14.4.3 *Chromosome 6*

The centromeric C-band of chromosome 6 is normally small and undistinguished, but a variant, $6ph^+$, has a larger C-band, extending into the short arms, which is negative with G- and Q-banding (Sofuni *et al.* 1974), and may occur in up to 9% of chromosomes (Madan & Bruinsma 1979). This C-band is not elongated by growing cells in the presence of 5-azaC (Section 11.2.3),

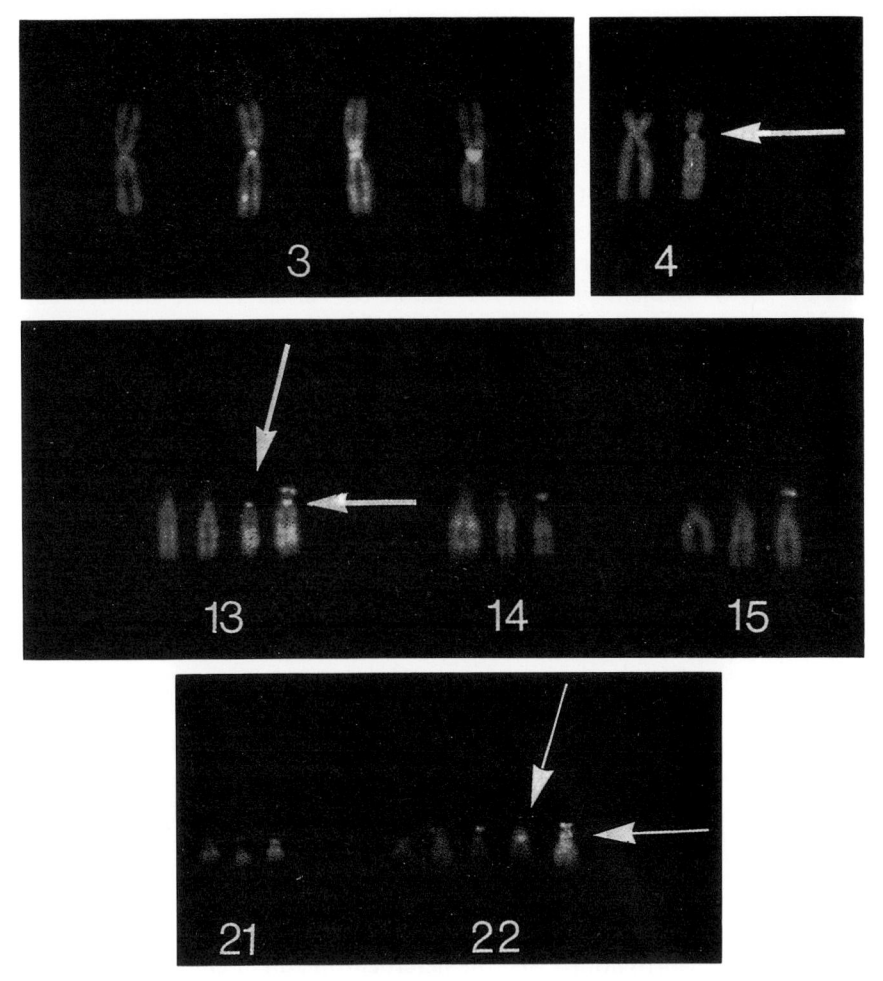

Figure 14.2 Quinacrine fluorescence variants of human chromosomes 3, 4, 13, 14, 15, 21 and 22. Note that in the acrocentric chromosomes the fluorescence of p1.1 and p1.3 can vary independently (arrows). Reproduced from Evans *et al.* (1971) by permission of the authors and Springer-Verlag.

and is particularly interesting because the increase in size is correlated with an increase in the amount of an alphoid satellite DNA which is specific to chromosome 6 (Jabs & Carpenter 1988).

14.4.4 Chromosome 9

The paracentromeric block of heterochromatin is one of the largest and most variable in the human genome. It is C-band positive, but G- and Q-band negative, and shows a relatively high frequency of partial or total pericentric inversions (Fig. 14.3) (Jacobs 1977). The C-band normally appears to be divided into two regions, a smaller proximal G-banded region, and a larger distal region that is stained by the G–11 technique (Donlon & Magenis 1981) and which is fluorescent with distamycin/DAPI (Buys *et al.* 1981a). Sometimes there may be complete deletion of the constitutive heterochromatin of chromosome 9, which then lacks both G–11 and distamycin/DAPI staining, and has only a very small C-band at the centromere itself, probably corresponding to centromere dots (Section 10.2.1) (Buys *et al.* 1979a). The C-band of chromosome 9 is also resistant to digestion by several restriction endonucleases (M. S. Bianchi *et al.* 1985, Babu & Verma 1986a, Babu 1988); digestion with some of these enzymes results in staining of the whole C-band, whereas others apparently digest a region equivalent to the G-banded segment.

The two segments of the C-band of chromosome 9 can apparently vary in size independently; according to Donlon & Magenis (1981) the G-banded region varies continuously in size, while the G–11 stained region appears to vary in discrete steps, larger blocks being two or three times the size of the smaller ones. These two types of heterochromatin can also behave independently in pericentric inversions. In total inversions, the G-band remains proximal and the G–11 band distal, as usual (Donlon & Magenis 1981); however, the situation in partial inversions is more complex. Donlon & Magenis (1981) found cases in which the G–11 material remained in its usual position on the long arm, but the G-band material on the short arm was increased; these authors doubted whether such an arrangement was the result of a partial inversion. Gosden *et al.* (1981b) also noted a similar arrangement, using distamycin/DAPI; all the fluorescent material remained in the long arm. However, other arrangements were found in which distamycin/DAPI fluorescence was only found in the short arm, or in which fluorescence occurred in both arms; all these cases appeared as partial pericentric inversions by C-banding.

Finally, it is worth noting that the heterochromatin of chromosome 9 has the almost unique property of forming 'parameres' (Fig. 14.4), multiple dense spherical chromatin bodies, at the pachytene stage of meiosis (Hungerford 1971, Page 1973). The significance of this remains unknown. The number of parameres varies between bivalents even from the same individual, and cannot therefore be used as a measure of the size of the block of heterochromatin (Sumner 1986a).

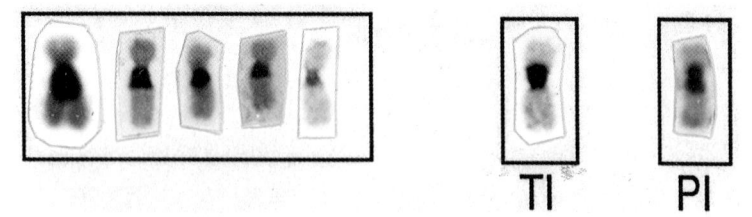

Figure 14.3 Variation of the heterochromatin of human chromosome 9. Left, size variations. Right, total (TI) and partial (PI) pericentric inversions.

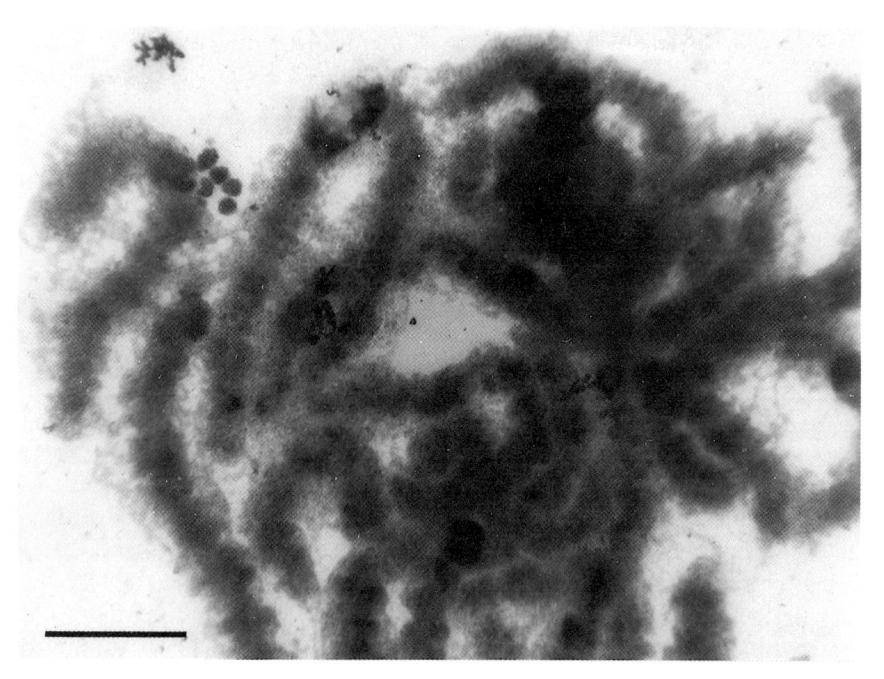

Figure 14.4 Transmission electron micrograph of a whole mount of human chromosome 9 from a male pachytene cell, showing the organization of the heterochromatin into 'parameres'. The bar represents 5 μm. Reproduced from Sumner (1986a) by permission of the author and Springer-Verlag.

14.4.5 Acrocentric chromosomes (Group D, 13–15 and Group G, 21–22)

The human acrocentric chromosomes show a particularly complex set of heteromorphisms, which, however, tend to show generally similar features in all these chromosomes. The short arms of the acrocentric chromosomes are divided into three segments (ISCN 1985) (Fig. 14.5): p1.1, the part of the short arm adjacent to the centromere; p1.2, the satellite stalk, which is the site of the 18S and 28S ribosomal genes (Section 9.1); and p1.3, the satellites. Each of these segments can vary in size, and p1.1 and p1.3 can vary in fluorescence properties with a number of stains; the fluorescence of p1.1 and p1.3 can vary independently (Fig. 14.2). Double satellites, that is, an extra satellite attached at the end of the arm beyond the usual one, occur as rare variants (e.g. Sofuni *et al.* 1980). Closer investigation may result in p1.1 being divided into a number of segments with different staining properties (Wachtler & Musil 1980).

The p1.1 segment of all the acrocentric chromosomes shows variation in size, in brightness of quinacrine fluorescence, in colour of fluorescence with acridine orange R-banding (Verma & Dosik 1980), presence or absence of concentrations of 5-methylcytosine (Okamoto *et al.* 1981), in density of

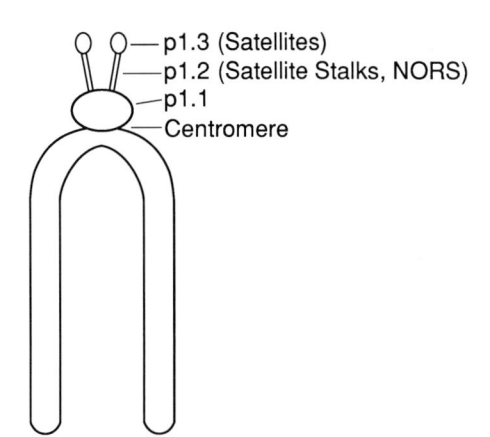

Figure 14.5 Diagram showing the arrangement of the short arms of the human acrocentric chromosomes.

- p1.3 (Satellites)
- p1.2 (Satellite Stalks, NORS)
- p1.1
- Centromere

C-banding (Jacobs 1977), and in resistance to digestion by various restriction endonucleases (Babu et al. 1988). Relationships between most of these heteromorphisms have not been established, but there is apparently no consistent relationship between the brightness of quinacrine fluorescence and the colour seen with acridine orange (Verma et al. 1977). Among the acrocentric chromosomes, there are differences in the amount of variability shown, chromosome 13 apparently showing most variation, at least with quinacrine (Jacobs 1977, Verma & Dosik 1980). In addition to these heteromorphisms, chromosome 15, which has the greatest amount of 5-methylcytosine on its short arms (Okamoto et al. 1981), also shows strong distamycin/DAPI fluorescence at a high frequency (Wachtler & Musil 1980), and indeed this fluorescence is regarded as diagnostic for chromosome 15 and its short arm. Wachtler & Musil (1980) also found a heteromorphic mithramycin positive region in the short arm of chromosome 15.

The satellite region (p1.3) of the acrocentric chromosomes shows a range of variation similar to, but rather more restricted than, that of the short arms proper (p1.1). The main differences are the absence of 5-methylcytosine and distamycin/DAPI staining, which in any case are largely restricted to chromosome 15. As mentioned earlier, the staining properties and sizes of p1.1 and p1.3 can vary independently on any one chromosome, but certain chromosomes tend to have a high frequency of particular variants; for examples, see Jacobs (1977) and Verma et al. (1977), and Section 14.4.9.

The satellite stalks, p1.2, show their own heteromorphism, but of a different type from that of the other segments. As the nucleolus-organizer regions, they show extensive variation in the degree of Ag-NOR staining (Section 9.4.3); numbers of silver stained NORs generally range between 4 and 9, and there is considerable variation in size also.

14.4.6 Chromosome 16

Chromosome 16 has one of the largest blocks of heterochromatin in the human genome, which is C-band and G-band positive, but negative with quinacrine or Giemsa-11. It is resistant to *Alu* I digestion, and all except the actual centromere is resistant to *Hae* III and stains with distamycin/DAPI (Babu & Verma 1986b). Inversions of the heterochromatin of chromosome 16, whether partial or total, are very scarce (Jacobs 1977, Phillips 1980, Erdtmann *et al.* 1981, Simi & Tursi 1982), but extensive size variations occur.

14.4.7 Y chromosome

The great variability in size of the human Y chromosome was already well known before the introduction of banding techniques. One of the earliest observations with banding methods on a human chromosome was the finding of a large segment on the long arm of the Y chromosome that showed brilliant quinacrine fluorescence (Zech 1969) (Fig. 7.8). It was soon found that the same region was also C-banded, and that the variation in size of the Y chromosome was due to variation in size of this segment (Bobrow *et al.* 1971). (In passing, it should be noted that, contrary to a number of statements in the literature, the Y chromosome has a small C-band at the centromere). As so often happens, the situation has turned out to be more complex than indicated by these early observations. For reviews of the organization of the Y chromosome see Bühler (1984) and especially Ibraimov & Mirrakhimov (1985), and see Figure 14.6.

Y chromosomes can vary in size between very large ones, larger than the F group chromosomes, and very small ones, which lack completely any visible heterochromatin in the long arm. While it is apparent at a glance, and has been confirmed by measurements, that most of this variability is due to variation in the size of the heterochromatic (fluorescent) segment (Fig. 7.8),

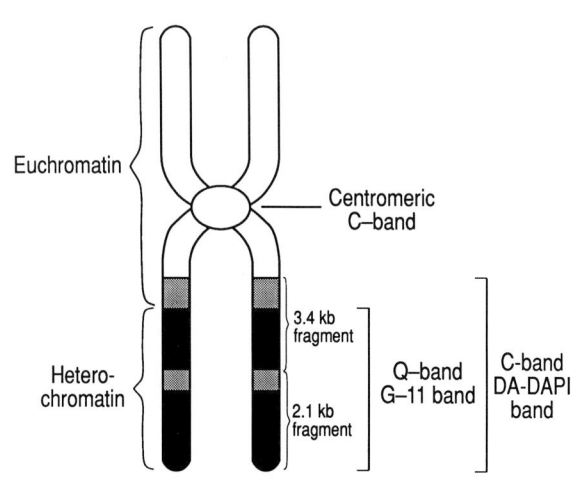

Figure 14.6 Diagram of the human Y chromosome, showing the distribution of the different types of repetitive DNA, and of the different types of banding. Note that the exact relationship between the distribution of DNA types and the different types of banding has not yet been established. Redrawn after Bühler (1984) by permission of the author and S. Karger AG.

careful measurements have shown that a small amount of variation in the length of the non-fluorescent segment of the Y also contributes to changes in the total length (Schnedl 1971b, Soudek et al. 1973, Verma et al. 1978). The heterochromatic block on the long arms is also not a single, simple block that is entirely positive with both C-banding and quinacrine. The whole block is apparently C-banded, and the same material is also intensely fluorescent with distamycin/DAPI; the quinacrine positive region is, however, slightly smaller, and can be completely deleted leaving a small amount of C-banded material (Soudek & Laraya 1976). The Q-banded material corresponds to that which can be stained by Giemsa-11 (Bühler 1984). A narrow dull band is frequently visible in the middle of the quinacrine-positive segment. Characteristic DNA sequences are present in the Y heterochromatin, and the distribution of these, as well as the staining patterns, are illustrated diagrammatically in Figure 14.6. One of the sequences, the 3.4 kb fragment, has been shown to increase in amount with increasing length of the Y heterochromatin, but the 2.1 kb fragment does not show a simple proportionality with length (McKay et al. 1978), and is apparently largely confined to the distal part of the heterochromatin (Schmidtke & Schmid 1980) (Fig. 14.6).

When C- or Q-banded Y chromosomes of different sizes are examined, particularly in preparations with relatively extended chromosomes, the heterochromatin is seen to be made up of a number of discrete blocks; Drets & Seuanez (1974) recorded up to five. Recent DNA measurements indicate that each of these blocks contains approximately the same amount of DNA; the size variation of the Y chromosome is therefore not continuous, but is in discrete steps (Wall & Butler 1989). It should be noted that, as well as straightforward variations in length, the Y chromosome shows a wide variety of structural aberrations, such as isochromosomes, dicentrics, rings, translocations, inversions and deletions, many of which are without phenotypic effects (Ibraimov & Mirrakhimov 1985). Since the chromosome is largely heterochromatic, and very few genes have been found even in the euchromatic segments, extensive structural alterations can presumably be tolerated without serious effects.

14.4.8 *Inheritance of human chromosome heteromorphisms*

A large number of studies have shown that human heterochromatin variants are heritable in strict Mendelian fashion. This is true of C-band variants, both for size (Craig-Holmes et al. 1975, Robinson et al. 1976, Carnevale et al. 1976) and position (Mayer et al. 1978, Phillips 1980), Q-band variants (Robinson et al. 1976, Ibraimov & Mirrakhimov 1985), acridine orange R-band variants (Verma & Lubs 1976b) and Ag-NOR variants (Mikelsaar et al. 1977b, Markovic et al. 1978, Taylor & Martin-de Leon 1981). Nevertheless, some exceptions, real or apparent, have been reported. The existence of somatic mosaicism as well as the occurrence in children of variants not present in the parents suggests that unequal crossing-over has occurred (Craig-Holmes et al. 1975, Simi & Tursi 1982). More intriguing are cases where preferential segregation of a particular variant seems to have happened. Both Robinson et al. (1976) and Carnevale et al. (1976) reported preferential inheritance of very large variants of chromosome 9, while Geraedts & Pearson (1974) and Robinson et al. (1976) described the preferential segregation of brilliant quinacrine variants of the acrocentric chromosomes. These and other examples quoted by Ibraimov & Mirrakhimov (1985) should not perhaps be taken as conclusive evidence yet for non-Mendelian inheritance of variants. Particularly in the case of fluorescence variants, there are serious technical difficulties, and these observations have not so far been confirmed by more extensive data.

14.4.9 Population studies of human chromosome heteromorphisms

Numerous studies have been carried out on the frequencies of heteromorphisms for human heterochromatin in various populations. Like studies on inheritance and on phenotypic effects, the population studies have been done mainly on C- and Q-band heteromorphisms, with limited work on variants shown by acridine orange R-banding and Ag-NOR staining. Other types of heteromorphisms, revealed by more esoteric methods, have not yet been studied in populations.

Studies of C-bands have concerned the major ones on chromosomes 1, 9, 16 and Y, and very few data are available for C-bands of other chromosomes. Such studies (Craig-Holmes et al. 1973, Müller et al. 1975, Buckton et al. 1976, Jacobs 1977) show that in the vast majority of chromosomes the C-bands fall into a normal or medium-sized category, with larger and smaller variants, and partial and total inversions, much less common.

Several studies have investigated differences in C-band size between races. Brazilian Indians were found to have larger autosomal C-bands than Caucasoids (Erdtmann et al. 1981), who in turn have larger C-bands than Japanese (Cavalli et al. 1985) or negroes (Zanenga et al. 1984). However, differences in C-band size have also been found between two different European populations (Berger et al. 1983). Frequencies of inversions of the C-bands of chromosomes 1 and 9 are highly variable between populations (reviewed by Potluri et al. 1985) and in particular the incidence of inversion in the 9 is high in the black population (Hsu et al. 1987).

Extensive studies on Q-band heteromorphisms were reviewed by Ibraimov & Mirrakhimov (1985). The highest levels of bright variants are found on chromosomes 3 and 13 (indeed, in chromosome 3 bright bands are commoner than the so-called 'normal' ones, and it is perhaps the latter which should be called the 'variants'). The same authors also summarize data showing that there are significant differences in the quinacrine heteromorphisms between different populations and races.

The Y chromosome, which can be studied by both C-banding and Q-banding, again shows extensive variation, but most fall into a medium-sized category (Craig-Holmes et al. 1973, Müller et al. 1975, Hsu et al. 1987). Racial differences occur in the Y chromosome also, and it has long been recognized that the Japanese have, on average, particularly long Y chromosomes (Cavalli et al. 1985). Further data not only confirm this, but also show extensive differences between populations and races in the frequency of large Y chromosomes (Ibraimov & Mirrakhimov 1985, Verma & Pandey 1987); Australian aborigines are reported to have particularly small Y chromosomes.

Acridine orange fluorescence heteromorphisms have been studied in Caucasian populations by Verma & Lubs (1975b) and Verma et al. (1977). Most colour variants fall into the intermediate orange-yellow or pale yellow classes, but the relative proportions of variants in these categories, as well as in the others, ranging between red and green, vary according to the chromosome.

Several studies have been done on the frequencies of silver-stained NORs in populations, although comparison is sometimes difficult because of differences in technique, and in expressing the results. Nevertheless, the surveys by Mikelsaar et al. (1977a) on a mixed German and Austrian population, by Mikelsaar & Ilus (1979) on an Estonian population, by Zakharov et al. (1982) on a Russian population, and by Dipierri & Fraisse (1983b) on a French population all give results which can be compared. All these studies, plus the study on East Indians by Verma et al. (1981) show a mean number of Ag-NOR stained chromosomes in the region of eight, out of a possible maximum of ten. Nevertheless, there are significant differences in the frequencies of Ag-NOR staining on the different acrocentric chromosomes, except in the French population. In the Austro-German population, chromosome 22 had the lowest frequency of Ag-NOR staining; in the Estonians, chromosome 14, although chromosome 22 was also low; while in the Muscovites, chromosome 15 was low, and chromosome 21 particularly high. Zakharov et al. (1982) also reported the degree of silver staining and frequency of NOR association, while Verma et al. (1981) reported Ag-NOR size, divided into five categories. In the East Indian population, most Ag-NORs were in the large and medium classes, but a substantial number were classified as very small, equivalent to being virtually absent. Apparently, as with other human heteromorphisms, there are racial differences in the frequencies of Ag-NOR staining on different chromosomes, but adequate data on Ag-NOR size are not yet available.

14.4.10 Phenotypic significance of human chromosome heteromorphisms

It is quite clear that, although there is a good deal of variation in size and staining properties of blocks of heterochromatin and NORs in the human genome, much of this variation is perfectly tolerable, as it occurs in normal people. Since (except for NORs) this variation occurs in heterochromatin, which is generally regarded as being genetically inert, it is not surprising that the variation has little immediately obvious effect; nevertheless, numerous studies have been carried out to see whether heterochromatin variants have any phenotypic effects, particularly of a deleterious nature.

Bobrow (1985) reviewed the literature on the possible association between C-band variants and reproductive failure, and emphasized the problems which make adequate comparisons of different studies difficult. These include selective ascertainment of the populations studied, lack of appropriate controls, and variable definition of what constitutes a chromosomal variant (including lack of objective assessment of the size of the heterochromatic blocks). Bobrow concluded that there was no good evidence that autosomal variants (including inversion of the chromosome 9 heterochromatin) had any significant effect; on the other hand, it seemed possible that Y heterochromatin variants did have some effect, Yq- being associated with infertility and Yq+ with recurrent abortion. It was concluded that further data were needed in these cases, a judgement supported in the review by Schwartz & Cohen (1985). Nothing seems to have been published since which affects this conclusion.

A possible association between heteromorphic variants and trisomy and other chromosome abnormalities has been investigated by a number of workers. Again results are often conflicting or inconclusive (see Erdtmann, 1982, for a review). More recently, Babu *et al.* (1987) found that in Edwards' syndrome (trisomy 18), the blocks of *Alu* I-resistant heterochromatin on chromosome 18 were usually in the large or very large categories. Particularly interesting is the possible association between trisomy of the acrocentric chromosomes, and large NORs. Since the acrocentric chromosomes associate to form a limited number of nucleoli, it has long been supposed that such associations could lead to non-disjunction, and that large NORs might promote such associations. Although a number of claims have been made that double NOR variants in the parents may predispose to non-disjunction, particularly of chromosome 21, a recent review has concluded that there is as yet no good evidence for the involvement of NOR variants in non-disjunction (Schwartz *et al.* 1989). There also appears to be no evidence of an association between heterochromatic variants and congenital abnormalities; Brown *et al.* (1980) in a careful study, found no systematic differences in heterochromatin between normal and abnormal babies. Studies of the relationship between heterochromatin variants and mental retardation generally give inconsistent results (Erdtmann 1982), and more recent papers do not shed much further light on the problem. In this connection the relationship between the length of the Y chromosome and serious behavioural problems is of particular interest, in view of a similar relationship in the case of XYY men. The paper of Brøgger *et al.* (1977), which also reviews earlier work, shows that there is no difference in Y chromosome length between normal males and those with severe behavioural problems.

In recent years there has been considerable and growing interest in possible relationships between heteromorphisms and malignancy. Much of this work has been reviewed by Atkin & Brito-Babapulle (1981, 1985), with emphasis on C-band heteromorphisms, particularly those of chromosome 1. The evidence indicates that there is a high degree of heteromorphism for C-band size, particularly that of chromosome 1, in patients with some, but not all, cancers. It would be impracticable to review all the now extensive literature on the subject, but a few examples will be given, taken more or less at random. Berger et al. (1985) found differences in size of C-bands on chromosomes 1, 9 and 16, and in the incidence of inversions of 1 and 9 in breast cancer patients compared with controls, and also summarized other findings in the literature. Shabtai et al. (1985) and Labal de Vinuesa et al. (1988a) looked at colon cancers, and found in particular an increase in inversions of chromosomes 1 and 9. Similar sorts of differences have been found in leukaemias (Adhvaryu et al. 1987) and lymphomas (Labal de Vinuesa et al. 1988b). Even allowing for all the difficulties in assessing the association of heteromorphisms with different conditions, there does seem to be considerable evidence for an association between malignancies and an increased level of heteromorphism of heterochromatin. Reasons for such an association have not been established, but various ideas were discussed by Atkin & Brito-Babapulle (1981). In particular they suggest that somatic pairing of chromosomes with unequal C-bands could result in unequal crossing-over, which could facilitate chromosomal mutations, and also result in homozygosity for mutant genes.

Not all studies of the phenotypic effects of heteromorphisms have been concerned with abnormalities. Yamada (1985) has presented evidence for a positive correlation between stature and length of the Y heterochromatin in Japanese men. Obviously this cannot be the only factor affecting human height, and in any case, although Japanese men have a high proportion of long Ys, they are not particularly tall, and this correlation seems unlikely to be universal. Ibraimov & Mirrakhimov (1985) have described extensive variation in Q-band heteromorphisms between populations and races, although a selective advantage of any pattern of heteromorphisms has yet to be demonstrated, and it seems likely that founder effects might play a large part in producing the ethnic differences seen.

Finally, it has been proposed that large blocks of heterochromatin can delay chromatid separation at anaphase, and that this could be a factor in inducing non-disjunction (Zhang et al. 1987).

Throughout this section, various effects that have been supposed to be mediated by heteromorphisms of heterochromatin have been described, and in most cases the connection has proved to be tenuous. The difficulties in performing a properly controlled study have already been alluded to. It should also be noted that only a very small part of the variation present in human heterochromatin has been investigated. Most work has been concerned with C-bands, plus a small amount on Q-heterochromatin and NORs. However, as described earlier (Sections 14.4.1–14.4.7) many C-bands can be divided into segments which show distinctive staining properties with various reagents, and it could well be in some cases that the total size of the heterochromatic block is not the important factor, but the size or presence of a sub-band within it. It is perhaps appropriate to conclude this section with a quotation from a recent paper by Hsu *et al.* (1987) in which they studied chromosome heteromorphisms from amniotic fluid specimens of no less than 6250 patients from four population groups: '... we are not aware of any deleterious phenotypic nor clinical effect of these chromosomal polymorphisms nor of any apparent association with fetal wastage.' If a large, well-controlled study such as this shows no obvious effects of heterochromatin variants on phenotype, it seems probable that such variation, even if not selectively neutral, can have only very minor deleterious effects.

14.4.11 Applications of human chromosome heteromorphisms

Heteromorphism of human chromosomes is not only a phenomenon of interest in its own right, but also a valuable tool for various purposes. Müller *et al.* (1975) remarked that each of the 376 newborns they investigated had a unique set of C- and Q-band heteromorphisms, so that these, and other banding heteromorphisms are valuable for identifying the karyotype of an individual. Olson *et al.* (1986) estimated that the probability of finding two people with the same set of quinacrine heteromorphisms was 0.0003, and if C-band variants were also included, the probability would no doubt be much lower. It is also possible, in many cases, to determine from which parent any chromosome carrying a heteromorphism has been derived.

Heteromorphisms have been used extensively to determine the origin of the extra chromosome in autosomal trisomies (Hassold *et al.* 1984), and of the extra chromosome sets in triploids (Jacobs *et al.* 1978) and tetraploids (Sheppard *et al.* 1982). In all trisomies, non-disjunction at the maternal first meiotic division is the commonest cause, whereas in polyploids the origin of the extra chromosome sets is usually paternal, often by dispermy. Use of restriction fragment length polymorphisms is now being used to supplement the information that can be gained from heteromorphisms (Hassold *et al.* 1987). The origin of *de novo* chromosome rearrangements can be determined in a similar way, provided that they involve chromosomes with distinctive heteromorphisms (Chamberlin & Magenis 1980). A particularly interesting use of heteromorphisms was in determining the origin of hydatidiform moles, products of abnormal pregnancies, which lack an embryo and have a 46,XX karyotype. Kajii & Ohama (1977) showed that the moles contain two identical sets of chromosomes of paternal origin, and are the result of development of an ovum under the influence of a spermatozoal nucleus or nuclei.

It was soon recognized that heteromorphisms could be valuable in determining paternity in disputed cases (Schnedl 1974). More recent work indicates that the method shows a high degree of reliability (Olsen *et al.* 1986), although chromosomal methods of paternity testing have now been superseded by 'genetic fingerprinting' using hypervariable minisatellite DNA sequences (Jeffreys et al. 1985).

Another field in which heteromorphisms have been applied is in gene assignment (Chaganti et al. 1975); indeed, the first gene to be assigned to a human autosome was the Duffy blood group gene, located on chromosome 1 by linkage with an unusually large block of heterochromatin (Donahue et al. 1968). Here again, alternative methods of gene assignment have proved superior, and the use of chromosomal heteromorphisms in such studies has been quite limited.

One situation in which chromosomal heteromorphisms are unlikely to be replaced by other methods is where there is a need to identify the origin of individual cells. Such a situation arises in prenatal diagnosis, where it is necessary to ensure that maternal cells do not contaminate amniotic fluid cell cultures (Barker et al. 1977) or chorionic villus cultures (Miny et al. 1985) derived from the fetus. A similar situation arises in transplants, for example those of bone marrow, where it is important to establish whether the marrow is repopulated with recipient or donor cells (Khokhar et al. 1987).

The human chromosome heteromorphisms have been described in considerable detail not only because of their practical importance, nor merely because they have been so intensively studied, but also as an example of the degree of variability that might be present in other organisms. We shall now turn to these, starting with man's closest relatives.

14.5 Heteromorphisms in primates

The great apes (gorilla, two species of chimpanzee, and orangutan) have karyotypes generally similar to that of man, differing in that $2n = 48$, the human chromosome 2 being represented by two separate acrocentrics in the apes. It is, therefore, not surprising that many of the chromosome heteromorphisms found in the great apes resemble those of man, although there are also some important differences. Much of the earlier work has been reviewed by Seuanez (1979) who points out the relatively small numbers of individuals examined compared with the thousands of humans. All the great apes show heteromorphism of the centromeric heterochromatin of their autosomes; however, the chromosomes with large centromeric C-bands in man (1, 9 and 16) often have only small C-bands in their ape homologues. In the chimpanzees and gorilla, the NOR-bearing acrocentric chromosomes also show heteromorphisms of brilliant quinacrine fluorescence on their short arms, generally with a much higher incidence than that found in man; the orangutan, however, has no brilliant fluorescence on its chromosomes.

As well as similarities, there are important differences between the heteromorphisms of the great apes and man. The gorilla, for example, shows intense distamycin/DAPI fluorescence on the short arms of all its acrocentric autosomes, and at the centromeres of its chromosomes 4, 17 and 18; the size of all these DA/DAPI blocks is variable (Haaf & Schmid 1987). Heteromorphic DA/DAPI blocks occur on the chromosomes of the chimpanzees (Schmid & Haaf 1984, Wienberg & Stanyon 1988) and of the orangutan (Schweizer et al. 1979) although the sites of these differ a good deal between species (Wienberg & Stanyon 1988). The chimpanzees and gorilla also have telomeric C-bands on many chromosomes (Seuanez 1979); these bands also show bright quinacrine fluorescence and are heteromorphic. Except in the gorilla, the Y chromosome of the great apes lacks a heterochromatic segment showing brilliant quinacrine fluorescence; even in the gorilla this segment is small, and has not been reported as heteromorphic (Seuanez 1979).

A particularly interesting chromosomal polymorphism has been described in orangutans from both Borneo and Sumatra (Seuanez et al. 1976a). The polymorphism involves a rearrangement of chromosome 9, and was found in 12 out of 23 specimens; two separate pericentric inversions, one inside the other, were required to derive one form of the chromosome from the other. This, therefore, qualifies as a genuine polymorphism, since there are two distinct forms of the chromosome, both at a high frequency. Since it occurs in two separate island populations, the polymorphism is presumably of ancient origin, and apparently has no adverse effect on fertility in heterozygotes.

Investigations on other primates have been less detailed, but it is clear that extensive heteromorphism does occur. For comparison with the situation in the orangutan, it is noteworthy that in the 44-chromosome gibbons of the genus *Hylobates* there are three forms of chromosome 8, which can be related to each other by pericentric inversions. All three forms of this chromosome have been found in three species, and two forms in another species; in those species where only one form of the chromosome was found, only very small numbers of individuals were available (Stanyon et al. 1987).

As an example of the more usual sort of variability, Matayoshi et al. (1987) reported widespread heteromorphism of heterochromatin in the platyrrhine monkey *Cebus apella*, and there is little reason to suppose such variability is not extensive in wild populations of different species of monkeys.

14.6 Heteromorphisms in other mammals

Investigations of heteromorphisms in other mammals have involved chiefly two groups: ungulates of agricultural importance, and various rodents, both wild and maintained in the laboratory.

Mayr et al. (1985) showed that there was considerable heteromorphism of centromeric heterochromatin in ten species of Bovidae, including cattle, yak, two species of buffalo, sheep and goat. In cattle, the sizes of the centromeric C-bands show little variation between breeds, although in one, the White

Park cattle of Great Britain, there is heteromorphism for the absence of a C-band on chromosome 27 (Royle 1986). As in humans, there is variation in the number and distribution of silver-stained NORs, which is characteristic for each animal (Mayr et al. 1987). There is also extensive polymorphism for Robertsonian fusions in cattle, certain breeds showing particularly high frequencies of the 1/29 translocation (Long 1985), and in certain breeds of sheep (Bruère 1974). Although cattle that carry these translocations tend to show some reduction in fertility, the sheep with Robertsonian translocations generally show good fertility.

Extensive variation in the size of blocks of heterochromatin (Christensen & Smedegard 1978, Switonski et al. 1983) and of silver-stained NORs (Czaker & Mayr 1982) has also been reported for domestic pigs. In addition, there is polymorphism for chromosome number in wild boar (Bosma 1976); the diploid chromosome number may be 36, 37 or 38, compared with $2n = 38$ in domestic pigs. The polymorphism is the result of Robertsonian fusion of the chromosomes corresponding to numbers 15 and 17 in the karyotype of the domestic pig.

Both laboratory and wild mice (*Mus musculus*) show considerable heteromorphism in the size of their centromeric heterochromatin; within each inbred strain there is a characteristic pattern of C-band size that is inherited over many generations (Davisson 1989).

In the black rat (*Rattus rattus*) there is heteromorphism, not only of C-bands, but also of structural rearrangements, both within and between populations (Yosida et al. 1974, Yosida & Sagai 1975). Different laboratory strains of the brown rat (*Rattus norvegicus*), like those of laboratory mice, show characteristic patterns not only of C-band size, but also in the size of the short arms and satellites on certain chromosomes (Sasaki et al. 1979).

Another group of rodents that has been studied extensively for chromosome heteromorphism is the North American deer mice of the genus *Peromyscus*. There are two main types of heteromorphism: pericentric inversions, resulting in either acrocentric or metacentric chromosomes, and presence or absence of heterochromatic short arms (Greenbaum et al. 1978b, Greenbaum & Reed 1984, Davis et al. 1986, Hale 1986). Similar differences characterize the karyotypes of different species. In the case of polymorphisms for pericentric inversions, there is no evidence for reduced fitness (Davis et al. 1986), and at pachytene the inverted segments do not form an inversion loop, but proceed directly to heterosynapsis. Crossing-over does not occur in the heterosynapsed segments, and thus unbalanced gametes are not produced, so that polymorphism for the pericentric inversions can be maintained in the population (Greenbaum & Reed 1984, Hale 1986).

The results described above could be supplemented by reports from many other species of rodents, although few if any have been studied as comprehensively as those just described. However, one further example of polymorphism in a rodent is worth mentioning. Ray et al. (1984) described polymorphism of heterochromatin on chromosome 9 of the Chinese hamster. In itself this would not be remarkable, but polymorphisms had not previously been reported in Chinese hamster chromosomes (perhaps because laboratory animals comprise an inbred population derived from a very small number of animals), and the polymorphism was detected using DNA measurement by flow cytometry. The polymorphism could scarcely be seen in C-banded preparations. More interestingly, the measurements indicate that the heterochromatin occurs as discrete blocks, and that the polymorphism is the result of the presence of one or two extra copies of this block.

14.7 Heteromorphisms in lower vertebrates

Reports of banding heteromorphisms in fishes, amphibia, reptiles and birds are inevitably fewer than in mammals, which have been studied much more intensively than these other groups. Nevertheless, it is clear that heteromorphisms are not uncommon in the lower vertebrates, and a few examples of these will be given.

Among birds, two main forms of heteromorphisms within species have been found, much as in mammals: differences in morphology due to pericentric inversions and variations in the amount of heterochromatin (Ansari & Kaul 1979, Christidis 1986 a,b). Similar sorts of chromosomal heteromorphisms have been found in reptiles, but the W chromosome is particularly variable in some species. Using G-, N- and C-banding, Moritz (1984) identified six different forms of the W chromosome in the gecko *Gehyra purpurascens*, although only two heteromorphisms were found in the whole of the rest of the karyotype. The various forms of the W chromosome differed by paracentric and pericentric inversions, and in one case by a centric shift.

Heteromorphisms of the heterochromatic segments are well documented in amphibians (King 1980), and perhaps the most remarkable of these occurs in the newts *Triturus cristatus* and *T. marmoratus* (Macgregor & Horner 1980). In these species, the number 1 chromosomes are always heteromorphic for their heterochromatic long arms, and this heterozygosity is maintained because the homozygotes, for either the larger or the smaller homologue, are lethal in the embryo. The genetical and molecular basis of this intriguing phenomenon has not yet been worked out.

Among fish, the salmonids have been studied a good deal because of their economic importance. Their chromosomes show heteromorphism for Robertsonian fusions (Hartley & Horne 1984) and also for heterochromatic segments. The latter can be demonstrated by C-banding (Hartley & Horne 1984), but when Q-banding is used, variations in fluorescence intensity as well as size can be seen (Phillips & Zajicek 1982); as with other heteromorphisms, the quinacrine heteromorphisms in salmonids are heritable (Phillips & Ihssen 1986). It has been suggested that the quinacrine heteromorphisms would be useful markers in the study of different populations (Phillips & Zajicek 1982, Phillips & Ihssen 1986, Pleyte *et al.* 1989). Heteromorphisms for NORs have also been reported (Phillips *et al.* 1988). It should be noted that in the absence of any euchromatic banding patterns, it is impossible to identify the chromosomes involved in Robertsonian polymorphisms. Polymorphism for presence or absence of quinacrine-positive bands has also been reported for chromosomes of a non-salmonid fish, the North American mudminnow, *Umbra limi* (Howell & Bloom 1973).

14.8 Heteromorphisms in insects

Two orders of insects in particular have been studied in some detail chromosomally, the Orthoptera and the Diptera, and banding heteromorphisms have also been looked at in both these groups.

Among various species of Australian grasshoppers there is extensive variation in chromosome morphology and C-banding both within and between populations of the same species (Shaw *et al.* 1976, John & King 1977, 1983, Webb *et al.* 1978). The morphological variation is largely, but not entirely, due to pericentric inversions, and to substantial variation in the amount of C-banded material. Particular C-bands may be present or absent, and show variation in size. In an extreme case, *Atractomorpha australis*, the variation is so great that each chromosome exists in between 10 and 50 distinct forms. Different C-bands differ in their intensity of staining in certain species (Webb *et al.* 1978), in their reactions with different base-specific fluorochromes (John *et al* 1985), and in the type of satellite DNA they contain (Arnold & Shaw 1985, John *et al.* 1986). These heteromorphic systems in grasshoppers are evidently very complex, and although there are clear differences in C-banding patterns and amounts between different populations within a species (Shaw *et al.* 1976, John & King 1977) there is no evidence that these have any phenotypic significance. However, Arnold & Shaw (1985) did suggest that an increase in heterochromatin might increase the cell cycle time and thereby reduce the size of the adult grasshopper.

Heteromorphism of heterochromatin has also been described in Spanish grasshoppers, and while less extensive than that described for Australian species, shows similar characteristics: differences between populations, and variability in staining characteristics of different bands (Sentis *et al.* 1986, Navas-Castillo *et al.* 1986, Sentis & Fernandez-Piqueras 1987), as well as an inversion polymorphism (Cabrero & Camacho 1987).

It should be noted that much of the evidence for an effect of heterochromatin on chiasma frequency comes from studies on grasshoppers (John & Miklos 1979, John & King 1985). Additional blocks of heterochromatin may alter chiasma frequency, or cause the redistribution of chiasmata away from heterochromatin.

Banding heteromorphisms are also common among the Diptera. Differences in quinacrine fluorescence patterns between individuals in laboratory stocks of *Drosophila melanogaster* were recognized in the early days of banding (Barr & Ellison 1971), and heteromorphisms of heterochromatin continue to be reported in *Drosophila* species (Hatsumi 1987). Extensive variation in the heterochromatin of the sex chromosomes has also been found in mosquitoes of various species of *Anopheles* (Bonaccorsi *et al.* 1980, Baimai & Traipakvasin 1987). It has been suggested that sex chromosome heteromorphism in *Anopheles* might affect sexual behaviour and fertility (Bonaccorsi *et al.* 1980).

14.9 Heteromorphisms in plants

Just as in animals, there is extensive variation of heterochromatin in plants. In some cases, C-banding patterns have been used to identify structural rearrangements in plant chromosomes, but obviously, in the absence of euchromatic banding patterns throughout the length of the chromosome arms, this cannot be done as easily as in mammals (Sections 14.4–14.6). Most of the work on heteromorphisms in plants has only used C-banding, although the limited work using various fluorochromes (see for example Tables 7.1, 8.2 and 8.4) indicates that considerably more information might be obtained in this way. Heteromorphism in plants shows some interesting features, and in particular it has been possible in some species to make correlations with the phenotype of the plants.

A substantial amount of information has been obtained for cereal crops, all of which appear to show variability for presence, size and position of C-bands. This has been shown for rye by various authors, who have also found that different cultivated varieties show characteristic patterns of C-banding (Vosa 1974, Weimarck 1975, Singh & Röbbelen 1975, Lelley *et al.* 1978). The heteromorphism of C-bands is not confined to the cultivated species *Secale cereale*, but also occurs in other species of the genus (Singh & Röbbelen 1975, Bennett *et al.* 1977). No clear phenotypic effect of heterochromatin appears to have been reported in rye, but Heneen & Brismar (1987) reported that in its hybrid with wheat, *Triticale*, strains that had more terminal heterochromatin on the chromosomes showed more pronounced grain shrivelling.

Maize (*Zea mays*) shows a particularly interesting situation, in which the amount of heterochromatin varies between different strains, and can be correlated not only with the total nuclear DNA content, but also with geographical location (Rayburn *et al.* 1985). Southern varieties, which mature more slowly, have higher nuclear DNA C-values and more heterochromatin than the more northerly varieties, which have to mature more quickly. These authors speculate that artificial selection for earlier maturation and larger plant size in varieties of maize may have resulted in reduced DNA content. This would result in a shorter cell cycle, and the production of more cells. It should be noted that even within inbred lines of maize, heteromorphisms of C-bands can be found (Rayburn *et al.* 1985).

In several species of onions (*Allium*), Vosa (1976b) described a number of different classes of heterochromatin, according to whether or not they were C-banded, and whether they showed increased or decreased fluorescence or no differentiation with quinacrine or Hoechst 33258. The heterochromatic segments varied in number and size, and were reported to 'replace, but may be in addition to, normal euchromatic segments.' In the absence of precise measurements it is difficult to be dogmatic on this point, but we shall return to it later in connection with other species.

The chromosomes of *Scilla sibirica* also show extensive heteromorphism in their heterochromatin, which is both C-banded and intensely fluorescent with Hoechst 33258 (Vosa 1973). Each chromosome is variable, with from four to nine different forms for each chromosome. Most of the heterochromatin is terminal or interstitial, and particular blocks may be present or absent, or vary in size with or without corresponding variation in the size of the chromosome. Thus in some cases heterochromatin appears to replace euchromatin. As a result of this extensive variability, all 20 plants studied had unique karyotypes.

A number of other examples of heteromorphism of C-bands in plants could be mentioned. G. E. Marks & Schweizer (1974) described heteromorphisms for both number and size of bands in *Anemone* spp., while Filion (1974) found large differences in banding patterns between different cultivars of *Tulipa*. Finally, Bentzer & Landström (1975) have described an homologous pair of chromosomes in *Leopoldia comosa* which have between three and eight C-bands, which also vary in width and staining intensity. With such a great degree of variation, it is hardly surprising that the majority of plants are heterozygous for this chromosome pair. There is no evidence that in this species heteromorphism is in any way deleterious. In related species, male fertility is not reduced in spite of substantial chromosomal heteromorphism.

Plant chromosomes clearly have banding heteromorphisms at least as extensive and complex as those of animals, even if, in most cases, they have not been studied in such great depth. Application of various fluorochromes would be expected to provide much more information, especially in conjunction with biochemical investigations of the type and amount of DNA in plant heterochromatin. Careful measurements are needed to solve many problems, particularly the question of whether euchromatin can be converted into heterochromatin. DNA studies would be valuable here also. Finally, plants seem, from the few examples given above, to be particularly favourable material for studying the phenotypic effects of banding heteromorphisms. Several correlations have been described between the total DNA amount per nucleus for various plants, and various phenotypic and environmental effects (Bennett 1985). If the results with maize, described above, can be regarded as typical, heteromorphism of heterochromatin is likely to be very much involved in the variations in total DNA which lead to these phenotypic effects.

14.10 Concluding remarks

The use of banding techniques has revealed an extraordinary amount of variability in the chromosomes of virtually all organisms that have been investigated adequately. This variability falls into three main classes: changes in amount of heterochromatin, changes in the size of NORs, and structural rearrangements, particularly pericentric inversions, present as polymorphisms in the population. Such differences occur between homologues within an individual, between individuals of the same population, and between different populations of the same species. As we shall see in Chapter 15, the same sort of differences can be found between related species, and it appears that heteromorphism within an individual and the differences in chromosomes that can be found between species are merely the extremes of a continuum. To be sure, these types of variation were known before chromosome banding came into use, but there can be no doubt that without the use of chromosome banding the extraordinarily high degree of variation in chromosomes would never have been appreciated. It is perhaps not surprising that extensive variation in the amount of heterochromatin in a chromosome or a genome is quite tolerable, although the evidence tends to be that such variation is not entirely without phenotypic effects, subtle though they may be in many cases. Additional well-controlled studies are needed to elucidate such phenotypic effects, even in the most intensively studied species. Accurate measurements of heterochromatin are required, in spite of the difficulties involved. This is essential, not only for any sort of study in this field, but also to resolve the, perhaps rather artificial, controversy over whether the variation in heterochromatin occurs continuously, or in discrete steps. In reality, the argument is perhaps only about how large the discrete steps are, but there is good evidence in a few mammals that the variation is in fact discrete, rather than continuous as is often claimed in the literature without adequate supporting evidence. Measurements are also needed to resolve the question of whether in some cases, heterochromatin can increase in size at the expense of euchromatin (euchromatin transformation, King 1980), rather than being additional to the euchromatin. If euchromatin transformation really does occur, it would be most interesting to study the changes (if any) in DNA and protein that accompany the process.

A further aspect of the variability of heterochromatin which has scarcely been investigated, except in humans, is its heterogeneity. It is clear in humans that not only do different blocks of heterochromatin contain different DNA fractions and have different staining reactions with, for example, fluorochromes, but also that such heterogeneity may occur within a single block of heterochromatin. Human heterochromatin may be unusually complicated compared with that of other species, but there is at present little evidence for such an assumption, and it could well be that particular phenotypic effects are due to a particular fraction of heterochromatin, rather than due to the total heterochromatin.

In the case of chromosomal polymorphisms due to structural rearrangements, chromosome banding plays an essentially supporting role, as an aid to the accurate identification of the type of rearrangement. Although euchromatic banding (e.g. G-, Q- or R-banding) is clearly the best for this purpose, heterochromatic bands can also be used, although their more restricted distribution inevitably means that the breakpoints of the rearrangement cannot be located precisely. Until satisfactory replication banding can be induced in organisms other than the higher vertebrates, however, studies on these species will remain dependent on heterochromatic bands. Whatever the means of detection, it is clear that rearrangements are quite common, particularly pericentric inversions. It was formerly supposed that in a heterozygous state, inversions would lead to a substantial degree of infertility, due to crossing-over within the inverted region producing unbalanced gametes. In fact the evidence is that in many cases inversions can occur as genuine polymorphisms within populations without any apparent reduction in fertility. The discovery that extensive variation in the genetic material of eukaryotes can be tolerated with little, if any, adverse effect is indeed one of the more remarkable findings resulting from the application of chromosome banding techniques.

15 Banding and chromosome evolution

15.1 Introduction

Banding can be used for the study of chromosome evolution in various ways. First, there is the purely descriptive approach, in which the differences in chromosomes between related species, as identified by banding techniques, are identified. In this way, the mechanisms of change in karyotype evolution can be deduced. Second, banding patterns can be used to deduce phylogenetic relationships. This can be a distinctly hazardous procedure, for several reasons. Clearly the banded karyotype is only one of many characters that can be used in deducing phylogeny, and it is evident that it has no especial virtue for this purpose; thus phylogenies based purely on karyotypes without reference to any other features of the organism may well conflict with phylogenies based on comparative anatomy, protein polymorphisms and such like (J. Marks 1983). Moreover, as more distantly related species are studied, with a greater number of rearrangements distinguishing their karyotypes, banding patterns alone may not be sufficient to confirm homology; in such cases it is valuable, indeed almost essential, to see whether the supposedly homologous chromosomal segments contain the same linkage groups (Satoh & Yoshida 1985, O'Brien & Seuanez 1988). It is also not usually possible to deduce from banding patterns alone anything about the direction of evolution. It has been pointed out that if a particular chromosome configuration or karyotype is common in a particular group of organisms, it does not necessarily mean that it is primitive (Qumsiyeh & Baker 1988); comparisons must be made with other groups of rather less closely related organisms to see if they also possess a similar chromosome configuration, and only then can it be regarded as primitive. Another point is that some groups are very conservative karyotypically, showing little, if any, difference between related species, while other groups show very extensive changes in their chromosomes, even between species that are morphologically almost indistinguishable; thus phylogenetic distances cannot be deduced from the amount of karyotypic change separating species. On a purely practical point, the absence of euchromatic bands in almost all organisms except higher vertebrates severely limits the information available in a chromosome segment; attempting to deduce phylogenies on the basis of changes in heterochromatic segments, which are inherently much more variable than euchromatic segments (Ch. 14) seems to be a particularly risky procedure.

A third aspect of banding and evolution is the origin of the bands themselves. The changes observed in evolution are largely reshuffling of existing bands, but at some stage the bands themselves must have come into being as longitudinal differentiations of the chromosomes. Although this is clearly an aspect of banding and chromosome evolution, it is more appropriately considered in Chapter 16. The present chapter will be restricted to an account of the types of evolutionary changes that can be observed in chromosomes, as deduced by banding, and to summarizing some of the phylogenetic studies made with banding.

On a point of terminology, chromosomes in different species that have the same banding pattern are often described as homoeologous, indicating a slightly lesser degree of resemblance than that indicated by the adjective homologous. However, in evolutionary studies, the term homology is used to indicate resemblance due to common evolutionary origin, and there seems to be no justification for using a special term to indicate homology of chromosomes between species, simply because the (almost) identical pairs of chromosomes within a diploid species are also called homologues. See J. Marks (1983) on some of the confusing terminology that has been used by cytogeneticists in this field. In this chapter the adjective homologous will be used throughout.

15.2 Mechanisms of evolutionary change as seen by chromosome banding

When related organisms are compared, their chromosomes may appear very similar, if not identical; or they may differ in the amount of heterochromatin on their chromosomes; or their chromosomes may differ by a variety of rearrangements, such as inversions and translocations. Most of these types of changes can, in fact, be found within species as heteromorphisms (Ch. 14), or as racial differences, although the connection between such intraspecific variation and interspecific differences is not clear. Since ancestral karyotypes of two species cannot be known before and at the time of speciation, it is not possible to tell whether the different chromosomal forms are prerequisites for speciation (in particular cases), are produced at the actual time of speciation, or develop subsequently, and independently of the speciation event. What is clear is that since many chromosomal variants such as size variations in heterochromatin, pericentric inversions and Robertsonian translocations are maintained in populations as heteromorphisms (Ch. 14), they do not necessarily cause reduced fertility and will therefore often fail to produce reproductive barriers leading to speciation. These points will not be pursued further here, and the reader is referred to the books by White (1973) and by Chiarelli & Capanna (1973) for detailed considerations of the connections between changes of karyotype, speciation and evolution. Nevertheless, both these books were written before banding had been applied to any significant extent to problems of chromosomal evolution, and in the subsequent years banding has helped us to see much more clearly the kinds and extent of chromosomal changes that distinguish species. Differences between species which were interpreted as being due to one type of chromosomal change in the pre-banding era were often found to be due to completely different causes when banding techniques were applied (Hsu & Arrighi 1966, Pathak et al. 1973).

15.2.1 Speciation with no apparent changes in banding pattern

Most species, even those that are very closely related, appear to differ in their karyotypes, however slightly. Nevertheless, there are several cases in which groups of related species have virtually identical karyotypes when investigated using a variety of banding methods. It should be noted that most evolutionary studies have used G- and C-banding as the primary methods, although R-banding has been used extensively in France.

Among mammals, the Felidae are perhaps the best studied group in which speciation has not necessarily involved karyotypic divergence (Wurster-Hill & Centerwall 1982). Certainly there are some members of this family with unique karyotypes, but many species can be put in one of five groups, containing between two and eight species, in which all species have identical karyotypes. The Phocidae (seals) also have very uniform karyotypes by G- and C-banding, falling into one of two groups with $2n=32$ and $2n=34$ (Arnason 1974a). Among primates, the karyotypes of two species of gibbons (*Hylobates*) are virtually identical by G-, Q- and C-banding (Tantravahi *et al.* 1975), as are those of the baboon *Papio papio* and the macaque *Macaca mulatta* by R-, G-, C- and Ag-NOR banding (Finaz *et al.* 1978); in the latter case, be it noted, the animals belong to different genera. Identical G-banded karyotypes have also been reported in hares (*Lepus* spp.) (Schröder *et al.* 1978), camelids (Bunch *et al.* 1985), in several species of marsupials (Rofe & Hayman 1985), and in certain bats (Bickham 1979a). The dasyurid marsupials of New Guinea and Australia have remarkably uniform karyotypes by both C- and G-banding (Westerman & Woolley 1990). Among birds, generally regarded as a karyotypically conservative group, four species of gulls (*Larus* spp.) were found to have indistinguishable G- and C-banded karyotypes (Ryttman *et al.* 1979).

Other examples could be given, yet it should be emphasized that complete identity of banded karyotypes between related species is comparatively rare. Moreover, none of the cases cited above has been studied using high resolution banding (Sections 5.5.8 and 11.3.5), and it is possible that the greater resolution afforded by such methods might reveal minor differences. There is, nevertheless, no reason to postulate that speciation necessarily requires any karyotypic change.

15.2.2 Differences in heterochromatin between species

Differences between species in quantity, position and type of heterochromatin are exceedingly common. The simplest type of variation is perhaps differences in size of homologous blocks of heterochromatin; that is, two related species have blocks of heterochromatin with similar staining properties at similar locations on homologous chromosomes. It must not be assumed in all cases that such segments contain identical or related DNA sequences, since such analyses have not been attempted in most cases. It would be tedious to list more than a few examples, but it should be noted that such quantitative changes are found in all groups of organisms: for example, various plant species (Singh & Röbbelen 1975, Bennett *et al*. 1977, Schweizer & Ehrendorfer 1983), insects (Holmquist 1975b, King & John 1980), amphibians (Schmid 1980b), lizards (Olmo *et al*. 1986), birds (Christidis 1986b), and, of course, many groups of mammals. Among mammals, simple differences in the amount of heterochromatin between related species have been reported for rats (Yosida 1975) and hamsters (Gamperl *et al*. 1976) among rodents; and among the Bovidae (Buckland & Evans 1978b), to give but a small selection of possible examples.

Differences between species in amount of heterochromatin are often restricted to specific sites, or associated with specific events. Quite commonly closely related species differ in the presence or absence of heterochromatic short arms; thus in one species the chromosomes appear acrocentric, but the homologous chromosomes in another species are biarmed, the second arm being wholly composed of heterochromatin. This situation is often found among rodents (Fig. 15.1) (Pathak *et al*. 1973, Voiculescu 1974), but has also been reported in birds (Christidis 1986a). The presence of heterochromatic short arms is commonly regarded as an addition of material to the karyotype, but it must be affirmed that banding studies alone cannot, in general, be used to determine the direction of evolution. Loss of heterochromatin has been reported in biarmed chromosomes formed by Robertsonian translocations in the Bovidae (Buckland & Evans 1978b) and in tandem fusions in cotton rats of the genus *Sigmodon* (Elder 1980). (Again evidence favouring fusion rather than fission of chromosomes is not strong in most cases.)

Differences in the amount of heterochromatin between species are often correlated with differences in the total amount of nuclear DNA (the 2C value). Close correlations have been found in certain rodents (Deaven *et al*. 1977, Sherwood & Patton 1982) and in gibbons (Pellicciari *et al*. 1988), in which changes in the amount of heterochromatin are responsible for essentially all the change in the 2C DNA amount. In other groups of species, although there is still a correlation between the amounts of heterochromatin and of nuclear DNA, it is clear that the heterochromatin is not the only source of variation. This is true when rodents as a whole are considered (Gamperl *et al*. 1982) and in the plant genera *Lolium* (Thomas 1981) and *Secale* (Bennett *et al*. 1977). In other groups (e.g. *Scilla*, Greilhuber *et al*.

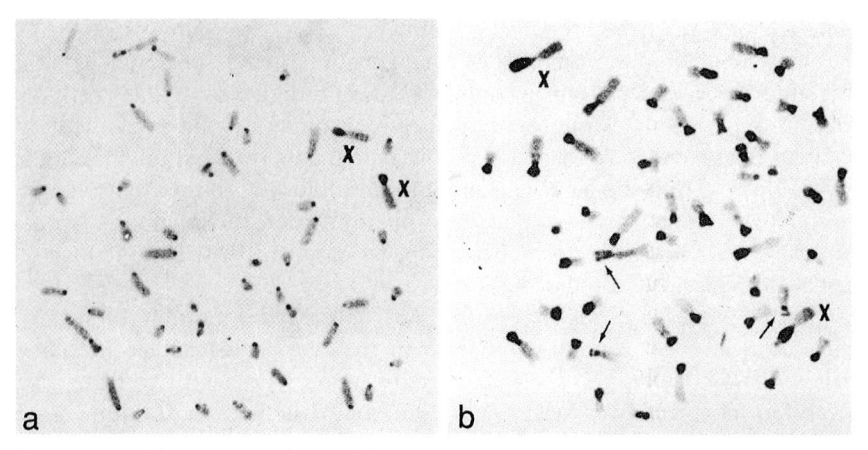

Figure 15.1 C-banded metaphases of (a) *Peromyscus crinitus* and (b) *P. eremicus*. In *Peromyscus crinitus*, almost all the autosomes lack short arms, whereas in *P. eremicus* all have short arms, which are wholly heterochromatic (except for those arrowed). Reproduced from Pathak *et al.* (1973) by permission of the authors and S. Karger AG.

1981, Acridoid grasshoppers, King & John 1980) there is no simple relationship between amount of heterochromatin and total nuclear DNA. This may well be the commonest situation, as there is no reason why amounts of euchromatin in the genome should not change as well as the amounts of heterochromatin. Cases where changes in total nuclear DNA are due solely to changes in heterochromatin should perhaps only be expected in closely related species.

One mechanism that would permit changes in the amount of heterochromatin without altering the total amount of nuclear DNA is *euchromatin transformation* (King 1980). In a genus of Australian frogs, he pointed out that: 'whole arm C-band blocks are present in the karyotype yet their inclusion has not produced a corresponding modification in relative chromosome size,' and described apparent examples of this phenomenon in other groups of frogs, in snakes and in grasshoppers. Similar observations have been reported in certain plants (Vosa 1973, 1976b) and in certain birds (Christidis 1986b). However, before the concept of euchromatin transformation can be accepted, a number of points must be established. First, it must be shown that the chromosomes that are being compared in the different species are truly homologous. This would presumably imply a good homology of DNA sequences throughout the chromosome, with similar sequences being in euchromatin in one species and in heterochromatin in the other. (This is, of course, a similar situation to that found with facultative heterochromatin.) So far homology has been based purely on morphology and banding patterns. Second, it must be recognized that no single banding technique, not even C-banding, will necessarily show all the heterochromatin in a species. Some supposed examples of euchromatin transformation could, therefore, be cases where the staining properties of the heterochromatin have changed. Finally, it must be demonstrated by measurements that the amount of heterochromatin has changed without affecting the size of the chromosome. Thus, although Vosa (1973) believed that, in *Scilla*, changes in amount of heterochromatin could occur without changes in chromosomal size, careful measurements by Greilhuber & Speta (1978) showed that the heterochromatin was additional to the euchromatin. In conclusion, therefore, while there is apparently nothing inherently impossible about euchromatin transformation, no case can yet be quoted which meets all the criteria discussed above.

Closely related species may differ not only in the amount of heterochromatin in their genomes, but also in the number of heterochromatic bands, their location and their staining properties. The first two categories are often inextricably mixed, since a different distribution of bands often implies a different number, and *vice versa*. At the same time, size differences can be found so that the heterochromatin of related species may differ simultaneously in several ways. A few examples of numerical differences in heterochromatic bands will be given. Many of the single C-bands in chromosomes of the newt *Triturus italicus* appear as pairs of bands in *T. vulgaris* (Nardi et al. 1973). Similarly, the plant *Brimeura amethystina* has groups of two or three Q-bands where other species of the same genus have only a single band (Vosa 1979), while in *Anemone* spp. the number of C-bands in the distal parts of certain chromosomes varies between none and four in different species (Marks & Schweizer 1974). A similar situation to the latter was also found in *Anacyclus* (Schweizer & Ehrendorfer 1976). Differences in both number and location of C-bands are found between species of tsetse flies, *Glossina* (Davies & Southern 1976); *G. morsitans* has groups of terminal bands on the long arms of its autosomes and X chromosome, which are absent in *G. austeni*, although the autosomes otherwise have similar patterns. Differences between species in the location of heterochromatin are very common, and sometimes quite complex. Only a few examples can be given. Often one species will have terminal C-bands on its chromosomes that are lacking in a closely related species; this has been found in rye (*Secale* spp.) (Singh & Röbbelen 1975), in lizards (Olmo et al. 1986), frogs (King 1980), and dolphins (Arnason 1980), although more complex differences can be found, as in *Scilla* (Greilhuber et al. 1981) and Acridoid grasshoppers (King & John 1980). At least three mechanisms can be envisaged to explain the appearance of a heterochromatic band in a new site. First, there is the movement, by translocation or inversion, of whole or part of an existing band. Such mechanisms seem plausible to explain the different patterns seen in *Lolium* spp. (Thomas 1981). Second, new bands could form by the process of euchromatin transformation (see above for a discussion of this mechanism). Third, a new band could form by amplification of an existing DNA sequence. Obviously a combination of mechanisms (e.g. translocation followed by amplification) could be involved in particular cases, but on the whole there is little or no evidence to distinguish between the various possible mechanisms.

Another way in which heterochromatin may differ between closely related species is in its staining properties. The heterochromatin of *Drosophila* species shows a particularly complex picture when patterns of C-, N- and Q-banding and fluorescence with Hoechst 33258 are compared (Holmquist 1975a, Gatti *et al.* 1976, Pimpinelli *et al.* 1976a, Lemeunier *et al.* 1978). Certain related species of mammals also show striking differences in the staining properties of their heterochromatin. It is well known that the heterochromatin of the domestic or laboratory mouse *Mus musculus* fluoresces weakly with quinacrine; that of *Mus cervicolor*, however, shows bright fluorescence (Dev *et al.* 1973). These two species have different satellite DNAs in their heterochromatin (that of *M. cervicolor* being less A+T-rich) but it is not established whether the difference in fluorescence is directly related to the sequences of the DNA in the heterochromatin. In another rodent genus, *Thomomys*, different species show various combinations of positive or negative C-banding and chromomycin fluorescence (Barros & Patton 1985). It must not be supposed that in such cases there is a wholesale coordinated transformation of the properties of a block of heterochromatin associated with speciation. Fry & Salser (1977) have proposed that rodents, and perhaps other mammals, share a library of satellite DNA sequences, and that in each species, individual sequences may be amplified to form major satellites, whereas others remain at a very low level. In another species, different sequences would be amplified from the same library. Although this account does not cover all aspects of Fry & Salser's hypothesis, it is nevertheless a useful conceptual framework for an understanding of how related species can have heterochromatin with different properties.

15.2.3 *Rules governing the evolution of heterochromatin*

Given the enormous variety of differences that can be found in patterns of heterochromatin distribution even between closely related species, it may be wondered whether there are any rules that govern the evolution of such patterns. In fact, the first such rule was proposed by Heitz (1933), that of equilocal heterochromatin distribution, which states that in any one species the constitutive heterochromatin tends to be located at similar sites on non-homologous chromosomes. Examples of this phenomenon have been described in plants (Greilhuber *et al.* 1981, Loidl 1983, Schweizer & Ehrendorfer 1983), grasshoppers (John *et al.* 1985) and fish (Mayr *et al.* 1988), and indeed the tendency for blocks of heterochromatin to occur in similar sites on different chromosomes of a particular species is widespread. In some cases, metacentric chromosomes show similar patterns of C-banding in both arms (Greilhuber *et al.* 1981). This equilocal distribution of heterochromatin has been explained in terms of chromosomal orientation at interphase or during meiotic prophase, so that repetitive DNA sequences in heterochromatin could be transferred between non-homologous chromosomes at stages when they are in close apposition (Loidl 1983, Schweizer & Ehrendorfer-

1983). These ideas have been developed further by Schweizer & Loidl (1987), who point out that telomeric C-bands tend to occur on short chromosomes or chromosome arms, whereas longer chromosomes or arms tend to have intercalary C-bands; in fact, the position of intercalary bands in longer chromosomes tends to be determined by the distance from the centromere to the telomeric band in the shorter chromosomes. According to these hypotheses, the positions of heterochromatin bands would be determined largely by mechanical considerations, and not by selective advantages. At present, it is too soon to judge the correctness or otherwise of these ideas, which at present must be regarded largely as descriptive.

15.2.4 Rearrangements of euchromatic parts of chromosomes

Whereas heterochromatic bands themselves may differ between closely related species, there is no discernible difference in individual euchromatic bands between related species; in fact, patterns of bands can be conserved even between distantly related species. Euchromatic bands can, therefore, be used to study chromosomal rearrangements that have occurred during evolution. Obviously, however, such studies must be largely confined to higher vertebrates, since these are the only organisms in which, at the present time, euchromatic bands are produced routinely. It should be noted, however, that good replication banding patterns have been obtained in the chromosomes of urodeles of the genus *Hynobius*, and have been used to make comparisons between species (Kuro-o *et al.* 1987). There can be little doubt that, as the techniques for replication banding are applied to various organisms that have not hitherto shown euchromatic bands, it should be possible to make more detailed evolutionary studies on them.

One of the commonest chromosomal changes occurring during evolution has been Robertsonian exchange, that is, the fusion of two acrocentric chromosomes to form a single metacentric chromosome, or alternatively the fission of a metacentric to form two acrocentrics. G-banding is valuable to confirm that morphological differences between the chromosomes of related species are due to this cause, and not to some other kind of rearrangement. Banding has confirmed the occurrence of Robertsonian exchanges in most orders of mammals, and also in crocodiles (King *et al.* 1986). Some particularly interesting examples of this process have been described. Although it is not normally possible to deduce the direction of chromosomal evolution from banding studies, there can be no doubt that in mice many populations have developed metacentric chromosomes by a process of fusion. The vast majority of populations of the mouse *Mus musculus* and its close relatives have 20 pairs of telocentric chromosomes. However, certain populations in the Alps and Apennines and some other localities in Europe and North Africa have karyotypes with a smaller number of chromosomes in which pairs of non-homologous telocentrics have fused to form single metacentric chromosomes (Gropp *et al.* 1972, Redi & Capanna 1988). In different populations, all the autosomes except chromosome 19 can be found fused into metacentrics, some populations having as many as nine pairs of metacentrics, with $2n=22$. One such form, the tobacco mouse with $2n=26$ and seven pairs of metacentrics, has been recognized as a separate species (*Mus poschiavinus*), the hybrids of which with laboratory mice show reduced fertility. Different mouse populations show fusions of different telocentrics, and it appears that in most populations the metacentrics were independently derived, although in the Rhaetian Alps several different populations share certain metacentrics.

A similar situation is found in the common shrew (*Sorex araneus*) in which different European populations have metacentric chromosomes made up of different combinations of acrocentrics (Searle 1984), all of which remain unfused in the closely related *S. granarius* (Wojcik & Searle 1988). G-banding is vital in these cases for identifying the chromosome arms which have fused in the different populations, thereby helping to elucidate the relationships between populations.

Robertsonian exchanges have also been important in the evolution of the Bovidae. A karyotype of 60 chromosomes, consisting of 29 pairs of acrocentric autosomes plus the sex chromosomes, is found in the ox (*Bos taurus*) and the goat (*Capra hircus*) (Evans *et al*. 1973). The karyotypes of many other species in this family can be derived from those of the ox or goat by a process of fusion to form metacentrics, although some other rearrangements have also been recognized (Buckland & Evans 1978a, Bunch 1978, Bunch & Nadler 1980). In the evolution of sheep there has been a progressive accumulation of homologous metacentrics as the chromosome number decreased (Bunch 1978, Bunch & Nadler 1980), but in the remainder of the Bovidae different species form their metacentrics from different combinations of acrocentrics. Even in the superfamilies Giraffoidea and Cervoidea, certain arms of metacentric chromosomes can be shown to be homologous, by G-banding, with individual acrocentrics of the ox (Buckland & Evans 1978a).

Another form of chromosome rearrangement that is common in vertebrate evolution is pericentric inversion. G-banding has been used to confirm that this type of change occurs between species of crocodiles (King *et al*. 1986); among birds, in several families of finches (Christidis 1983, 1986a,b); and in several groups of mammals. The predominance of reports from the primates (e.g. Garcia *et al*. 1976, Dutrillaux *et al*. 1978, Rumpler *et al*. 1987) and from rodents (e.g. Deaven *et al*. 1977, Greenbaum *et al*. 1978a, Vistorin *et al*. 1978) probably reflects the frequency with which these groups are studied, rather than a tendency for chromosomal evolution to occur by pericentric inversion in these, but not in other, groups.

Other types of chromosomal rearrangement evidently occur less frequently in the evolution of higher vertebrates; nevertheless, paracentric inversions (Olert & Schmid 1978, Viegas-Péquignot *et al*. 1983b, Christidis 1983), reciprocal translocations (Olert & Schmid 1978, Viegas-Péquignot *et al.* 1983b) and tandem fusions (Shi *et al.* 1980, Elder 1980, Christidis 1983) have all been confirmed by detailed banding studies. One of the most remarkable chromosomal differences between closely related species involves extensive tandem fusions to produce what has been called 'The Muntjac Scandal' (Capanna 1973): the fact that two very closely related species which are very similar morphologically, and can produce hybrids which are occasionally fertile, have completely different chromosome numbers. The Chinese muntjac (*Muntiacus reevesi*) has $2n = 46$, all chromosomes being acrocentric, while the Indian muntjac (*M. muntjak vaginalis*) has $2n = 6(♀)$ or $7(♂)$. Shi *et al.* (1980) have shown that the latter karyotype can be derived from the former by a series of Robertsonian fusions and tandem fusions (Fig. 15.2). In the process a substantial amount of DNA has been lost (Wurster & Atkin 1972), and the complexity of the process is emphasized by the presence of multiple kinetochores occupying a very elongated centromeric constriction (Brinkley *et al.* 1984).

15.3 Use of banding to study chromosome phylogenies

The earlier sections of this chapter have described the different types of changes in banding patterns that can occur in chromosomal evolution: either actual changes in the bands themselves, if heterochromatic, or in the case of euchromatic bands changes in the patterns of bands that indicate different types of chromosomal rearrangements. Obviously, it is also possible, in principle, to deduce evolutionary relationships by comparing the banding patterns of the chromosomes of different organisms. There are, however, many problems and pitfalls in making such deductions. One of the major problems is that the majority of organisms do not show euchromatic banding patterns on their chromosomes, so that comparisons can only be made of heterochromatin and NORs. Apart from the greatly reduced amount of information that is available from such types of banding, the fact

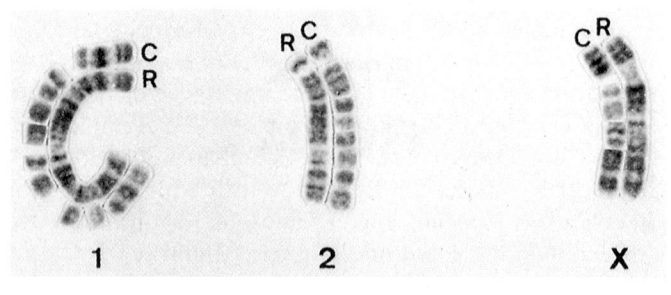

Figure 15.2 Comparison of the G-banding patterns of the chromosomes of *Muntiacus reevesi* with those of *M. muntjak vaginalis* to show how the chromosomes of the latter can be derived from those of the former by Robertsonian and tandem fusions. Reproduced from Shi *et al.* (1980) by permission of the authors and S. Karger AG.

that heterochromatin has, on the whole, little known function (Section 4.2.4) and can vary substantially within (Ch. 14) and between closely related species (Section 15.2.2) makes it of very limited value in working out evolutionary relationships. Euchromatic bands are, therefore, more useful for this purpose, and for this reason most studies of banding phylogeny have been done on higher vertebrates. Even here things are not straightforward, since chromosomal changes have occurred at vastly different rates in different groups of organisms. It must be realized that banding patterns are only one of a large number of characters which can be used in studies of phylogeny, and do not have any pre-eminence in determining evolutionary relationships. Rather, they should be used in conjunction with many other types of characters.

It is perhaps more interesting to take a group of organisms whose evolutionary relationships are well established, and examine how their chromosomes have changed during evolution. Obviously it is impossible to determine the karyotype of an extinct ancestral form directly, but reasonable deductions can be made about the ancestral karyotype, and about the types of chromosomal changes that would have led to the present day arrangements. Some examples of such studies will be described; for a general review of chromosomal evolution in mammals, see O'Brien & Seuanez (1988).

Numerous comparative studies have been made on the chromosomes of primates, and especially those of the great apes. All these have shown a high degree of homology of banding patterns. Results for the great apes have been reviewed in several publications (D. A. Miller 1977, de Grouchy et al. 1978, Mitchell & Gosden 1978, Seuanez 1979), but cytogenetics has not clarified the relationships of the great apes. High resolution banding, while confirming the very close homologies between man, gorilla and chimpanzee, and locating breakpoints of rearrangements more precisely, has not elucidated relationships further (Yunis et al. 1980, Yunis & Prakash 1982). Reasons for the failure of banding studies to clarify matters are discussed by J. Marks (1983), but it must be realized that problems are found with all other approaches so far made (R. Holmquist et al. 1988), and that there is as yet no agreement on higher primate phylogeny, whatever characters are used to deduce it.

Extensive homologies have also been found between human chromosome banding patterns, and those of many lower primates (e.g. Dutrillaux et al. 1978, 1980a, Clemente et al. 1987). Detailed comparative studies have also been carried out in lemurs (e.g. Rumpler & Dutrillaux 1979, Rumpler et al. 1986). Dutrillaux (1979) used banding studies from more than 60 species of primates, ranging from the lemur *Microcebus murinus* to man to deduce a chromosome phylogeny for primates, which involved some 150 rearrangements; nevertheless, at least 70% of the bands were shown to be common to Simians and lemurs. The quantitative variation in the karyotypes was confined to the heterochromatin.

The rodents are a group that shows extensive karyotypic variability, and virtually all the possible types of chromosomal rearrangements (addition of heterochromatin, Robertsonian and tandem fusion, pericentric and paracentric inversions and reciprocal translocation) can be identified in closely related species (e.g. Mascarello & Hsu 1976, Elder 1980, Baker *et al.* 1988, Baverstock *et al.* 1983). In some cases there is little if any obvious banding homology between related species (Gamperl *et al.* 1978, Baverstock *et al.* 1983). It has, nevertheless, been possible to deduce ancestral karyotypes in certain groups, and to identify chromosomes and banding patterns that have been conserved over considerable evolutionary distances. In *Peromyscus*, the primitive karyotype was probably composed mainly of acrocentric chromosomes, and during evolution changes have occurred by addition of heterochromatic short arms, and by pericentric inversions (Greenbaum & Baker 1978). In *Neotoma* there is good conservation of G-banding patterns in spite of extensive variation in the form of the chromosomes, resulting from addition of heterochromatin, Robertsonian fusion, and in one case perhaps fission (Mascarello & Hsu 1976). Chromosome banding homologies have also been reported for many chromosomes between more distantly related species of rodents (Mascarello *et al.* 1974, Viegas-Péquignot *et al.* 1985), and such homologies have been supported by gene mapping and the establishment of common syntenies in the mouse and Chinese hamster (Satoh & Yoshida 1985). Koop *et al.* (1984) have used comparisons of banding patterns of several genera of Cricetidae and Muridae to deduce primitive banding patterns for the chromosomes of the former family. The fact that it is possible even to attempt such studies indicates that there is substantial conservation of banding patterns even though extensive morphological alterations to the chromosomes have occurred.

The Carnivora appear to be a much more conservative group karyotypically, with very similar banding patterns in the Procyonidae, the Viverridae, and particularly in the Felidae (Wurster-Hill & Gray 1975). The chromosomes of the Mustelidae and a hyena also show several banding homologies with felids, viverrids and procyonids (Wurster-Hill & Centerwall 1982). Interestingly, several characteristic carnivoran chromosomes can also be identified in the Pinnipedia (seals) which split off quite early from the carnivores (Árnason 1974a, Wurster-Hill & Gray 1975). The bears (Ursidae) (Wurster-Hill & Bush 1980) and the Canidae, however, tend to show higher chromosome numbers than other Carnivora, and only limited homology of their banding patterns with those of the groups already described. The chromosomal evolution of the Canidae has recently been examined in detail, and it has been concluded that chromosome fission has been important in the evolution of karyotypes with large numbers of chromosomes (Wayne *et al.* 1987a, b). The Carnivora as a whole are remarkable for having very little heterochromatin in their karyotypes, but the karyotype of the blue fox (*Alopex lagopus*) differs from that of the silver fox (*Vulpes fulva*) mainly in the development of large heterochromatic blocks on ten pairs of chromosomes (Yoshida *et al.* 1983).

One outstanding problem in carnivore phylogeny has been the position of the giant panda (*Ailuropoda melanoleuca*). Earlier G-banding studies showed only a few homologies with the chromosomes of either bears (Ursidae) or the lesser panda (Procyonidae) (Wurster-Hill & Bush 1980). The question has been re-investigated more recently (O'Brien *et al.* 1985) using a variety of molecular techniques as well as high resolution G-banding. These studies showed that nearly every large chromosome of the brown bear could be homologized to a giant panda chromosome arm. Thus giant panda chromosomes seem to be composed largely of bear chromosomes joined together by Robertsonian fusion, although there were few homologies between the chromosomes of bears and giant panda on the one hand, and of procyonids and the lesser panda on the other. The chromosomal data, therefore, confirm conclusions obtained from molecular studies that the giant panda has diverged from the ursid line, whereas the lesser panda is much more closely allied to the procyonids, which diverged from the ursids much earlier than the giant panda. This study not only emphasizes the importance of using banding patterns as only one of several characters in deducing evolutionary relationships, but is also a striking illustration of the extra information that can be obtained using high-resolution banding.

Several other groups of mammals show great karyotypic conservatism, particularly the Cetacea (Árnason 1974b, 1980, Árnason *et al.* 1977) and the Pinnipedia (Árnason 1977). In both orders the G-banding patterns are almost identical throughout, with a minimum of chromosome rearrangement, and most of the variability is due to changes in heterochromatin. It had been proposed that both the Cetacea and the Pinnipedia might be diphyletic, but the karyotypic uniformity in both these groups is strong evidence that each is monophyletic.

Another group showing great conservatism in its G-banding patterns is the Bovidae, in which chromosomal evolution has been mainly by Robertsonian exchange, with individual chromosome arms in general retaining their banding patterns largely unchanged (Evans *et al.* 1973, Buckland & Evans 1978a, Bunch 1978, Bunch & Nadler 1980). Nevertheless, similarity of banding patterns may obscure other differences between genomes. The karyotypes of the ox (*Bos taurus*) and goat (*Capra hircus*) are almost identical, with $2n=60$, and all the autosomes are acrocentric, mostly with indistinguishable G-banding patterns, and similar amounts of heterochromatin in the two species. Yet the ox has about 14% more DNA in its genome than the goat, and it must be concluded that the extra DNA is evenly distributed in small packets throughout the genome, rather than being concentrated at a few distinctive sites (Sumner & Buckland 1976). Dasyurid marsupials also have very similar G-banded karyotypes with similar amounts of heterochromatin, but a wide range of nuclear DNA values (Westerman & Woolley 1990).

Conservation of G-banding patterns has also been described in marsupials (Rofe & Hayman 1985, Westerman & Woolley 1990). These animals typically have a small number of large chromosomes, and the most common chromosome number is $2n = 14$, which is found in most groups of marsupials. The G-banding patterns in those species with $2n = 14$ are very similar, and it is argued that this must be similar to the ancestral karyotype, rather than being derived by fusion from karyotypes with greater chromosome numbers. In the latter case, it is unlikely that the same fusions would have occurred in distantly related species; rather, the higher chromosome numbers must have been derived from the $2n = 14$ karyotype by fission.

Some groups of mammals, e.g. bats, show conservatism in some lines, whereas others have evolved rapidly, with a large number of chromosome rearrangements which are identifiable by banding studies (Bickham 1979b, Haiduk & Baker 1982). A high level of chromosome change during evolution is also found in the Equidae (horses, donkeys and zebras) (Ryder et al. 1978), in which the diploid chromosome number varies from 32 to 66, and the number of chromosome arms from 62 to 102. G-banding studies show that while some chromosomes are common to the whole group of seven species, others are confined to subgroups such as the donkeys or the zebras. Although the G-banded karyotypes of the zebras *Equus grevyi* ($2n = 46$) and *E. burchelli* ($2n = 44$) are quite similar, that of *E. zebra hartmannae* ($2n = 32$) seems to show only limited homology with the other species. Evidently numerous chromosomal rearrangements of many different types must have occurred during evolution of the Equidae, and it has been calculated that this family has the highest rate of chromosomal evolution among the mammals.

In view of the extensive chromosomal differences among the Equidae, it is remarkable that viable hybrids have been produced between almost all pairs of species in this family (Ewart 1899, Ryder et al. 1978). Except for the hybrid between Przewalski's horse and the domestic horse, which is fertile (Short et al. 1974), such hybrids are normally sterile. The best known of these hybrids is, of course, the mule (*E. asinus* ♂ × *E. caballus* ♀), and it should be noted that while the diploid numbers of these species are 64 and 62 respectively, there is little obvious homology between their karyotypes. It is, therefore, even more remarkable that earlier anecdotal records of fertile mules (Ewart 1899, Chandley 1981) have recently been confirmed by cytogenetic and other studies of animals from China (Rong et al. 1988). Viability, and even more fertility, imply a balanced karyotype and genome, and detailed study of such hybrids should provide further insights into chromosomal homologies among the Equidae.

Similarities of euchromatic (G-) banding patterns between even distantly related species appear, in fact, to be the rule rather than the exception. Obviously one would not expect to find similarities in complete G-banded karyotypes in different orders of mammals, yet there are clearly close resemblances of the banding patterns of certain chromosomal segments. Such resemblances have been described between chromosomes of rodents and primates (Dutrillaux et al. 1979, Petit et al. 1984), between lagomorphs and primates (Dutrillaux *et al.* 1980b) and between rodents and carnivores (Petit *et al.* 1984). In some cases the banding homology is supported by evidence that the chromosomes involved carry the same genes and have similar linkage groups (O'Brien & Seuanez 1988), thus greatly strengthening the conclusion that these chromosomal segments are truly homologous. Such combined studies of banding patterns and linkage groups have been reported for comparisons of carnivores and primates (Nash & O'Brien 1982, Dutrillaux & Couturier 1983) and for comparison of mouse and man (Sawyer & Hozier 1986). These observations show quite clearly the constancy during evolution of euchromatic segments of chromosomes, although it is clear that the content of the bands need not remain the same (e.g. Sumner & Buckland 1976), and the overall pattern of banding can be extensively modified by chromosomal rearrangements.

Inevitably, data on chromosome phylogenies of birds and reptiles based on banding are much fewer than the data from mammals. Moreover, comparisons are made more difficult by the large numbers of microchromosomes in these organisms; such chromosomes are too small to give useful banding patterns, and comparisons are therefore generally restricted to a small proportion of the karyotype, the large macrochromosomes.

Birds have been regarded as a karyotypically conservative group (Shields 1982), and to some extent this has been confirmed by banding studies (Takagi & Sasaki 1974, Stock *et al.* 1974, Carlenius *et al.* 1981). Takagi & Sasaki (1974) even claimed G-banding homology between the three largest bird macrochromosomes and the largest chromosomes of the freshwater turtle *Geoclemys reevesii*, species which had a very distant common ancestor. However, Stock & Mengden (1975) stated that there was no banding homology between birds, snakes, turtles or amphibia (*Xenopus*), although homology within the groups could be found. Although certain chromosomes are conserved with little or no change across several orders of birds, it has become clear that in other groups of birds extensive chromosomal rearrangements have taken place (Christidis 1986a,b); nevertheless, the degree of modification is not so great that the banding patterns cannot be used to identify the types of rearrangements involved.

Reptiles, like birds, include groups that are karyotypically conservative, and others that are highly variable. Both crocodiles (King *et al.* 1986) and snakes (Mengden & Stock 1980) show many chromosomal differences between species when studied with banding. In snakes, although a $2n=36$ karyotype is common to many families, G-band patterns are not homologous (Mengden & Stock 1980). Turtles, on the other hand, show conservation of G-banding patterns over substantial evolutionary distance and time. This conservatism enabled Bickham (1981) to deduce the primitive karyotypes of eight families of turtles, work out by what rearrangements they had been derived from the primitive karyotype of the suborder Cryptodira, and describe the rather limited number of rearrangements that had given rise to the karyotypes of present-day species. Moreover, by comparison with the age of the suborder, families and genera calculated from the fossil record, it was possible to show that in the Mesozoic radiation of the turtles the rate of karyotypic change was more than twice as great as in Tertiary and modern turtles. Certain chromosomes have apparently been conserved for some 200 million years.

As remarked frequently hitherto, detailed euchromatic banding patterns are not obtained routinely on chromosomes of lower vertebrates (amphibia and fish), of invertebrates or of plants. These organisms do not, therefore, lend themselves to the detailed karyotypic comparisons that can be made in higher vertebrates, and it is not possible to identify the same chromosome in different organisms separated by substantial evolutionary distances. In most cases, comparisons have to be made between closely related species, and the only bands that can be compared are heterochromatic bands and NORs. The former are particularly variable, and are not, in themselves, of much value for chromosome identification. Thus, while many comparisons have been made of the distribution and staining properties of heterochromatin in closely related plant species (e.g. Singh & Röbbelen 1975, Schweizer & Ehrendorfer 1976, 1983, Greilhuber et al. 1981, Deumling & Greilhuber 1982, Linde-Laursen et al. 1986), insects (e.g. Holmquist 1975b, Gatti et al. 1976, Pimpinelli et al. 1976a, King & John 1980), fish (Hartley 1987), and amphibia (reviewed by Schmid 1980b, Birstein 1982), such studies are incapable of yielding the detailed information on banding homologies and chromosomal rearrangements that can be provided by euchromatic banding. Nevertheless, some interesting examples can be found, where banding methods have been used to study particular evolutionary problems.

Both among grasshoppers and frogs there are distinct species that have apparently arisen as a result of hybridization between two other species. The Australian parthenogenetic grasshopper *Warramaba virgo* consists of a number of clones with relatively minor differences in C-banding patterns, and which differ also by a few inversions, translocations or fusions. Comparison of its C-banding patterns with those of other species confirms that it probably originated by hybridization between the as yet unnamed sexually reproducing species 'P196' and 'P169' (Webb et al. 1978). Further studies using base-specific fluorochromes show that the staining patterns of certain populations of *W. virgo* are significantly different, and probably arose by independent crosses between 'P169' and 'P196' (Schweizer et al. 1983). Another example of a hybrid species occurs in the European water frogs of the *Rana esculenta* complex. It now appears that *R. esculenta* is essentially a hybrid between *R. ridibunda* and *R. lessonae*, although the complete situation is quite complex (Heppich 1978). Study of the C-banded karyotypes of these species showed that the chromosomes of *R. ridibunda* and *R. lessonae* had the same banding patterns, except for chromosome 11; *R. esculenta* possessed one of each type of chromosome 11, thus supporting other evidence that it is a hybrid (Heppich 1978).

15.4 Evolution of sex chromosomes

The mechanism of sex determination has been a subject of great interest since antiquity, and the existence of morphologically distinct sex chromosomes was recognized in early studies of chromosomes (McClung 1901). Nevertheless, not all organisms with separate sexes have heteromorphic sex chromosomes, and a complete range can be identified, from species with no morphologically recognizable sex chromosomes to extreme differentiation of a sex chromosome pair. Some organisms have greatly modified sex chromosomes, or more complex sex determining mechanisms involving several chromosomes. Use of chromosome banding has helped greatly to clarify both the evolution of normal sex chromosome systems, and the understanding of more complex systems.

A sequence for the evolution of vertebrate sex chromosomes has been proposed by Ohno (1967, 1983), and the following paragraphs will be built on this framework. Although there is extensive information on insect sex chromosome systems (White 1973), no comparable theory of their evolution has, apparently, been proposed, and in any case, studies with banding have been much more restricted. Most higher plants are monoecious, and therefore do not have sex chromosomes. The principles of Ohno's hypothesis are that the sex chromosomes (X and Y in the case of male heterogamety, or Z and W in female heterogamety) were originally homomorphic and carried numerous genes, and that allelic differences at only a few genes determined the sex. Such a situation can be found in most fish and amphibia, and in many reptiles. In the course of evolution, however, the Y or the W chromosomes have been modified until they retain only one or two genes, which are concerned with sex determination, and have either become very small, or are largely heterochromatic. The X (in mammals) or Z chromosomes (in birds) have retained their original sizes and a full complement of genes not connected with sex determination. Intermediate stages in this evolution can be found in snakes; in the early stages, some mechanism is required to inhibit crossing-over, otherwise differentiation of the sex chromosome pair could not occur. Ohno (1967) reported that in some species the Z and W chromosomes differ by a pericentric inversion but are similar in length, and believed that this was a first stage in their differentiation. Ohno *et al.* (1964) also proposed that the majority of mammalian X chromosomes were similar in size (about 5% of the haploid genome) and should be expected to carry the same set of genes in different species.

Differentiation of sex chromosomes in fishes appears to be rare, whether they are studied without or with banding. A few cases have been described in Salmonidae (Hartley 1987), but earlier reports of sex chromosomes in eels have been refuted by Wiberg (1983) using a variety of banding techniques. In amphibia, however, the use of banding methods has revealed a number of examples of sex chromosome differentiation (Schmid 1983, Sessions 1980, Green & Sharbel 1988). In some species the sex chromosomes turn out to be different in size or morphology, but even here banding is important for precise identification of the chromosomes. The situation in the frog *Rana esculenta* is particularly interesting (Schempp & Schmid 1981); the sex chromosomes are essentially monomorphic, but study of replication patterns shows that one homologue of chromosome 4 in males alone had a very late replicating segment in the long arms. This chromosome showed no heteromorphism with any other banding technique. It should be mentioned that banding studies have demonstrated both XX/XY and ZZ/ZW sex chromosome systems, and that sex chromosome differentiation with banding has been reported both in Urodela and Anura. It should also be noted that sex chromosomes may be differentiated in one species, but not in another, closely related, species (Green & Sharbel 1988).

Differentiated sex chromosomes occur sporadically in many reptiles (e.g. Carr & Bickham 1981, Moritz 1984), but the most detailed investigations have been on snakes. In snakes the chromosomal sex determining system is always a ZZ♂/ZW♀ system, and a complete range can be found from the primitive Boidae with normally no morphological or banding differentiation, to the advanced families in which the W chromosome has been reduced to a very small, largely heterochromatic element. In between are the Colubridae, in which the Z and W chromosomes are distinguished by a pericentric inversion. 'By creating a need for permanent isolation, a pericentric inversion in the Y or the W facilitates further differentiation of sex chromosomes' (Ohno 1967). Use of banding techniques combined with studies of DNA composition have, however, shown that the differentiation is not simply shrinkage of the W or Y. Snake W chromosomes, even when large, are typically largely heterochromatic (Singh *et al.* 1976, Mengden & Stock 1980) and contain a substantial amount of a highly conserved satellite DNA. Jones & Singh (1985) have proposed that heterochromatinization is an important early event in sex chromosome differentiation, and that genes (other than sex-determining genes) on the W or Y chromosomes are not simply lost gradually but are actively replaced by heterochromatin. It should be noted that the W chromosome of snakes often does not consist of a single large undifferentiated block of heterochromatin, but shows longitudinal differentiation with C-banding, and in some cases distinctive G-banding patterns (Mengden & Stock 1980, Mengden 1981).

Most birds, like the more advanced snakes, have a strongly heterochromatic W chromosome, but the ratites (emus, rheas, etc.) show a different situation (Ansari *et al.* 1988). Their W chromosomes are slightly smaller than the Z chromosomes, but are euchromatic and show extensive homology with the Z chromosomes by a variety of banding techniques, including replication banding. It is not clear whether the ratite condition represents a more primitive stage in differentiation of the sex chromosomes than that found in carinate birds.

An interesting parallel to the ratites is found in the primitive monotreme mammals, in which the long arms of the X and Y chromosomes are largely

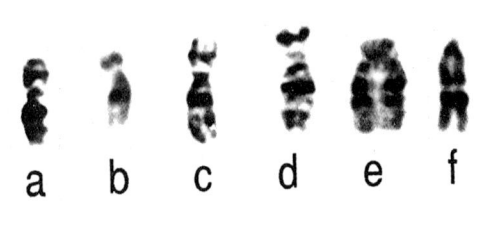

Figure 15.3 G-bands of X chromosomes from various mammals, showing the presence of two major bands in all species. (a) *Atilax paludinosus*, (b) *Ursus americanus*, (c) *Homo sapiens*, (d) *Xerus rutilis*, (e) *Sigmodon hispidus*, and (f) *Clethrionomys rutilis*. Reproduced from Pathak & Stock (1974) by permission of the authors and the Genetics Society of America.

homologous by G-banding (Wrigley & Graves 1988). It is again tempting to regard this as a primitive condition compared with that found in both marsupials and the Eutheria, in which the Y chromosome is commonly reduced to a very small chromosome, largely heterochromatic. The eutherian X chromosome not only carries a highly conserved linkage group (Lalley *et al.* 1988), as suggested by Ohno (1967), but also a conserved G-banding pattern. Pathak & Stock (1974) showed that two major bands were always present in normal-sized eutherian X chromosomes, although their relative position depends on whether the chromosome is acrocentric or metacentric (Fig. 15.3). Eutherian Y chromosomes seem to be more variable, although since they are usually so small, it is not possible to obtain detailed banding patterns. It is, however, clear that there can be substantial differences between the Y chromosomes of related species due to heterochromatin, as for example in the great apes (D. A. Miller 1977, Seuanez 1979).

Although the X chromosomes of most mammals comprise about 5% of the haploid genome, some species have much larger X chromosomes. Ohno *et al.* (1964) and Ohno (1967) supposed that these had arisen literally by duplication, triplication or even quadruplication of the chromosome, but C-banding shows that the extra material is, in fact, heterochromatin, as for example in hamsters (Vistorin *et al.* 1977), *Microtus agrestis* (Pera 1972) and certain whales (Arnason 1974b, Arnason *et al.* 1977). Interestingly, in the rodents at least, the Y chromosome is also enlarged, and it seems unlikely that any simple multiplication of the X chromosome occurs.

In other species, an enlarged X chromosome occurs because an autosome has become translocated on to the X, producing an $XY_1Y_2\male$, where Y_1 is the original Y chromosome, and Y_2 the homologue of the translocated autosome. Banding permits the identification of the autosome involved, as in the Indian muntjac (Fredga 1971). More complex modifications of the X due to translocations of autosomes can be identified in gerbils (Wahrman *et al.* 1983), and the reverse process, forming an X_1X_2Y system by translocation of an autosome on to the Y, has been identified in certain bovids using G-banding (Buckland & Evans 1978a). Other quite extraordinary sex determining mechanisms have been identified in certain rodents with the aid of banding. In the wood lemming (*Myopus schisticolor*) both XX and XY females are found, as well as XY males. Careful studies, including G-banding, showed that in fact there were three types of females: XX, X^*X, and X^*Y. The X^* chromosome is

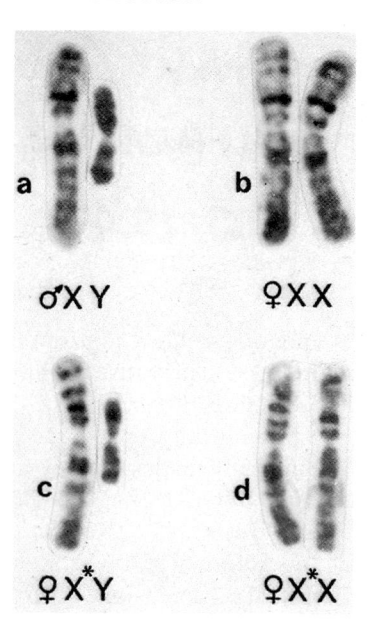

Figure 15.4 The two different types of X chromosome in the wood lemming (*Myopus schisticolor*): (a) XY ♂ with a normal X; (b) XX ♀; (c) X*Y ♀ with modified X* chromosome; (d) X*X ♀. Reproduced from Herbst *et al.* (1978) by permission of the authors and Springer-Verlag.

evidently a modified X chromosome, with a shorter short arm than the normal X, and a different G-banding pattern (Fig. 15.4), that in some way suppresses the male determining factor on the Y (Herbst *et al.* 1978). Finally, perhaps the most extraordinary situation is that found in *Ellobius lutescens*, with $2n = 17$ in both sexes. Vogel *et al.* (1988) have shown conclusively that the unpaired chromosome 9 is in fact the X chromosome, and that both sexes have indistinguishable chromosome complements with no cytologically visible trace of a Y chromosome.

It was proposed many years ago that the X and Y chromosomes consist of an homologous pairing segment, and non-pairing differential segments. Banding of metaphase chromosomes fails to show homology between the X and the Y in most cases, probably because the homologous pairing segments are too small to reveal adequate banding patterns. It is, therefore, interesting that Müller & Schempp (1982) showed homologous early replication patterns in the pairing segment of the short arms of the human X and Y chromosomes, which can also be found in the great apes (Weber *et al.* 1986), but not in a marsupial (Müller *et al.* 1988). This segment has presumably been conserved since the time when eutherian X and Y chromosomes first began to differentiate.

Most of the studies described in this section inevitably refer to vertebrates, and mammals in particular. Nevertheless, observations have been made on the sex chromosomes of two groups of insects, grasshoppers and Diptera (especially *Drosophila* spp.), although comparisons depend on morphology as much as on banding patterns. In grasshoppers, C-band patterns have been used to identify rearrangements of the X in different races of *Caledia captiva* (Shaw *et al.* 1976). In the Acrididae, which normally have an XX ♀/XO ♂

sex determining mechanism, neo X-neo Y sex chromosomes produced, as in vertebrates, by translocation of an autosome on to the X, can be identified and interpreted using C-banding (Cardoso & Dutra 1979, King & John 1980). In *Drosophila orena*, increase in size of the sex chromosomes by accumulation of heterochromatin has been described (Lemeunier *et al.* 1978). On the whole, however, the limited information provided by C-banding and other stains for heterochromatin is of rather limited use for tracing evolutionary changes in sex chromosomes, except those of the more dramatic kind.

15.5 Concluding remarks

Although it is probably true to say that use of chromosome banding has not demonstrated any fundamentally new mechanisms of chromosomal evolution, its power clearly lies in the precision with which the chromosomal changes that occur in evolution can be defined. This is, of course, especially true of higher vertebrates, in which the use of G-banding and other euchromatic banding techniques can be used to identify precisely the fate of different chromosomal segments. This is true even of quite distantly related species belonging to different orders of one class, for example birds or mammals. In such cases the supporting evidence of homology provided by common linkage groups is valuable. Except in mammals, work of this type is so far wholly inadequate, in spite of important efforts by many workers, and it is to be hoped that many more comprehensive studies of chromosome phylogeny will be made in lower vertebrates, invertebrates and plants. The restrictive factor here is, of course, the general absence of euchromatic bands in such organisms, but the evidence that detailed patterns of replication banding can be induced in their chromosomes (Section 11.3) gives hope that this problem can be overcome.

The other major type of chromosomal evolution, changes in heterochromatin, has also been illuminated by the use of specific banding techniques. Obviously, it is important to know that changes in chromosome morphology are due to addition or loss of heterochromatin, rather than to rearrangement of euchromatic segments. Essentially, the role of banding in the study of chromosome evolution falls into two parts: the use of euchromatic banding to identify chromosomal segments and their rearrangements, and the study of changes in heterochromatin. While the variability of the latter has proved to be extraordinarily great, the degree of stability of euchromatic bands, in spite of alterations in their gross patterns, has been one of the more remarkable discoveries in this field.

16 Genome organization in the light of chromosome banding

16.1 Introduction

Since chromosome banding methods came into general use twenty years or so ago, they have become indispensable for the study of the cytogenetics of both plants and animals (including man), and have proved to be invaluable in such diverse fields as evolutionary studies and molecular biology. Many applications are recounted in earlier chapters of this book, and if chromosome banding were used only as a practical tool for identifying chromosomes and parts of chromosomes, it would be of very great importance. There is, however, another aspect to chromosome banding. In the author's view, chromosome banding is equally important for directing attention to the different properties of different chromosome segments. In previous chapters considerable emphasis has been put on mechanisms of chromosome banding with the intention of understanding the structural and functional entities that are revealed as longitudinal differentiation of chromosomes by special staining techniques. In fact, banding is an expression of the ways in which the eukaryotic genome is organized. Instead of being uniform rods with a centromeric constriction, chromosomes have turned out to be highly differentiated structures. Although there is still much to be learnt about this level of organization, it is now possible to put together a reasonably coherent picture of it. The purpose of this chapter, therefore, is to review genome organization in the context of the longitudinal differentiation of chromosomes revealed by banding.

Four classes of chromosome bands can be recognized (Section 2.2). Two of these, nucleolar organizers and kinetochores, are clearly defined structural entities whose functions are relatively well understood. They have been described in Chapters 9 & 10 respectively, and will not be considered further here. The other two types of banding are heterochromatic and euchromatic bands, and it is with these, especially the latter, that this chapter will be concerned.

16.2 Heterochromatic bands

Although heterochromatin has been recognized for over 60 years (Heitz 1928), it required specific staining methods to appreciate several aspects of heterochromatin, and indeed to make it easily accessible for study. The main features of heterochromatin demonstrated by specific banding techniques are its universality (Ch. 4), its diversity (Ch. 5, 7, 8, 11 & 12), and its variability (Ch. 14). Almost all chromosomes in almost all eukaryotes have some heterochromatin; its staining properties and the nature of the DNA it contains vary extensively both within and between species, and in general blocks of heterochromatin are heteromorphic for size, and sometimes for staining properties. The ability to stain heterochromatin in specific ways, and to determine the nature of its DNA has settled a long-standing controversy. Although Heitz regarded heterochromatin as a substance, later authors regarded it as a state of the chromosome; different chromosomal regions might or might not be heterochromatic at different stages of development (Brown 1966, Lima-de-Faria & Jaworska 1968). Although it is obviously true that facultative heterochromatin, such as the inactive X in female mammals, or the paternal set of chromosomes in mealy bugs, must be a state of chromatin, since there is always a homologous euchromatic chromosome or set of chromosomes, this is not true of constitutive heterochromatin. The evidence that constitutive heterochromatin, as well as having distinctive staining properties, also contains, in most cases that have been examined, distinctive types of DNA (Table 4.1) implies that it is a distinct substance or substances.

No general function has been satisfactorily ascribed to constitutive heterochromatin, although several effects have been described (Section 4.2, John 1988). The distinction between functions and effects is an important one. The word 'function' implies that the heterochromatin is present specifically to produce the result observed, whereas the word 'effect' is more neutral and does not imply such a direct functional correlation.

Heterochromatin has long been regarded as inert, and evidence from its composition and variability supports this view. The frequent, but far from invariable, occurrence in heterochromatin of highly repeated simple sequence DNA, which cannot possibly code for any meaningful protein, seems a clear confirmation of its essential inertness. So does the high degree of variability in amount of heterochromatin in a species without any obvious phenotypic effects. It must, nevertheless, be remembered that heterochromatin is highly variable in DNA composition, and does not always contain very highly repeated DNAs. It is thus possible that different types of heterochromatin have different functions, or no function at all. In fact, closer examination of heterochromatin has shown clearly that it is not necessarily without effect. In *Drosophila* spp. it has been established that heterochromatin contains a very low concentration of typical genes (Hilliker *et al*. 1980), as well as other less conventional functional entities, some of them of considerable size (Pimpinelli *et al.* 1986). Among these are the fertility factors on the Y chromosome (Hennig 1985), the ABO heterochromatin (Pimpinelli *et al.* 1985) and the Responder locus (Wu *et al.* 1989), none of which appears to be a conventional gene. No other organism has been investigated as thoroughly as *Drosophila*, but recent evidence shows that constitutive heterochromatin and highly repetitive DNA can be transcribed in plants (Nagl & Schmitt 1985), amphibia (Varley *et al.* 1980) and mammals (Sperling *et al.* 1987), although the functional significance, if any, of such transcription is not yet understood.

In some cells in some organisms, heterochromatin is evidently dispensable. For example, in *Drosophila* polytene chromosomes, heterochromatin is grossly under-replicated compared with euchromatin. Organisms as diverse as *Ascaris* (Goday & Pimpinelli 1984), copepods (Beermann 1977) and hagfish (Nakai & Kohno 1987) eliminate C-banded heterochromatin in somatic cells. While these observations might be taken to support the idea of the lack of function of heterochromatin, it is noteworthy that the heterochromatin is retained in its full amount in germ line cells; in other words, the organisms appear to be capable of getting rid of heterochromatin where it is not required, and therefore apparently deliberately retain it in germ line cells. The necessity for retention of heterochromatin in the germ line is not understood, although the presence of fertility factors on the *Drosophila* Y chromosome, which has no somatic effects, has already been mentioned. However, one important and widespread effect of heterochromatin in germ cells of both plants and animals is its influence on distribution of chiasmata (Miklos & Nankivell 1976, Rhoades 1978, Miklos & John 1979, Loidl 1982), although other cases are known in which heterochromatin apparently has no effect on chiasmata (Attia & Lelley 1987). An effect on chiasma distribution is clearly of some importance in influencing the generation of genetic diversity in the organism. It should be noted that chiasmata are almost unknown in heterochromatin. The possible importance of the development of heterochromatin combined with suppression of crossing-over as a mechanism for the differentiation of sex chromosomes has already been mentioned (Section 15.4).

Another apparent effect of heterochromatin which would be of importance in the germ line has been described by Vig and his colleagues (reviewed by Vig 1987). Essentially, it is argued that at anaphase, sister chromatids of different chromosomes do not separate simultaneously, and that the amount of centromeric heterochromatin influences the time of separation. Premature separation resulting from too little heterochromatin could, therefore, lead to aneuploidy. In fact, observations made by scanning electron microscopy show that in mitotic metaphase chromosomes, when the euchromatic arms are clearly separated into chromatids, the centromeric heterochromatin apparently remains undivided (Fig. 16.1). It may be, therefore, that centromeric heterochromatin has a function in holding sister chromatids together until anaphase, although terminal and interstitial blocks of heterochromatin would probably not share this function.

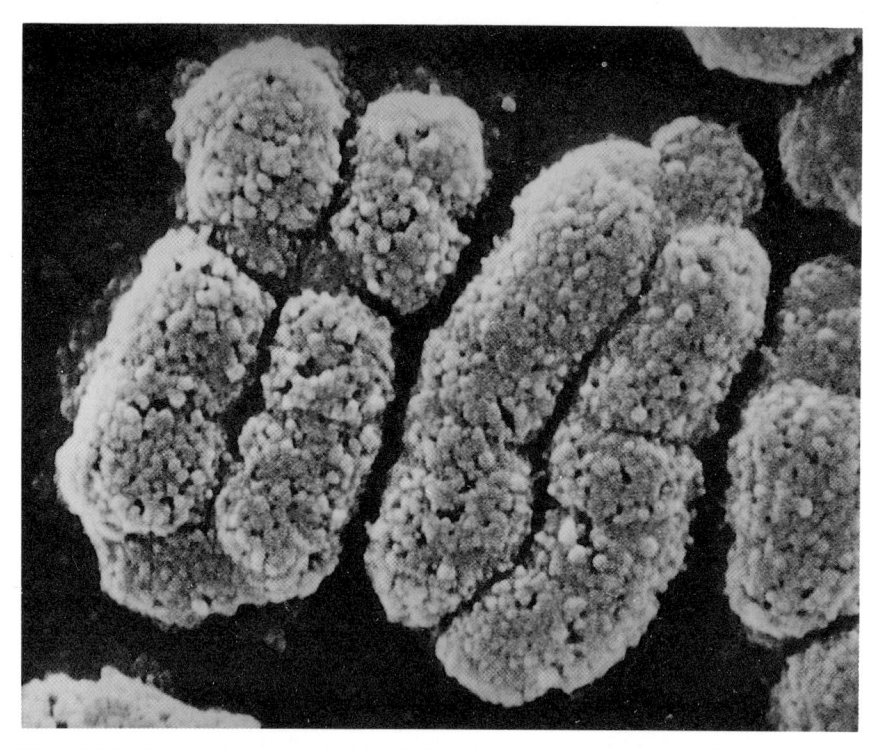

Figure 16.1 Scanning electron micrograph of mouse chromosomes, showing that the heterochromatin remains undivided after the arms have separated into two distinct chromatids. Reproduced from Sumner (1989) *Scanning Microscopy* suppl 3, 87–99, by permission of Scanning Microscopy International.

Although the influence of heterochromatin in the germ line is clear in some cases, it is certainly not true that it is without somatic effect. Cavalier-Smith (1978) has argued that the function of much of the nuclear DNA, including heterochromatin, is not genic, but instead regulates nuclear volume, an important parameter in cell physiology. Bennett (1985) has described several such 'nucleotypic' effects in plants; that is, parameters such as pollen grain volume, duration of meiosis and latitude of cultivation, that are closely correlated with the amount of nuclear DNA. An involvement of C-banded heterochromatin in such nucleotypic effects can be seen clearly in maize, as described by Rayburn *et al.* (1985). In this species, strains adapted to more northerly climates, which have to grow and mature more quickly, have less heterochromatin and less total DNA than the more southerly, more slowly growing strains. The amount of heterochromatin is believed to affect the growth rate through its effect on cell cycle time in this species. On the other hand, Nagl (1974) found that in various plants, the total nuclear DNA can be increased by addition of heterochromatin without increasing the cell cycle time.

In fact, the whole study of heterochromatin is beset with contradictions, and it may be unprofitable to seek universal functions for heterochromatin. The diversity of heterochromotin composition shown by banding techniques and DNA analysis could well reflect a diversity of function. An organism such as *Drosophila*, with a very small genome, may well need to incorporate various genic or quasi-genic functions in its heterochromatin, whereas an organism with a large genome may be able to afford a genically inert type of heterochromatin which would exert only nucleotypic and structural effects. It must be admitted that heterochromatin still remains mysterious, and in most cases lacks any assigned functions. Nevertheless, from the effects that have been described, it appears unwise to dismiss it merely as 'junk' or 'selfish' DNA whose only purpose seems to be to reproduce itself. Whatever the eventual truth about heterochromatin may turn out to be, C-banding has proved to be an invaluable label for defining it, and many other banding methods have helped to emphasize its diversity.

16.3 Euchromatic bands

The most remarkable discovery made with the introduction of G-, Q- and R-banding was that the euchromatic part of the genome, in higher vertebrates at least, was segregated into two distinct phases, distinguishable by their intensity of staining. This differentiation of euchromatin was only dimly foreshadowed by autoradiographic studies of chromosome replication and is a discovery due essentially to chromosome banding. A number of correlations between euchromatic banding patterns and other features of chromosome organization, such as base composition, time of DNA replication and pachytene chromomeres, have been mentioned in previous chapters. In this chapter, such correlations will be considered in more detail, and their functional, structural and evolutionary implications assessed. It should be noted that nearly all the relevant work has been done on mammalian chromosomes; this work and its implications will, therefore, be considered before going on to the situations in non-mammalian species, and the relevance of such findings for the evolution and functional significance of different types of banding.

16.3.1 *Molecular correlations with euchromatic banding*

In comparing features of the genome studied at the molecular level with the microscopically visible chromosome bands, it is essential to have some idea of scale. We know quite a lot about how genomic activities are organized at the molecular level, and also quite a lot about bands, but without some idea of the size ratio between them, it is not possible to make satisfactory correlations. On the basis that up to 2000 bands can be recognized in the human haploid genome (Yunis 1981), it has been calculated that each band, on average, contains between 1250 and 1640 kb (kilobases) of DNA (Holmquist *et al.* 1982, Holmquist 1988a). If, therefore, banding is to be explained on a molecular basis, the molecular phenomena must show a periodicity comparable with this range.

The close correlation between G-, Q- and R-banding patterns on the one hand, and patterns of chromosomal replication on the other, have already been noted (Section 11.3). At the molecular level, replication occurs in independent units called replicons, consisting of 50–330 Kb per replicon, and clustered into groups of 25–100 replicons; all the replicons within a cluster initiate and terminate replication at a specific time in S phase (Hand 1978). Thus the size of replicon clusters is comparable with, but generally rather larger than, that of the finest bands. Given the uncertainties in these estimates, there can be little doubt that a replicon cluster corresponds to a replication band at the finest level of resolution. However, Holmquist (1988a) has pointed out that at a lower level of resolution, less than 500 bands/genome, an individual band takes much longer to replicate than a cluster of synchronous replicons should, and concludes that at this level of resolution each band must consist of several less synchronous replication clusters.

The difference in base composition between early and late-replicating DNA fractions that was demonstrated by centrifugation studies (Section 5.3) found a cytological parallel in the banding patterns obtained using base-specific fluorochromes (Sections 8.3 & 8.4). The differentiation of mammalian DNA into A+T-rich and G+C-rich fractions has been investigated in more detail by Bernardi and his colleagues (Bernardi *et al.* 1985, Thiery *et al.* 1976) and their conclusions confirmed by others (Ikemura & Aota 1988). The main points of their work are that the DNA (excluding satellites) of mammals and other higher vertebrates is made up of a number of fractions of different base composition, that the same fraction can be found in different chromosomes, and different fractions in the same chromosome, and that the overall base composition and codon usage of individual genes is correlated with the overall base composition of the DNA fraction in which the gene is embedded. Segments of DNA having a fairly homogeneous composition are known as *isochores*; these are believed to average much greater than 200 kb in length, and are regarded as the molecular basis for base-specific chromosome banding (Bernardi *et al.* 1985, Holmquist 1988a). Although a relationship between isochores and bands has not yet been demonstrated directly, by *in situ* hybridization for example, there is no reason to doubt the connection. Other data also support this connection, as well as providing an intriguing view of chromosome organization.

There has been considerable interest in the distribution of genes on chromosomes. Earlier, circumstantial evidence suggested that most genes were probably in negative G-bands (Ganner & Evans 1971, Hoehn 1975, Wahrman *et al.* 1976, Korenberg *et al.* 1978). Study of the distribution of genes in chromosomal DNA fractionated by replication time (Goldman *et al.* 1984, Goldman 1988, Hatton *et al.* 1988) or by base composition (Bernardi *et al.* 1985, Ikemura & Aota 1988) has tended to confirm and refine this conclusion. Put very simply, 'housekeeping' genes (i.e., those that are constitutively active in all cells) are always early replicating and are found in G+C-rich positive R-bands. Tissue specific genes are usually in the late-replicating, A+T-rich positive G-bands, but when such genes are active, they replicate early. However, such genes are not translocated to early replicating bands, and presumably remain as cytologically invisible early replicating segments within a late replicating band.

Other data concerning gene distribution on chromosomes tend to support those just cited, but may indicate a more extreme heterogeneity. There are now sufficient genes that have been mapped to single bands on human chromosomes to enable one to make quite reliable statements about gene distribution, and it turns out that most genes, both housekeeping and tissue specific, are in negative G-bands (Rodionov 1985, Gardiner *et al.* 1988, Bickmore & Sumner 1989). Another source of information is the distribution of CpG islands, which are associated with the 5' regions of all housekeeping genes, and with many tissue-specific genes (Bird 1986, Gardiner-Garden & Frommer 1987). It is now possible, using pulsed field gel electrophoresis, to map these CpG islands in DNA fragments of a length comparable with the amount of DNA in a band. The result is that CpG islands are many fold more abundant in negative than in positive G-bands (Gardiner *et al.* 1988, Bickmore & Sumner 1989). Although, as suggested by Burmeister *et al.* (1987) this non-random distribution may merely mean that genes with CpG islands are in negative G-bands, while genes without CpG islands are in positive G-bands, it seems possible that rather few genes actually lack CpG islands. As mentioned above, many tissue-specific genes do have them. Moreover, the total number of CpG islands in the human genome, about 30 000, is very similar to the estimated number of genes (Bird 1986), although the large uncertainty in these estimates could still allow for a substantial proportion of genes that lacked CpG islands and could be located in positive G-bands. Nevertheless, most of the evidence now seems to be indicating that the great majority of genes, whether housekeeping or tissue-specific, are in negative G-bands.

The intermediate repetitive DNA sequences of mammalian genomes also show a non-random distribution between positive and negative G-bands. Such sequences fall into one or other of two classes: the long interspersed repeated sequences (LINES), usually a few thousand base pairs long, and short interspersed repeated sequences (SINES), which are much shorter. Such sequences fall into several families, although one or two sequences form a major part of each category. Together, the LINES and SINES constitute about 20% of the human genome (Holmquist & Caston 1986), and are therefore very abundant sequences. Both *in situ* hybridization (Manuelidis & Ward 1984, Korenberg & Rykowski 1988) and studies of replication (Holmquist & Caston 1986) have shown that LINES are concentrated into late-replicating DNA, or positive G-bands, while SINES occur predominantly in the negative, early-replicating G-bands. Since SINES are

G+C-rich, and LINES A+T-rich, it has been suggested that they are largely responsible for the base-specific fluorochrome banding of human (and other mammalian) chromosomes (Korenberg & Rykowski 1988). However, the evidence already discussed in this section indicates that positive and negative G-bands as a whole have A+T- and G+C-rich DNA respectively, so it seems more probable that both SINES and LINES, even if they show a more extreme bias in base composition, are actually embedded in DNA segments that show a similar bias in base composition. Curiously, both SINES and LINES have turned out to be mobile elements or retroposons, that can move about from one part of the genome to another (Holmquist 1988b). It is, therefore, quite remarkable that the intermediate repetitive sequences of whatever class normally appear to integrate into regions of the chromosome having a similar base composition, and thus the segregation of the genome into A+T- and G+C-rich fractions is maintained. Holmquist (1988b) suggests that general loss or accumulation of retroposons (i.e. SINES and LINES) might be a cause of differences in amount of DNA between species that do not affect chromosome morphology or banding patterns. On the other hand, selective accumulation or loss of retroposons could alter banding patterns by shrinking or swelling specific bands.

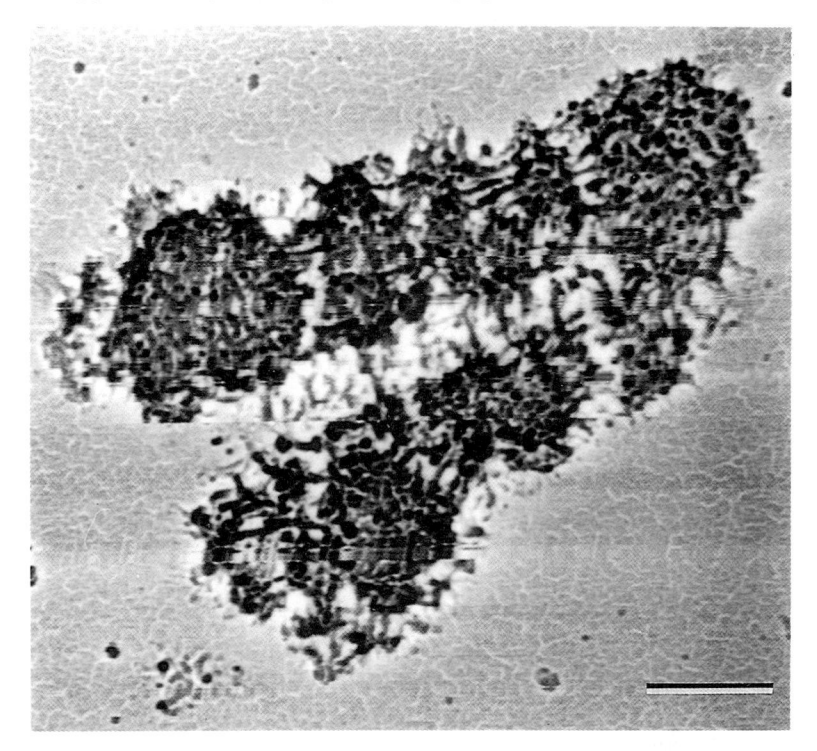

Figure 16.2 Electron micrograph of a mouse metaphase chromosome showing condensations of randomly orientated fibres connected by fibres parallel to the axis of the chromosome. The bar represents 1 μm.

16.3.2. Structural correlations with euchromatic banding

Current ideas on chromosome structure have been reviewed in Chapter 3, and it will be realized that although the scaffold model has gained general acceptance, it does not yet appear to provide a complete explanation of how chromatin fibres are arranged within chromosomes. Other features of chromosomes, such as chromomeres and coiling, are not generally taken into account by the scaffold model, although they are not necessarily incompatible with it. Can banding throw any light on the structural organization of chromosomes, and conversely, can knowledge of chromosome structure help to explain banding?

One aspect of chromosome structure that shows a strong, but not perfect, correlation with banding is the chromomeres of meiotic pachytene chromosomes (Ferguson-Smith & Page 1973, Okada & Comings 1974, Luciani *et al.* 1975, Hungerford & Hungerford 1978, 1979, Jagiello & Fang 1980, Jhanwhar & Chaganti 1981, Ambros & Sumner 1987). Nevertheless, although the pattern of chromomeres matches quite well the pattern of positive G-bands, the pachytene chromosomes that have chromomeres cannot be G-banded satisfactorily, while mitotic prophase chromosomes do not have chromomeres, but can easily be banded. Quite clearly, chromomeres are not simply the structural basis of G-banding, although both could be alternative manifestations of the same underlying phenomenon. The pachytene chromomere represents a more highly condensed region of the chromosome, and adjacent chromomeres apparently grow and fuse as the chromosome contracts (Lima-de-Faria 1975). Similarly, positive G-bands appear to grow and fuse as the chromosomes contract (Seabright *et al.* 1975, Sumner 1976, Yunis *et al.* 1978, Sen & Sharma 1985).

In fact, there is extensive evidence from electron microscopy that mitotic chromosomes are organized into dense aggregations of apparently randomly arranged fibres (Fig. 16.2), which are connected by a number of parallel fibres (Golomb & Bahr 1974b, Yunis & Bahr 1979, Mullinger & Johnson 1980, 1983, 1987). In interphase nuclei, there are about 8000 of these aggregations of chromatin fibres per haploid genome, and it has been suggested that these correspond to the ultimate chromosome bands (Yunis & Bahr 1979, Holmquist 1988a). As the chromosomes contract, the aggregations of fibres eventually become closely apposed, with no room for connecting fibres between them (Mullinger & Johnson 1983, 1987). Meiotic prophase chromosomes also consist of aggregations of fibres (Fig. 16.3) connected by parallel fibres (Sumner 1986a). At this level, therefore, there seems to be a good agreement between the structural principles of mitotic and meiotic chromosomes, but it remains true that G-bands can be induced on mitotic chromosomes that show no hint of longitudinal structural differentiation.

Harrison and Allen and their colleagues have proposed an alternative relationship between G-banding and chromosome structure (Harrison *et al.*

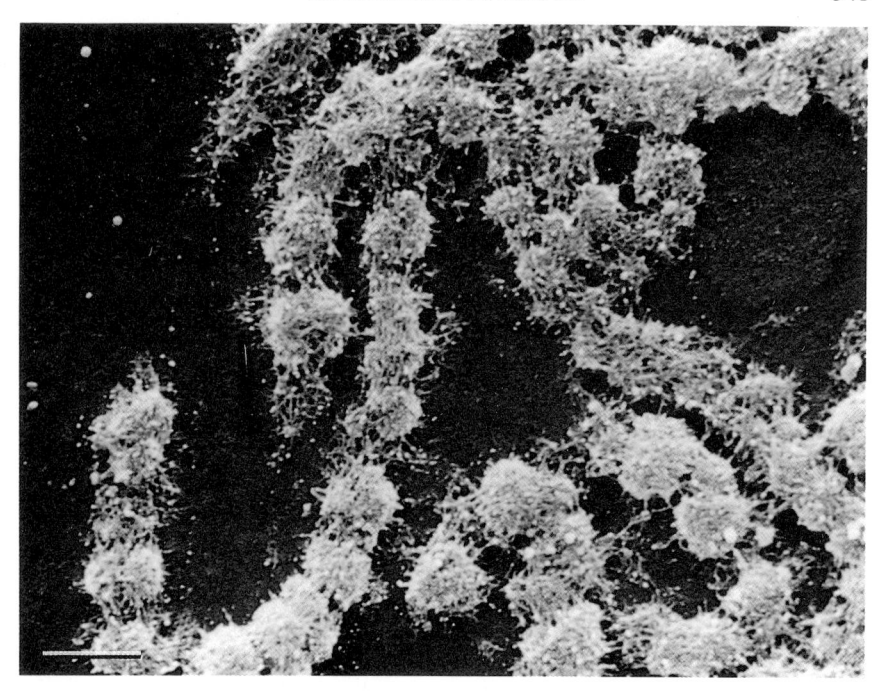

Figure 16.3 Scanning electron micrograph of human male pachytene chromosomes, showing parallel chromatin fibres connecting the chromomeres. The bar represents 2 μm. Reproduced from Sumner (1986a) by permission of the author and Springer-Verlag.

1981, 1985b). They prepared metaphase chromosomes by fixation in methanol-acetic acid in the standard way, treated them with trypsin to induce G-bands, and examined them by scanning electron microscopy. They observed that the chromosomes were divided into a number of segments by 'circumferential grooves' (Fig. 16.4). By comparison with the banding patterns, they deduced that the grooves corresponded to G-bands. Even if the proposed relationship between these grooves and G-banding is accepted (and the relationship does not always appear as clear as the original authors claim, see also Mullinger & Johnson 1987), it appears to be implicit that, in this model, chromomeres would correspond to negative G-bands, contrary to all other evidence.

Relationships between chromosome coiling and G-bands have also been claimed (Kato & Yosida 1972, Hatami-Monazah 1974, Goradia & Davis 1977, Johnson *et al.* 1981). A fundamental problem here is how the regularly spaced coils can be correlated with the variably spaced bands. Takayama (1976) claimed that by progressive trypsin treatment, coils could be converted to bands; some coils became closer together to form positive bands, while others became more widely separated to form negative bands. This would imply that in the banded chromosome, but not in the fixed unbanded chromosome, there would be an uneven distribution of chromatin corresponding to the banding pattern; this point has apparently not been tested. A serious objection to the idea of an equivalence between coils and bands or any other subdivision of chromosomes is that the interface between adjacent coils must necessarily be at an acute angle to the axis of the chromatid, while the interfaces between bands, or between chromomeres, are perpendicular to the axis, as are the circumferential grooves. It must be concluded that at present the relationships between bands and the various structural features of chromosomes that have been described are as yet very imperfectly understood, and indeed will continue to be until an unambiguous account of chromosome structure is available.

16.3.3 Taxonomic distribution of bands, and its implications

It has already been remarked that euchromatic bands are rare or absent in lower vertebrates, invertebrates and plants. In fact, the situation, as summarized in Table 16.1, is quite complicated, with different types of banding being present or absent in different groups. Consideration of the distribution of the different types of bands should help to illuminate the relationships between them, and to solve some of the problems raised by the consideration of the situation in mammalian chromosomes, considered in Sections 16.3.1 and 16.3.2.

Three main points emerge from Table 16.1. First, almost all eukaryotes apparently have chromomeres in their pachytene chromosomes, and rep-

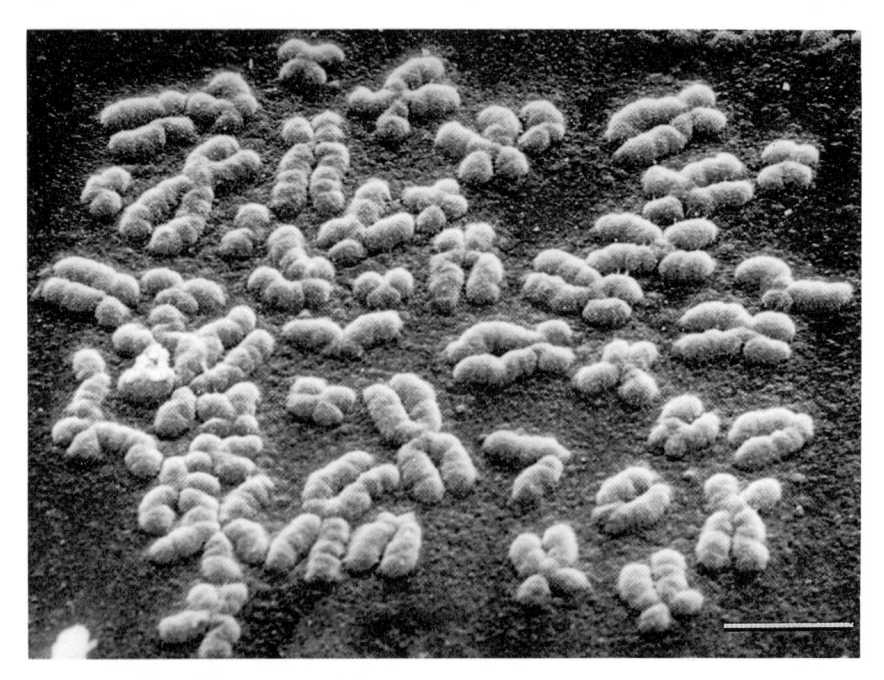

Figure 16.4 Scanning electron micrograph showing 'circumferential grooves' on human chromosomes. The bar represents 5 μm.

Table 16.1 Distribution of different types of banding in different groups of eukaryotes.

	Pachytene chromomeres	Replication bands	G-bands	Base-specific fluorochromes		G+C-rich isochores
				Q-bands (A+T-rich DNA)	Chromomycin R-bands (G+C-rich DNA)	
Mammals	Yes	Yes	Good	Good	Good	Yes (1)
Birds	Yes (2)	Yes (3)	Good	Moderate(4)	Moderate (4)	Yes (1)
Reptiles	Yes (5)	Yes (6)	Good	Poor (4)	Poor or absent (4)	Poor (7)
Amphibia	Yes (8)	Yes (9,10)	No	No (4)	No (4)	No (7)
Xenopus			Good (11)			No (7)
Fish	Yes (12)	Yes (13)	No	No (4,16)	No (4)	No (14)
Eels			Good (15)	Poor (16)	No	Yes (16)
Thermophilic spp.						Yes (17)
Insects						
Drosophila	Yes (18)		No	No	No	
Plants			No	No	No	No (7)
Monocotyledons	Yes (18)	Yes (19)	No	No	No	Yes (20)
Dicotyledons	Yes (18)	Yes (21)	No	No	No	No (20)

References
1 Cuny et al. 1981
2 Stahl et al. 1974
3 Carlenius et al. 1981
4 Schmid & Guttenbach 1988
5 Bull 1978
6 Yonenaga-Yassuda et al. 1988
7 Thiery et al. 1976
8 Sessions & Kezer 1987
9 Cuny & Malacinski 1985
10 Schempp & Schmid 1981
11 Stock & Mengden 1975
12 Schmid et al. 1982
13 Delany & Bloom 1984
14 Hudson et al. 1980
15 Wiberg 1983
16 Medrano et al. 1988
17 Bernardi & Bernardi 1986
18 Lima-de-Faria 1975
19 Cortes & Escalza 1986
20 Salinas et al. 1988
21 Schweizer et al. 1990

lication bands are almost universal also. Second, among vertebrates, there is a good correlation between the quality of banding with base-specific fluorochromes, and the presence or absence of G+C-rich isochores. Third, the distribution of G-banding is neither as extensive as that of chromomeres and replication banding, nor as limited as the distribution of banding with base-specific fluorochromes. Inevitably, all these points need some qualification. Obviously the taxonomic distribution of replication bands is, at present, based on a very restricted number of organisms, and the assumption that they may be almost universal comes from the finding that, wherever appropriate technical procedures have been worked out, good quality, reproducible banding has been obtained. Regarding isochores and base-specific banding, it must be noted that the correlation only holds good for vertebrates. G+C-rich isochores have now been reported for monocotyledonous plants (Salinas *et al.* 1988), which are generally regarded as lacking euchromatic bands in their chromosomes. The weight of evidence seems to indicate that this absence of banding is genuine, although it could be due to purely technical factors (Section 5.3), or to the high degree of condensation reported for certain plant chromosomes (Greilhuber 1977b). On the other hand, it is worth recalling the apparently exceptional finding of Q-bands throughout the length of the chromosomes in *Lilium* spp. (Holm 1976, Kongsuwan & Smyth 1977, 1980). It may indeed turn out that the correlation between isochores and base-specific banding will hold for plants also, if only the technical details of chromosome preparation and staining can be optimized. Finally, the distribution of G-banding given in Table 16.1 must also be regarded as tentative. Reports are continually published of G-banding in plants (Section 5.3), but the quality is frequently poor, and repeatable patterns of banding do not seem to be obtained. It is, nevertheless, possible that good quality G-banding will some day be obtained in organisms other than those listed in Table 16.1, although the continued passage of time without such reports makes it unlikely that G-banding is much more widely distributed.

Whereas in mammals there is good correlation between all the types of euchromatic banding, it is clear from Table 16.1 that there are, in fact, several different independent classes of euchromatic bands, which nevertheless tend to have, presumably for important functional reasons, similar distributions. Evidently the forms of longitudinal differentiation of the euchromatic parts of chromosomes that evolved first were the pachytene chromomeres and the replication bands. Next must have been the G-bands, which may well have arisen independently on several occasions, judging from their patchy distribution in lower vertebrates (see also Holmquist 1988a). Finally came the differentiation by DNA base composition, manifested as isochores at the molecular level, and as base-specific banding at the cytogenetic level. Again, this type of differentiation has probably evolved independently more than once, in monocotyledons (Salinas *et al.* 1988), in certain fish (Bernardi & Bernardi 1986, Medrano *et al.* 1988), as well as in higher vertebrates.

The different types of banding have presumably evolved in response to a variety of selective pressures. Evidently the same selective pressures have not acted on all eukaryotes, or if they have, the chromosomes of different groups have responded in different ways. At present we have few clues as to what the selective pressures might have been, or indeed as to why the euchromatin should be almost universally divided into two fractions. In any case, it appears that, whatever the cause, new types of longitudinal differentiation can only be superimposed on the existing basic pattern, that of the pachytene chromomeres or of replication.

16.3.4 Some speculations on the significance of euchromatic banding

A proper understanding of the significance of euchromatic banding must include an explanation of why the euchromatin should be divided into two fractions at all. It is difficult to imagine that the mitotic metaphase chromosome requires such a division; indeed, it is a remarkably homogeneous body, in which, when fully contracted, banding is difficult to discern. It, therefore, seems more likely that euchromatic banding is a reflection of some requirement of interphase chromatin organization; indeed, replication banding is in fact an expression during mitosis of a purely interphase phenomenon, that of DNA replication. Note, however, that the chromomeric organization of meiotic prophase chromosomes could be due to a requirement to retain some chromosomal segments in an extended state to facilitate crossing-over.

If euchromatic banding is a consequence of some interphase requirements, then what are the most plausible forms of differentiation during interphase that might appear as bands in chromosomes? Cavalier-Smith (1978) has proposed that euchromatic DNA consists of genic DNA (G-DNA), which is that fraction which contains the genes, and nucleoskeletal DNA (S-DNA), whose primary function is to regulate nuclear size, and thereby affect cell size, cell cycle time, the growth rate of the organism and several other parameters. (A full account of the relationship between the quantity of DNA per nucleus, and many other parameters, while fascinating, is not immediately relevant here, but the subject is reviewed by Cavalier-Smith (1978, 1985)). It is postulated that the amount of skeletal (S-) DNA is determined by that type of cell which requires the largest nucleus; if a tissue requires smaller nuclei, then the surplus S-DNA is condensed (Cavalier-Smith 1985). Here, then, is a basic division of the euchromatic genome into a transcriptionally active, uncondensed fraction, and an inactive fraction showing a strong tendency to condense, each fraction showing, in fact, properties in common with negative and positive G-bands respectively. Chromomeres would, therefore, be a reflection of the interphase state, and are found in prophase chromosomes, first because such segments of chromatin have a natural tendency to condense, and second because they are not required for transcription, and so can be condensed into a recognizable chromosome at the earliest opportunity. Thuriaux (1977) has argued that meiotic crossing-over is restricted to the vicinity of the genes; in other words, according to the present proposals, to the gene-rich, interchromomeric DNA, which might well need to remain in an uncondensed state to permit crossing-over. This would explain why chromomeres are much more prominent at meiotic than at mitotic prophase. Note also that Chandley (1986) has proposed that meiotic pairing and recombination occur at early replicating sites along chromosomes, which correspond to the interchromomeric regions. It is also worth noting that chromosome breakage, due to a variety of causes, is concentrated in negative G- or Q-bands (Morad *et al.* 1973, San Roman & Bobrow 1973, Buckton 1976, Nakagome & Chiyo 1976); these would correspond to the interchromomeric G-DNA, which is uncondensed in interphase, and might therefore be more difficult to repair because of the ease with which the broken ends of the DNA molecule could drift apart. A prediction of this hypothesis would be that in closely related species having different amounts of DNA, the difference would be largely in the S-DNA, manifest as chromomeres in meiotic prophase. Good quantitative comparisons of this sort are not known to the author, and an added complication is the changes in number and size of the chromomeres that occur during chromosome condensation.

It is not entirely clear why asynchronous replication should occur in chromosomal DNA; that is, why there should be distinct early and late replicating segments. Indeed, it is known that when extremely rapid cell multiplication is required, DNA synthesis can occur in all chromosomal segments simultaneously. There is, nevertheless, strong evidence that late replication is associated with transcriptional inactivity, which is also associated with chromatin condensation (Goldman 1988). Various reasons can be suggested for the development of asynchronous replication. It might be that the enzymes required for replication are not available in sufficient quantity to replicate the whole genome simultaneously. Alternatively, the condensation of part of the genome might inhibit replication due to purely mechanical factors. Neither of these explanations appears to offer the quite precise control of replication time actually seen. It may be that if control of nuclear volume is critical, it would be necessary to delay replication of S-DNA as long as possible, to maintain the optimum size of the nucleus. On the other hand, Goldman (1988) has proposed that replication early in S phase results in assembly of chromatin in a state that is accessible to the transcriptional machinery. Genes in early replicating chromatin are not necessarily transcribed, but late replicating genes would never be transcribable. Comparison of the replication times of mammalian genes (Goldman 1988) shows that early replicating genes are normally expressed (38 of 41), whereas late replicating genes are never expressed (15 of 15). It may well be that chromatin replicated at different stages of the cell cycle is in different states, so that not only would early replication result in a transcriptionally competent form of chromatin, but that late replication would necessarily produce a condensed, transcriptionally incompetent form of chromatin. The close relationship between replication time and condensation would then be a causal one, and the two would have evolved together early in the evolution of eukaryotes. Of course, if it is necessary to replicate early for a gene to be expressed, certain genes must be able to switch from late to early replication. This need not affect patterns of condensation or replication at the chromosomal level, although small differences in replication patterns between different tissues have in fact been reported (Sheldon & Nichols 1981a, b). It should also be noted that if different cell types decondense their S-DNA to different extents, depending on the nuclear volume required, such decondensation could also be associated with the activation of genes specific for that cell type.

If the basic types of euchromatic banding, chromomeres and replication banding, are so closely connected, and concerned with various aspects of control of gene expression, why should any other type of euchromatic banding have evolved? G-banding remains essentially unexplained, since although there is clearly some connection with chromosome condensation, it has a much more limited taxonomic distribution than chromomeres. If G-banding is simply an expression of condensation, then it must be some aspect of condensation that is far from universal. Possibly those organisms that have G-bands need to stabilize their chromosome condensation in some more secure way than that required by most organisms. Perhaps there is a parallel here with the stabilization of sperm heads by disulphide bonds in higher vertebrates and certain other organisms. However, until the true mechanism of G-banding is elucidated, further speculation would be unprofitable. It should be noted, however, that in spite of the suggestion by van Duijn *et al.* (1985) G-banding does not appear to be a consequence of variation in DNA base composition along the chromosome; excellent G-banding can be produced in lower vertebrates such as *Xenopus* and reptiles, in which there is no large scale variation in base composition (Table 16.1). G-banding and base-specific banding are evidently independent phenomena, even if their patterns are similar. Perhaps their patterns are less similar than is generally supposed. In prophase, there are numerous G-bands which grow and fuse together as the chromosomes contract, until there is a very small number of bands in the fully contracted chromosome; indeed, I have argued (Sumner 1976) that a fully contracted chromosome consists of a single G-band. On the other hand, base composition of chromosomes cannot change during the cell cycle, and thus $G+C$-rich and $A+T$-rich bands are fixed and unchanging.

This brings us to the question of the reason for the origin of isochores and base-specific banding. Bernardi & Bernardi (1986) have argued that the evolution of $G+C$-rich isochores confers greater thermal stability on those stretches of DNA in warm-blooded vertebrates; remarkably, proteins coded by $G+C$-rich genes also show greater thermal stability. This explanation appears sound as far as it goes, and the prediction that any organism living at elevated temperatures might possess $G+C$-rich isochores has been confirmed, at least for certain thermophilic fish (Bernardi & Bernardi 1986), although it has not yet been shown that such species also have base-specific banding. However, it does not explain why portions of the genome remain in the less stable, A + T-rich form. All the evidence indicates that such regions are poor in genes; moreover, such genes are generally tissue-specific and can, therefore, be maintained in a more stable condensed state for much of the time. Indeed, it could be that it was the decondensed state of chromatin containing active genes that necessitated the evolution of G + C-rich, gene-rich isochores in warm-blooded species. Thus the similarity in pattern between chromomere distribution and base-specific banding would be explained.

The existence of chromosome banding, and of euchromatic banding in particular, has been one of the most remarkable discoveries in cytogenetics and molecular biology in recent years. The occurrence of several distinct, although related, types of euchromatic banding is not so widely realized, most authors apparently regarding them all as one, perhaps lumped together as G-bands. Not until the distinctions between the different types of bands are clearly recognized can banding be not merely a vital tool for cytogeneticists, but also the way towards understanding the organization of chromosomes.

References

Abraham, R., Weinberg, J. & Schnedl W. 1983. Analysis of heterochromatin and identification of mitotic chromosomes in *Drosophila virilis* by GC- and AT-specific fluorochromes. *Mikroskopie* **40**, 4–8.

Adams, R. L. P., Burdon, R. H. & Fulton, J. 1983. Methylation of satellite DNA. *Biochemical and Biophysical Research Communications* **113**, 695–702.

Adhvaryu, S. G., Dave, B. J., Trivedi, A. H., Jani, K. H. & Vyas, R. C. 1987. Heteromorphism of C-band positive chromosomal regions in CML patients. *Cancer Genetics and Cytogenetics* **27**, 33–8.

Adkisson, K. P., Perreault, W. J. & Gay, H. 1971. Differential fluorescent staining of *Drosophila* chromosomes with quinacrine mustard. *Chromosoma* **34**, 190–205.

Adolph, K. W. 1981. A serial sectioning study of the structure of human mitotic chromosomes. *European Journal of Cell Biology* **24**, 146–53.

Adolph, K. W. 1988. *Chromosomes and chromatin*. 3 Vols. Boca Raton: CRC Press.

Adolph, K. W., Cheng, S. M., Paulson, J. R. & Laemmli, U. K. 1977. Isolation of a protein scaffold from mitotic HeLa cell chromosomes. *Proceedings of the National Academy of Sciences* **74**, 4937–41.

Adolph, S. 1988. *In situ* nick translation distinguishes between C-band positive regions on mouse chromosomes. *Chromosoma* **96**, 102–6.

Adolph, S. & Hameister, H. 1985. *In situ* nick translation of metaphase chromosomes with biotin-labelled d-UTP. *Human Genetics* **69**, 117–21.

Aghamohammadi, S. Z. & Savage, J. R. K. 1990. BrdU pulse/reverse staining protocols for investigating chromosome replication. *Chromosoma* **99**, 76–82.

Ahnström, G. & Natarajan, A. T. 1974. Localization of repetitive DNA in mouse cells by *in situ* hybridization and dye binding techniques. *Hereditas* **76**, 316–20.

Aleixandre, C., Miller, D. A., Mitchell, A. R., Warburton, D. A., Gerson, S. L., Disteche, C. & Miller, O. J. 1987. p82H identifies sequences at every human centromere. *Human Genetics* **77**, 46–50.

Alfi, O. S., Donnell, G. N. & Derencsenyi, A. 1973. C-banding of human chromosomes produced by DNase. *Lancet* **2**, 505.

Alfi, O. S. & Menon, R. 1973. A rapid C-band staining technique for chromosomes. *Journal of Laboratory Clinical Medicine* **82**, 692–4.

Alhadeff, B., Velivasakis, M. & Siniscalco, M. 1977. Simultaneous identification of chromatid replication and of human chromosomes in metaphase of man-mouse somatic cell hybrids. *Cytogenetics and Cell Genetics* **19**, 236–9.

Alitalo, K., Schwab, M., Lin, C. C., Varmus, H. E. & Bishop, J. M. 1983. Homogeneously staining chromosomal regions contain amplified copies of an abundantly expressed cellular oncogene (c-myc) in malignant neuroendocrine cells from a human colon carcinoma. *Proceedings of the National Academy of Sciences* **80**, 1707–11.

Allen, J. W., Liang, J. C., Carrano, A. V. & Preston, R. J. 1986. Review of literature on chemical-induced aneuploidy in mammalian male germ cells. *Mutation Research* **167**, 123–37.

Ambros, P. F. & Schweizer, D. 1987. Non-radioactive *in situ* hybridization and chromosome banding: methods and applications. *Annales of University Sarav. Med.* suppl. **7**, 1–6.

Ambros, P. F. & Sumner, A. T. 1987. Correlation of pachytene chromomeres and metaphase bands of human chromosomes, and distinctive properties of telomeric regions. *Cytogenetics and Cell Genetics.* **44**, 223–8.

Amemiya, C. T. & Gold, J. R. 1987. Chromomycin staining of vertebrate chromosomes: enchancement of banding patterns by NaOH. *Cytobios* **49**, 147–52.

Andersen, O. & Rønne, M. 1978. Effect of histidinol and parafluorophenylalanine on metaphase chromosome structure in human lymphoid cells. *Hereditas* **88**, 197–201.

Andersen, O. & Rønne, M. 1983. Quantitation of spindle-inhibiting effects of metal compounds by chromosome length measurements. *Hereditas* **98**, 215–8.

Andersen, O., Rønne, M. & Nordberg, G. F. 1983. Effects of inorganic metal salts on chromosome length in human lymphocytes. *Hereditas* **98**, 65–70.

Anderson, L. K., Stack, S. M. & Mitchell, J. B. 1982. An investigation of the basis of a current hypothesis for the lack of G-banding in plant chromosomes. *Experimental Cell Research* **138**, 433–6.

Andrulis, I. L., Duff, C., Evans-Blackler, S., Worton, R. & Siminovitch, L. 1983. Chromosomal alterations associated with overproduction of asparagine synthetase in albizzin-resistant Chinese hamster ovary cells. *Molecular and Cellular Biology* **3**, 391–8.

Angelier, N., Hernandez-Verdun, D. & Bouteille, M. 1982. Visualization of Ag-NOR proteins on nucleolar transcriptional units in molecular spreads. *Chromosoma* **86**, 661–72.

Angell, R. R. & Jacobs, P. A. 1975. Lateral asymmetry in human constitutive heterochromatin. *Chromosoma* **51**, 301–10.

Angell, R. R. & Jacobs, P. A. 1978. Lateral asymmetry in human constitutive heterochromatin: frequency and inheritance. *American Journal of Human Genetics* **30**, 144–52.

Anon. 1987. NORs – a new method for the pathologist. *Lancet* **1**, 1413–4.

Ansari, H. A. & Kaul, D. 1979. Inversion polymorphism in common green pigeon, *Treron phoenicoptera* (Latham) (Aves). *Japanese Journal of Genetics* **54**, 197–202.

Ansari, H. A., Takagi, N. & Sasaki, M. 1988. Morphological differentiation of sex chromosomes in three species of ratite birds. *Cytogenetics and Cell Genetics* **47**, 185–8.

Appels, R., Driscoll, C. & Peacock, W. J. 1978. Heterochromatin and highly repeated DNA sequences in rye (*Secale cereale*). *Chromosoma* **70**, 67–89.

Archidiacono, N., de Capoa, A., Ferraro, M., Pelliccia, F., Rocchi, A. & Rocchi, M. 1977. Nucleolus organizer and N-band distribution in morphologic and fluorescence variants of human chromosomes. *Human Genetics* **37**, 285–9.

Armada, J. L. & Seuanez, H. N. 1984. Late DNA replication in rhesus monkey (*Macaca mulatta*) chromosomes. *Cytobios* **41**, 95–103.

Armstrong, K. C. 1982. N-banding in *Triticum aestivum* following Feulgen hydrolysis. *Theoretical and Applied Genetics* **61**, 337–9.

Árnason, Ú 1974a. Comparative chromosome studies in Pinnepedia. *Hereditas* **76**, 179–226.

Árnason, Ú. 1974b. Comparative chromosome studies in Cetacea. *Hereditas* **77**, 1–36.

Árnason, Ú. 1977. The relationship between the four principal pinniped karyotypes. *Hereditas* **87**, 227–42.

Árnason, Ú. 1980. C- and G-banded karyotypes of three delphinids: *Stenella clymene, Lagenorhynchus albirostris* and *Phocoena phocoena*. *Hereditas* **92**, 179–87.

Árnason, Ú., Benirschke, K., Mead, J. G. & Nichols, W. W. 1977. Banded karyotypes of three whales: *Mesoplodon europaeus, M. carlhubbsi* and *Balaenoptera acutorostrata*. *Hereditas* **87**, 189–200.

Árnason, Ú., Purdom, I. F. & Jones, K. W. 1978. Conservation and chromosomal localization of DNA satellites in Balaenopterid whales. *Chromosoma* **66**, 141–59.

Arndt-Jovin, D. J., Robert-Nicoud, M., Baurschmidt, P. & Jovin, T. M. 1985. Immunofluorescence localization of Z-DNA in chromosomes: quantitation by scanning microphotometry and computer-assisted image analysis. *Journal of Cell Biology* **101**, 1422–33.

Arnold, M. L. & Shaw, D. D. 1985. The heterochromatin of grasshoppers from the *Caledia captiva* species complex. II. Cytological organization of tandemly repeated DNA sequences. *Chromosoma* **93**, 183–90.

Arrighi, F. E. & Hsu, T. C. 1965. Experimental alteration of metaphase chromosome morphology. Effect of actinomycin D. *Experimental Cell Research* **39**, 305–8.
Arrighi, F. E. & Hsu, T. C. 1971. Localization of heterochromatin in human chromosomes. *Cytogenetics* **10**, 81–6.
Arrighi, F. E., Hsu, T. C., Pathak, S. & Sawada, H. 1974. The sex chromosomes of the Chinese hamster: constitutive heterochromatin deficient in repetitive DNA sequences. *Cytogenetics and Cell Genetics* **13**, 268–74.
Arrighi, F. E., Hsu, T. C., Saunders, P. & Saunders, G. F. 1970. Localization of repetitive DNA in the chromosomes of *Microtus agrestis* by means of *in situ* hybridization. *Chromosoma* **32**, 224–36.
Arruga, M. V. & Monteaguido, L. V. 1989. Evidence of Mendelian inheritance of the nucleolar organizer regions in the Spanish common rabbit. *Journal of Heredity* **80**, 85–6.
Atkin, N. B. & Brito-Babapulle, V. 1981. Heterochromatin polymorphism and human cancer. *Cancer Genetics and Cytogenetics* **3**, 261–72.
Atkin, N. B. & Brito-Babapulle, V. 1985. Chromosome 1 heterochromatin variants and cancer: a reassessment. *Cancer Genetics and Cytogenetics* **18**, 325–331.
Attardi, G. & Amaldi, F. 1970. Structure and synthesis of ribosomal RNA. *Annual Review of Biochemistry* **39**, 183–226.
Attia, T. & Lelley, T. 1987. Effects of constitutive heterochromatin and genotype on frequency and distribution of chiasmata in the seven individual rye bivalents. *Theoretical and Applied Genetics* **74**, 527–30.
Ayer, L. M. & Fritzler, M. J. 1984. Anticentromere antibodies bind to trout testis histone 1 and a low molecular weight protein from rabbit thymus. *Molecular Immunology* **21**, 761–70.

Babu, A. 1988. Heterogeneity of heterochromatin of human chromosomes as demonstrated by restriction endonuclease treatment. In *Heterochromatin*, R. S. Verma (ed.), 250–75. Cambridge: Cambridge University Press.
Babu, A. & Verma, R. S. 1986a. Characterization of human chromosomal constitutive heterochromatin. *Canadian Journal of Genetics and Cytology* **28**, 631–44.
Babu, A. & Verma, R. S. 1986b. Cytochemical heterogeneity of the C-band in human chromosome 1. *Histochemical Journal* **18**, 329–33.
Babu, A. & Verma, R. S. 1986c. Expression of heterochromatin by restriction endonuclease treatment and distamycin A/DAPI staining of Indian muntjac (*Muntiacus muntjak*) chromosomes. *Cytogenetics and Cell Genetics* **41**, 96–100.
Babu, A. & Verma, R. S. 1986d. Heteromorphic variants of human chromosome 4. *Cytogenetics and Cell Genetics* **41**, 60–1.
Babu, A., Agarwal, A. K. & Verma, R. S. 1988. A new approach in recognition of heterochromatic regions of human chromosomes by means of restriction endonucleases. *American Journal of Human Genetics* **42**, 60–5.
Babu, A., Verma, R. S. & Patil, S. R. 1987. *Alu* I-resistant chromatin of chromosome 18: classification, frequencies and implications. *Chromosoma* **95**, 163–6.
Bahr, G. F. & Larsen, P. M. 1974. Structural 'bands' in human chromosomes. *Advances in Cell and Molecular Biology* **3**, 191–212.
Bahr, G. F., Mikel, U. & Engler, W. F. 1973. Correlates of chromosomal banding at the level of ultrastructure. In *Chromosome identification. Nobel Symposium 23*, T. Caspersson & L. Zech (eds), 280–9. New York: Academic Press.
Baimai, V. & Traipakvasin, A. 1987. Intraspecific variation in sex heterochromatin of species B of the *Anopheles dirus* complex in Thailand. *Genome* **29**, 401–4.
Baker, J. R. 1958. *Principles of biological microtechnique*. London: Methuen.
Baker, R. J., Qumsiyeh, M. B. & Rautenbach, I. L. 1988. Evidence for eight tandem and five centric fusions in the evolution of the karyotype of *Aethomys namaquensis* A. Smith (Rodentia: Muridae). *Genetica* **76**, 161–9.
Balaban-Malenbaum, G. & Gilbert, F. 1980. The proposed origin of double minutes from

homogeneously staining region (HSR)-marker chromosomes in human neuroblastoma hybrid cell lines. *Cancer Genetics and Cytogenetics* **2**, 339–48.

Balczon, R. D. & Brinkley, B. R. 1987. Tubulin interaction with kinetochore proteins: analysis by *in vitro* assembly and chemical cross-linking. *Journal of Cell Biology* **105**, 855–62.

Baldini, G., Doglia, S., Dolci, S. & Sassi, G. 1981. Fluorescence-determined preferential binding of quinacrine to DNA. *Biophysical Journal* **36**, 465–77.

Baldini, A., Felli, M. P., Ravenna, L., Camaioni, A., de Capoa, A., Spadoni, G. L., Cianfarani, S. & Boscherini, B. 1988. Differential ribosomal gene responsiveness to human growth hormone is visualized by selective silver staining. *Cytogenetics and Cell Genetics* **47**, 22–5.

Balíček, P., Žižka, J. & Skalská, H. 1977. Length of human constitutive heterochromatin in relation to chromosomal contraction. *Human Genetics* **38**, 189–93.

Balíček, P., Žižka, J. & Skalská, H. 1978. Variability and familial transmission of constitutive heterochromatin of human chromosomes evaluated by the method of linear measurement. *Human Genetics* **42**, 257–65.

Baranovskaya, L. T., Zakharov, A. F., Dutrillaux, B., Carpentier, S., Prieur, M. & Lejeune, J. 1972. Différenciation des chromosomes X par les méthodes de déspiralisation au 5 bromodéoxyuridine (BUDR) et de dénaturation thermique ménagée. *Annales de Génétique* **15**, 271–4.

Barker, P. E., Lau, Y-F. & Hsu, T. C. 1980. A heterochromatic homogenously staining region (HSR) in the karyotype of a human breast carcinoma cell line. *Cancer Genetics and Cytogenetics* **1**, 311–19.

Barker, P. E., Mohandas, T. & Kanack, M. M. 1977. Chromosome polymorphisms in karyotypes from amniotic fluid cell cultures. *Clinical Genetics* **11**, 243–8.

Barlow, P. & Vosa, C. G. 1970. The Y chromosome in human spermatozoa. *Nature* **226**, 961–2.

Barnes, I. C. S. & Maltby, E. L. 1986. Prometaphase chromosome analysis as a routine diagnostic technique. *Clinical Genetics* **29**, 378–83.

Barnes, S. R., James, A. M., & Jamieson, G. 1985. The organization, nucleotide sequence, and chromosomal distribution of a satellite DNA from *Allium cepa*. *Chromosoma* **92**, 185–192.

Barr, H. J. & Ellison, J. R. 1971. Quinacrine staining of chromosomes and evolutionary studies in *Drosophila*. *Nature* **233**, 190–1.

Barros, M. A. & Patton, J. L. 1985. Genome evolution in pocket gophers (genus *Thomomys*). III. Fluorochrome-revealed heterochromatin heterogeneity. *Chromosoma* **92**, 337–43.

Barsacchi-Pilone, G., Batistoni, R., Andronico, F., Vitelli, L. & Nardi, I. 1986. Heterochromatic DNA in *Triturus* (Amphibia, Urodela). I. A satellite DNA component of the pericentric C-bands. *Chromosoma* **93**, 435–46.

Barsanti, P., Marazia, T. & Maggini, F. 1981. Q-banding and A-T rich DNA in *Ornithogalum montanum* (Liliaceae). *Experientia* **37**, 467–8.

Bastide, P., de Rocca Serra, M. F., Fellmann, N., Jaffray, J. Y., Geneix, A., Malet, P. & Turchini, J. P. 1981. Action de divers agents pharmacologiques sur les structures nucleaires et chromosomiques de lymphocytes humains. *Cytologia* **46**, 387–91.

Bath, D. W. & Gendel, B. R. 1973. Giemsa banding of meiotic chromosomes. *Lancet* **2**, 455.

Baumann, T. W. 1971. Heterochromatin und DNS-Replikation bei *Scilla sibirica*. *Expermiental Cell Research* **64**, 323–30.

Baverstock, P. R., Watts, C. H. S., Gelder, M. & Jahnke, A. 1983. G-banding homologies of some Australian rodents. *Genetica* **60**, 105–17.

Beatty, R. A. 1977. F-bodies as Y chromosome markers in mature human sperm heads: a quantitative approach. *Cytogenetics and Cell Genetics* **18**, 33–49.

Beçak, M. L., dos Santos, R. de C. S., Soares-Scott, M. D., Batistic, R. F. & Costa, H. 1988. Chromosome structure in man and Amphibia-Anura, restriction enzymes. *Rev. Brasil. Genet.* **11**, 939–48.

Beermann, S. 1977. The diminution of heterochromatic chromosomal segments in *Cyclops* (Crustacea, Copepoda). *Chromosoma* **60**, 297–344.

Behr, W., Honikel, K. & Hartmann, G. 1969. Interaction of the RNA polymerase inhibitor chromomycin with DNA. *European Journal of Biochemistry* **9**, 82–92.

Beil, B. & Limon, J. 1975. The effect of cacodylate buffer on morphological differentiation of human chromosomes. *Genet. Polonica* **16**, 117–22.

Beltran, I. C., Robertson, F. W. & Page, B. M. 1979. Human Y chromosome variation in normal and abnormal babies and their fathers. *Annals of Human Genetics* **42**, 315–25.

Bennett, M. D. 1985. Intraspecific variation in DNA amount and the nucleotypic dimension in plant genetics. In *Plant genetics*, M. Freeling (ed.), 283–302. New York: Alan R. Liss.

Bennett, M. D., Gustafson, J. P. & Smith, J. B. 1977. Variation in nuclear DNA in the genus *Secale*. *Chromosoma* **61**, 149–76.

Bennett, M.D., Heslop-Harrison, J. S., Smith, J. B. & Ward, J. P. 1983. DNA density in mitotic and meiotic metaphase chromosomes of plants and animals. *Journal of Cell Science* **63**, 173–9.

Bentzer, B. & Landstrom, T. 1975. Polymorphism in chromosomes of *Leopoldia comosa* (Liliaceae) revealed by Giemsa staining. *Hereditas* **80**, 219–32.

Berger, R. 1972a. Effet de l'hydroxyde de sodium sur les chromosomes. *Comptes rendus de l'Academie des Sciences (de Paris* D), **275**, 1511–12.

Berger, R. 1972b. Étude du caryotype du porc avec une nouvelle technique. *Experimental Cell Research* **75**, 298–300.

Berger, R., Bernheim, A. & Schaison, G. 1981. Discrepancy between G and R bands. Example of an acute non-lymphocytic leukaemia. *Human Genetics* **59**, 84–6.

Berger, R., Bernheim, A., Kristoffersson, U., Mineur, A. & Mitelman, F. 1983. Differences in human C-band pattern between two European populations. *Hereditas* **99**, 147–9.

Berger, R., Bernheim, A., Kristoffersson, U., Mitelman, F. & Olsson, H. 1985. C-band heteromorphism in breast cancer patients. *Cancer Genetics and Cytogenetics* **18**, 37–42.

Beridze, T. 1986. *Satellite DNA*. Berlin: Springer.

Berman, E., Brown, S. C., James, T. L. & Shafer, R. H. 1985. NMR studies of chromomycin A_3 interaction with DNA. *Biochemistry* **24**, 6887–93.

Bernardi, G. & Bernardi, G. 1986. Compositional constraints and genome evolution. *Journal of Molecular Evolution* **24**, 1–11.

Bernardi, G., Olofsson, B., Filipski, J., Zerial, M., Salinas, J., Cuny, G., Meunier-Rotival, M. & Rodier, F. 1985. The mosaic genome of warm-blooded vertebrates. *Science* **228**, 953–8.

Bernheim, A. & Berger, R. 1981. A simple method for improving the reproducibility of the R-banding technique. *Human Genetics* **57**, 432–3.

Bernheim, A., Berger, R. & Szabo, P. 1984. Localization of actin-related sequences by *in situ* hybridization to R-banded human chromosomes. *Chromosoma* **89**, 163–7.

Bhasin, M. K. & Foerster, W. 1972. A simple banding technique for identification of human metaphase chromosomes. *Humangenetik* **14**, 247–50.

Bianchi, M. S., Bianchi, N. O., Pantelias, G. E. & Wolff, S. 1985. The mechanism and pattern of banding induced by restriction endonucleases in human chromosomes. *Chromosoma* **91**, 131–6.

Bianchi, N. O., Bianchi, M. S. & Cleaver, J. E. 1984. The action of ultraviolet light on the patterns of banding induced by restriction endonucleases in human chromosomes,. *Chromosoma* **90**, 133–8.

Bianchi, N. O, Vidal-Rioja, L. & Cleaver, J. E. 1986. Direct visualization of the sites of DNA methylation in human, and mosquito chromosomes. *Chromosoma* **94**, 362–6.

Bianchi, N. O, Bianchi, M. S., Cleaver, J. E. & Wolff, S. 1985. The pattern of restriction enzyme-induced banding in the chromosomes of chimpanzee, gorilla, and orangutan and its evolutionary significance. *Journal of Molecular Evolution* **22**, 323–33.

Bickham, J. W. 1979a. Banded karyotypes of 11 species of American bats (genus *Myotis*). *Cytologia* **44**, 789–97.

Bickham, J. W. 1979b. Chromosomal variation and evolutionary relationships of vespertilionid bats. *Journal of Mammalogy* **60**, 350–63.

Bickham, J. W. 1981. Two-hundred-million-year-old chromosomes: deceleration of the rate of karyotypic evolution in turtles. *Science* **212**, 1291–3.

Bickmore, W. A. & Sumner, A. T. 1989. Mammalian chromosome banding – an expression of genome organization. *Trends in Genetics* **5**, 144–8.

Biedler, J. L. 1982. Evidence for transient or prolonged extrachromosomal existence of amplified DNA sequences in antifolate-resistant, vincristine-resistant, and human neuroblastoma cells. In *Gene amplification*, R. T. Schimke (ed.), 39–45. New York: Cold Spring Harbor Laboratory.

Biedler, J. L. & Spengler, B. A. 1976. Metaphase chromosome anomaly: association with drug resistance and cell-specific products. *Science* **191**, 185–7.

Biedler, J. L., Melera, P. W. & Spengler, B. A. 1980. Specifically altered metaphase chromosomes in antifolate-resistant Chinese hamster cells that overproduce dihydrofolate reductase. *Cancer Genetics and Cytogenetics* **2**, 47–60.

Bigger, T. R. L. 1975. Karyotypes of some Lepidoptera chromosomes and changes in their holokinetic organization as revealed by new cytological techniques. *Cytologia* **40**, 713–26.

Bigger, T. R. L. & Savage, J. R. K. 1975. Mapping G-bands on human prophase chromosomes. *Cytogenetics and Cell Genetics* **15**, 112–21.

Bignone, F. A., Panarello, C. & Gimelli, G. 1983. G-bands without pretreatment of slides, in chemically defined conditions. *Human Genetics* **63**, 63–6.

Bird, A. P. 1986. CpG-rich islands and the function of DNA methylation. *Nature* **321**, 209–13.

Bird, A. P. & Taggart, M. H. 1980. Variable patterns of total DNA and rDNA methylation in animals. *Nucleic Acids Research* **8**, 1485–97.

Birnstiel, M. L., Chipchase, M. & Speirs, J. 1971. The ribosomal RNA cistrons. *Progress in Nucleic Acid Research and Molecular Biology* **11**, 351–89.

Biro, P. A., Carr-Brown, A., Southern, E. M. & Walker P. M. B. 1975. Partial sequence analysis of mouse satellite DNA: evidence for short range periodicities. *Journal of Molecular Biology* **94**, 71–86.

Birstein, V. J. 1982. Structural characteristics of genome organization in Amphibians: differential staining of chromosomes and DNA structure. *Journal of Molecular Evolution* **18**, 73–91.

Bishop, D. T., Williamson, J. A. & Skolnick, M. H. 1983. A model for restriction fragment length distributions. *American Journal of Human Genetics* **35**, 795–815.

Black, M. M. & Ansley, H. R. 1964. Histone staining with ammoniacal silver. *Science* **143**, 693–5.

Black, M. M. & Ansley, H. R. 1966. Histone specificity revealed by ammoniacal silver staining. *Journal of Histochemistry and Cytochemistry* **14**, 177–81.

Blackburn, E. H. & Szostak, J. W. 1984. The molecular structure of centromeres and telomeres. *Annual Review of Biochemistry* **53**, 163–94.

Blakey, D. H. & Filion, W. G. 1976. Differential Giemsa staining in plants. II. Morphological changes in chromosome structure observed with interference contrast and scanning electron microscopy. *Chromosoma* **56**, 191–7.

Blaxhall, P. C. 1983. Chromosome karyotyping of fish using conventional and G-banding methods. *Journal of Fish Biology* **22**, 417–24.

Bloom, S. E. & Goodpasture, C. 1976. An improved technique for selective silver staining of nucleolar organizer regions in human chromosomes. *Human Genetics* **34**, 199–206.

Blumenfeld, M., Orf, J. W., Sina, B. J., Kreber, R. A., Callahan, M. A., Mullins, J. I & Snyder, L. A. 1978. Correlation between phophorylated H1 histones and satellite DNAs in *Drosophila virilis*. *Proceedings of the National Academy of Sciences* **75**, 866–70.

Bobrow, M. 1974. Acridine orange and the investigation of chromosome banding. *Cold Spring Harbor Symposia on Quantitative Biology* **38**, 435–40.

Bobrow, M. 1985. Heterochromatic chromosome variation and reproductive failure. *Experimental and Clinical Immunogenetics* **2**, 97–105.

Bobrow, M. & Cross, J. 1974. Differential staining of human and mouse chromosomes in interspecific cell hybrids. *Nature* **251**, 77–9.

Bobrow, M. & Madan, K. 1973. The effects of various banding procedures on human chromosomes studied with acridine orange. *Cytogenetics and Cell Genetics* **12**, 145–56.

Bobrow, M., Collacott, H. E. A. C. & Madan, K. 1972a. Chromosome banding with acridine orange. *Lancet* **2**, 1311.

Bobrow, M., Madan, K. & Pearson, P. L. 1972b. Staining of some specific regions of human chromosomes, particularly the secondary constriction of no. 9. *Nature New Biology* **238**, 122–4.

Bobrow, M., Pearson, P. L., Pike, M. C. & El Alfi, O. S. 1971. Length variation in the quinacrine-binding segment of human Y chromosomes of different sizes. *Cytogenetics* **10**, 190–8.

Bonaccorsi, S., Pimpinelli, S. & Gatti, M. 1981. Cytological dissection of sex chromosome heterochromatin of *Drosophila hydei*. *Chromosoma* **84**, 391–403.

Bonaccorsi, S., Santini, G., Gatti, M., Pimpinelli, S. & Coluzzi, M. 1980. Intraspecific polymorphism of sex chromosome heterochromatin in two species of the *Anopheles gambiae* complex. *Chromosoma* **76**, 57–64.

Bontemps, J. & Fredericq, E. 1974. Comparative binding study of the interaction of quinacrine and ethidium bromide with DNA and nucleohistone. *Biophysical Chemistry* **2**, 1–22.

Bontemps, J., Houssier, C. & Fredericq, E. 1974. Optical- and electro-optical properties of the complexes of dibutylproflavine with DNA and nucleohistone. *Biophysical Chemistry* **2**, 301–5.

Boothroyd, E. R. 1953. The reaction of *Trillium* pollen-tube chromosomes to cold treatment during mitosis. *Journal of Heredity* **44**, 3–9.

Boothroyd, E. R. & Lima-de-Faria, A. 1964. DNA synthesis and differential reactivity in the chromosomes of *Trillium* at low temperature. *Hereditas* **52**, 122–6.

Bosma, A. A. 1976. Chromosomal polymorphism and G-banding patterns in the wild boar (*Sus scrofa* L.) from the Netherlands. *Genetica* **46**, 391–9.

Bosman, F. T. & Nakane, P. K. 1982. Immunoelectronmicroscopy of metaphase chromosomes. *Histochemistry* **74**, 341–6.

Bosman, F. T. & Schaberg, A. 1973. A new G-banding modification for metaphase chromosomes. *Nature New Biology* **241**, 216–17.

Bosman, F. T., van der Ploeg, M. & Geraedts, J. P. M. 1977a. *Histochemical Journal* **9**, 31–42.

Bosman, F. T., van der Ploeg, M., van Duijn, P. & Schaberg, A. 1977b. Photometric determination of the DNA distribution in the 24 human chromosomes. *Experimental Cell Research* **105**, 301–11.

Bostock, C. J. 1980. The organization of DNA sequences in chromosomes. In *Cell Biology*, Vol. 3, L. Goldstein & D. M. Prescott (eds), 1–59. New York: Academic Press.

Bostock, C. J. & Christie, S. 1974. Quinacrine fluorescence staining of chromosomes and its relationship to DNA base composition. *Experimental Cell Research* **86**, 157–61.

Bostock, C. J. & Christie, S. 1976. Analysis of the frequency of sister chromatid exchange in different regions of chromosomes of the kangaroo rat (*Dipodomys ordii*). *Chromosoma* **56**, 275–87.

Bostock, C. J. & Clark, E. M. 1980. Satellite DNA in large marker chromosomes of methotrexate-resistant mouse cells. *Cell* **19**, 709–15.

Bostock, C. J. & Prescott, D. M. 1971a. Buoyant density of DNA synthesized at different stages of S phase in Chinese hamster cells. *Experimental Cell Research* **64**, 481–4.

Bostock, C. J. & Prescott, D. M. 1971b. Buoyant density of DNA synthesized at different stages of S phase of mouse L cells. *Experimental Cell Research* **64**, 267–74.

Bostock, C. J. & Prescott, D. M. 1971c. Shift in buoyant density of DNA during the synthetic period and its relation to euchromatin and heterochromatin in mammalian cells. *Journal of Molecular Biology* **60**, 151–62.

Bostock, C. J. & Sumner, A. T. 1978. *The eukaryotic chromosome*. Amsterdam: North-Holland.

Bostock, C. J., Prescott, D. M. & Hatch, F. T. 1972. Timing of the replication of the satellite and main band DNAs in cells of the kangaroo rat (*Dipodomys ordii*). *Experimental Cell Research* **74**, 487–95.

Bostock, D. E., Crocker, J., Harris, K & Smith, P. 1989. Nucleolar organizer regions as indicators of post-surgical prognosis in canine spontaneous mast cell tumours. *British Journal of Cancer* **59**, 915–8.

Boué, A., Boué, J. & Gropp, A. 1985. Cytogenetics of pregnancy wastage. *Advances in Human Genetics* **14**, 1–57.

Boy de la Tour, E. & Laemmli, U.K. 1988. The metaphase scaffold is helically folded: sister chromatids have predominantly opposite helical handedness. *Cell* **55**, 937–44.

Bradbury, E. M., Inglis, R. J., Matthews, H. R. & Sarner, N. 1973. Phosphorylation of very-lysine-rich histone in *Physarum polycephalum*. Correlation with chromosome condensation. *European Journal of Biochemistry* **33**, 131–9.

Brandriff, B., Gordon, L., Ashworth, L., Watchmaker, G., Moore, D., Wyrobek, A. J. & Carrano, A. V. 1985. Chromosomes of human sperm: variability among normal individuals. *Human Genetics* **70**, 18–24.

Brasch, J. M. & Smyth, D. R. 1987. Silver bands in chronic granulocytic leukaemia. II. The Philadelphia chromosome. *Cancer Genetics and Cytogenetics* **25**, 131–9.

Braselton, J. P. 1973. The ultrastructure of cold-modified segments of metaphase chromosomes. *Protoplasma* **76**, 97–101.

Braselton, J. P. 1975. Ribonucleoprotein staining of *Allium cepa* kinetochores. *Cytobiologie* **12**, 148–51.

Breg, W. R. 1972. Quinacrine fluorescence for identifying metaphase chromosomes, with special reference to photomicrography. *Stain Technology* **47**, 87–93.

Brenner, S., Pepper, D., Berns, M. W., Tan, E. & Brinkley, B. R. 1981. Kinetochore structure, duplication, and distribution in mammalian cells: analysis by human autoantibodies from scleroderma patients. *Journal of Cell Biology* **91**, 95–102.

Brinkley, B. R., Valdivia, M. M., Tousson, A. & Brenner, S. S. 1984. Compound kinetochores of the Indian muntjac. *Chromosoma* **91**, 1–11.

Brinkley, B. R., Brenner, S. L., Hall, J. M., Tousson, A., Balczon, R. D. & Valdivia, M. M. 1986. Arrangements of kinetochores in mouse cells during meiosis and spermiogenesis. *Chromosoma* **94**, 309–17.

Brito-Babapulle, V. 1981. Lateral asymmetry in human chromosomes 1, 3, 4, 15 and 16. *Cytogenetics and Cell Genetics* **29**, 198–202.

Brody, T. 1974. Histones in cytological preparations. *Experimental Cell Research* **85**, 255–63.

Brøgger, A., Urdal, T., Larsen, F. B. & Lavik, N. J. 1977. No evidence for a correlation between behaviour and the size of the Y chromosome. *Clinical Genetics* **11**, 349–58.

Brothman, A. R. 1987. Identification of metaphase chromosomes after *in situ* hybridization with radiolabelled probes. *Stain Technology* **62**, 128–30.

Brown, J. E. & Jones, K. W. 1972. Localization of satellite DNA in the microchromosomes of the Japanese quail by *in situ* hybridization. *Chromosoma* **38**, 313–8.

Brown, P. A. & Loughman, W. D. 1980. Visible light observations on the kinetochore of the Indian muntjac, *Muntiacus muntjac*, Z. *Cytogenetics and Cell Genetics* **27**, 123–8.

Brown, S. W. 1966. Heterochromatin. *Science* **151**, 417–25.

Brown, T., Robertson, F. W., Dawson, B. M., Hanlin, S. J. & Page, B. M. 1980. Individual variation of centric heterochromatin in man. *Human Genetics* **55**, 367–73.

Bruère, A. N. 1974. The segregation patterns and fertility of sheep heterozygous for three different Robertsonian translocations. *Journal of Reproduction and Fertility* **41**, 453–64.

Bruère, A. N., Zartman, D. L. & Chapman, H. M. 1974. The significance of the G-bands and C-bands of three different Robertsonian translocations of domestic sheep (*Ovis aries*). *Cytogenetics and Cell Genetics* **13**, 479–88.

Bruère, A. N., Evans, E. P., Burtenshaw, M. D. & Brown, B. B. 1978. Centric fusion polymorphism in Romney Marsh sheep of England. *Journal of Heredity* **69**, 8–10.

Brum-Zorrilla, N. & Postiglioni, A. 1980. Karyological studies on Uruguayan spiders. I. Banding pattern in chromosomes of *Lycosa* species (Araneae – Lycosidae). *Genetica* **54**, 149–53.

Buckland, R. A. & Evans, H. J. 1978a. Cytogenetic aspects of phylogeny in the Bovidae. I. G-banding. *Cytogenetics and Cell Genetics* **21**, 42–63.

Buckland, R. A., & Evans, H. J. 1978b. Cytogenetic aspects of phylogeny in the Bovidae. II. C-banding. *Cytogenetics and Cell Genetics* **21**, 64–71.

Buckland, R. A., Evans, H. J. & Sumner, A. T. 1971. Identifying mouse chromosomes with the ASG technique. *Experimental Cell Research* **69**, 231–6.

Buckton, K. E. 1976. Identification with G and R banding of the position of breakage points induced in human chromosomes by *in vitro* X-irradiation. *International Journal of Radiation Biology* **29**, 475–88.
Buckton, K. E., Spowart, G., Newton, M. S. & Evans, H. J. 1985. Forty-four probands with an additional 'marker' chromosome. *Human Genetics* **69**, 353–70.
Buckton, K. E., O'Riordan, M. L., Jacobs, P. A., Robinson, J. A., Hill, R. & Evans, H. J. 1976. C- and Q-band polymorphisms in the chromosomes of three human populations. *Annals of Human Genetics* **40**, 99–112.
Buhariwalla, F. & Blecher, S. R. 1983. A rapid DAPI-distamycin method for identification of the mouse Y chromosome. *Cytogenetics* and *Cell Genetics* **35**, 78–9.
Bühler, E. M. 1984. Formal analysis of the Y chromosome. In *Aspects of human genetics*, C. San Roman Cos-Gayon & A. McDermott (eds), 106–18. Basel: Karger.
Bühler, E. M. & Malik, N. J. 1988. DA/DAPI heteromorphisms in acrocentric chromosomes other than 15. *Cytogenetics and Cell Genetics* **47**, 104–5.
Bühler, E. M., Tsuchimoto, T., Jurik, L. P. & Stalder, G. R. 1975. Satellite DNA III and alkaline Giemsa staining. *Humangenetik* **26**, 329–33.
Bulatova, N. S. & Radjabli, S. I. 1974. Trypsin-banding procedure for avian chromosomes. *Journal of Heredity* **65**, 188.
Bull, J. 1978. Sex chromosome differentiation: an intermediate stage in a lizard. *Canadian Journal of Genetics and Cytology* **20**, 205–9.
Bullerdiek, J., Dittmer, J., Faehre, A., Bartnitzke, S., Kasche, V. & Schloot, W. 1985 A new banding pattern of human chromosomes by *in situ* nick translation using ECO RI and biotin-dUTP. *Clinical Genetics* **28**, 173–6.
Bullerdiek, J., Dittmer, J., Faehre, A. & Bartnitzke, S. 1986a. An improved method for *in situ* nick translation of human chromosomes with biotin 11-labelled dUTP detected by biotinylated alkaline phosphatase. *Cytobios* **45**, 35–43.
Bullerdiek, J., Dittmer, J., Faehre, A. & Bartniotzke, S. 1986b. Mechanisms of *in situ* nick translation of chromosomes using restriction endonucleases. *Cytobios* **47**, 33–44.
Bultmann, H. & Mezzanotte, R. 1987. Characterization and origin of extrachromosomal DNA granules in *Sarcophaga bullata*. *Journal of Cell Science* **88**, 327–34.
Bunch, T. D. 1978. Fundamental karyotype in domestic and wild species of sheep. *Journal of Heredity* **69**, 77–80.
Bunch, T. D. & Nadler, C. F. 1980. Giemsa-band patterns of the tahr and chromosomal evolution of the tribe Caprini. *Journal of Heredity* **71**, 110–6.
Bunch, T. D., Foote, W. C. & Maciulis, A. 1985. Chromosome banding pattern homologies and NORs for the Bactrian camel, guanaco, and llama. *Journal of Heredity* **76**, 115–8.
Burkholder, G. D. 1974. Electron microscopic visualization of chromosomes banded with trypsin. *Nature* **247**, 292–4.
Burkholder, G. D. 1975. The ultrastructure of G- and C-banded chromosomes. *Experimental Cell Research* **90**, 269–78.
Burkholder, G. D. 1978. Reciprocal Giemsa staining of late DNA replicating regions produced by low and high pH sodium phosphate. *Experimental Cell Research* **111**, 489–92.
Burkholder, G. D. 1979. An investigation of the mechanism of the reciprocal differential staining of BUdR-substituted and unsubstituted chromosome regions. *Experimental Cell Research* **121**, 209–19.
Burkholder, G. D. 1981. The ultrastructure of R-banded chromosomes. *Chromosoma* **83**, 473–80.
Burkholder, G. D. 1982. The mechanisms responsible for reciprocal BrdU-Giemsa staining. *Experimental Cell Research* **141**, 127–37.
Burkholder, G. D. 1989. Morphological and biochemical effects of endonucleases on isolated mammalian chromosomes *in vitro*. *Chromosoma* **97**, 347–55.
Burkholder, G. D. & Comings, D. E. 1972. Do the Giemsa-banding patterns of chromosomes change during embryonic development? *Experimental Cell Research* **75**, 268–71.
Burkholder, G. D. & Duczek, L. L. 1980a. Proteins in chromosome banding. I. Effect of G-banding treatments on the proteins of isolated nuclei. *Chromosoma* **79**, 29–41.

Burkholder, G. D. & Duczek, L. L. 1980b. Proteins in chromosome banding. II. Effect of R- and C-banding treatments on the proteins of isolated nuclei. *Chromosoma* **79**, 43–51.

Burkholder, G. D. & Duczek, L. L. 1982a. The effect of the chromosome banding techniques on the histone and nonhistone proteins of isolated chromatin. *Canadian Journal of Biochemistry* **60**, 328–37.

Burkholder, G. D. & Duczek, L. L. 1982b. The effect of chromosome banding techniques on the proteins of isolated chromosomes. *Chromosoma* **87**, 425–35.

Burkholder, G. D. & Schmidt, G. J. 1986. Endonuclease banding of isolated mammalian metaphase chromosomes. *Experimental Cell Research* **164**, 379–87.

Burkholder, G. D. & Weaver, M. G. 1975. Differential accessibility of DNA in extended and condensed chromatin to pancreatic DNase I. *Experimental Cell Research* **92**, 518–22.

Burkholder, G. D. & Weaver, M. G. 1977. DNA-protein interactions and chromosome bandings. *Experimental Cell Research* **110**, 251–62.

Burkholder, G. D., Latimer, L. J. P. & Lee, J. S. 1988. Immunofluorescent staining of mammalian nuclei and chromosomes with a monoclonal antibody to triplex DNA *Chromosoma* **97**, 185–92.

Burmeister, M., Monaco, A. P., Gillard, E. F., van Ommen, G-J. B., Affara, N. A., Ferguson-Smith, M. A., Kunkel, L. M. & Lehrach, H. 1987. A 10-megabase physical map of human Xp21, including the Duchenne muscular dystrophy gene. *Genomics* **2**, 189–202.

Busch, H., Lischwe, M. A., Michalik, J., Chan, P-K. & Busch, R. K. 1982. Nucleolar proteins of special interest: silver-staining proteins B23 and C23 and antigens of human tumour nucleoli. In *The nucleolus*, E. G. Jordan & C. A. Cullis (eds), 43–71. Cambridge: Cambridge University Press.

Bustin, M. 1979. Immunological approaches to chromatin and chromosome structure and function. *Current Topics in Microbiology and Immunology* **88**, 105–42.

Bustin, M., Yamasaki, H., Goldblatt, D., Shani, M., Huberman, E. & Sachs, L. 1976. Histone distribution in chromosomes revealed by antihistone sera. *Experimental Cell Research* **97**, 440–4.

Buys, C. H. C. M. & Osinga, J. 1980a. Abundance of protein-bound sulphydryl and disulphide groups at chromosomal nucleolus organizing regions. *Chromosoma* **77**, 1–11.

Buys, C. H. C. M. & Osinga, J. 1980b. The mechanism of differential sister chromatid fluorescence as studied with the GC-specific DNA-ligand mithramycin. *Experimental Cell Research* **125**, 105–9.

Buys, C. H. C. M. & Osinga, J. 1981. The role of chromosomal proteins in the induction of a differential staining of sister chromatids by light. *Histochemical Journal* **13**, 735–46.

Buys, C. H. C. M. & Osinga, J. 1982. A relation between G-, C-, and N-band patterns as revealed by progressive oxidation of chromosomes and a note on the nature of N-bands. *Genetica* **58**, 3–9.

Buys, C. H. C. M. & Osinga, J. 1984. Selective staining of the same set of nucleolar phosphoproteins by silver and Giemsa. A combined biochemical and cytochemical study on staining of NORs. *Chromosoma* **89**, 387–96.

Buys, C. H. C. M., Aanstoot, G. H. & Nienhaus, A. J. 1984a. The Giemsa-11 technique for species-specific chromosome differentiation. A simple stain modification leading to dependable direct and sequential staining procedures. *Histochemistry* **81**, 465–8.

Buys, C. H. C. M., Koerts, T. & van der Veen, A. Y. 1984b. Banding of unfixed mitotic chromosomes in suspension after release from human lymphocytes and fibroblasts. *Human Genetics* **66**, 361–4.

Buys, C. H. C. M., Osinga, J. & Anders, G. J. P. A. 1979b. Age-dependent variability of ribosomal RNA-gene activity in man as determined from frequencies of silver staining nucleolus organizing regions on metaphase chromosomes of lymphocytes and fibroblasts. *Mechanisms of Ageing and Development* **11**, 55–75.

Buys, C. H. C. M., Osinga, J. & Stienstra, S. 1981b. Rapid irradiation procedure for obtaining permanent differential staining of sister chromatids and aspects of its underlying mechanism. *Human Genetics* **57**, 35–8.

Buys, C. H. C. M., Osinga, J. & van der Veen, A. 1982. Effects on chromosomal proteins in sister chromatid differentiation by incorporation of 5-bromodeoxyuridine into DNA. *Experimental Cell Research* **137**, 452–5.

Buys, C. H. C. M., Gouw, W. L., Blenkers, J. A. M. & van Dalen, C. H. 1981a. Heterogeneity of human chromosome 9 constitutive heterochromatin as revealed by sequential distamycin A/DAPI staining and C-banding. *Human Genetics* **57**, 28–30.

Buys, C. H. C. M., Osinga, J., Gouw, W. L. & Anders, G. J. P. A. 1978. Rapid identification of chromosomes carrying sliver-stained nucleolus-organizing regions. *Human Genetics* **44**, 173–80.

Buys, C. H. C. M., Anders, G. J. P. A., Gouw, W. L., Borkent-Ypma, J. M. M. & Blenkers-Platter, J. A. M. 1979a. A comparison of constitutive heterochromatin staining methods in two cases of familial heterochromatin deficiencies. *Human Genetics* **52**, 133–8.

Buys, C. H. C. M., Aten, J. A., Koerts, T., Osinga, J. & van der Veen, A. 1984c. Isolated metaphase chromosomes stabilized by DNA-intercalation or polyamine addition: a comparison. *Cell Biology International Reports* **8**, 273.

Cabrero, J. & Camacho, J. P. M. 1987. Population cytogenetics of *Chorthippus vagans*. I. Polymorphisms for pericentric inversion and for heterochromatic deletion. *Genome* **29**, 280–4.

Cabrero, J., Alché, J. D. & Camacho, J. P. M. 1987. Effects of B chromosomes on the activity of nucleolar organizer regions in the grasshopper *Eyprepocnemis plorans*: activation of a latent nucleolar organizer region on a B chromosome fused to an autosome. *Genome* **29**, 116–21.

Caizzi, R. & Bostock, C. J. 1982. Gene amplification in methotrexate-resistant mouse cells. IV. Different DNA sequences are amplified in different resistant cell lines. *Nucleic Acids Research* **10**, 6597–618.

Calderon, D. & Schnedl, W. 1973. A comparison between quinacrine fluorescence banding and ^{3}H-thymidine incorporation patterns in human chromosomes. *Humangenetik* **18**, 63–70.

Callan, H. G. 1942. Heterochromatin in *Triton*. *Proceedings of the Royal Society of London* B, **130**, 324–35.

Camacho, J. P. M., Viseras, E., Navas, J. & Cabrero, J. 1984. C-heterochromatin content of supernumerary chromosome segments of grasshoppers: detection of an euchromatic extra segment. *Heredity* **53**, 167–75.

Cameron, I. L., Smith, N. K. R. & Pool, T. B. 1979. Element concentration changes in mitotically active and postmitotic enterocytes. An X-ray microanalysis study. *Journal of Cell Biology* **80**, 444–50.

Capanna, E. 1973. Concluding remarks. In *Cytotaxonomy and vertebrate evolution*, A. B. Chiarelli & E. Capanna (eds), 681–95. London: Academic Press.

Capanna, E., Cristaldi, M., Perticone, P. & Rizzoni, M. 1975. Identification of chromosomes involved in the 9 Robertsonian fusions of the Apennine mouse with a 22-chromosome karyotype. *Experientia* **31**, 294–6.

Caratzali, A., Roman, I. C., Cirnu-Georgian, L. & Geormaneanu, C. 1972. Identification des chromosomes humains par chromomérisation *in vitro*. *Comptes Rendus de l'Academie des Sciences (Paris)* D, **274**, 1198–9.

Carbon, J. 1984. Yeast centromeres: structure and function. *Cell* **37**, 351–3.

Cardoso, H. & Dutra, A. 1979. The neo-X neo-Y sex pair in Acrididae, its structure and association. *Chromosoma* **70**, 323–36.

Carlenius, C., Ryttman, H., Tegelstrom, H. & Jansson, H. 1981. R-, G- and C-banded chromosomes in the domestic fowl (*Gallus domesticus*). *Hereditas* **94**, 61–6.

Carleton, H. M. 1920. Observations on an intra-nucleolar body in columnar epithelial cells of the intestine. *Quarterly Journal of Micr. Science* **64**, 329–43.

Carlin, C. R. & Rao, K. W. 1982. Sequential staining with Hoechst 33258 and quinacrine mustard for the identification of human chromosomes in somatic cell hybrids. *Experimental Cell Research* **138**, 466–9.

Carnevale, A., Ibañez, B. B. & del Castillo, V. 1976. The segregation of C-band polymorphisms on chromosomes 1, 9, and 16. *American Journal of Human Genetics* **28**, 412–6.

Carpentier, S., Dutrillaux, B. & Lejeune, J. 1972. Effet du milieu ionique sur la dénaturation thermique ménagée des chromosomes humains. *Annales de Génétique* **15**, 203–5.

Carr, J. L. & Bickham, J. W. 1981. Sex chromosomes of the Asian black pond turtle, *Siebenrockiella crassicollis* (Testudines: Emydidae). *Cytogenetics and Cell Genetics* **31**, 178–83.

Caspersson, T. 1973. Cytochemistry in chromosome analysis. In *Chromosome identification. Nobel Symposium 23*, T. Caspersson & L. Zech (eds), 25–6. New York: Academic Press.

Caspersson, T., Lomakka, G. & Zech, L. 1971a. The 24 fluorescence patterns of the human metaphase chromosomes – distinguishing characters and variability. *Hereditas* **67**, 89–102.

Caspersson, T., Zech, L. & Johansson, C. 1970a. Differential binding of alkylating fluorochromes in human chromosomes. *Experimental Cell Research* **60**, 315–9.

Caspersson, T., de la Chapelle, A., Schröder, J. & Zech, L. 1972. Quinacrine fluorescence of metaphase chromosomes. Identical patterns in different tissues. *Experimental Cell Research* **72**, 56–9.

Caspersson, T., Hultén, M., Lindsten, J. & Zech, L. 1971b. Identification of chromosome bivalents in human male meiosis by quinacrine mustard fluorescence analysis. *Hereditas* **67**, 147–9.

Caspersson, T., Zech, L., Johansson, C. & Modest, E. J. 1970c. Identification of human chromosomes by DNA-binding fluorescent agents. *Chromosoma* **30**, 215–27.

Caspersson, T., Zech, L., Harris, H., Wiener, F. & Klein, G. 1971c. Identification of human chromosomes in a mouse/human hybrid by fluorescence techniques. *Experimental Cell Research* **65**, 475–8.

Caspersson, T., Zech, L., Johansson, C., Lindsten, J. & Hulten, M. 1970b. Fluorescent staining of heteropycnotic chromosome regions in human interphase nuclei. *Experimental Cell Research* **61**, 472–4.

Caspersson, T., Zech, L., Modest, E. J. Foley, G. E., Wagh, U. & Simonsson, E. 1969a. Chemical differentiation with fluorescent alkylating agents in *Vicia faba* metaphase chromosomes. *Experimental Cell Research* **58**, 128–40.

Caspersson, T., Zech, L., Modest, E. J., Foley, G. E., Wagh, U. & Simonsson, E. 1969b. DNA-binding fluorochromes for the study of the organization of the metaphase nucleus. *Experimental Cell Research* **58**, 141–52.

Caspersson, T., Farber, S., Foley, G. E., Kudynowski, J., Modest, E. J., Simonsson, E., Wagh, U. & Zech, L. 1968. Chemical differentiation along metaphase chromosomes. *Experimental Cell Research* **49**, 219–22.

Castleman, K. R. & Wall, R. J. 1973. Automatic systems for chromosome identification. In *Chromosome identification. Nobel Symposium 23*, T. Caspersson & L. Zech (eds), 77–84. New York: Academic Press.

Catala, A., Vidal-Rioja, L. & Bianchi, N. O. 1981. Liver chromatin fractions in *Mus* and *Akodon*. The concept of constitutive heterochromatin. *Molecular and Cellular Biochemistry* **36**, 135–41.

Cau, A., Salvadori, S., Deiana, A. M., Bella, J. L. & Mezzanotte, R. 1988. The characterization of *Muraena helena* L. mitotic chromosomes: karyotype, C-banding, nucleolar organizer regions, and *in situ* digestion with restriction endonucleases. *Cytogenetics and Cell Genetics* **47**, 223–6.

Cavalier-Smith, T. 1978. Nuclear volume control by nucleoskeletal DNA, selection for cell volume and cell growth rate, and the solution of the DNA C-value paradox. *Journal of Cell Science* **34**, 247–78.

Cavalier-Smith, T. 1985. *The evolution of genome size*. Chichester: Wiley.

Cavalli, I. J., Mattevi, M. S., Erdtmann, B., Sbalqueiro, I. J. & Maia, N. A. 1985. Equivalence of the total constitutive heterochromatin content by an interchromosomal compensation in the C-band sizes of chromosomes 1, 9, 16, and Y in Caucasian and Japanese individuals. *Human Heredity* **35**, 379–87.

Cavatorta, P., Masotti, L. & Szabo, A. G. 1985. A time-resolved fluorescence study of 4′, 6′-diamidine-2-phenylindole dihydrochloride binding to polynucleotides. *Biophysical Chemistry* **22**, 11–16.
Cawood, A. H. 1981. Chromosome replication in fibroblasts of the Syrian hamster (*Mesocricetus auratus*). *Chromosoma* **83**, 711–20.
Cawood, A. H. & Savage, J. R. K. 1985. Uninterrupted DNA synthesis during S-phase in untransformed diploid hamster fibroblasts. *Chromosoma* **91**, 164–6.
Cermeno, M. C. & Lacadena, J. R. 1985. Nucleolar organizer competition in *Aegilops*-rye hybrids. *Canadian Journal of Genetics and Cytology* **27**, 479–83.
Červenka, J., Jacobson, D. E. & Gorlin, R. J. 1971. Fluorescing structures of human metaphase chromosomes. Detection of Y body. *American Journal of Human Genetics* **23**, 317–24.
Chaganti, R. S. K., Weitkamp, L. R., Fleming, J., Miller, G. & German, J. 1975. Polymorphic chromosomes and gene assignment in man. *Cytogenetics and Cell Genetics* **14**, 263–8.
Chamberlin, J. & Magenis, R. E. 1980. Parental origin of *de novo* chromosome rearrangements. *Human Genetics* **53**, 343–7.
Chamla, Y. & Ruffié, M. 1976. Production of C and T bands in human mitotic chromosomes after heat treatment. *Human Genetics* **34**, 213–6.
Chandley, A. C. 1981. Does 'affinity' hold the key to fertility in the female mule? *Genetical Research* **37**, 105–9.
Chandley, A. C. 1986. A model for effective pairing and recombination at meiosis based on early replicating sites (R-bands) along chromosomes. *Human Genetics* **72**, 50–7.
Chandley, A. C. & Fletcher, J. M. 1973. Centromere staining at meiosis in man. *Humangenetik* **18**, 247–52.
Chandley, A. C. & McBeath, S. 1987. DNase I hypersensitive sites along the XY bivalent at meiosis in man include the XpYp pairing region. *Cytogenetics and Cell Genetics* **44**, 22–31.
Chaudhuri, J. P., Vogel, W., Voiculescu, I. & Wolf, U. 1971. A simplified method of demonstrating Giemsa-band pattern in human chromosomes. *Humangenetik* **14**, 83–4.
Chen, T. R. 1974. A simple method to sequentially reveal Q- and C-bands on the same metaphase chromosomes. *Chromosoma* **47**, 147–56.
Chen, T. R. 1985. The Chinese hamster Don cell line differs from normal diploidy by one chromosome band. *Cytogenetics and Cell Genetics* **39**, 57–63.
Chernay, P. R., Hsu, L. Y. F., Streicher, H. & Hirschhorn, K. 1971. Human chromosome identification by differential staining: G group (21–22–Y). *Cytogenetics* **10**, 219–24.
Cherry, L. M. & Johnston, D. A. 1987. Size variation in kinetochores of human chromosomes. *Human Genetics* **75**, 155–8.
Cherry, L. M. & Shah, S. A. 1987. A technique for simultaneous antikinetochore immunofluorescence staining and Q-banding in chromosomes from human lymphocytes. *Stain Technology* **62**, 221–5.
Cherry, L. M., Faulkner, A. J., Grossberg, L. A. & Balczon, R. 1989. Kinetochore size variation in mammalian chromosomes: an image analysis study with evolutionary implications. *Journal of Cell Science* **92**, 281–9.
Cheung, S. W., Crane, J. P., Johnson, A., Simms, L. & Reid, J. 1987. A simple method for preparing prometaphase chromosomes from amniotic fluid cell cultures. *Prenat. Diag.* **7**, 383–8.
Chiarelli, A. B. & Capanna, E. 1973. *Cytotaxonomy and vertebrate evolution*. London: Academic Press.
Christensen, K. & Smedegard, K. 1978. Chromosome marker in domestic pigs. C-band polymorphism. *Hereditas* **88**, 269–72.
Christidis, L. 1983. Extensive chromosomal repatterning in two congeneric species: *Pytilia melba*, L. and *Pytilia phoenicoptera* Swainson (Estrildidae: Aves). *Cytogenetics and Cell Genetics* **36**, 641–8.
Christidis, L. 1986a. Chromosomal evolution within the family Estrildidae (Aves). I. The Poephilae. *Genetica* **71**, 81–97.

Christidis, L. 1986b. Chromosomal evolution within the family Estrildidae (Aves). II. The Lonchurae. *Genetica* **71**, 99–113.

Christman, J. K. 1984. DNA methylation in Friend erythroleukaemia cells: the effects of chemically induced differentiation and of treatment with inhibitors of DNA methylation. *Current Topics in Microbiology and Immunology* **108**, 49–78.

Chuprevich, T. W., Meisner, L. F., Inhorn, S. L. & Indriksons, A. 1973. Chromosomal protein and C-banding. *Lancet* **1**, 1453–4.

Cionini, P. G., Bassi, P., Cremonini, R. & Cavallini, A. 1985. Cytological localization of fast renaturing and satellite DNA sequences in *Vicia faba*. *Protoplasma* **124**, 106–11.

Citoler, P., Gropp, A. & Natarajan, A. T. 1972. Timing of DNA replication of autosomal heterochromatin in the hedgehog. *Cytogenetics* **11**, 53–62.

Clapham, L. & Östergren, G. 1978. A fixation method for the observation of kinetochores. *Hereditas* **89**, 89–106.

Clavaguera, A., Querol, E., Coll, D., Genesca, J. & Egozcue, J. 1983. Cytochemical studies on the nature of NOR (nucleolus organizer region) silver stainability. *Cellular and Molecular Biology* **29**, 255–9.

Clemente, I. C., Garcia, M., Ponsa, M. & Egozcue, J. 1987. High-resolution chromosome banding studies in *Cebus apella*, *Cebus albifrons*, and *Lagothrix lagothricha*: comparison with the human karyotype. *American Journal of Primatology* **13**, 23–36.

Cohen, M. M. & Shaw, M. W. 1964. Effects of mitomycin C on human chromosomes. *Journal of Cell Biology* **23**, 386–95.

Cohen, M. M., Shaw, M. W. & Craig, A. P. 1963. The effects of streptonigrin on cultured human leukocytes. *Proceedings of the National Academy of Sciences* **50**, 16–24.

Coleman, A. W., Maguire, M. J. & Coleman, J. R. 1981. Mithramycin- and 4'-6'-diamidino-2-phenylindole (DAPI)-DNA staining for fluorescence microspectrophotometric measurement of DNA in nuclei, plastids, and virus particles. *Journal of Histochemistry and Cytochemistry* **29**, 959–68.

Comings, D. E. 1971. Heterochromatin of the Indian muntjac. Replication, condensation, DNA ultracentrifugation, fluorescent and heterochromatin staining. *Experimental Cell Research* **67**, 441–60.

Comings, D. E. 1972. The structure and function of chromatin. *Advances in Human Genetics* **3**, 237–431.

Comings, D. E. 1975a. Mechanisms of chromosome banding. IV. Optical properties of the Giemsa dyes. *Chromosoma* **50**, 89–110.

Comings, D. E. 1975b. Mechanisms of chromosome banding. VIII. Hoechst 33258-DNA interaction. *Chromosoma* **52**, 229–43.

Comings, D. E. & Avelino, E. 1974. Mechanisms of chromosome banding. II. Evidence that histones are not involved. *Experimental Cell Research* **86**, 202–6.

Comings, D. E. & Avelino, E. 1975. Mechanisms of chromosome banding. VII. Interaction of methylene blue with DNA and chromatin. *Chromosoma* **51**, 365–79.

Comings, D. E. & Drets, M. E. 1976. Mechanisms of chromosome banding. IX. Are variations in DNA base composition adequate to account for quinacrine, Hoechst 33258 and daunomycin banding? *Chromosoma* **56**, 199–211.

Comings, D. E. & Harris, D. C. 1975. Nuclear proteins. I. Electrophoretic comparison of mouse nucleoli, heterochromatin, euchromatin and contractile proteins. *Experimental Cell Research* **96**, 161–79.

Comings, D. E. & Okada, T. A. 1976. Fine structure of the heterochromatin of the kangaroo rat, *Dipodomys ordii*, and examination of the possible role of actin and myosin in heterochromatin condensation. *Journal of Cell Science* **21**, 465–77.

Comings, D. E. & Wyandt, H. E. 1976. Reverse banding of Japanese quail microchromosomes. *Experimental Cell Research* **99**, 183–5.

Comings, D. E., Avelino, E., Okada, T. A. & Wyandt, H. E. 1973. The mechanism of C- and G-banding of chromosomes. *Experimental Cell Research* **77**, 469–93.

Comings, D. E., Kovacs, B. S., Avelino, E. & Harris, D. C. 1975. Mechanisms of chromosome banding. V. Quinacrine banding. *Chromosoma* **50**, 111–45.
Comings, D. E., Limon, J., Ledochowski, A. & Tsou, K. C. 1978. Mechanisms of chromosome banding. XI. The ability of various acridine derivatives to cause Q-banding. *Experimental Cell Research* **117**, 451–5.
Committee for a standardized karyotype of *Rattus norvegicus* 1973. Standard karyotype of the Norway rat, *Rattus norvegicus*. *Cytogenetics and Cell Genetics* **12**, 199–205.
Committee for standardization of chromosomes of *Peromyscus* 1977. Standardized karyotype of deer mice, *Peromyscus* (Rodentia). *Cytogenetics and Cell Genetics* **19**, 38–43.
Committee for standardized karyotype of *Oryctolagus cuniculus*. 1981. Standard karyotype of the laboratory rabbit, *Oryctolagus cuniculus*. *Cytogenetics and Cell Genetics* **31**, 240–8.
Committee for the standard karyotype of *Alopex lagopus*, 1985a. The standard karyotype of the blue fox (*Alopex lagopus* L.) *Hereditas* **103**, 33–38.
Committee for standardized karyotype of *Ovis aries*. 1985b. Standard nomenclature for the G-band karyotype of the domestic sheep (*Ovis aries*). *Hereditas* **103**, 165–70.
Committee for the standard karyotype of *Vulpes fulvus* Desm. 1985c. The standard karyotype of the silver fox (*Vulpes fulvus* Desm.). *Hereditas* **103**, 171–6.
Cooke, H. J., Schmidtke, J. & Gosden, J. R. 1982. Characterization of a human Y chromosome repeated sequence and related sequences in higher primates. *Chromosoma* **87**, 491–502.
Cooper, D. N. 1983. Eukaryotic DNA methylation. *Human Genetics* **64**, 315–33.
Cordeiro-Stone, M. & Lee, C. S. 1976. Studies on the satellite DNAs of *Drosophila nasutoides*: their buoyant densities, reassociation rates and localizations in polytene chromosomes. *Journal of Molecular Biology* **104**, 1–24.
Cortes, F. & Escalza, P. 1986. Analysis of different banding patterns and late replicating regions in chromosome of *Allium cepa*, *A. sativum* and *A. nigrum*. *Genetica* **71**, 39–46.
Cortes, F., Gonzalez-Gil, G. & Lopez-Saez, J. F. 1980. Differential staining of late replicating DNA-rich regions in *Allium cepa* chromosomes. *Caryologia* **33**, 193–202.
Couturier, J. & Lejeune, J. 1976. Modification des chromatides avant fixation: production de bandes G par certains colorants. *Annales de Génétique* **19**, 217–18.
Cowden, R. R. & Curtis, S. K. 1973. A study of grasshopper meiotic chromosomes by a combination of light and electron microscopic procedures. *Histochemical Journal* **5**, 225–38.
Cowden, R. R. & Curtis, S. K. 1981. Microfluorometric investigations of chromatin structure. I. Evaluation of nine DNA-specific fluorochromes as probes of chromatin organization. *Histochemistry* **72**, 11–23.
Cowell, J. K. 1982. Double minutes and homogeneously staining regions: gene amplification in mammalian cells. *Annual Reviews of Genetics* **16**, 21–59.
Cowell, J. K. & Franks, L. M. 1980. A rapid method for accurate DNA measurements in single cells *in situ* using a simple microfluorimeter and Hoechst 33258 as a quantitative fluorochrome. *Journal of Histochemistry and Cytochemistry* **28**, 206–10.
Cox, J. V. & Olmsted, J. B. 1984. Kinetochore antigens: complexity, synthesis, and evolutionary conservation. In *Molecular biology of the cytoskeleton*, G. G. Borisy, D. W. Cleveland & D. B. Murphy (eds), 71–7. New York: Cold Spring Harbor Laboratory.
Cox, J. V., Schenk, E. A. & Olmsted, J. B. 1983. Human anticentromere antibodies: distribution, characterization of antigens, and effect on microtubule organization. *Cell* **35**, 331–9.
Craig-Holmes, A. P., Moore, F. B. & Shaw, M. W. 1973. Polymorphism of human C-band heterochromatin. I. Frequency of variants. *American Journal of Human Genetics* **25**, 181–92.
Craig-Holmes, A. P., Moore, F. B. & Shaw, M. W. 1975. Polymorphism of human C-band heterochromatin. II. Family studies with suggestive evidence for somatic crossing-over. *American Journal of Human Genetics* **27**, 178–89.
Cramer, H. & Hansen, S. 1972. On the Y fluorescence of human male fibroblasts. *Humangenetik* **17**, 23–8.
Crissman, H. A., Stevenson, A. P., Orlicky, D. J. & Kissane, R. J. 1978. Detailed studies on the application of three fluorescent antibiotics for DNA staining in flow cytometry. *Stain Technology* **53**, 321–30.

Crossen, P. E. 1972. Giemsa banding patterns of human chromosomes. *Clinical Genetics* **3**, 169–79.

Crossen, P. E. 1973. Factors influencing Giemsa band formation of human chromosomes. *Histochemie* **35**, 51–62.

Crossen, P. E. 1974. Unusual chromosome bands revealed by aging. *Humangenetik* **21**, 197–202.

Crossen, P. E., Pathak, S. & Arrighi, F. E. 1975. A high resolution study of the DNA replication patterns of Chinese hamster chromosomes using sister chromatid differential staining technique. *Chromosoma* **52**, 339–47.

Cuny, G., Soriano, P., Macaya, G. & Bernardi, G. 1981. The major components of the mouse and human genomes. 1. Preparation, basic properties and compositional heterogeneity. *European Journal of Biochemistry* **115**, 227–33.

Cuny, R. & Malacinski, G. M. 1985. Banding differences between tiger salamander and axolotl chromosomes. *Canadian Journal of Genetics and Cytology* **27**, 510–14.

Curtis, D. J. & Horobin, R. W. 1975. Staining banded human chromosomes with Romanowsky dyes: some practical consequences of the nature of the stain. *Humangenetik* **26**, 99–104.

Curtis, D. J. & Horobin, R. W. 1982. Chromosome banding: specification of structural features of dyes giving rise to G-banding. *Histochemical Journal* **14**, 911–28.

Czaker, R. & Mayr, B. 1980. Ag-G staining, a rapid technique for producing combined silver staining and Giemsa banding in mammalian chromosomes. *Experientia* **36**, 625–6.

Czaker, R. & Mayr, B. 1982. Comparative studies on the polymorphism of nucleolus organizer regions (NORs) in four breeds of domestic pigs (*Sus scrofa domestica* L.) with special emphasis on the development of breeds. *Zeitschrift Syst. Evolutionsforsch.* **20**, 233–41.

D'Alisa, R. M., Korf, B. R. & Gershey, E. L. 1979, T antigen banding on chromosomes of simian virus 40 infected muntjac cells. *Cytogenetics and Cell Genetics* **24**, 27–36.

Dallapiccola, B. & Ricci, N. 1975. Observations on specific Giemsa staining of the Y and on selective oil destaining of the chromosomes. *Humangenetik* **26**, 251–5.

Daniel, A. 1979. Single Cd band in dicentric translocations with one suppressed centromere. *Human Genetics* **48**, 85–92.

Darlington, C. D. & LaCour, L. 1938. Differential reactivity of the chromosomes. *Annals of Botany* N.S. **2**, 615–25.

Darlington, C. D. & LaCour, L. 1940. Nucleic acid starvation in chromosomes in *Trillium*. *Journal of Genetics* **40**, 185–213.

Darzynkiewicz, Z. 1979. Acridine orange as a molecular probe in studies of nucleic acids *in situ*. In *Flow cytometry and sorting*, M. R. Melamed, P. F. Mullaney & M. L. Mendelsohn (eds), 285–316. New York: Wiley.

Darzynkiewicz, Z. & Traganos, F. 1988. Unstainable DNA in cell nuclei. *Anal. Quant. Cytol. Histol.* **10**, 462–6.

Darzynkiewicz, Z., Traganos, F., Kapuscinski, J., Staiano-Coico, L. & Melamed, M. R. 1984. Accessibility of DNA *in situ* to various fluorochromes: relationship to chromatin changes during erythroid differentiation of Friend leukaemia cells. *Cytometry* **5**, 355–63.

Das, B. C., Rani, R., Mitra, A. B. & Luthra, U. K. 1986. The number of silver-staining NORs (rDNA) in lymphocytes of newborns and its relationship to human development. *Mechanism of Ageing and Development* **36**, 117–23.

Daskal, Y., Mace, M. & Busch, H. 1978. Demonstration of membranous patches on isolated chromosomes. *Experimental Cell Research* **111**, 472–5.

Davidson, M. W., Griggs, B. G., Lopp, I. G., Boykin, D. W. & Wilson, W. D. 1978. Nuclear magnetic resonance investigation of the interaction of a ^{13}C-labelled quinacrine derivative with DNA. *Biochemistry* **17**, 4220–5.

Davidson, N. R. 1973. Photographic techniques for recording chromosome banding patterns. *Journal of Medical Genetics* **10**, 122–6.

Davies, E. D. G. & Southern, D. I. 1976. Giemsa C-banding within the genus *Glossina* (Diptera, Glossinidae). *Genetica* **46**, 413–8.

Davis, K. M., Smith, S. A. & Greenbaum, I. F. 1986. Evolutionary implications of chromosomal polymorphisms in *Peromyscus boylii* from southwestern Mexico. *Evolution* **40**, 645–9.

Davisson, M. T. 1989. Centromeric heterochromatin variants. In *Genetic variants and strains of the laboratory mouse*, M. F. Lyon & A. G. Searle (eds), 617–8. Oxford: Oxford University Press.

Davisson, M. T. & Roderick, T. H. 1973. Chromosome banding patterns of two paracentric inversions in mice. *Cytogenetics and Cell Genetics* **12**, 398–403.

Deaven, L. L. & Petersen, D. F. 1973. The chromosomes of CHO, an aneuploid Chinese hamster cell line: G-band, C-band, and autoradiographic analyses. *Chromosoma* **41**, 129–44.

Deavan, L. L., Vidal-Rioja, L., Jett, J. H. & Hsu, T. C. 1977. Chromosomes of *Peromyscus* (Rodentia, Cricetidae). VI. The genomic size. *Cytogenetics and Cell Genetics* **19**, 241–9.

de Braekeleer, M., Keushnig, M. & Lin, C. C. 1985. Synchronization of human lymphocyte cultures by fluorodeoxyuridine. *Canadian Journal of Genetics and Cytology* **27**, 622–5.

de Braekeleer, M., Keushnig, M. & Lin, C. C. 1986. A high-resolution C-banding technique. *Canadian Journal of Genetics and Cytology* **28**, 317–22.

de Capoa, A., Ferraro, M., Lavia, P., Pelliccia, F. & Finazzi-Agro, A. 1982. Silver staining of the nucleolus organizer regions (NOR) requires clusters of sulphydryl groups. *Journal of Histochemistry and Cytochemistry* **30**. 908–11.

de Capoa, A., Pelliccia, F., Markelaj, P., Ciofi-Luzzato, A. R. & Buongiorno Nardelli, M. 1983. Silver positivity of the NORs during embryonic development of *Xenopus laèvis*. *Experimental Cell Research* **147**, 472–8.

de Grouchy, J. & Turleau, C. 1986. Microcytogenetics. *Experientia* **42**, 1090–7.

de Grouchy, J., Turleau, C & Finaz, C. 1978. Chromosomal phylogeny of the primates. *Annual Review of Genetics* **12**, 289–328.

de la Chapelle, A., Schröder, J., Selander, R-K. & Strenstrand, K. 1973. Differences in DNA composition along mammalian metaphase chromosomes. *Chromosoma* **42**, 365–82.

de la Maza, L. M. & Sanchez, O. 1976. Simultaneous G and C banding of human chromosomes. *Journal of Medical Genetics* **13**, 235–6.

Delany, M. E. & Bloom, S. E. 1984. Replication banding patterns in the chromosomes of the rainbow trout. *Journal of Heredity* **75**, 431–4.

del Mazo, J., Martin-Sempere, M. J., Kremer, L. & Avila, J. 1986. Centromere pattern in different mouse seminiferous tubule cells. *Cytogenetics and Cell Genetics* **43**, 201–6.

den Nijs, J. I., Gonggrijp, H. S., Augustinus, E. & Leeksma, C. H. W. 1985. Hot bands: a simple G-banding method for leukaemic metaphases. *Cancer Genetics and Cytogenetics* **15**, 373–4.

Denton, T. E., Brooke, W. R. & Howell, W. M. 1977. A technique for the simultaneous staining of both nucleolar organizer regions and kinetochores of human chromosomes with silver. *Stain Technology* **52**, 311–13.

Denton, T. E., Howell, W. M. & Barrett, J. V. 1976. Human nucleolar organizer chromosomes: satellite associations. *Chromosoma* **55**, 81–4.

Denton, T. E., Liem, S. L., Cheng, K. M. & Barrett, J. V. 1981. The relationship between aging and ribosomal gene activity in humans as evidenced by silver staining. *Mechanism of Ageing and Development* **15**, 1–7.

Dervan, P. B. 1986. Design of sequence-specific DNA-binding molecules. *Science* **232**, 464–71.

de Torres, M. L. & Abrisqueta, J. A. 1978. Study of human male meiosis. II. Q-banding in pachytene bivalents. *Human Genetics* **42**, 283–9.

Deumling, B. 1981. Sequence arrangement of a highly methylated satellite DNA of a plant, *Scilla*: a tandemly repeated inverted repeat. *Proceedings of the National Academy of Sciences* **78**, 338–42.

Deumling, B. & Greilhuber, J. 1982. Characterization of heterochromatin in different species of the *Scilla siberica* group (Liliaceae) by *in situ* hybridization of satellite DNAs and fluorochrome banding. *Chromosoma* **84**, 535–55.

Dev. V. G., Miller, D. A., Allderdice, P. W. & Miller, O. J. 1972a. Method for locating the centromeres of mouse meiotic chromosomes and its application to T163H and T70H translocations. *Experimental Cell Research* **73**, 259–62.

Dev. V. G., Miller, D. A., Miller, O. J., Marshall, J. T. & Hsu, T. C. 1973. Quinacrine fluorescence of *Mus cervicolor* chromosomes. Bright centromeric heterochromatin. *Experimental Cell Research* **79**, 475–9.

Dev, V. G., Warburton, D., Miller, O. J., Miller, D. A., Erlanger, B. F. & Beiser, S. M. 1972c. Consistant pattern of binding of nati-adenosine antibodies to human metaphose chromosomes. *Experimental Cell Research* **74**, 288–93.

Dev. V. G., Warburton, D. & Miller, O. J. 1972b. Giemsa banding of chromosomes. *Lancet* **1**, 1285.

Dhaliwal, M. K., Pathak, S., Shirley, L. R. & Flanagan, J. P. 1988. Ag-NOR staining in the Bennett wallaby, *Macropus rufogriseus*: evidence for dosage compensation. *Cytobios* **56**, 29–38.

di Castro, M., Prantera, G., Marchetti, E. & Rocchi, A. 1979. Characterization of the chromatin of *Asellus aquaticus* (Crust. Isop.) by treatment *in vivo* with BrdU and Hoechst 33258. *Caryologia* **32**, 81–8.

Dick, C. & Johns, E. W. 1968. The effect of two acetic acid containing fixatives on the histone content of calf thymus deoxyribonucleoprotein and calf thymus tissue. *Experimental Cell Research* **51**, 626–32.

di Paolo, J. A. & Popescu, N. C. 1974. Chromosome bands induced in human and Syrian hamster cells by chemical carcinogens. *British Journal of Cancer* **30**, 103–8.

Dipierri, J. E. & Fraisse, J. 1983a. A simple combined Ag-I/RHG technique for human metaphase chromosomes. *Human Genetics* **64**, 286–7.

Dipierri, J. E. & Fraisse, J. 1983b. Polymorphisme des bandes NOR dans une population francaise normale. *Annals de Genetique* **26**, 215–9.

Disney, J. E., Johnson, K. R., Magnuson, N. S., Sylvester, S. R. & Reeves, R. 1989. High-mobility group protein HMG-I localizes to G/Q- and C-bands of human and mouse chromosomes. *Journal of Cell Biology* **109**, 1975–82.

Distèche, C. & Bontemps, J. 1974. Chromosome regions containing DNAs of known base composition, specifically evidenced by 2,7-di-t-butyl proflavine. Comparison with the Q-banding and relation to dye-DNA interactions. *Chromosoma* **47**, 263–81.

Distèche, C., Bontemps, J., Houssier, C., Frederic, J. & Fredericq, E. 1980. Quantitative analysis of fluorescence profiles of chromosomes. Influence of DNA base composition on banding. *Experimental Cell Research* **125**, 251–64.

Döbel, P., Schubert, I. & Rieger, R. 1978. Distribution of heterochromatin in a reconstructed karyotype of *Vicia faba* as identified by banding and DNA-late replication patterns. *Chromosoma* **69**, 193–209.

Donahue, R. P., Bias, W. B., Renwick, J. H. & McKusick, V. A. 1968. Probable assignment of the Duffy blood group locus to chromosome 1 in man. *Proceedings of the National Academy of Sciences* **61**, 949–55.

Donlon, T. A. & Magenis, R. E. 1981. Structural organization of the heterochromatic region of human chromosome 9. *Chromosoma* **84**, 353–63.

Donlon, T. A. & Magenis, R. E. 1983. Methyl green is a substitute for distamycin A in the formation of distamycin A/DAPI C-bands. *Human Genetics* **65**, 144–6.

Douvas, A. S., Harrington, C. A. & Bonner, J. 1975. Major nonhistone proteins of rat liver chromatin: preliminary identification of myosin, actin, tubulin, and tropomyosin. *Proceedings of the National Academy of Sciences* **72**, 3902–6.

Dreskin, S. C. & Mayall, B. H. 1974. Deoxyribonucleic acid cytophotometry of stained human leukocytes. III. Thermal denaturation of chromatin. *Journal of Histochemistry and Cytochemistry* **22**, 120–6.

Dresser, M., Pisetsky, D., Warren, R., McCarty, G. & Moses, M. 1987. A new method for the cytological analysis of autoantibody specificities using whole-mount, surface-spread meiotic nuclei. *Journal of Immunological Methods* **104**, 111–21.

Drets, M. E. & Seuanez, H. 1974. Quantitation of heterogeneous human heterochromatin: microdensitometric analysis of C- and G-bands. In *Physiology and genetics of reproduction* part A, E. M. Coutinho & F. Fuchs (eds), 29–52. New York: Plenum Press.

Drets, M. E. & Shaw, M. W. 1971. Specific banding patterns of human chromosomes. *Proceedings of the National Academy of Sciences* **68**, 2073–7.

Drewry, A. 1982. G-banded chromosomes in *Pinus resinosa*. *Journal of Heredity* **73**, 305–6.
Drouin, R. & Richer, C-L. 1989. High-resolution R-banding at the 1250-band level. II. Schematic representation and nomenclature of human RBG-banded chromosomes. *Genome* **32**, 425–39.
Drouin, R., Lemieux, N. & Richer, C-L. 1988a. High-resolution R-banding at the 1250-band level. I. Technical considerations on cell synchronization and R-banding. *Cytobios* **56**, 107–25.
Drouin, R., Messier, P-E. & Richer, C-L. 1988b. Human chromosome banding specific for electron microscopy. *Cytogenetics and Cell Genetics* **47**, 117–20.
Drwinga, H. L. & Pathak, S. 1982. G bands of prematurely condensed mouse diplotene bivalents. *Cytogenetics and Cell Genetics* **33**, 264–6.
Duhamel-Maestracci, N., Simard, R., Harbers, K. & Spencer, J. H. 1979. Localization of satellite DNAs in the chromosomes of the guinea pig. *Chromosoma* **75**, 63–74.
Duijndam, W. A. L. & van Duijn, P. 1975. The influence of chromatin compactness on the stoichiometry of the Feulgen-Schiff procedure studied in model films. II. Investigations on films containing condensed or swollen chicken erythrocyte nuclei. *Journal of Histochemistry and Cytochemistry* **23**, 891–900.
Dunsmuir, P. 1976. Satellite DNA in the kangaroo *Macropus rufogriseus*. *Chromosoma* **56**, 111–25.
DuPraw, E. J. 1970. *DNA and chromosomes*. New York: Holt, Rinehart & Winston.
Dutrillaux, B. 1973. Nouveau système de marquage chromosomique: les bandes T. *Chromosoma* **41**, 395–402.
Dutrillaux, B. 1975. Obtention simultanée de plusieurs marquages chromosomiques sur les mêmes préparations, après traitement par le BrdU. *Humangenetik* **30**, 297–306.
Dutrillaux, B. 1979. Chromosomal evolution in primates: tentative phylogeny from *Microcebus murinus* (Prosimian) to man. *Human Genetics* **48**, 251–314.
Dutrillaux, B. & Couturier, J. 1983. The ancestral karyotype of Carnivora: comparison with that of platyrrhine monkeys. *Cytogenetics and Cell Genetics* **35**, 200–8.
Dutrillaux, B. & Covic, M. 1974. Étude de facteurs influençant la dénaturation thermique menagée des chromosomes. *Experimental Cell Research* **85**, 143–53.
Dutrillaux, B. & Lejeune, J. 1971. Sur une nouvelle technique d'analyse du caryotype humain. *Comptes Rendus de l'Academie des Sciences (Paris)* D, **272**, 2638–40.
Dutrillaux, B. & Viegas-Péquignot, E. 1981. High resolution R- and G-banding on the same preparation. *Human Genetics* **57**, 93–5.
Dutrillaux, B., Aurias, A. & Lombard, M. 1979. Présence de chromosomes communs chez un rongeur (*Eliomys quercinus*, Lerot) et chez les primates. *Annals of Génétics* **22**, 21–4.
Dutrillaux, B., Couturier, J. & Fosse, A-M. 1980a. The use of high resolution banding in comparative cytogenetics: comparison between man and *Lagothrix lagotricha* (Cebidae). *Cytogenetics and Cell Genetics* **27**, 45–51.
Dutrillaux, B., Viegas-Péquignot, E. & Couturier, J. 1980b. Très grande analogie de marquage chromosomique entre le lapin (*Oryctolagus cuniculus*) et les primates, dont l'homme. *Annales de Génétiques* **23**, 22–5.
Dutrillaux, B., de Grouchy, J., Finaz, C. & Lejeune, J. 1971. Mise en évidence de la structure fine des chromosomes humains par digestion enzymologique (pronase en particulier). *Comptes Rendus de l'Academie des Sciences (Paris)* D, **273**, 587–8.
Dutrillaux, B., Laurent, C., Couturier, J. & Lejeune, J. 1973a. Coloration des chromosomes humains par l'acridine orange après traitement par le 5 bromodeoxyuridine. *Comptes Rendus de l'Academie des Sciences (Paris)* D, **276**, 3179–81.
Dutrillaux, B., Rethore, M-O, Prieur, M. & Lejeune, J. 1973b. Analyse de la structure fine des chromosomes du Gorille (*Gorilla gorilla*). *Humangenetik* **20**, 343–54.
Dutrillaux, B., Viegas-Péquignot, E., Dubos, C. & Masse, R. 1978. Complete or almost complete analogy of chromosome banding between the baboon (*Papio papio*) and man. *Human Genetics* **43**, 37–46.
Dutrillaux, B., Rethoré, M-O., Prieur, M., Raoul, O., Berger, R. & Lejeune, J. 1972. Reconnaissance des chromosomes du groupe G par la méthode de dénaturation ménagée. *Experimental Cell Research* **70**, 453–5.

Dyer, A. F. 1963. Allocyclic segments of chromosomes and the structural heterozygosity that they reveal. *Chromosoma* **13**, 545–76.

Dzhemilev, Z. A. & Ataeva, D. M. 1975. Comparison of late incorporation patterns of thymidine-H^3 with distribution of G-segments in easily identifiable chromosomes of *Macaca mulatta. Byull, Eksp. Biol. Med.* **79**, 101–4.

Earnshaw, W. C. 1988, Mitotic chromosome structure. *Bioessays* **9**, 147–50.

Earnshaw, W. C. & Heck, M. M. S. 1985. Localization of topoisomerase II in mitotic chromosomes. *Journal of Cell Biology* **100**, 1716–25.

Earnshaw, W. C. & Migeon, B. R. 1985. Three related centromere proteins are absent from the inactive centromere of a stable isodicentric chromosome. *Chromosoma* **92**, 290–6.

Earnshaw, W. C. & Rothfield, N. 1985. Identification of a family of human centromere proteins using autoimmune sera from patients with scleroderma. *Chromosoma* **91**, 313–21.

Earnshaw, W. C., Ratrie, H. & Stetten, G. 1989. Visualization of centromere proteins CENP-B and CENP-C on a stable dicentric chromosome in cytological spreads. *Chromosoma* **98**, 1–12.

Earnshaw, W. C., Halligan, N., Cooke, C. & Rothfield, N. 1984. The kinetochore is part of the metaphase chromosome scaffold. *Journal of Cell Biology* **98**, 352–7.

Earnshaw, W. C., Machlin, P. S., Bordwell, B. J., Rothfield, N. F. & Cleveland, D. W. 1987a. Analysis of anticentromere autoantibodies using cloned autoantigen CENP-B. *Proceedings of the National Academy of Science* **84**, 4979–83.

Earnshaw, W. C., Sullivan, K. F., Machlin, P. S., Cooke, C. A., Kaiser, D. A., Pollard, T. D., Rothfield, N. F. & Cleveland, D. W. 1987b. Molecular cloning of cDNA for CENP-B, the major human centromere autoantigen. *Journal of Cell Biology* **104**, 817–29.

Eckhardt, R. A., & Gall, J. G. 1971. Satellite DNA associated with heterochromatin in *Rhynchosciara. Chromosoma* **32**, 407–27.

Eiberg, H. 1973. G, R and C banding patterns of human chromosomes produced by heat treatment in organic and inorganic salt solutions. *Clinical Genetics* **4**, 556–62.

Eiberg, H. 1974. New selective Giemsa technique for human chromosomes, C_d staining. *Nature* **248**, 55.

Einck, L. & Bustin, M. 1985. The intracellular distribution and function of the high mobility group chromosomal proteins. *Experimental Cell Research* **156**, 295–310.

Elder, F. F. B. 1980. Tandem fusion, centric fusion, and chromosomal evolution in the cotton rats, genus *Sigmodon. Cytogenetics and Cell Genetics* **26**, 199–210.

Elgin, S. C. R. 1988. The formation and function of DNase I hypersensitive sites in the process of gene activation. *Journal of Biological Chemistry* **263**, 19259–62.

Elgin, S. C. R. & Weintraub, H. 1975. Chromosomal proteins and chromatin structure. *Annual Review of Biochemistry* **44**, 725–74.

Ellison, J. R. & Barr, H. J. 1971. Differences in the quinacrine staining of the chromosomes of a pair of sibling species: *Drosophila melanogaster* and *Drosophila simulans. Chromosoma* **34**, 424–35.

Ellison, J. R. & Barr, H. J. 1972. Quinacrine fluorescence of specific chromosome regions. Late replication and high A : T content in *Samoaia leonensis. Chromosoma* **36**, 375–90.

Emanuel, B. S. 1978. Compound lateral asymmetry in human chromosome 6: BrdU-dye studies of 6q12→6q14. *American Journal of Human Genetics* **30**, 153–9.

Emanuel, B. S., Balaban, G., Boyd, J. P., Grossman, A., Negishi, M., Parmiter, A., & Glick, M. C. 1985. N-myc amplification in multiple homogeneously staining regions in two human neuroblastomas. *Proceedings of the National Academy of Science* **82**, 3736–40.

Endo, T. R. & Gill, B. S. 1984. Somatic karyotype, heterochromatin distribution, and nature of chromosome differentiation in common wheat, *Triticum aestivum* L. em Thell. *Chromosoma* **89**, 361–9.

Endow, S. A. & Gall, J. G. 1975. Differential replication of satellite DNA in polyploid tissues of *Drosophila virilis. Chromosoma* **50**, 175–92.

Ennis, T. J. 1975. Feulgen hydrolysis and chromosome banding in *Chilocorus* (Coleoptera: Coccinellidae). *Canadian Journal of Genetics and Cytology* **17**, 75–80.

Epplen, J. T., Siebers, J-W & Vogel, W. 1975. DNA replication patterns of human chromosomes from fibroblasts and amniotic fluid cells revealed by a Giemsa staining technique. *Cytogenetics and Cell Genetics* **15**, 177–85.

Erdtmann, B. 1982. Aspects of evaluation, significance, and evolution of human C-band heteromorphism. *Human Genetics* **61**, 281–94.

Erdtmann, B., Salzano, F. M., Mattevi, M. S. & Flores, R. Z. 1981. Quantitative analysis of C-bands in chromosomes 1, 9, and 16 of Brazilian Indians and Caucasoids. *Human Genetics* **57**, 58–63.

Ericsson, R. J. & Glass, R. H. 1982. Functional differences between sperm bearing the X- or Y-chromosome. In *Prospects for sexing mammalian sperm*, R. P. Amann & G. E. Seidel (eds), 201–11. Boulder: Colorado Associated University Press.

Esponda, P. 1978. Cytochemistry of kinetochores under electron microscopy. *Experimental Cell Research* **114**, 247–252.

Evans, E. P. 1989. Standard normal chromosomes. In *Genetic variants and strains of the laboratory mouse*, M. F. Lyon & A. G. Searle (eds). 576–81. Oxford: Oxford University Press.

Evans, H. J. & Ross, A. 1974. Spotted centromeres in human chromosomes. *Nature* **249**, 861–2.

Evans, H. J., Buckland, R. A. & Sumner, A. T. 1973. Chromosome homology and heterochromatin in goat, sheep and ox studied by banding techniques. *Chromosoma* **42**, 383–402.

Evans, H. J., Buckton, K. E. & Sumner, A. T. 1971. Cytological mapping of human chromosomes: results obtained with quinacrine fluorescence and the acetic-saline-Giemsa techniques. *Chromosoma* **35**, 310–25.

Ewart, J. C. 1899. *The Penycuik experiments*. London: A & C Black.

Faccio Dolfini, S. 1987. The effect of distamycin A on heterochromatin condensation of *Drosophila* chromosomes. *Chromosoma* **95**, 57–62.

Faccio Dolfini, S. & Bonifazio Razzini, A. 1983. High resolution of heterochromatin of *Drosophila melanogaster* by distamycin A. *Experientia* **39**, 1402–4.

Fakan, S. & Hernandez-Verdun, D. 1986. The nucleolus and the nucleolar organizer regions. *Biology of the Cell* **56**, 189–206.

Fantes, J. A., Green, D. K. & Cooke, H. J. 1983. Purifying human Y chromosomes by flow cytometry and sorting. *Cytometry* **4**, 88–91.

Fantes, J. A., Green, D. K., Malloy, P. & Sumner, A. T. 1989. Flow cytometry measurements of human chromosome kinetochore labelling. *Cytometry* **10**, 134–42.

Farber, R. A. & Liskay, R. M. 1974. Karyotypic analysis of a near-diploid established mouse cell line. *Cytogenetics and Cell Genetics* **13**, 384–96.

Faust, J. & Vogel, W. 1974. Are 'N bands' selective staining of specific heterochromatin? *Nature* **249**, 352–3.

Ferguson-Smith, M. A. & Page, B. M. 1973. Pachytene analysis in a human reciprocal (10;11) translocation. *Journal of Medical Genetics* **10**, 282–7.

Ferguson-Smith, M. A., Ferguson-Smith, M. E., Ellis, P. M. & Dickson, M. 1962. The sites and relative frequencies of secondary constrictions in human somatic chromosomes. *Cytogenetics* **1**, 325–43.

Ferguson-Smith, M. A., Newman, B. F., Ellis, P. M., Thomson, D. M. G. & Riley, I. D. 1973. Assignment by deletion of human red cell acid phosphatase gene locus to the short arm of chromosome 2. *Nature New Biology* **243**, 271–4.

Fernandez-Gomez, M. E., Sanchez-Pina, M-A., Risueño, M. C. & Medina, F-J. 1983. Differential staining of the nucleolar organising region (NOR) and nucleolar components by a new silver technique in plants. *Cellular and Molecular Biology* **29**, 181–7.

Ferraro, M. & Lavia, P. 1983. Activation of human ribosomal genes by 5-azacytidine. *Experimental Cell Research* **145**, 452–7.

Ferraro, M. & Lavia, P. 1985. Differential gene activity visualized on sister chromatids after replication in the presence of 5-azacytidine. *Chromosoma* **91**, 307–12.

Ferraro, M. & Prantera, G. 1988. Human NORs show correlation between transcriptional activity, DNase I sensitivity, and hypomethylation. *Cytogenetics and Cell Genetics* **47**, 58–61.

Ferraro, M., Lavia, P., Pelliccia, F. & de Capoa, A. 1981. Clonal inheritance of rRNA gene activity: cytological evidence in human cells. *Chromosoma* **84**, 345–51.

Ferraro, M., Lavia, P., Pelliccia, F. & de Capoa, A. 1982. Effects of potassium cyanide on silver stainability of specific cell structures. *Stain Technology* **57**, 259–63.

Ferrucci, L. & Mezzanotte, R. 1982. A cytological approach to the role of guanine in determining quinacrine fluorescence response in eukaryotic chromosomes. *Journal of Histochemistry and Cytochemistry* **30**, 1289–92.

Ferrucci, L. & Sumner, A. T. 1984. Further evidence for the DNA base specificity of light-induced banding. *Histochemical Journal* **16**, 521–8.

Ferrucci, L., Romano, E. & de Stefano, G. F. 1987. The *Alu* I-induced bands in great apes and man: implication for heterochromatin characterization and satellite DNA distribution. *Cytogenetics and Cell Genetics* **44**, 53–7.

Filion, W. G. 1974. Differential Giemsa staining in plants. I. Banding patterns in three cultivars of *Tulipa*. *Chromosoma* **49**, 51–60.

Finaz, C., Cochet, C. & de Grouchy, J. 1978. Identité des caryotypes de *Papio papio* et *Macaca mulatta* en bandes R, G, C et Ag-NOR. *Annales de Génétique* **21**, 149–51.

Finch, J. T. & Klug, A. 1976. Solenoidal model for superstructure in chromatin. *Proceedings of the National Academy of Sciences* **73**, 1897–901.

Fiskesjö, G. 1974. Two types of constitutive heterochromatin made visible in *Allium* by a rapid C-banding method. *Hereditas* **78**, 153–6.

Fiskesjö, G. 1988. The *Allium* test – an alternative in environmental studies: the relative toxicity of metal ions. *Mutation Research* **197**, 243–60.

Flamm, W. G., McCallum, M. & Walker, P. M. B. 1967. The isolation of complementary strands from a mouse DNA fraction. *Proceedings of the National Academy of Sciences* **57**, 1729–34.

Flamm, W. G., Walker, P. M. B. & McCallum, M. 1969. Some properties of the single strands isolated from the DNA of the nuclear satellite of the mouse (*Mus musculus*). *Journal of Molecular Biology* **40**, 423–43.

Fletcher, J. M. 1979. Light microscope analysis of meiotic prophase chromosomes by silver staining. *Chromosoma* **72**, 241–8.

Folle, G. A. & López de Griego, S. 1987. Clastogenicity of methylated p-benzoquinones: chemical warfare in nature? In *Cytogenetics*, G. Obe & A. Basler (eds), 361–78. Berlin: Springer.

Fontana, F. & Goldoni, D. 1986. Actinomycin D effects on termite chromosomes: induction of high resolution banding patterns with silver staining in mitotic stages. *Stain Technology* **61**, 361–6.

Ford, C. E., Pollock, D. L. & Gustavsson, I. 1980. Proceedings of the first international conference for the standardization of banded karyotypes of domestic animals. *Hereditas* **92**, 145–62.

Ford, E. B. 1960. *Mendelism and evolution*, 7th ed. London: Methuen.

Ford, E. H. R., Thurley, K. & Woollam, D. H. M. 1968. Electron-microscopic observations on whole human mitotic chromosomes. *Journal of Anatomy* **103**, 143–50.

Fox, D. P. & Santos, J. L. 1985. N-bands and nucleolus expression in *Schistocerca gregaria* and *Locusta migratoria*. *Heredity* **54**, 333–41.

Francke, U. & Oliver, N. 1978. Quantitative analysis of high-resolution trypsin-Giemsa bands on human prometaphase chromosomes. *Human Genetics* **45**, 137–65.

Franke, W. W., Deumling, B. & Zentgraf, H. 1973. Losses of material during cytological preparation of nuclei and chromosomes. *Experimental Cell Research* **80**, 445–9.

Fredga, K. 1971. Idiogram and fluorescence pattern of the chromosomes of the Indian muntjac. *Hereditas* **68**, 332–7.

Frediani, M., Mezzanotte, R., Vanni, R., Pignone, D. & Cremonini, R. 1987. The biochemical and cytological characterization of *Vicia faba* DNA by means of *Mbo* I, *Alu* I and *Bam* HI restriction endonucleases. *Theoretical and Applied Genetics* **75**, 46–50.

Freeman, M. V. R., Beiser, S. M., Erlanger, B. F. & Miller, O. J. 1971. Reaction of antinucleoside antibodies with human cells in vitro. *Experimental Cell Research* **69**, 345–55.

Friebe, B. 1977. Identification of heterochromatic regions in plant chromosomes by use of Giemsa banding techniques. *Microscopica Acta* **80**, 53–6.

Friedrich, U. & Therkelsen, A. J. 1982. An attempt to define !qh +, 9qh +, and 16qh +. *Human Genetics* **60**, 139–44.

Friend, K. K., Dorman, B. P., Kucherlapati, R. S. & Ruddle, F. H. 1976. Detection of interspecific translocations in mouse-human hybrids by alkaline Giemsa staining. *Experimental Cell Research* **99**, 31–6.

Fritschi, S. & Stranzinger, G. 1985. Fluorescent chromosome banding in inbred chicken: quinacrine bands, sequential chromomycin and DAPI bands, *Theoretical and Applied Genetics* **71**, 408–12.

Frommer, M., Prosser, J. & Vincent, P. C. 1984. Human satellite I sequences include a male specific 2.47 kb tandemly repeated unit containing one Alu family member per repeat. *Nucleic acids Research* **12**, 2887–900.

Fry, K. & Salser, W. 1977. Nucleotide sequences of HS-α satellite DNA from kangaroo rat *Dipodomys ordii* and characterization of similar sequences in other rodents. *Cell* **12**, 1069–84.

Fucik, W., Michaelis, A. & Rieger, R. 1970. On the induction of segment extension and chromatid structural changes in *Vicia faba* chromosomes after treatment with 5-azacytidine and 5-azadeoxycytidine. *Mutation Research* **9**, 599–606.

Fukuda, I. & Grant, W. F. 1980. Chromosome variation and evolution in *Trillium grandiflorum*. *Canadian Journal of Genetics and Cytology* **22**, 81–91.

Funaki, K., Matsui, S-I. & Sasaki, M. 1975. Location of nucleolar organizers in animal and plant chromosomes by means of an improved N-banding technique. *Chromosoma* **49**, 357–70.

Fussell, C. P. 1975. The position of interphase chromosomes and late replicating DNA in centromere and telomere regions of *Allium cepa* L. *Chromosoma* **50**, 201–10.

Gagné, R. 1981. DNase I digestion of fixed chromosomes specifically reveals NORs in man. *Experimental Cell Research* **131**, 476–8.

Gagné, R. & Laberge, C. 1972. Specific cytological recognition of the heterochromatic segment of number 9 chromosome in man. *Experimental Cell Research* **73**, 239–42.

Gagné, R., Laberge, C. & Tanguay, R. 1973. Aspect cytologique et localisation intranucléaire de l'hétérochromatine constitutive des chromosomes C9 chez l'homme. *Chromosoma* **41**, 159–66.

Gagné, R., Tanguay, R. & Laberge, C. 1971. Differential staining patterns of heterochromatin in man. *Nature New Biology* **232**, 29–30.

Gall, J. G., Cohen, E. H. & Polan, M. L. 1971. Repetitive DNA sequences in *Drosophila*. *Chromosoma* **33**, 319–44.

Gallagher, A., Hewitt, G. & Gibson, I. 1973. Differential Giemsa staining of heterochromatic B-chromosomes in *Myrmeleotettix maculatus* (Thunb.) (Orthoptera: Acrididae). *Chromosoma* **40**, 167–72.

Gallimore, P. H. & Richardson, C. R. 1973. An improved banding technique exemplified in the karyotype analysis of two strains of rat. *Chromosoma* **41**, 259–63.

Galloway, S. M. & Buckton, K. E. 1978. Aneuploidy and ageing: chromosome studies on a random sample of the population using G-banding. *Cytogenetics and Cell Genetics* **20**, 78–95.

Galloway, S. M. & Evans, H. J. 1975. Asymmetrical C-bands and satellite DNA in man. *Experimental Cell Research* **94**, 454–9.

Gallyas, F. 1982. Physico-chemical mechanism of the argyrophil III reaction. *Histochemistry* **74**, 409–21.

Gamperl, R., Ehmann, C. & Bachmann, K. 1982. Genome size and heterochromatin variation in rodents. *Genetica* **58**, 199–212.

Gamperl, R., Vistorin, G. & Rosenkranz, W. 1976. A comparative analysis of the karyotypes of *Cricetus cricetus* and *Cricetulus griseus*. *Chromosoma* **55**, 259–65.

Gamperl, R., Vistorin, G. & Rosenkranz, W. 1978. Comparison of chromosome banding patterns in five members of Cricetinae with comments on possible relationships. *Caryologia* **31**, 343–53.

Ganner, E. & Evans, H. J. 1971. The relationship between patterns of DNA replication and of quinacrine fluorescence in the human chromosome complement. *Chromosoma* **35**, 326–41.
Gao, X. & Patel, D. J. 1989. Solution structure of the chromomycin-DNA complex. *Biochemistry* **28**, 751–62.
García, M., Freitas, L., Miró, R. & Egozcue, J. 1976. Banding patterns of the chromosomes of *Cebus albifrons*. *Folia Primatologica* **25**, 313–9.
Gardiner, K., Watkins, P., Munke, M., Drabkin, H., Jones, C. & Patterson, D. 1988. Partial physical map of human chromosome 21. *Somatic Cell and Molecular Genetics* **14**, 623–38.
Gardiner-Garden, M. & Frommer, M. 1987. CpG islands in vertebrate genomes. *Journal of Molecular Biology* **196**, 261–82.
Gasser, S. M., Laroche, T., Falquet, J., Boy de la Tour, E. & Laemmli, U. K. 1986. Metaphase chromosome structure. Involvement of topoisomerase II. *Journal of Molecular Biology* **188**, 613–29.
Gatti, M., Pimpinelli, S. & Santini, G. 1976. Characterization of *Drosophila* heterochromatin. I. Staining and decondensation with Hoechst 33258 and quinacrine. *Chromosoma* **57**, 351–75.
Gazit, B., Cedar, H., Lerer, I. & Voss, R. 1982. Active genes are sensitive to deoxyribonuclease I during metaphase. *Science* **217**, 648–50.
Gendel, S. & Fosket, D.E. 1978. The role of chromosomal proteins in the C-banding of *Allium cepa* chromosomes. *Cytobios* **22**, 155–68.
George, D. L. 1984. Amplification of cellular proto-oncogenes in tumours and tumour cell lines. *Cancer Surveys* **3**, 497–513.
George, D. L. & Francke, U. 1980. Homogeneously staining chromosome regions and double minutes in a mouse adrenocortical tumour cell line. *Cytogenetics and Cell Genetics* **28**, 217–26.
George, K. P. 1970. Cytochemcial differentiation along human chromosomes. *Nature* **226**, 80–1.
George, K. P. 1971. Quinacrine mustard – a selective fluorescent stain for the Y chromosome in human tissues for routine cytogenetic screening. *Stain Technology* **46**, 34–6.
Geraedts, J. P. M. & Pearson, P. L. 1974. Fluorescent chromosome polymorphisms: frequencies and segregations in a Dutch population. *Clinical Genetics* **6**, 247–57.
Geraedts, J. P. M., Pearson, P. L., van der Ploeg, M. & Vossepoel, A. M. 1975. Polymorphisms for human chromosomes 1 and Y. *Experimental Cell Research* **95**, 9–14.
Gerlach, W. L. 1977. N-banded karyotypes of wheat species. *Chromosoma* **62**, 49–56.
Ghosh, P. K., Nand, R. & Rani, R. 1981. Restrictive distribution of asymmetric C-bands in the chromosome complement of the Rhesus (*Macaca mulatta*). *Experimental Cell Research* **135**, 424–7.
Ghosh, S. 1976. The nucleolar structure. *International Reviews in Cytology* **44**, 1–28.
Giles, V., Thode, G. & Alvarez, M. C. 1988. Early replication bands in two scorpion fishes, *Scorpaena porcus* and *S. notata* (order Scorpaneiformes). *Cytogenetics and Cell Genetics* **47**, 80–3.
Gill, D. 1979. Inhibition of fading in fluorescence microscopy of fixed cells. *Experientia* **35**, 400–1.
Gill, D. & Nadjar, G. 1977. Daunomycin-bands are similar to Q-bands on chromosomes of *Vicia faba*. *Experientia* **33**, 756–7.
Gill, J. E., Jotz, M. M., Young, S. G., Modest, E. J. & Sengupta, S. K. 1975. 7-amino-actinomycin D as a cytochemical probe. I. Spectral properties. *Journal of Histochemistry and Cytochemistry* **23**, 793–9.
Gledhill, B. L. 1983. Control of mammalian sex ratio by sexing sperm. *Fertility and Sterility* **40**, 572–4.
Goday, C. & Pimpinelli, S. 1984. Chromosome organization and heterochromatin elimination in *Parascaris*. *Science* **224**, 411–3.
Godward, M. B. E. 1985. The kinetochore. *International Reviews in Cytology* **94**, 77–105.
Goessens, G. 1984. Nucleolar structure. *International Reviews in Cytology* **87**, 107–58.

Goessens, G. & Lepoint, A. 1979. The nucleolus-organising regions (NORS): recent data and hypotheses. *Biologie Cellulaire* **35**, 211–20.
Gold, J. R. & Ellison, J. R. 1983. Silver staining for nucleolar organizing regions of vertebrate chromosomes. *Stain Technology* **58**, 51–5.
Goldman, M. A. 1988. The chromatin domain as a unit of gene regulation. *Bioessays* **9**, 50–5.
Goldman, M. A., Holmquist, G. P., Gray, M. C., Caston, L. A. & Nag, A. 1984. Replication timing of genes and middle repetitive sequences. *Science* **224**, 686–92.
Gollin, S. M., Wray, W., Hanks, S. K., Hittelman, W. N. & Rao, P. N. 1984. The ultrastructural organization of prematurely condensed chromosomes. *Journal of Cell Science* suppl. **1**, 203–21.
Golomb, H. M. & Bahr, G. F. 1974a. Correlation of the fluorescent banding pattern and ultrastructure of a human chromosome. *Experimental Cell Research* **84**, 121–6.
Golomb, H. M. & Bahr, G. F. 1974b. Human chromatin from interphase to metaphase. *Experimental Cell Research* **84**, 79–87.
Goodpasture, C. & Bloom, S. E. 1975. Visualization of nucleolar organizer regions in mammalian chromosomes using silver staining. *Chromosoma* **53**, 37–50.
Goradia, R. Y. & Davis, B. K. 1977. Banding and spiralization of human metaphase chromosomes. *Human Genetics* **36**, 155–60.
Gormley, I. P. & Ross, A. 1972. Surface topography of human chromosomes examined at each stage during ASG banding procedure. *Experimental Cell Research* **74**, 585–7.
Gormley, I. P. & Ross, A. 1976. Studies on the relationship of a collapsed chromosomal morphology to the production of Q- and G-bands. *Experimental Cell Research* **98**, 152–8.
Gosálvez, J., Bella, J. L., López-Fernández, C. & Mezzanotte, R. 1987. Correlation between constitutive heterochromatin and restriction enzyme resistant chromatin in *Arcyptera tornosi* (Orthoptera). *Heredity* **59**, 173–80.
Gosálvez, J., de la Torre, J., Garcia de la Vega, C. & López-Fernández, C. 1986. The effect of double-strength standard saline citrate on silver staining. I. Nucleoli and micronucleoli in the somatic and germ line of the grasshopper *Arcyptera fusca* (Orthoptera). *Canadian Journal of Genetics and Cytology* **28**, 219–26.
Gosden, J. R., Lawrie, S. S. & Cooke, H. J. 1981a. A cloned repeated DNA sequence in human chromosome heteromorphisms. *Cytogenetics and Cell Genetics* **29**, 32–9.
Gosden, J. R., Spowart, G. & Lawrie, S. S. 1981b. Satellite DNA and cytological staining patterns in heterochromatic inversions of human chromosome 9. *Human Genetics* **58**, 276–8.
Gosden, J. R., Buckland, R. A., Clayton, R. P. & Evans, H. J. 1975a. Chromosomal localization of DNA sequences in condensed and dispersed human chromatin. *Experimental Cell Research* **92**, 138–47.
Gosden, J. R., Mitchell, A. R. Buckland, R. A., Clayton, R. P. & Evans, H. J. 1975b. The location of four human satellite DNAs on human chromosomes. *Experimental Cell Research* **92**, 148–58.
Goto, K., Akematsu, T., Shimazu, H. & Sugiyama, T. 1975. Simple differential Giemsa staining of sister chromatids after treatment with photosensitive dyes and exposure to light and the mechanism of staining. *Chromosoma* **53**, 223–30.
Gottesfeld, J. M., Bonner, J., Radda, G. K. & Walker, I. O. 1974. Biophysical studies on the mechanism of quinacrine staining of chromosomes. *Biochemistry* **13**, 2937–45.
Goyanes, V. J. 1985. Electron microscopy of chromosomes: toward an ultrastructural cytogenetics? *Cancer Genetics and Cytogenetics* **15**, 349–67.
Gray, J. W., Langlois, R. G., Carrano, A. V., Burkhart-Schulte, K. & van Dilla, M. A. 1979. High resolution chromosome analysis: one and two parameter flow cytometry. *Chromosoma* **73**, 9–27.
Green, D. M. & Sharbel, T. F. 1988. Comparative cytogenetics of the primitive frog, *Leiopelma archeyi* (Anura, Leiopelmatidae). *Cytogenetics and Cell Genetics* **47**, 212–6.
Greenbaum, I. F. & Baker, R. J. 1978. Determination of the primitive karyotype for *Peromyscus*. *Journal of Mammalogy* **59**, 820–34.

Greenbaum, I. F. & Reed, M. J. 1984. Evidence for heterosynaptic pairing of the inverted segment in pericentric inversion heterozygotes of the deer mouse (*Peromyscus maniculatus*). *Cytogenetics and Cell Genetics* **38**, 106–11.

Greenbaum, I. F., Baker, R. J. & Bowers, J. H. 1978a. Chromosomal homology and divergence between sibling species of deer mice: *Peromyscus maniculatus* and *P. melanotis* (Rodentia, Cricetidae). *Evolution* **32**, 334–341.

Greenbaum, I. F., Baker, R. J. & Ramsey, P. R. 1978b. Chromosomal evolution and the mode of speciation in three species of *Peromyscus*. *Evolution* **32**, 646–54.

Greilhuber, J. 1973. Differential staining of plant chromosomes after hydrochloric acid treatments (Hy bands). *Österr. Bot. Z.* **122**, 333–51.

Greilhuber, J. 1975. Heterogeneity of heterochromatin in plants: comparison of Hy- and C-bands in *Vicia faba*. *Plant Syst. Evol.* **124**, 139–56.

Greilhuber, J. 1977a. Nuclear DNA and heterochromatin contents in the *Scilla hohenackeri* group, *S. persica*, and *Puschkinia scilloides* (Liliaceae). *Plant Syst. Evol.* **128**, 243–57.

Greilhuber, J. 1977b. Why plant chromosomes do not show G-bands. *Theoretical and Applied Genetics* **50**, 121–4.

Greilhuber, J. 1979. C-band distribution, DNA content, and base composition in *Adoxa moschatellina* (Adoxaceae), a plant with cold-sensitive chromosome segments. *Plant Syst. Evol.* **131**, 243–59.

Greilhuber, J. & Speta, F. 1978. Quantitative analyses of C-banded karyotypes, and systematics in the cultivated species of the *Scilla siberica* group (Liliaceae). *Plant Syst. Evol.* **129**, 63–109.

Greilhuber, J., Deumling, B. & Speta, F. 1981. Evolutionary aspects of chromosome banding, heterochromatin, satellite DNA, and genome size in *Scilla* (Liliaceae). *Ber. Deutsch. Bot. Ges.* **94**, 249–66.

Griffith, J. D. 1975. Chromatin structure: deduced from a minichromosome. *Science* **187**, 1202–3.

Gropp, A., Winking, H., Zech, L. & Müller, H. 1972. Robertsonian chromosomal variation and identification of metacentric chromosomes in feral mice. *Chromosoma* **39**, 265–88.

Gross, D. S. & Garrard, W. T. 1988. Nuclease hypersensitive sites in chromatin. *Annual Reviews of Biochemistry* **57**, 159–97.

Guldner, H. H., Lakomek, H-J. & Bautz, F A. 1984. Human anti-centromere sera recognize a 19.5 kD non-histone chromosomal protein from HeLa cells. *Clinical and Experimental Immunology* **58**, 13–20.

Guo, X-C., Morgan, W. F., & Cleaver, J. E. 1986. Hoechst 33258 dye generates DNA-protein cross-links during ultraviolet light-induced photolysis of bromodeoxyuridine in replicated and repaired DNA. *Photochemistry and Photobiology* **44**, 131–6.

Gurley, L. R., Walters, R. A. & Tobey, R. A. 1973. Histone phosphorylation in late interphase and mitosis. *Biochemical and Biophysical Research Communications* **50**, 744–50.

Gurley, L. R., Walters, R. A. & Tobey, R. A. 1974. Cell cycle-specific changes in histone phosphorylation associated with cell proliferation and chromosome condensation. *Journal of Cell Biology* **60**, 356–64.

Gustavsson, I. 1980. Banding techniques in chromosome analysis of domestic animals. *Advances in Veterinary Science and Comparative Medicine* **24**, 245–89.

Gustavsson, I. 1988. Standard karyotype of the domestic pig. *Hereditas* **109**, 151–7.

Haaf, T. & Schmid, M. 1984. An early stage of ZW/ZZ sex chromosome differentiation in *Poecilia sphenops* var. *melanistica* (Poeciliidae, Cyprinodontiformes). *Chromosoma* **89**, 37–41.

Haaf, T. & Schmid, M. 1987. Chromosome heteromorphisms in the gorilla karyotype. *Journal of Heredity* **78**, 287–92.

Haaf, T. & Schmid, M. 1989. 5-azadeoxycytidine induced undercondensation in the giant X chromosomes of *Microtus agrestis*. *Chromosoma* **98**, 93–8.

Haaf. T., Muller, H. & Schmid, M. 1986a. Distamycin A/DAPI staining of heterochromatin in male meiosis of man. *Genetica* **70**, 179–85.

Haaf, T., Ott, G. & Schmid, M. 1986b. Differential inhibition of sister chromatid condensation induced by 5-azadeoxycytidine in human chromosomes. *Chromosoma* **94**, 389–94.
Haaf, T., Ott, G. & Schmid, M. 1988a. Inhibition of condensation in the late-replicating X chromosome induced by 5-azadeoxycytidine in human lymphocyte cultures. *Human Genetics* **79**, 18–23.
Haaf, T., Reimer, G. & Schmid, M. 1988b. Immunocytogenetics: localization of transcriptionally active rRNA genes in nucleoli and nucleolus organizer regions by use of human autoantibodies to RNA polymerase I. *Cytogenetics and Cell Genetics* **48**, 35–42.
Haaf, T., Weis, H., Schindler, D. & Schmid, M. 1984. Specific silver staining of experimentally undercondensed chromosome regions. *Chromosoma* **90**, 149–55.
Haaf, T., Feichtinger, W., Guttenbach, M., Sanchez, L., Muller, C. R. & Schmid, M. 1989. Berenil-induced undercondensation in human heterochromatin. *Cytogenetics and Cell Genetics* **50**, 27–33.
Hadlaczky, Gy., Sumner, A. T. & Ross, A. 1981a. Protein-depleted chromosomes. I. Structure of isolated protein-depleted chromosomes. *Chromosoma* **81**, 537–55.
Hadlaczky, Gy., Sumner, A. T. & Ross, A. 1981b. Protein-depleted chromosomes. II. Experiments concerning the reality of chromosome scaffolds. *Chromosoma* **81**, 557–67.
Hadlaczky, Gy., Went, M. & Ringertz, N. R. 1986. Direct evidence for the non-random localization of mammalian chromosomes in the interphase nucleus. *Experimental Cell Research* **167**, 1–15.
Hadlaczky, Gy., Praznovsky, T., Rasko, I. & Kereso, J. 1989. Centromere proteins. I. Mitosis specific centromere antigen recognized by anti-centromere autoantibodies. *Chromosoma* **97**, 282–8.
Hägele, K. 1977. N-banding in polytene chromosomes of *Chironomus* and *Drosophila*. *Chromosoma* **63**, 79–88.
Hägele, K. 1979. Characterization of heterochromatin in *Schistocerca gregaria* by C- and N-banding methods. *Chromosoma* **70**, 239–50.
Haiduk, M. W. & Baker, R. J. 1982. Cladistical analysis of G-banding chromosomes of nectar feeding bats (Glossophaginae: Phyllostomidae). *Syst. Zool.* **31**, 252–65.
Hajduk, S. L. 1976. Demonstration of kinetoplast DNA in dyskinetoplastic strains of *Trypanosoma equiperdum*. *Science* **191**, 858–9.
Hale, D. W. 1986. Heterosynapsis and suppression of chiasmata within heterozygous pericentric inversions of the Sitka deer mouse. *Chromosoma* **94**, 425–32.
Halleck, M. S. & Gurley, L. R. 1980. Histone H2A subfractions and their phosphorylation in cultured *Peromyscus* cells. *Experimental Cell Research* **125**, 377–88.
Halleck, M. S. & Gurley, L. R. 1981. Histone acetylation and heterochromatin content of cultured *Peromyscus* cells. *Experimental Cell Research* **132**, 201–13.
Halleck, M. S. & Schlegel, R. A. 1983. C-banding of *Peromyscus* constitutive heterochromatin persists following histone hyperacetylation. *Experimental Cell Research* **147**, 269–79.
Haluska, F. G., Tsujimoto, Y. & Croce, C. M. 1987. Oncogene activation by chromosome translocation in human malignancy. *Annual Review of Genetics* **21**, 321–45.
Hamada, S. & Fujita, S. 1983. Fluorescence enhancement in DNA-DAPI complexes. *Acta Histochemica et Cytochemica* **16**, 606–9.
Hamilton, L. D., Barclay, R. K., Wilkins, M. H. F., Brown, G. L., Wilson, H. R., Marvin, D. A., Ephrussi-Taylor, H. & Simmons, N. S. 1959. Similarity of the structure of DNA from a variety of sources. *Journal of Biophysical and Biochemical Cytology* **5**, 397–404.
Hamlin, J. L., Milbrandt, J. D., Heintz, N. H. & Azizkhan, J. C. 1984. DNA sequence amplification in mammalian cells. *International Review of Cytology* **90**, 31–82.
Hampel, K. E. & Levan, A. 1964. Breakage in human chromosomes induced by low temperature. *Hereditas* **51**, 315–43.
Hancock, J. M. & Burns, R. G. 1987. Specificity and biological significance of microtubule-associated protein-DNA interactions in chick. *Biochimica Biophysica Acta* **927**, 163–9.
Hancock, J. M. & Sumner, A. T. 1982. The role of proteins in the production of different types of chromosome bands. *Cytobios* **35**, 37–46.

Hand, R. 1978. Eucaryotic DNA: organization of the genome for replication. *Cell* **15**, 317–25.
Hanlon, S., Johnson, R. S., Wolf, B. & Chan, A. 1972. Mixed conformations of deoxyribonucleic acid in chromatin: a preliminary report. *Proceedings of the National Academy of Sciences* **69**, 3263–7.
Hansen, K. M. 1972a. Bovine chromosomes identified by quinacrine mustard and fluorescence microscopy. *Hereditas* **70**, 225–34.
Hansen, K. M. 1972b. The karyotype of the pig (*Sus scrofa domestica*), identified by quinacrine mustard staining and fluorescence microscopy. *Cytogenetics* **11**, 286–94.
Hansen-Melander, E., Melander, Y. & Olin, M. L. 1974. Chromosome preparation by air drying at low temperature and Giemsa banding procedures. *Hereditas* **76**, 35–40.
Harbers, K. & Spencer, J. H. 1974. Nucleotide clusters in deoxyribonucleic acids. Pyrimidine oligonucleotides of mouse L-cell satellite deoxyribonucleic acid and main-band deoxyribonucleic acid. *Biochemistry* **13**, 1094–101.
Harnden, D. G. 1974. Skin culture and solid tumor technique. In *Human chromosome methodology*, 2nd ed., J. J. Yunis (ed.), 167–84. New York: Academic Press.
Harris, P., Boyd, E., Young, B. D. & Ferguson-Smith, M. A. 1986. Determination of the DNA content of human chromosomes by flow cytometry. *Cytogenetics and Cell Genetics* **41**, 14–21.
Harrison, C. J., Britch, M., Allen, T. D. & Harris, R. 1981. Scanning electron microscopy of the G-banded human karyotype. *Experimental Cell Research* **134**, 141–53.
Harrison, C. J., Jack, E. M., Allen, T. D. & Harris, R. 1985a. Investigation of human chromosome polymorphisms by scanning electron microscopy. *Journal of Medical Genetics* **22**, 16–23.
Harrison, C. J., Jack, E. M., Allen, T. D. & Harris, R. 1985b. Light and scanning electron microscopy of the same human metaphase chromosomes. *Journal of Cell Science* **77**, 143–53.
Harshman, K. D. & Dervan, P. B. 1985. Molecular recognition of B-DNA by Hoechst 33258. *Nucleic Acids Research* **13**, 4825–35.
Hartley, S. E. 1987. The chromosomes of Salmonid fishes. *Biological Reviews* **62**, 197–214.
Hartley, S. E. & Horne, M. T. 1984. Chromosome polymorphism and constitutive heterochromatin in the Atlantic salmon, *Salmo salar*. *Chromosoma* **89**, 377–80.
Hartung, M., Mirre, C. & Stahl, A. 1979. Nucleolar organizers in human oocytes at meiotic prophase I, studied by the silver-NOR method and electron microscopy. *Human Genetics* **52**, 295–308.
Hassold, T. J. 1986. Chromosome abnormalities in human reproductive wastage. *Trends in Genetics* **2**, 105–10.
Hassold, T., Chiu, D. & Yamane, J. A. 1984. Parental origin of autosomal trisomies. *Annals of Human Genetics* **48**, 129–44.
Hassold, T., Jacobs, P. A., Leppert, M. & Sheldon, M. 1987. Cytogenetic and molecular studies of trisomy 13. *Journal of Medical Genetics* **24**, 725–32.
Hassold, T., Kumlin, E., Takaesu, N. & Leppert, M. 1985. The use of restriction fragment length polymorphisms to study the origin of human aneuploidy. *Annals of the New York Academy of Sciences* **450**, 179–89.
Hatami-Monazah, H. 1974. Banding and chromatid separation in Chinese hamster chromosomes. *Nature* **249**, 827–8.
Hatfield, J. M. R., Peden, K. W. C. & West, R. M. 1975. Binding of quinacrine to the human Y chromosome. *Chromosoma* **52**, 67–71.
Hatsumi, M. 1987. Karyotype polymorphism in *Drosophila albomicans*. *Genome* **29**, 395–400.
Hatton, K. S., Dhar, V., Brown, E. H., Iqbal, M. A., Stuart, S., Didamo, V. T. & Schildkraut, C. L. 1988. Replication program of active and inactive multigene families in mammalian cells. *Molecular and Cellular Biology* **8**, 2149–58.
Hauser-Urfer, I., Leeman, U. & Ruch, F. 1982. Cytofluorometric determination of the DNA base content in human chromosomes with quinacrine mustard, Hoechst 33258, DAPI, and mithramycin. *Experimental Cell Research* **142**, 455–9.

Hayman, D. & Sharp, P. 1981. Hoechst 33258 induced undercondensed sites in marsupial chromosomes. *Chromosoma* **83**, 249–62.

Hazen, M. W., Kuo, M. T. & Arrighi, F. E. 1977. Genome analysis of *Peromyscus* (Rodentia, Cricetidae), VII. Localization of satellite DNA sequences and cytoplasmic poly(A) RNA sequences of *P. eremicus* on metaphase chromosomes. *Chromosoma* **64**, 133–42.

Hegde, U. C., Shastry, P. R. & Rao, S. S. 1978. Use of dithiothreitol for improved visibility of the F-body in human Y-bearing spermatozoa. *Journal of Reproduction and Fertility* **53**, 403–5.

Heim, & Mitelman, F. 1987. *Cancer cytogenetics.* New York: Alan R. Liss.

Heitz, E. 1928. Das Heterochromatin der Moose. I. *Jahrb. Wiss. Bot.* **69**, 762–818.

Heitz, E. 1933. Die Herkunft der Chromocentren. *Planta* **18**, 571–636.

Henderson, A. S., Warburton, D. & Atwood, K. C. 1972. Location of ribosomal DNA in the human chromosome complement. *Proceedings of the National Academy of Sciences* **69**, 3394–8.

Henderson, L. M. & Bruère, A. N. 1980. Inheritance of Ag-stainability of the nucleolus organiser regions in domestic sheep, *Ovis aries*. *Cytogenetics and Cell Genetics* **26**, 1–6.

Heneen, W. K. 1975. Ultrastructure of the prophase kinetochore in cultured cells of rat-kangaroo (*Potorous tridactylis*). *Hereditas* **79**, 209–20.

Heneen, W. K. & Brismar, K. 1987. Rye heterochromatin in the somatic chromosomes of triticale in relation to grain shrivelling. *Hereditas* **107**, 137–45.

Heneen, W. K. & Caspersson, T. 1973. Identifying the chromosomes of rye by distribution patterns of DNA. *Hereditas* **74**, 259–72.

Hennig, W. 1972. Highly repetitive DNA sequences in the genome of *Drosophila hydei*. I. Preferential localization in the X chromosomal heterochromatin *Journal of Molecular Biology* **71**, 407–17.

Hennig, W. 1985. Y chromosome function and spermatogenesis in *Drosophila hydei*. *Advances in Genetics* **23**, 179–234.

Henry, S. M., Weinfeld, H. & Sandberg, A. A. 1980. Effects of divalent cations and glucose on mitotic-like events in fused interphase-metaphase cells. *Experimental Cell Research* **125**, 351–62.

Heppich, S. 1978. Hybridogenesis in *Rana esculenta*: C-band karyotypes of *Rana ridibunda*, *Rana lessonae* and *Rana esculenta*. *Z. zool. Systematik Evolutionsforsch.* **16**, 27–39.

Herbst, E. W., Fredga, K., Frank, F., Winking, H. & Gropp, A. 1978. Cytological identification of two X chromosome types in the wood lemming (*Myopus schisticolor*). *Chromosoma* **69**, 185–91.

Hernandez-Verdun, D., Hubert, J., Bourgeois, C. A. & Bouteille, M. 1980. Ultrastructural localization of Ag-NOR stained proteins in the nucleolus during the cell cycle and in other nucleolar structures. *Chromosoma* **79**, 349–62.

Herrera, A. H. & Olson, M. O. J. 1986. Association of protein C23 with rapidly labelled nucleolar RNA. *Biochemistry* **25**, 6258–64.

Herrero, P. & Garcia de la Vega, C. 1986. Intermediate heterochromatic zones in *Triturus* (Amphibia: Caudata). *Cytobios* **46**, 115–18.

Herrero, P. & Gosálvez, J. 1985. Presence of interchromatidic bands in mitotic chromosomes of *Triturus marmoratus* (Amphibia: Caudata). *Experientia* **41**, 683–4.

Heyting, C., Kroes, W., Kriek, E., Meyer, I. & Slater, R. 1985. Hybridisation of N-acetoxy-N-acetyl-2-aminofluorene-labelled RNA to Q-banded metaphase chromosomes. *Acta Histochemica* **77**, 177–84.

Hill, R. J. & Stollar, B. D. 1983. Dependence of Z-DNA antibody binding to polytene chromosomes on acid fixation and DNA torsional strain. *Nature* **305**, 338–40.

Hilliker, A. J., Appels, R. & Schalet, A. 1980. The genetic analysis of *D. melanogaster* heterochromatin. *Cell* **21**, 607–19.

Hilwig, I. & Gropp, A. 1972. Staining of constitutive heterochromatin in mammalian chromosomes with a new fluorochrome. *Experimental Cell Research* **75**, 122–6.

Hilwig, I. & Gropp, A. 1973. Decondensation of constitutive heterochromatin in L cell chromosomes by a benzimidazole compound ('Hoechst 33258'). *Experimental Cell Research* **81**, 474–7.

Hilwig, I. & Gropp, A. 1975. pH-dependent fluorescence of DNA and RNA in cytologic staining with 'Hoechst 33258'. *Experimental Cell Research* **91**, 457–60.

Hizume, M., Sato, S. & Tanaka, A. 1980. A highly reproducible method of nucleolus organizing regions staining in plants. *Stain Technology* **55**, 87–90.

Hoehn, H. 1975. Functional implications of differential chromosome banding. *American Journal of Human Genetics* **27**, 676–85.

Hoehn, H. & Martin, G. M. 1972. Heritable alteration of human constitutive heterochromatin induced by mitomycin C. *Experimental Cell Research* **75**, 275–8.

Hofgärtner, F. J., Schmid, M., Krone, W., Zenzes, M. T. & Engel, W. 1979. Pattern of activity of nucleolus organizers during spermatogenesis in mammals as analyzed by silver-staining. *Chromosoma* **71**, 197–216.

Holden, J. J. A., Reimer, D. L., Higgins, M. J., Roder, J. C. & White, B. N. 1985. Amplified sequences from chromosome 15, including centromeres, nucleolar organizer regions, and centromeric heterochromatin, in homogeneously staining regions in the human melanoma cell line MeWo. *Cancer Genetics and Cytogenetics* **14**, 131–46.

Hollander, D. H., Litton, L. E. & Liang, Y. W. 1976. Ethidium bromide counterstain for differentiation of quinacrine stained interphase bodies and brilliant metaphase bands. *Experimental Cell Research* **99**, 174–5.

Holm, P. B. 1976. The C and Q banding patterns of the chromosomes of *Lilium longiflorum* (Thunb.). *Carlsberg Research Communications* **41**, 217–24.

Holmberg, M. & Jonasson, J. 1973. Preferential location of X-ray induced chromosome breakage in the R-bands of human chromosomes. *Hereditas* **74**, 57–68.

Holmquist, G. 1975a. Hoechst 33258 fluorescent staining of *Drosophila* chromosomes. *Chromosoma* **49**, 333–56.

Holmquist, G. 1975b. Organisation and evolution of *Drosophila virilis* heterochromatin. *Nature* **257**, 503–6.

Holmquist, G. 1979. The mechanism of C-banding: depurination and β-elimination. *Chromosoma* **72**, 203–24.

Holmquist, G. 1988a. DNA sequences in G-bands and R-bands. In *Chromosomes and chromatin*, Vol. II, K. W. Adolph (ed.), 76–121. Boca Raton: CRC Press.

Holmquist, G. P. 1988b. Mobile genetic elements in G-band and R-band DNA. In *The cytogenetics of mammalian autosomal rearrangements*, A. Daniel (ed.), 803–33. New York: Alan R. Liss.

Holmquist, G. P. & Caston, L. A. 1986. Replication time of interspersed repetitive DNA sequences in hamsters. *Biochimica Biophysica Acta* **868**, 164–77.

Holmquist, G. P. & Comings, D. E. 1975. Sister chromatid exchange and chromosome organization based on a bromodeoxyuridine Giemsa-C-banding technique (TC-banding). *Chromosoma* **52**, 245–59.

Holmquist, G., Gray, M., Porter, T. & Jordan, J. 1982. Characterization of Giemsa dark- and light-band DNA. *Cell* **31**, 121–9.

Holmquist, R., Miyamoto, M. M. & Goodman, M. 1988. Higher-primate phylogeny – why can't we decide? *Molecular Biology and Evolution* **5**, 201–16.

Hong, Y. & Zhou, T. 1985. Chromosome banding in fishes. I. An improved BrdU-Hoechst-Giemsa method for revealing DNA-replication bands in fish chromosomes (In Chinese). *Acta Genetica Sinica* **12**, 67–71.

Hoo, J. J. 1986. Routine application of high resolution chromosome analysis. *American Journal of Medical Genetics* **24**, 533–7.

Hook, E. B. & Hamerton, J. L. 1977. The frequency of chromosome abnormalities detected in consecutive newborn studies. In *Population cytogenetics*, E. B. Hook & I. H. Porter (eds), 63–79. New York: Academic Press.

Horobin, R. W. & Walter, K. J. 1987. Understanding Romanowsky staining. I. The Romanowsky-Giemsa effect in blood smears. *Histochemistry* **86**, 331–6.

Hors-Cayla, M. C., Heuertz, S. & Frezal, J. 1983. Coreactivation of four inactive X genes in a hamster × human hybrid and persistence of late replication of reactivated X chromosome. *Somatic Cell Genetics* **9**, 645–57.

Houghton, J. A. & Houghton, S. E. 1978. Satellite associations of human chromosomes. *Science Progr.* **65**, 331–41.
Howell, W. M. 1982. Selective staining of nucleolus organizer regions (NORs). In *The cell nucleus*, Vol. 11, H. Busch & L. Rothblum (eds), 89–142. New York: Academic Press.
Howell, W. M. & Black, D. A. 1978. A rapid technique for producing silver-stained nucleolus organizer regions and trypsin-Giemsa bands on human chromosomes. *Human Genetics* **43**, 53–6.
Howell, W. M. & Black, D. A. 1980. Controlled silver-staining of nucleolus organizer regions with a protective colloidal developer: a 1-step method. *Experientia* **36**, 1014–15.
Howell, W. M. & Bloom, S. E. 1973. Sex-associated differential fluorescence of mudminnow chromosomes and spermatozoa. *Nature* **245**, 261–3.
Howell, W. M. & Denton, T. E. 1974. An ammoniacal-silver stain technique for satellite III DNA regions on human chromosomes. *Experientia* **30**, 1364–5.
Howell, W. M. & Denton, T. E. 1976. Negative silver staining in A-T and satellite DNA-rich regions of human chromosomes. *Chromosoma* **57**, 165–9
Howell, W. M. & Hsu, T. C. 1979. Chromosome core structure revealed by silver staining. *Chromosoma* **73**, 61–6.
Howell, W. M., Denton, T. E. & Diamond, J. R. 1975. Differential staining of the satellite regions of human acrocentric chromosomes. *Experientia* **31**, 260–2.
Hozier, J. C., Furcht, L. T. & Wendelshafer-Grabb, G. 1981. Structure of human chromosomes visualized at the electron microscope level. *Chromosoma* **82**, 55–64.
Hsieh, T. & Brutlag, D. L. 1979. A protein that preferentially binds *Drosophila* satellite DNA. *Proceedings of the National Academy of Sciences* **76**, 726–30.
Hsu, L. Y. F., Benn, P. A., Tannenbaum, H. L., Perlis, T. E. & Carlson, A. D. 1987. Chromosomal polymorphisms of 1, 9, 16 and Y in 4 major ethnic groups: a large prenatal study. *American Journal of Medical Genetics* **26**, 95–101.
Hsu, T. C. 1971. Heterochromatin pattern in metaphase chromosomes of *Drosophila melanogaster*. *Journal of Heredity* **62**, 285–7.
Hsu, T. C. 1974. Longitudinal differentiation of chromosomes. *Annual Review of Genetics* **7**, 153–76.
Hsu, T. C. 1975. A possible function of constitutive heterochromatin: the bodyguard hypothesis. *Genetics* **79**, 137–50.
Hsu, T. C. 1976. Editorial. *Mammalian Chromosomes Newsletter* **17**, pt. 3, 1–3.
Hsu, T. C. & Arrighi, F. E. 1966. Chromosomal evolution in the genus *Peromyscus* (Cricetidae, Rodentia). *Cytogenetics* **5**, 355–9.
Hsu, T. C. & Somers, C. E. 1961. Effect of 5-bromodeoxyuridine on mammalian chromosomes. *Proceedings of the National Academy of Sciences* **47**, 396–403.
Hsu, T. C., Arrighi, F. E. & Saunders, G. F. 1972. Compositional heterogeneity of human heterochromatin. *Proceedings of the National Academy of Sciences* **69**, 1464–66.
Hsu, T. C., Pathak, S. & Shafer, D. A. 1973. Induction of chromosome crossbanding by treating cells with chemical agents before fixation. *Experimental Cell Research* **79**, 484–7.
Hsu, T. C., Spirito, S. E. & Pardue, M. L. 1975. Distribution of 18+28S ribosomal genes in mammalian genomes. *Chromosoma* **53**, 25–36.
Htun, H. & Dahlberg, J. E. 1989. Topology and formation of triple-stranded H-DNA. *Science* **243**, 1571–6.
Hubbell, H. R. 1985. Silver staining as an indicator of active ribosomal genes. *Stain Technology* **60**, 285–94.
Hubbell, H. R. & Hsu, T. C. 1977. Identification of nucleolus organizer regions (NORs) in normal and neoplastic cells by the silver-staining technique. *Cytogenetics and Cell Genetics* **19**, 185–96.
Hubbell, H. R., Rothblum, L. I. & Hsu, T. C. 1979. Identification of a silver binding protein associated with the cytological silver staining of actively transcribing nucleolar regions. *Cell Biology International Reports* **3**, 615–22.

Hudson, A. P., Cuny, G., Cortadas, J., Haschemeyer, A. E. V. & Bernardi, G. 1980. An analysis of fish genomes by density gradient centrifugation. *European Journal of Biochemistry* **112**, 203–10.

Hungerford, D. A. 1971. Chromosome structure and function in man. I. Pachytene mapping in the male, improved methods and general discussion of initial results. *Cytogenetics* **10**, 23–32.

Hungerford, D. A. & Hungerford, A. M. 1978. Chromosome structure and function in man. VI. Pachytene chromomere maps of 16, 17 and 18; pachytene as a reference standard for metaphase banding. *Cytogenetics and Cell Genetics* **21**, 212–30.

Hungerford, D. A. & Hungerford, A. M. 1979. A pachytene chromomere map of chromosome 10. *Human Genetics* **52**, 153–5.

Hutchison, N. & Weintraub, H. 1985. Localization of DNAase I-sensitive sequences to specific regions of interphase nuclei. *Cell* **43**, 471–82.

Ibraimov, A. I. & Mirrakhimov, M. M. 1985. Q-band polymorphism in the autosomes and the Y chromosome in human populations. In *The Y chromosome*, part A, A. A. Sandberg (ed.), 213–87. New York: Alan R. Liss.

Ibraimov, A. I., Mirrakhimov, M. M., Nazarenko, S. A., Axenrod, E. I. & Akbanova, G. A. 1982. Human chromosomal polymorphism. I. Chromosomal Q polymorphism in Mongoloid populations of Central Asia. *Human Genetics* **60**, 1–7.

Ieshima, A., Yorita, T. & Takeshita, K. 1984. A simple R-banding technique by BrdU-Hoechst 33258 treatment and Giemsa staining following heating and ultraviolet exposure. *Human Genetics* **29**, 133–8.

Ikemura, T. & Aota, S-I. 1988. Global variation in G+C content along vertebrate genome DNA. Possible correlation with chromosome band structures. *Journal of Molecular Biology* **203**, 1–13.

Ikeuchi, T. 1984. Inhibitory effect of ethidium bromide on mitotic chromosome condensation and its application to high-resolution chromosome banding. *Cytogenetics and Cell Genetics* **38**, 56–61.

Ikeuchi, T. & Sasaki, M. 1979. Accumulation of early mitotic cells in ethidium bromide-treated human lymphocyte cultures. *Proceedings of the Japanese Academy.* series B, **55**, 15–18.

ISCN 1985. *An international system for human cytogenetic nomenclature*, D. G. Harnden & H. P. Klinger (eds). Basel: Karger.

Islam, M. Q. & Levan, G. 1987. A new fixation procedure for improved quality G-bands in routine cytogenetic work. *Hereditas* **107**, 127–30.

Iverson, G. M., Bianchi, D. W., Cann, H. M. & Herzenberg, L. A. 1981. Detection and isolation of fetal cells from maternal blood using the fluorescence-activated cell sorter (FACS). *Prenatal Diagnosis* **1**, 61–73.

Jabs, E. W. & Carpenter, N. 1988. Molecular cytogenetic evidence for amplification of chromosome-specific alphoid sequences at enlarged C-bands on chromosome 6. *American Journal of Human Genetics* **43**, 69–74.

Jabs, E. W., Wolf, S. F. & Migeon, B. R. 1984. Characterization of a cloned DNA sequence that is present at centromeres of all human autosomes and the X chromosome and shows polymorphic variation. *Proceedings of the National Academy of Sciences* **81**, 4884–8.

Jack, E. M., Harrison, C. J., Allen, T. D. & Harris, R. 1985. The structural basis for C-banding. A scanning electron microscope study. *Chromosoma* **91**, 363–8.

Jack, E. M., Harrison, C. J., Allen, T. D. & Harris, R. 1986. A structural basis for R- and T-banding: a scanning electron microscope study. *Chromosoma* **94**, 395–402.

Jacobs, P. A. 1977. Human chromosome heteromorphisms (variants). *Progress in Medical Genetics*, new series, **2**, 251–74.

Jacobs, P. A., Angell, R. R., Buchanan, I. M., Hassold, T. J., Matsuyama, A. M. & Manuel, B. 1978. The origin of human triploids. *Annals of Human Genetics* **42**, 49–57.

Jacobs, P. A., Buckton, K. E., Cunningham, C. & Newton, M. 1974. An analysis of the break points of structural rearrangements in man. *Journal of Medical Genetics* **11**, 50–64.

Jagiello, G. & Fang, J-S. 1980. A pachytene map of the mouse oocyte. *Chromosoma* **77**, 113–21.
Jalal, S. M., Clark, R. W., Hsu, T. C. & Pathak, S. 1974. Cytological differentiation of constitutive heterochromatin. *Chromosoma* **48**, 391–403.
Jalal, S. M., Markvong, A. & Hsu, T. C. 1975. Differential chromosome fluorescence with Hoechst 33258. *Experimental Cell Research* **90**, 443–4.
James, T. C. & Elgin, S. C. R. 1986. Identification of a nonhistone chromosomal protein associated with heterochromatin in *Drosophila melanogaster* and its gene. *Molecular and Cellular Biology* **6**, 3862–72.
Jan, K. Y., Sheng, W-W. & Wen, W-N. 1982. Contradictory differential staining results with Coomassie brilliant blue and silver carbonate on sister chromatids. *Experientia* **38**, 853–4.
Jan, K. Y., Su, P-F. & Lee, T-C. 1985. Reverse differential staining of sister chromatids induced by Hoechst plus black light and endonuclease. *Experimental Cell Research* **157**, 307–14.
Jan, K. Y., Yang, J. L. & Su, P. F. 1984. DNA lysis is involved in the simplified fluorescence plus Giemsa method for differential staining of sister chromatids. *Chromosoma* **89**, 76–8.
Jeffreys, A. J., Wilson, V. & Thein, S. L. 1985. Hypervariable 'minisatellite' regions in human DNA. *Nature* **314**, 67–73.
Jeppesen, P. & Nicol, L. 1986. Non-kinetochore directed autoantibodies in scleroderma/CREST. *Molecular Biology and Medicine* **3**, 369–84.
Jewell, D. C. 1981. Recognition of two types of positive staining chromosomal material by manipulation of critical steps in the N-banding technique. *Stain Technology* **56**, 227–34.
Jhanwar, S. C. & Chaganti, R. S. K. 1981. Pachytene chromomere maps of Chinese hamster autosomes. *Cytogenetics and Cell Genetics* **31**, 70–6.
John, B. 1976. Myths and mechanisms of meiosis. *Chromosoma* **54**, 295–325.
John, B. 1988. The biology of heterochromatin. In *Heterochromatin*, R. S. Verma (ed.), 1–147. Cambridge: Cambridge University Press.
John, B. & King, M. 1977. Heterochromatin variation in *Cryptobothrus chrysophorus*. II. Patterns of C-banding. *Chromosoma* **65**, 59–79.
John, B. & King, M. 1983. Population cytogenetics of *Atractomorpha similis*. I. C-band variation. *Chromosoma* **88**, 57–68.
John, B. & King, M. 1985. The inter-relationship between heterochromatin distribution and chiasma distribution. *Genetica* **66**, 183–94.
John, B. & Miklos, G. L. G. 1979. Functional aspects of satellite DNA and heterochromatin. *International Review of Cytology* **58**, 1–114.
John, B., Appels, R. & Contreras, N. 1986. Population cytogenetics of *Atractomorpha similis*. II. Molecular characterisation of the distal C-band polymorphisms. *Chromosoma* **94**, 45–58.
John, B., King, M., Schweizer, D. & Mendelak, M. 1985. Equilocality of heterochromatin distribution and heterochromatin heterogeneity in acridid grasshoppers. *Chromosoma* **91**, 185–200.
Johns, E. W. 1982. *The HMG chromosomal proteins*. London: Academic Press.
Johnson, P. W., Barnett, R. I., Gray, V. A. & MacKinnon, E. A. 1981. Spirals and G-bands as a function of chromosome length. *Cytobios* **32**, 7–14.
Johnston, F. P., Jorgenson, K. F., Lin, C. C. & van de Sande, J. H. 1978. Interaction of anthracyclines with DNA and chromosomes. *Chromosoma* **68**, 115–29.
Jones, K. W. 1970. Chromosomal and nuclear location of mouse satellite DNA in individual cells. *Nature* **225**, 912–15.
Jones, K. W. & Robertson, F. W. 1970. Localisation of reiterated nucleotide sequences in *Drosophila* and mouse by *in situ* hybridization of complementary RNA. *Chromosoma* **31**, 331–45.
Jones, K. W. & Singh, L. 1985. Snakes and the evolution of sex chromosomes. *Trends in Genetics* **1**, 55–61.
Jones, R. L., Davidson, M. W. & Wilson, W. D. 1979. Comparative viscometric analysis of the interaction of chloroquine and quinacrine with superhelical and sonicated DNA. *Biochemica Biophysica Acta* **561**, 77–84.

Jordan, E. G. 1984. Nucleolar nomenclature. *Journal of Cell Science* **67**, 217–20.
Jordan, G. 1987. At the heart of the nucleolus. *Nature* **329**, 489–90.
Jorgenson, K. F., van de Sande, J. H. & Lin, C. C. 1978. The use of base pair specific DNA binding agents as affinity labels for the study of mammalian chromosomes. *Chromosoma* **68**, 287–302.
Jorgenson, K. F., Varshney, U. & van de Sande, J. H. 1988. Interaction of Hoechst 33258 with repeating synthetic DNA polymers and natural DNA. *J. biomolec. Struct. Dynam.* **5**, 1005–23.
Joseph, A. M., Gosden, J. R. & Chandley, A. C. 1984. Estimation of aneuploidy levels in human spermatozoa using chromosome specific probes and *in situ* hybridization. *Human Genetics* **66**, 234–8.
Jotterand-Bellomo, M. 1983. Les effets de la distamycine A sur les cellules du liquide amniotique cultivées *in vitro*. *Annales de Génétique* **26**, 27–30.
Jovin, T. M., McIntosh, L. P., Arndt-Jovin, D. J., Zarling, D. A., Robert-Nicoud, M., van de Sande, J. H., Jorgenson, K. F. & Eckstein, F. 1983. Left-handed DNA: from synthetic polymers to chromosomes. *J. biomolec. Struct. Dynam.* **1**, 21–57.

Kaback, M. M., Saksela, E. & Mellman, W. J. 1964. The effect of 5-bromodeoxyuridine on human chromosomes. *Experimental Cell Research* **34**, 182–6.
Kaelbling. M., Miller, D. A. & Miller, O. J. 1984. Restriction enzyme banding of mouse metaphase chromosomes. *Chromosoma* **90**, 128–32.
Kaiserman, M. Z. & Burkholder, G. D. 1980. Silver stained core-like structures in Chinese hamster metaphase chromosomes. *Canadian Journal of Genet. Cytol.* **22**, 627–32.
Kajii, T. & Ohama, K. 1977. Androgenetic origin of hydatidiform mole. *Nature* **268**, 633–4.
Kanda, N. 1973. A new differential technique for staining the heteropycnotic X-chromosome in female mice. *Experimental Cell Research* **80**, 463–7.
Kanda, N. 1976. Banding pattern observed in human chromosomes by the modified BSG technique. *Human Genetics* **31**, 283–92.
Kanda, N. 1978. Differential staining in the secondary constriction regions of human chromosomes A1, C9 and E16 by a heat-hypotonic treatment technique. *Human Genetics* **45**, 19–24.
Kao, F-T. 1983. Somatic cell genetics and gene mapping. *International Review of Cytology* **85**, 109–46.
Kao, F-T. 1985. Human genome structure. *International Review of Cytology* **96**, 51–88.
Kao, Y. S., Whang-Peng, J. & Lee, E. 1983. A simple, rapid, high-resolution chromosome technique for lymphocytes. *American Journal of Clinical Pathology* **79**, 481–3.
Kapp, R. W. & Jacobson, C. B. 1980. Analysis of human spermatozoa for Y chromosomal nondisjunction. *Teratogenesis Carcinogenesis and Mutagenesis* **1**, 193–211.
Kapuściński, J. & Skoczylas, B. 1978. Fluorescent complexes of DNA with DAPI 4',6-diamidine-2-phenyl indole.2HCl or DCI 4',6-dicarboxyamide-2-phenyl indole. *Nucleic Acids Research* **5**, 3775–99.
Kapuściński, J. & Szer, W. 1979. Interactions of 4',6-diamidine-2-phenylindole with synthetic polynucleotides. *Nucleic Acids Research* **6**, 3519–34.
Kapuściński, J. & Yanagi, K. 1979. Selective staining by 4',6-diamidine-2-phenylindole of nanogram quantities of DNA in the presence of RNA on gels. *Nucleic Acids Research* **6**, 3535-42.
Karube, T. & Watanabe, S. 1988. Analysis of the chromosomal DNA replication pattern using the bromodeoxyuridine labelling method. *Cancer Research* **48**, 219–22.
Kato, H. & Moriwaki, K. 1972. Factors involved in the production of banded structures in mammalian chromosomes. *Chromosoma* **38**, 105–20.
Kato, H. & Yosida, T. H. 1972. Banding patterns of Chinese hamster chromosomes revealed by new techniques. *Chromosoma* **36**, 272–80.
Kato, H., Tsuchiya, K. & Yosida, T. H. 1974. Constitutive heterochromatin of Indian muntjac chromosomes revealed by DNase treatment and a C-banding technique. *Canadian Journal of Genetics and Cytology* **16**, 273–80.
Kaufmann, B. P., Gay, H. & McDonal, M. R. 1960. Organizational patterns within chromosomes. *International Review of Cytology* **9**, 77–127.

Kawano, T., Simoes, L. C. G. & Toledo, L. F. de A. 1987. Nucleolar organizer region in three species of the genus *Biomphalaria* (Mollusca, Gastropoda). *Revista Brasileira Genetica* **10**, 695–707.

Kerem, B-S., Goitein, R., Diamond, G., Cedar, H. & Marcus, M. 1984. Mapping of DNAase I sensitive regions on mitotic chromosomes. *Cell* **38**, 493–9.

Kerem, B-S., Goitein, R., Richler, C., Marcus, M. & Cedar, H . 1983. *In situ* nick translation distinguishes between active and inactive X chromosomes. *Nature* **304**, 88–90.

Kernell, A. M. & Ringertz, N. R. 1972. Cytochemical characterization of deoxyribonucleoprotein by UV-microspectrophotometry on heat denatured cell nuclei. *Experimental Cell Research* **72**, 240–51.

Kezer, J., Leon, P. E. & Sessions, S. K. 1980. Structural differentiation of the meiotic and mitotic chromosomes of the salamander, *Ambystoma macrodactylum*. *Chromosoma* **81**, 177–97.

Khokhar, M. T., Lawler, S. D., Powles, R. L. & Millar, J. L. 1987. Cytogenetic studies using Q-band polymorphisms in patients with AML receiving marrow from like-sex donors. *Human Genetics* **76**, 176–80.

Kim, M. A. 1975. Fluorometrical detection of thymine base differences in complementary strands of satellite DNA in human metaphase chromosomes. *Humangenetik* **28**, 57–63.

Kim, M. A. & Grzeschik, K. H. 1974. A method for discriminating murine and human chromosomes in somatic cell hybrids. *Experimental Cell Research* **88**, 406–10.

King, M. 1980. C-banding studies on Australian hylid frogs: secondary constriction structure and the concept of euchromatin transformation. *Chromosoma* **80**, 191–217.

King, M. & John, B. 1980. Regularities and restrictions governing C-band variation in acridoid grasshoppers. *Chromosoma* **76**, 123–50.

King, M., Honeycutt, R & Contreras, N. 1986. Chromosomal repatterning in crocodiles: C, G and N-banding and the *in situ* hybridization of 18S and 26S rRNA cistrons. *Genetica* **70**, 191–201.

Kingwell, B. & Rattner, J. B. 1987. Mammalian kinetochore/centromere composition: a 50kDa antigen is present in the mammalian kinetochore/centromere. *Chromosoma* **95**, 403–7.

Kitchin, R. M. & Loudenslager, E. J. 1976. An *in vivo* Giemsa chromosome banding technique. *Stain Technology* **50**, 371–4.

Klasen, M. & Schmid, M. 1981. An improved method for Y-body identification and confirmation of a high incidence of YY sperm nuclei. *Human Genetics* **58**, 156–61.

Klevecz, R. R., Keniston, B. A. & Deaven, L. L. 1975. The temporal structure of S phase. *Cell* **5**, 195–203.

Klevit, R. E., Wemmer, D. E. & Reid, B. R. 1986. ^{1}H NMR studies on the interaction between distamycin A and a symmetrical DNA dodecamer. *Biochemistry* **25**, 3296–303.

Knuutila, S. & Teerenhovi, L. 1989. Immunophenotyping of aneuploid cells. *Cancer Genetics and Cytogenetics* **41**, 1–7.

Kodama, Y., Yoshida, M. C. & Sasaki, M. 1980. An improved silver staining technique for nucleolus organizer regions by using nylon cloth. *Japanese Journal of Human Genetics* **25**, 229–33.

Kohlstaedt, L. A., Sung, E. C., Fujishige, A. & Cole, R. D. 1987. Non-histone chromosomal protein HMG 1 modulates the histone H1-induced condensation of DNA. *Journal of Biological Chemistry* **262**, 524–6.

Kongsuwan, K. & Smyth, D. 1977. Q-bands in *Lilium* and their relationship to C-banded heterochromatin. *Chromosoma* **60**, 169–78.

Kongsuwan, K. & Smyth, D. 1978. DNA loss during C-banding of chromosomes of *Lilium longiflorum*. *Chromosoma* **68**, 59–72.

Kongsuwan, K. & Smyth, D. 1980. Late labelled regions in relation to Q- and C-bands in chromosomes of *Lilium longiflorum* and *L. pardalinum*. *Chromosoma* **76**, 151–64.

Koop, B. F., Baker, R. J., Haiduk, M. W. & Engstrom, M. D. 1984. Cladistical analysis of primitive G-band sequences for the karyotype of the ancestor of the Cricetidae complex of rodents. *Genetica* **64**, 199–208.

Kopnin, B. P., Massino, J. S. & Gudkov, A. V. 1985. Regular pattern of karyotypic alterations accompanying gene amplification in Djungarian hamster cells: study of colchicine, adriablastin, and methotrexate resistance. *Chromosoma* **92**, 25–36.

Kopp, E., Mayr, B. & Schleger, W. 1986. Species-specific non-expression of ribosomal RNA genes in a mammalian hybrid, the mule. *Chromosoma* **94**, 346–52.

Korenberg, J. R. & Engels, W. R. 1978. Base ratio, DNA content, and quinacrine-brightness of human chromosomes. *Proceedings of the National Academy of Sciences* **75**, 3382–6.

Korenberg, J. R. & Freedlender, E. F. 1974. Giemsa technique for the detection of sister chromatid exchanges. *Chromosoma* **48**, 355–60.

Korenberg, J. R. & Rykowski, M. C. 1988. Human genome organization: Alu, Lines, and the molecular structure of metaphase chromosome bands. *Cell* **53**, 391–400.

Korenberg, J. R., Therman, E. & Denniston, C. 1978. Hot spots and functional organization of human chromosomes. *Human Genetics* **43**, 13–22.

Korf, B. R., Schuh, B. E. & Salwen, M. J. 1975. Optimum pH for nuclear sex identification using quinacrine. *Clinical Genetics* **8**, 145–8.

Korf, B. R., Schuh, B. E., Salwen, M. J., Warburton, D. & Miller, O. J. 1976. The role of trypsin in the pre-treatment of chromosomes for Giemsa banding. *Human Genetics* **31**, 27–33.

Korson, R. 1964. A silver stain for deoxyribonucleic acid. *Journal of Histochemistry and Cytochemistry* **12**, 875–9.

Kosztolányi, G. & Bühler, E. M. 1978. The effect of SH-SS transition in the structural organization of mitotic chromosomes. *Human Genetics* **42**, 83–8.

Krajca, J. B. & Wray, W. 1977. Banding isolated metaphase chromosomes by a sequential G/Q technique. *Histochemistry* **51**, 103–11.

Kram, R., Botchan, M. & Hearst, J. E. 1972. Arrangement of the highly reiterated DNA sequences in the centric heterochromatin of *Drosophila melanogaster*. Evidence for interspersed spacer DNA. *Journal of Molecular Biology* **64**, 103–17.

Kubista, M., Akerman, B. & Norden, B. 1987. Characterization of interaction between DNA and 4′,6-diamidino-2-phenylindole by optical spectroscopy. *Biochemistry* **26**, 4545–53.

Kuo, M. T. 1982. Comparison of chromosomal structures isolated under different conditions. *Experimental Cell Research* **138**, 221–9.

Kuo, M. T. & Hsu, T. C. 1978. Studies in heterochromatin DNA: characterization of transcripts synthesized *in situ* from C-banded preparations. *Chromosoma* **65**, 325–34.

Kuo, M. T. & Plunkett, W. 1985. Nick-translation of metaphase chromosomes: *in vitro* labelling of nuclease-hypersensitive regions in chromosomes. *Proceedings of the National Academy of Sciences* **82**, 854–8.

Kurnit, D. M. 1974. DNA helical content during the C-banding procedure. *Cytogenetics and Cell Genetics* **13**, 313–29.

Kurnit, D. M. 1979. Satellite DNA and heterochromatin variants: the case for unequal mitotic crossing over. *Human Genetics* **47**, 169–86.

Kurnit, D. M. & Maio, J. J. 1973. Subnuclear redistribution of DNA species in confluent and growing mammalian cells. *Chromosoma* **42**, 23–36.

Kurnit, D. M. & Maio, J. J. 1974. Variable satellite DNAs in the African green monkey *Cercopithecus aethiops*. *Chromosoma* **45**, 387–400.

Kurnit, D. M., Brown, F. L. & Maio, J. J. 1978. Mammalian repetitive DNA sequences in a stable Robertsonian system. *Cytogenetics and Cell Genetics* **21**, 145–67.

Kurnit, D. M., Shafit, B. R. & Maio, J. J. 1973. Multiple satellite deoxyribonucleic acids in the calf and their relation to the sex chromosomes. *Journal of Molecular Biology* **81**, 273–84.

Kuro-o, M., Ikebe, C. & Kohno, S. 1986. Cytogenetic studies of Hynobiidae (Urodela). IV. DNA replication bands (R-banding) in the genus *Hynobius* and the banding karyotype of *Hynobius nigrescens* Stejneger. *Cytogenetics and Cell Genetics* **43**, 14–18.

Kuro-o, M., Ikebe, C. & Kohno, S. 1987. Cytogenetic studies of Hynobiidae (Urodela). VI. R-banding patterns in five pond-type *Hynobius* from Korea and Japan. *Cytogenetics and Cell Genetics* **44**, 69–75.

Labal de Vinuesa, M., Mudry de Pargament, M., Slavutsky, I., Meiss, R., Chopita, N. &

Larripa, I. 1988a. Heterochromatic variants and their association with neoplasias: IV. Colon adenomas and carcinomas. *Cancer Genetics and Cytogenetics* **31**, 171–4.
Labal de Vinuesa, M., Slavutsky, I., Mudry de Pargament, M. & Larripa, I. 1988b. Heterochromatic variants and their association with neoplasias: V. Non-Hodgkin's lymphomas. *Cancer Genetics and Cytogenetics* **31**, 175–8.
LaCour, L. F. & Wells, B. 1974. Fine structure and staining behaviour of heterochromatic segments in two plants. *Journal of Cell Science* **14**, 505–21.
Lakhotia, S. C. & Kumar, M. 1978. Heterochromatin in mitotic chromosomes of *Drosophila nasuta*. *Cytobios* **21**, 79–89.
Lakhotia, S. C., Roy, J. K. & Kumar, M. 1979. A study of heterochromatin in *Drosophila nasuta* by the 5-bromodeoxyuridine-Giemsa staining technique. *Chromosoma* **72**, 249–55.
Lalley, P. A., Davisson, M. T., Graves, J. A. M., O'Brien, S. J., Roderick, T. H., Doolittle, D. P. & Hillyard, A. L. 1988. Report of the committee on comparative mapping. *Cytogenetics and Cell Genetics* **49**, 227–35.
Lambiase, S., Maraschio, P. & Zuffardi, O. 1984. The Cd technique identifies a specific structure related to centromeric function. *Human Genetics* **67**, 214–15.
Langlois, R. G. & Jensen, R. H. 1979. Interactions between pairs of DNA-specific fluorescent stains bound to mammalian cells. *Journal of Histochemistry and Cytochemistry* **27**, 72–9.
Langlois, R. G., Carrano, A. V., Gray, J. W. & van Dilla, M. A. 1980. Cytochemical studies of metaphase chromosomes by flow cytometry. *Chromosoma* **77**, 229–51.
Larsen, J. K., Munch-Petersen, B., Christiansen, J. & Jørgensen, K. 1986. Flow cytometric discrimination of mitotic cells: resolution of M, as well as G_1, S, and G_2 phase nuclei with mithramycin, propidium iodide, and ethidium bromide after fixation with formaldehyde. *Cytometry* **7**, 54–63.
Latos-Bielenska, A., Hameister, H. & Vogel, W. 1987. Detection of BrdUrd incorporation in mammalian chromosomes by a BrdUrd antibody. III. Demonstration of replication patterns in highly resolved chromosomes. *Human Genetics* **76**, 293–5.
Latt, S. A. 1973. Microfluorometric detection of deoxyribonucleic acid replication in human metaphase chromosomes. *Proceedings of the National Academy of Sciences* **70**, 3395–9.
Latt, S. A. 1974a. Microfluorometric analysis of deoxyribonucleic acid replication kinetics and sister chromatid exchanges in human chromosomes. *Journal of Histochemistry and Cytochemistry* **22**, 478–91.
Latt, S. A. 1974b. Microfluorometric analysis of DNA replication in human X chromosomes. *Experimental Cell Research* **86**, 412–15.
Latt, S. A. 1975. Fluorescence analysis of late DNA replication in human metaphase chromosomes. *Somatic Cell Genetics* **1**, 293–321.
Latt, S. A. 1977. Fluorescent probes of chromosome structure and replication. *Canadian Journal of Genetics and Cytology* **19**, 603–23.
Latt, S. A. & Gerald, P. S. 1973. Staining of human metaphase chromosomes with fluorescent conjugates of polylysine. *Experimental Cell Research* **81**, 401–6.
Latt, S. A. & Wohlleb, J. C. 1975. Optical studies of the interaction of 33258 Hoechst with DNA, chromatin, and metaphase chromosomes. *Chromosoma* **52**, 297–316.
Latt, S. A., Brodie, S. & Munroe, S. H. 1974a. Optical studies of complexes of quinacrine with DNA and chromatin: implications for the fluorescence of cytological preparations. *Chromosoma* **49**, 17–40.
Latt, S. A., Sahar, E. & Eisenhard, M. E. 1979. Pairs of fluorescent dyes as probes of DNA and chromosomes. *Journal of Histochemistry and Cytochemistry* **27**, 65–71.
Latt, S. A., Barell, E. F., Dougherty, C. P. & Lazarus, H. 1981. Patterns of late replication in X chromosomes of human lymphoid cells. *Cancer Genetics and Cytogenetics* **3**, 171–81.
Latt, S. A., Davidson, R. L., Lin, M. S. & Gerald, P. S. 1974b. Lateral asymmetry in the fluorescence of human Y chromosomes stained with 33258 Hoechst. *Experimental Cell Research* **87**, 425–9.
Latt, S. A., Sahar, E., Eisenhard, M. E. & Juergens, L. A. 1980b. Interactions between pairs of DNA-binding dyes: results and implications for chromosome analysis. *Cytometry* **1**, 2–12.

Latt, S. A., Juergens, L. A., Matthews, D. J., Gustashaw, K. M. & Sahar, E. 1980a. Energy transfer-enhanced chromosome banding. An overview. *Cancer Genetics and Cytogenetics* **1**, 187–96.

Latt, S. A., Stetten, G., Juergens, L. A., Willard, H. F. & Scher, C. D. 1975. Recent developments in the detection of deoxyribonucleic acid synthesis by 33258 Hoechst fluorescence. *Journal of Histochemistry and Cytochemistry* **23**, 493–505.

Lau, Y-F. & Hsu, T. C. 1977. Variable modes of Robertsonian fusions. *Cytogenetics and Cell Genetics* **19**, 231–5.

Lau, Y-F., Arrighi, F. E. & Chuang, C. R. 1977. Studies of the squirrel monkey, *Saimiri sciureus*, genome. II. C-band characterization and DNA replication patterns. *Cytogenetics and Cell Genetics* **19**, 14–25.

Lau, Y-F., Pfeiffer, R. A., Arrighi, F. E. & Hsu, T. C. 1978. Combination of silver and fluorescent staining for metaphase chromosomes. *American Journal of Human Genetics* **30**, 76–9.

Lavania, U. C. & Sharma, A. K. 1979. Trypsin-orcein banding in plant chromosomes. *Stain Technology* **54**, 261–3.

Lawler, S.D. & Reeves, B. R. 1976. Chromosome studies in man; past achievements and recent advances. *Journal of Clinical Pathology* **29**, 569–82.

Lawrie, S. S. & Gosden, J. R. 1980. The identification of human chromosomes by quinacrine fluorescence after hybridization *in situ*. *Human Genetics* **53**, 371–3.

LeBeau, M. M. & Rowley, J. D. 1984. Heritable fragile sites in cancer. *Nature* **308**, 607–8.

Ledbetter, D. H., Riccardi, V. M., Au, W. W., Wilson, D. P. & Holmquist, G. P. 1980. Ring chromosome 15: phenotype, Ag-NOR analysis, secondary aneuploidy, and associated chromosome instability. *Cytogenetics and Cell Genetics* **27**, 111–22.

Lee, C. L. Y., Welch, J. P. & Lee, S. H. S. 1973. Banding of human chromosomes by protein denaturation. *Nature New Biology* **241**, 142–3.

Lee, J. C. K. & Bahr, G. F. 1983. Microfluorometric studies on chromosomes. Quantitative determination of protein content of Chinese hamster chromosome 1 *in situ* with and without trypsin digestion. *Chromosoma* **88**, 374–6.

Leeman, U. & Ruch, F. 1978. Selective excitation of mithramycin or DAPI fluorescence on double-stained cell nuclei and chromosomes. *Histochemistry* **58**, 329–34.

Leeman, U. & Ruch, F. 1982. Cytofluorometric determination of DNA base content in plant nuclei and chromosomes by the fluorochromes DAPI and chromomycin A3. *Experimental Cell Research* **140**, 275–82.

Leeman, U. & Ruch, F. 1983. DNA and base content in the nuclei and the sex chromatin of *Rumex acetosa*. *Bot. Helvetica* **93**, 77–83.

Leeman, U. & Ruch, F. 1984. Cytofluorometry and visualisation of DNA base composition in the chromosomes of *Drosophila melanogaster*. *Chromosoma* **90**, 6–15.

Lejeune, J. 1973. Scientific impact of the study of fine structure of chromatids. In *Chromosome identification*. Nobel Symposium 23, T. Caspersson & L. Zech (eds), 16–24. New York: Academic Press.

Lejeune, J., Dutrillaux, B., Rethoré, M-O. & Prieur, M. 1973. Comparaison de la structure fine des chromatides d'*Homo sapiens* et de *Pan troglodytes*. *Chromosoma* **43**, 423–44.

Lelley, T., Josifek, K. & Kaltsikes, P. J. 1978. Polymorphism in the Giemsa C-banding pattern of rye chromosomes. *Canadian Journal of Genetics and Cytology* **20**, 307–12.

Lemeunier, F., Dutrillaux, B. & Ashburner, M. 1978. Relationships within the *melanogaster* subgroup species of the genus *Drosophila* (*Sophophora*). III. The mitotic chromosomes and quinacrine fluorescent patterns of the polytene chromosomes. *Chromosoma* **69**, 349–61.

LePecq, J-B. & Paoletti, C. 1967. A fluorescent complex between ethidium bromide and nucleic acids. Physical-chemical characterization. *Journal of Molecular Biology* **27**, 87–106.

LePecq, J-B., le Bret, M., Barbet, J. & Roques, B. 1975. DNA polyintercalating drugs: DNA binding of diacridine derivatives. *Proceedings of the National Academy of Sciences* **72**, 2915–9.

Lester, D. S., Crozier, R. H. & Shipp, E. 1979. G-banding patterns of the housefly, *Musca domestica*, autosomes and sex chromosomes. *Experientia* **35**, 174–5.

Levan, A. 1945. Cytological reactions induced by inorganic salt solutions. *Nature* **156**, 751–2.

Levan, A. 1946. Heterochromaty in chromosomes during their contraction phase. *Hereditas* **32**, 449–68.
Levan, A., Levan, G. & Mandahl, N. 1978. A new chromosome type replacing the double minutes in a mouse tumor. *Cytogenetics and Cell Genetics* **20**, 12–23.
Levan, G., Mandahl, N., Bregula, U., Klein, G. & Levan, A. 1976. Double minute chromosomes are not centromeric regions of the host chromosomes. *Hereditas* **83**, 83–90.
Lica, L. & Hamkalo, B. 1983. Preparation of centromeric heterochromatin by restriction endonuclease digestion of mouse L929 cells. *Chromosoma* **88**, 42–9.
Lichtenberger, M. J. 1983. Quick and reversible staining methods for G- and Q-bands of chromosomes. *Stain Technology* **58**, 185–8.
Lima-de-Faria, A. 1975. The relationship between chromomeres, replicons, operons, transcription units, genes, viruses and palindromes. *Hereditas* **81**, 249–84.
Lima-de-Faria, A. & Jaworska, H. 1968. Late DNA synthesis in heterochromatin. *Nature* **217**, 138–42.
Lima-de-Faria, A., Isaksson, M. & Olsson, E. 1980. Action of restriction endonucleases on the DNA and chromosomes of *Muntiacus muntjak*. *Hereditas* **92**, 267–73.
Limon, J. & Gibas, Z. 1985. Lateral asymmetry of the Y chromosome and its significance. In *The Y chromosome*, part A, A. A. Sandberg (ed.), 317–26. New York: Alan R. Liss.
Limon, J., Babińska, M. & Ledóchowski, A. 1975. Studies on the structure-fluorescence relationships of chromatin-bound 9-amino acridine derivatives. *Experimental Cell Research* **92**, 299–306.
Lin, C. C. & van de Sande, J. H. 1975. Differential fluorescent staining of human chromosomes with daunomycin and adriamycin – the D-bands. *Science* **190**, 61–3.
Lin, C. C., Chiarelli, B., de Boer, L. E. M. & Cohen, M. M. 1973. A comparison of the fluorescent karyotypes of the chimpanzee (*Pan troglodytes*) and man. *Journal of Human Evolution* **2**, 311–21.
Lin, C. C., Jorgensen, K. F. & van de Sande, J. H. 1980. Specific fluorescent bands on chromosomes produced by acridine orange after prestaining with base specific non-fluorescent DNA ligands. *Chromosoma* **79**, 271–86.
Lin, C. C., van de Sande, H., Smink, W. K. & Newton, D. R. 1975. Quinacrine fluorescence and Q-banding patterns of human chromosomes. I. Effects of varying factors. *Canadian Journal of Genetics and Cytology* **17**, 81–92.
Lin, M. S. & Davidson, R. L. 1974. Centric fusion, satellite DNA, and DNA polarity in mouse chromosomes. *Science* **185**, 1179–81.
Lin, M. S., Alfi, O. S. & Donnell, G. N. 1976. Differential fluorescence of sister chromatids with 4'-6-diamidino-2-phenylindole. *Canadian Journal of Genetics and Cytology* **18**, 545–7.
Lin, M. S., Comings, D. E. & Alfi, O. S. 1977. Optical studies of the interaction of 4'-6-diamidino-2-phenylindole with DNA and metaphase chromosomes. *Chromosoma* **60**, 15–25.
Lin, M. S., Latt, S. A. & Davidson, R. L. 1974. Microfluorometric detection of asymmetry in the centromeric region of mouse chromosomes. *Experimental Cell Research* **86**, 392–5.
Lin, S-Y., Lin, D. & Riggs, A. D. 1976. Histones bind more tightly to bromodeoxyuridine-substituted DNA than to normal DNA. *Nucleic Acids Research* **3**, 2183–91.
Linde-Laursen, I., von Bothmer, R. & Jacobsen, N. 1986. Giemsa C-banded karyotypes of *Hordeum* taxa from North America. *Canadian Journal of Genetics and Cytology* **28**, 42–62.
Lischwe, M. A., Richards, R. L., Busch, R. K. & Busch, H. 1981. Localization of phosphoprotein C23 to nucleolar structures and to the nucleolus organizer regions. *Experimental Cell Research* **136**, 101–9.
Lischwe, M. A., Smetana, K., Olson, M. O. J. & Busch, H. 1979. Proteins C23 and B23 are the major silver staining nucleolar proteins. *Life Sciences* **25**, 701–8.
Liu, L. 1988. Analysis of high-resolution G-banding pattern in fish chromosomes treated by BrdU (In Chinese). *Acta Genetica Sinica* **15**, 117–21.
Lloyd, M. A. & Thorgaard, G. H. 1988. Restriction endonuclease banding of rainbow trout chromosomes. *Chromosoma* **96**, 171–7.

Löber, G. 1975. On the spectroscopic basis of acridine-induced fluorescence banding patterns in chromosomes. *Studia Biophysica* **48**, 109–23.

Löber, G., Kleinwächter, V. & Koudelka, J. 1976. Staining of chromosomes with basic dyes. *Studia Biophysica* **55**, 49–56.

Löber, G., Beensen, V., Zimmer, C. & Hanschmann, H. 1978. Changes of quinacrine staining on human chromosomes by the competitive binding of AT- and GC-specific substances. *Studia Biophysica* **69**, 237–8.

Loidl, J. 1979. C-band proximity of chiasmata and absence of terminalization in *Allium flavum* (Liliaceae). *Chromosoma* **73**, 45–51.

Loidl, J. 1982. Further evidence for a heterochromatin-chiasma correlation in some *Allium* species. *Genetica* **60**, 31–5.

Loidl, J. 1983. Some features of heterochromatin in wild *Allium* species. *Plant Syst. Evol.* **143**, 117–31.

Loidl, J. 1985. Banding of *Allium* chromosomes protected against DNase digestion by DNA-binding drugs. *Stain Technology* **60**, 13–19.

Long, E. O. & Dawid, I. B. 1980. Repeated genes in eukaryotes. *Annual Review of Biochemistry* **49**, 727–64.

Long, S. E. 1985. Centric fusion translocations in cattle: a review. *Veterinary Record* **116**, 516–18.

Lopetegui, P. H. 1980. 1, 9, and 16 C-band heteromorphisms in parents of Down's syndrome patients: distribution and etiological significance. *Japanese Journal of Human Genetics* **25**, 29–37.

Lopez-Fernandez, C., Gosalvez, J. & Mezzanotte, R. 1989. Heterochromatin heterogeneity in *Oedipoda germanica* (Orthoptera) detected by *in situ* digestion with restriction endonucleases. *Heredity* **62**, 269–77.

Lopez-Fernandez, C., Gosalvez, J., Suja, J. A. & Mezzanotte, R. 1988. Restriction endonuclease digestion of meiotic and mitotic chromosomes in *Pyrgomorpha conica* (Orthoptera: Pyrgomorphidae). *Genome* **30**, 621–6.

Low, C. M. L., Fox, K. R. & Waring, M. J. 1986. DNA sequence selectivity of three biosynthetic analogues of the quinoxaline antibiotics. *Anti-Cancer Drug Design* **1**, 149–60.

Lubit, B. W., Pham, T. D., Miller, O. J. & Erlanger, B. F. 1976. Localization of 5-methylcytosine in human metaphase chromosomes by immunoelectron microscopy. *Cell* **9**, 503–9.

Lubs, H. A., Kimberling, W. J., Hecht, F., Patil, S. R., Brown, J., Gerald, P. & Summitt, R. L. 1977. Racial differences in the frequency of Q and C chromosomal heteromorphisms. *Nature* **268**, 631–3.

Lubs, H. A., McKenzie, W. H. & Merrick, S. 1973. Comparative methodology and mechanisms of banding. In *Chromosome identification. Nobel Symposium 23*, T. Caspersson & L. Zech (eds), 315–22. New York: Academic Press.

Luciani, J. M., Morazzani, M-R. & Stahl, A. 1975. Identification of pachytene bivalents in human male meiosis using G-banding technique. *Chromosoma* **52**, 275–82.

Lundsteen, C., Kristoffersen, L. & Ottosen, F. 1974. Studies on the mechanism of chromosome banding with trypsin. *Humangenetik* **24**, 67–9.

McClung, C. E. 1901. Notes on the accessory chromosome. *Anatomischer Anzeiger* **20**, 220–6.

Macgregor, H. C. & Horner, H. 1980. Heteromorphism for chromosome 1, a requirement for normal development in crested newts. *Chromosoma* **76**, 111–12.

Macgregor, H. C. & Kezer, J. 1971. The chromosomal localization of a heavy satellite DNA in the testis of *Plethodon c. cinereus*. *Chromosoma* **33**, 167–82.

Macgregor, H. C. & Varley, J. M. 1988. *Working with animal chromosomes*, 2nd ed. Chichester: Wiley.

McKay, R. D. G. 1973. The mechanism of C and G banding in mammalian metaphase chromosomes. *Chromosoma* **44**, 1–14.

McKay, R. D. G., Bobrow, M. & Cooke, H. J. 1978. The identification of a repeated DNA sequence involved in the karyotype polymorphism of the human Y chromosome. *Cytogenetics and Cell Genetics* **21**, 19–32.

McKenzie, W. H. & Lubs, H. A. 1973. An analysis of the technical variables in the production of C bands. *Chromosoma* **41**, 175–82.

McKeon, F. D., Tuffanelli, D. L., Kobayashi, S. & Kirschner, M. W. 1984. The redistribution of a conserved nuclear envelope protein during the cell cycle suggests a pathway for chromosome condensation. *Cell* **36**, 83–92.

McLean, I. W. & Nakane, P. K. 1974. Periodate-lysine-paraformaldehyde fixative. A new fixative for immunoelectron microscopy. *J. Histochem. Cytochem.* **22**, 1077–83.

McNeilage, L. J., Whittingham, S., McHugh, N. & Barnett, A. J. 1986. A highly conserved 72,000 dalton centromeric antigen reactive with autoantibodies from patients with progressive systemic sclerosis. *Journal of Immunology* **137**, 2541–7.

Mace, M. L., Tevethia, S. S. & Brinkley, B. R. 1972. Differential immunofluorescent labelling of chromosomes with antisera specific for single strand DNA. *Experimental Cell Research* **75**, 521–3.

Madan, K. & Bruinsma, A. H. 1979. C-band polymorphism in human chromosome no. 6. *Clinical Genetics* **15**, 193–7.

Maes, A., Staessen, C., Hens, L., Vamos, E., Kirsch-Volders, M., Lauwers, M. C., Defrise-Gussenhoven, E. & Susanne, C. 1983. C heterochromatin variation in couples with recurrent early abortions. *Journal of Medical Genetics* **20**, 350–6.

Magaud, J-P., Rimokh, R., Brochier, J., Lafage, M. & Germain, D. 1985. Chromosomal R-banding with a monoclonal antidouble-stranded DNA antibody. *Human Genetics* **69**, 238–42.

Magenis, R. E., Donlon, T. A. & Wyandt, H. E. 1978. Giemsa-11 staining of chromosome 1: a newly described heteromorphism. *Science* **202**, 64–5.

Magenis, R. E., Overton, K. M., Chamberlin, J., Brady, T. & Lovrien, E. 1977. Parental origin of the extra chromosome in Down's syndrome. *Human Genetics* **37**, 7–16.

Maio, J. J. & Schildkraut, C. L. 1969. Isolated mammalian metaphase chromosomes. II. Fractionated chromosomes of mouse and Chinese hamster cells. *Journal of Molecular Biology* **40**, 203–16.

Mäkinen, T., Stenstrand, K. & Selander, R-K. 1975. The effect of ionic strength on G-bands. *Humangenetik* **28**, 71–3.

Mamaev, N., Mamaeva, S., Liburkina, I., Kozlova, T., Medvedeva, N. & Makarkina, G. 1985. The activity of nucleolar organizer regions of human bone marrow cells studied with silver staining. I. Chronic myelocytic leukaemia. *Cancer Genetics and Cytogenetics* **16**, 311–20.

Manolov, G. & Manolova, Y. 1972. Marker band in one chromosome 14 from Burkitt lymphomas. *Nature* **237**, 33–4.

Manton, I. 1950. The spiral structure of chromosomes. *Biological Reviews* **25**, 486–508.

Manuelidis, L. 1978. Chromosomal localization of complex and simple repeated human DNAs. *Chromosoma* **66**, 23–32.

Manuelidis, L. & Ward, D. C. 1984. Chromosomal and nuclear distribution of the Hind III 1.9-kb human DNA repeat segment. *Chromosoma* **91**, 28–38.

Manzini, G., Barcellona, M. L., Avitabile, M. & Quadrifoglio, F. 1983. Interaction of diamidino-2-phenylindole (DAPI) with natural and synthetic nucleic acids. *Nucleic Acids Research* **11**, 8861–76.

Maraschio, P., Zuffardi, O. & Lo Curto, F. 1980. Cd bands and centromeric function in dicentric chromosomes. *Human Genetics* **54**, 265–7.

Marchi, A. & Mezzanotte, R. 1988. Restriction endonuclease digestion and chromosome banding in the mosquito, *Culiseta longiareolata* (Diptera: Culicidae). *Heredity* **60**, 21–6.

Marcus, M., Goitein, R. & Gropp, A. 1979a. Condensation of all human chromosomes in phase G2 and early mitosis can be drastically inhibited by 33258–Hoechst treatment. *Human Genetics* **51**, 99–105.

Marcus, M., Nielsen, K., Goitein, R. & Gropp, A. 1979b. Pattern of condensation of mouse and Chinese hamster chromosomes in G2 and mitosis of 33258-Hoechst-treated cells. *Experimental Cell Research* **122**, 191–201.

Marcus, M., Nattenberg, A., Goitein, R., Nielsén, K. & Gropp. A. 1979c. Inhibition of condensation of human Y chromosome by the fluorochrome Hoechst 33258 in a mouse-human cell hybrid. *Human Genetics* **46**, 193–8.

Markovic, V. D., Worton, R. G. & Berg, J. M. 1978. Evidence for the inheritance of silver-stained nucleolus organizer regions. *Human Genetics* **41**, 181–7.

Marks, G. E. 1975. The Giemsa-staining centromeres of *Nigella damascena*. *Journal of Cell Science* **18**, 19–25.

Marks, G. E. 1983. Feulgen banding of heterochromatin in plant chromosomes. *Journal of Cell Science* **62**, 171–6.

Marks, G. E. & Schweizer, D. 1974. Giemsa banding: karyotype differences in some species of *Anemone* and in *Hepatica nobilis*. *Chromosoma* **44**, 405–16.

Marks, J. 1983. Hominoid cytogenetics and evolution. *Yearbook Phys. Anthropol.* **26**, 131–59.

Marsden, M. P. F. & Laemmli, U. K. 1979. Metaphase chromosome structure: evidence for a radial loop model. *Cell* **17**, 849–58.

Marshall, C. J. 1975. A method for analysis of chromosomes in hybrid cells employing sequential G-banding and mouse specific C-banding. *Experimental Cell Research* **91**, 464–9.

Marshall, P. N. & Galbraith, W. 1984. On the nature of the purple coloration of leucocyte nuclei stained with azure B-eosin Y. *Histochemical Journal* **16**, 793–7.

Martin, P. K. & Rowley, J. D. 1983. An improved technique for sequential R-, Q- and C-banding of bone marrow chromosomes. *Stain Technology* **58**, 7–12.

Martin, R. F. & Holmes, N. 1983. Use of an ^{125}I-labelled DNA ligand to probe DNA structure. *Nature* **302**, 452–4.

Martin, R. H., Balkan, W., Burns, K., Rademaker, A. W., Lin, C. C. & Rudd, N. L. 1983. The chromosome constitution of 1000 human spermatozoa. *Human Genetics* **63**, 305–9.

Mascarello, J. T. & Hsu, T. C. 1976. Chromosome evolution in woodrats, genus *Neotoma* (Rodentia: Cricetidae). *Evolution* **30**, 152–69.

Mascarello, J. T., Stock. A. D. & Pathak, S. 1974. Conservatism in the arrangement of genetic material in rodents. *J. Mammal.* **55**, 695–704.

Mason, D., Lauder, I., Rutovitz, D. & Spowart, G. 1975. Measurement of C-bands in human chromosomes. *Computers in Biology and Medicine* **5**, 179–201.

Masumoto, H., Sugimoto, K. & Okazaki, T. 1989. Alphoid satellite DNA is tightly associated with centromere antigens in human chromosomes throughout the cell cycle. *Experimental Cell Research* **181**, 181–96.

Matayoshi, T., Howlin, E., Nasazzi, N., Nagle, C., Gadow, E. & Seuanez, H. N. 1986. Chromosome studies of *Cebus apella*: the standard karyotype of *Cebus apella paraguayanus*, Fischer, 1829. *American Journal of Primatology* **10**, 185–93.

Matayoshi, T., Seuanez, H. N., Nasazzi, N., Nagle, C., Armada, J. L., Freitas, L., Alves, G., Barroso, C. M. & Howlin, E. 1987. Heterochromatic variation in *Cebus apella* (Cebidae, Platyrrhini) of different geographic regions. *Cytogenetics and Cell Genetics* **44**, 158–62.

Matsubara, T. & Nakagome, Y. 1983. High-resolution banding by treating cells with acridine orange before fixation. *Cytogenetics and Cell Genetics* **35**, 148–51.

Matsui, S. 1974. Structural proteins associated with ribosomal cistrons in *Xenopus laevis* chromosomes. *Experimental Cell Research* **88**, 88–94.

Matsui, S-I & Sasaki, M. 1973. Differential staining of nucleolus organizers in mammalian chromosomes. *Nature* **246**, 148–50.

Matsui, S-I. & Sasaki, M. 1975. The mechanism of Giemsa-banding of mammalian chromosomes, with special attention to the role of non-histone proteins. *Japanese Journal of Genetics* **50**, 189–204.

Matsui, S-I., Fuke, M., Chai, L., Sandberg, A.A. & Elassouli, S. 1986. N-band proteins of nucleolar organizers: chromosomal mapping, subnucleolar localization and rDNA binding. *Chromosoma* **93**, 231–42.

Matsukuma, S. & Utakoji, T. 1976. Uneven extraction of protein in Chinese hamster chromosomes during G-staining procedures. *Experimental Cell Research* **97**, 297–303.

Matthes, E., Fenske, H., Eichhorn, I., Langen, P. & Lindigkeit, R. 1977. Altered histone-DNA interactions in rat liver chromatin containing 5-bromodeoxyuridine-substituted DNA. *Cell Differentiation* **6**, 241–51.

Maul, G. G., French, B. T., van Venrooij, W. J. & Jimenez, S. A. 1986. Topoisomerase I identified by scleroderma 70 antisera: enrichment of topoisomerase I at the centromere in mouse mitotic cells before anaphase. *Proceedings of the National Academy of Sciences* **83**, 5145–9.

Mayer, M., Matsuura, J. & Jacobs, P. 1978. Inversions and other unusual heteromorphisms detected by C-banding. *Human Genetics* **45**, 43–50.

Mayfield, J. E. & Ellison, J. R. 1975. The organization of interphase chromatin in Drosophilidae. *Chromosoma* **52**, 37–48.

Mayr, B., Kalat, M. & Rab, P. 1988. Heterochromatins and band karyotypes in three species of salmonids. *Theoretical and Applied Genetics* **76**, 45–53.

Mayr, B., Schleger, W. & Auer, H. 1987. Frequency of Ag-stained nucleolus organizer regions in the chromosomes of cattle. *Journal of Heredity* **78**, 206–7.

Mayr, B., Schweizer, D. & Schleger, W. 1983. Characterization of the canine karyotype by counterstain-enhanced chromosome banding. *Canadian Journal of Genetics and Cytology* **25**, 616–21.

Mayr, B., Lambrou, M., Schweizer, D. & Kalat, M. 1989. An inversion polymorphism of a DA/DAPI-positive chromosome pair in *Anas platyrhynchos* (Aves). *Cytogenetics and Cell Genetics* **50**, 132–4.

Mayr, B., Geber, G., Auer, H., Kalat, M. & Schleger, W. 1986. Heterochromatin composition and nucleolus organizer activity in four canid species. *Canadian Journal of Genetics and Cytology* **28**, 744–53.

Mayr, B., Schweizer, D., Mendelak, M., Krutzler, J., Schleger, W., Kalat, M. & Auer, H. 1985. Levels of conservation and variation of heterochromatin and nucleolus organizers in the Bovidae. *Canadian Journal of Genetics and Cytology* **27**, 665–82.

Medina, F. J., Risueño, M. C., Sánchez-Pina, M. A. & Fernández-Gomez, M. E. 1983. A study on nucleolar silver staining in plant cells. The role of argyrophilic proteins in nucleolar physiology. *Chromosoma* **88**, 149–55.

Medina, F. J., Solanilla, E. L., Sánchez-Pina, M. A., Fernández-Gómez, M. E. & Risueño, M. C. 1986. Cytological approach to the nucleolar functions detected by silver staining. *Chromosoma* **94**, 259–66.

Medrano, L., Bernardi, G., Couturier, J., Dutrillaux, B. & Bernardi, G. 1988. Chromosome banding and genome compartmentalization in fishes. *Chromosoma* **96**, 178–83.

Meer, B., Hameister, H. & Cerrillo, M. 1981. Early and late replication patterns of increased resolution in human lymphocyte chromosomes. *Chromosoma* **82**, 315–9.

Mehra, R. C., Brekrus, S. & Butler, M. G. 1985. A simple two-step procedure for silver staining nucleolus organizer regions in plant chromosomes. *Canadian Journal of Genetics and Cytology* **27**, 255–7.

Meisner, L. F., Chuprevich, T. W. & Inhorn, S. L. 1973a. Chromosome banding in G_2 with tetracycline. *Lancet* **1**, 1509–10.

Meisner, L. F., Chuprevich, T. W., Inhorn, S. L. Indriksons, A. & Peterson, G. G. 1973b. Microanalysis of chromosomes with X-ray energy dispersion. *Lancet* **2**, 561.

Meisner, L. F., Chuprevich, T. W., Inhorn, S. L. & Johnson, C. B. 1974. Dye-nucleoprotein interactions in Giemsa banding. *Journal of Cell Biology* **61**, 248–52.

Mello, M. L. S. & Vidal, B. C. 1980. Acid lability of deoxyribonucleic acids of some polytene chromosome regions of *Rhynchosciara americana*. *Chromosoma* **81**, 419–29.

Meltzer, P. S., Cheng, Y-C. & Trent, J. M. 1985. Analysis of dihydrofolate reductase gene amplification in a methotrexate-resistant human tumor cell line. *Cancer Genetics and Cytogenetics* **17**, 289–300.

Mendelsohn, J. 1974. Studies of isolated mammalian metaphase chromosomes. In *The cell nucleus*, Vol. 2, H. Busch (ed.), 123–47. New York: Academic Press.

Mendelsohn, M. L., Mayall, B. H., Bogart, E., Moore, D. H. & Perry, B. H. 1973. DNA content and DNA-based centromeric index of the 24 human chromosomes. *Science* **179**, 1126–9.

Mengden, G. A. 1981. Linear differentiation of the C-band pattern of the W chromosome in snakes and birds. *Chromosoma* **83**, 275–87.

Mengden, G. A. & Stock. A. D. 1980. Chromosomal evolution in Serpentes: a comparison of G and C chromosome banding patterns of some colubrid and boid genera. *Chromosoma* **79**, 53–64.

Merrick, S., Ledley, R. S. & Lubs, H. A. 1973. Production of G and C banding with progressive trypsin treatment. *Pediatric Research* **7**, 39–44.

Merry, D. E., Pathak, S., Hsu, T. C. & Brinkley, B. R. 1985. Anti-kinetochore antibodies: use as probes for inactive centromeres. *American Journal of Human Genetics* **37**, 425–30.

Metzner, R. 1894. Beitrage zur Granulalehre. I Kern und Kerntheilung. *Archiv fur Physiologie* **18**, 309–48.

Meulenbroek, G. H. M. & Geraedts, J. P. M. 1982. Parental origin of chromosome abnormalities in spontaneous abortions. *Human Genetics* **62**, 129–33.

Meyne, J., Bartholdi, M. F., Travis, G. & Cram, L. S. 1984. Counterstaining human chromosomes for flow karyology. *Cytometry* **5**, 580–3.

Mezzanotte, R. 1978. Differential banding induced in polytene chromosomes of *Drosophila melanogaster* stained with acridine orange. *Experientia* **34**, 322–3.

Mezzanotte, R. 1986. The selective digestion of polytene and mitotic chromosomes of *Drosophila melanogaster* by the Alu I and Hae III restriction endonucleases. *Chromosoma* **93**, 249–55.

Mezzanotte, R. & Ferrucci, L. 1981. The differential banding pattern produced by actinomycin-D/acridine orange counterstaining in metaphase chromosomes of *Drosophila melanogaster*. *Experientia* **37**, 822–3.

Mezzanotte, R. & Ferrucci, L. 1983. The metaphase banding pattern produced in the chromosomes of *Culiseta longiareolata* (Diptera: Culicidae) by Alu I restriction endonuclease or actinomycin-D treatment followed by acridine orange staining. *Genetica* **62**, 47–50.

Mezzanotte, R. & Ferrucci, L. 1984. Alterations induced in mouse chromosomes by restriction endonucleases. *Genetica* **64**, 123–8.

Mezzanotte, R., Bianchi, U. & Marchi, A. 1987. *In situ* digestion of *Drosophila virilis* polytene chromosomes by Alu I and Hae III restriction endonucleases. *Genome* **29**, 630–4.

Mezzanotte, R., Ferrucci, L. & Marchi, A. 1979. Light-induced banding (LIB) in metaphase chromosomes of *Culiseta longiareolata* (Diptera: Culicidae) *Genetica* **51**, 149–52.

Mezzanotte, R., Ferrucci, L. & Marchi, A. 1981. Methylene-blue/coriphosphine-O/acridine-orange, chromosomal DNA and visible light: interaction and cytological effect in *Drosophila melanogaster*. *Genetica* **55**, 203–7.

Mezzanotte, R., Bianchi, U., Vanni, R. & Ferrucci, L. 1983a. Chromatin organization and restriction endonuclease activity on human metaphase chromosomes. *Cytogenetics and Cell Genetics* **36**, 562–6.

Mezzanotte, R., Ferrucci, L., Marchi, A. & Bianchi, U. 1982. The longitudinal differentiation produced by photo-oxidation and acridine-orange staining in eukaryote chromosomes: role and involvement of DNA base composition. *Cytogenetics and Cell Genetics* **33**, 277–84.

Mezzanotte, R., Ferrucci, L., Vanni, R. & Bianchi, U. 1983b. Selective digestion of human metaphase chromosomes by Alu I restriction endonuclease. *Journal of Histochemistry and Cytochemistry* **31**, 553–6.

Mezzanotte, R., Ferrucci, L., Vanni, R. & Sumner, A. T. 1985. Some factors affecting the action of restriction endonucleases on human metaphase chromosomes. *Experimental Cell Research* **161**, 247–53.

Mezzanotte, R., Peretti, D., Ennas, M. G., Vanni, R. & Sumner, A. T. 1989. Analysis of human metaphase chromosomes using antibodies to double-stranded and single-stranded DNA: staining patterns are related to DNA conformation. *Cytogenetics and Cell Genetics* **50**, 54–8.

Mezzanotte, R., Vanni, R., Flore, O., Ferrucci, L. & Sumner, A. T. 1988. Ageing of fixed cytological preparations produces degradation of chromosomal DNA. *Cytogenetics and Cell Genetics* **48**, 60–2.

Michelson, A. M., Monny, C. & Kovoor, A. 1972. Action of quinacrine mustard on polynucleotides. *Biochimie* **54**, 1129–36.
Mihalakis, N., Miller, O. J. & Erlanger, B. F. 1976. Antibodies to histones and histone-histone complexes: immunochemical evidence for secondary structure in histone I. *Science* **192**, 469–71.
Mikelsaar, A-V. & Ilus, T. 1979. Population polymorphisms in silver staining of nucleolus organizer regions (NORs) in human acrocentric chromosomes. *Human Genetics* **51**, 281–5.
Mikelsaar, A-V. & Schwarzacher, H. G. 1978. Comparison of silver staining of nucleolus organizer regions in human lymphocytes and fibroblasts. *Human Genetics* **42**, 291–9.
Mikelsaar, A.-V., Ilus, T. & Kivi, S. 1978. Variant chromosome 3 (inv 3) in normal newborns and their parents, and in children with mental retardation. *Human Genetics* **41**, 109–13.
Mikelsaar, A.-V., Schmid, M., Krone, W., Schwarzacher, H. G. & Schnedl, W. 1977a. Frequency of Ag-stained nucleolus organizer regions in the acrocentric chromosomes of man. *Human Genetics* **37**, 73–7.
Mikelsaar, A.-V., Schwarzacher, H. G., Schnedl, W. & Wagenbichler, P. 1977b. Inheritance of Ag-stainability of nucleolus organizer regions. *Human Genetics* **38**, 183–8.
Miklos, G. L. G. & John, B. 1979. Heterochromatin and satellite DNA in man: properties and prospects. *American Journal of Human Genetics* **31**, 264–80.
Miklos, G. L. G. & Nankivell, R. N. 1976. Telomeric satellite DNA functions in regulating recombination. *Chromosoma* **56**, 143–67.
Miller, D. A. 1977. Evolution of primate chromosomes. *Science* **198**, 1116–24.
Miller, D. A., Choi, Y–C. & Miller, O. J. 1983. Chromosome localization of highly repetitive human DNAs and amplified ribosomal DNA with restriction enzymes. *Science* **219**, 395–7.
Miller, D. A., Allderdice, P. W., Miller, O. J. & Breg, W. R. 1971. Quinacrine fluorescence patterns of human D group chromosomes. *Nature* **232**, 24–7.
Miller, D. A., Dev, V. G., Tantravahi, R. & Miller, O. J. 1976. Suppression of human nucleolus organizer activity in mouse-human somatic hybrid cells. *Experimental Cell Research* **101**, 235–43.
Miller, D. A., Gosden, J. R., Hastie, N. D. & Evans, H. J. 1984. Mechanism of endonuclease banding of chromosomes. *Experimental Cell Research* **155**, 294–8.
Miller, D. A., Okamoto, E., Erlanger, B. F. & Miller, O. J. 1982. Is DNA methylation responsible for mammalian X chromosome inactivation? *Cytogenetics and Cell Genetics* **33**, 345–9.
Miller, D. A., Tantravahi, R., Dev, V. G. & Miller, O. J. 1977. Frequency of satellite association of human chromosomes is correlated with amount of Ag-staining of the nucleolus organizer region. *American Journal of Human Genetics* **29**, 490–502.
Miller, D. A., Firschein, I. L. Dev, V. G., Tantravahi, R. & Miller, O. J. 1974. The gorilla karyotype: chromosome lengths and polymorphisms. *Cytogenetics and Cell Genetics* **13**, 536–50.
Miller, O. J. 1970. Autoradiography in human cytogenetics. *Advances in Human Genetics* **1**, 35–130.
Miller, O. J. & Erlanger, B. F. 1975. Immunochemical probes of human chromosome organization. *Pathobiology Annual* 1975., 71–103.
Miller, O. J., Schnedl, W., Allen, J. & Erlanger, B. F. 1974. 5-methylcytosine localised in mammalian constitutive heterochromatin. *Nature* **251**, 636–7.
Miller, O. J., Miller, D. A., Dev, V. G., Tantravahi, R. & Croce, C. M. 1976. Expression of human and suppression of mouse nucleolus organizer activity in mouse-human somatic cell hybrids. *Proceedings of the National Academy of Sciences* **73**, 4531–5.
Miller, O. J., Miller, D. A., Kouri, R. E., Allderdice, P. W., Dev, V. G., Grewal, M. S. & Hutton, J. J. 1971. Identification of the mouse karyotype by quinacrine fluorescence, and tentative assignment of seven linkage groups. *Proceedings of the National Academy of Sciences* **68**, 1530–3.
Miller, O. L. 1981. The nucleolus, chromosomes, and visualization of genetic activity. *Journal of Cell Biology* **91**, 15s–27s.
Miny, P., Holzgreve, W., Basaran, S., Gerbaulet, K. H., Beller, F. K. & Pawlowitski, I-H. 1985. Maternal cell contamination in chorionic villi cultures – exclusion by chromosomal fluorescence polymorphisms. *Clinical Genetics* **28**, 262–3.

Misawa, S., Takino, T., Morita, M., Abe, T. & Ashihara, T. 1977. Staining properties of a benzimidazol derivative '33258 Hoechst' and a simplified staining method for chromosome banding. *Japanese Journal of Human Genetics* **22**, 1–9.

Mitchell, A. R. & Gosden, J. R. 1978. Evolutionary relationships between man and the great apes. *Science Progr.* **65**, 273–93.

Mitchison, T. J. 1988. Microtubule dynamics and kinetochore function in mitosis. *Annual Review of Cell Biology* **4**, 527–49.

Mitchison, T. J. & Kirschner, M. W. 1985. Properties of the kinetochore *in vitro*. I. Microtubule nucleation and tubulin binding. *Journal of Cell Biology* **101**, 755–65.

Mitelman, F. 1984. Restricted number of chromosomal regions implicated in aetiology of human cancer and leukaemia. *Nature* **310**, 325–7.

Modest, E. J. & Sengupta, S. K. 1973. Chemical correlates of chromosome banding. In *Chromosome identification. Nobel Symposium 23*, T. Caspersson & L. Zech (eds), 327–34. New York: Academic Press.

Modest, E. J. & Sengupta, S. K. 1974. 7-substituted actinomycin D (NSC-3053) analogues as fluorescent DNA-binding and experimental antitumor agents. *Cancer Chemotherapy Reports*. **58**, 35–48.

Moens, P. B., Heyting, C., Deitrich, A. J. J., van Raamsdonk, W. & Chen, Q. 1987. Synaptonemal complex antigen location and conservation. *Journal of Cell Biology* **105**, 93–103.

Mogford, D. J. 1977. Chromosome associations in onion root tip nuclei. *Journal of South African Botany* **43**, 97–102.

Monaco, P. J. & Rasch, E. M. 1982. Differences in staining with DNA-specific fluorochromes during spermiogenesis. *Journal of Histochemistry and Cytochemistry* **30**, 585.

Montgomery, K. T., Biedler, J. L., Spengler, B. A. & Melera, P. W. 1983. Specific DNA sequence amplification in human neuroblastoma cells. *Proceedings of the National Academy of Sciences* **80**, 5724–8.

Moorhead, P. S., Nowell, P. C., Mellman, W. J., Battips, D. M. & Hungerford, D. A. 1960. Chromosome preparations of leucocytes cultured from human peripheral blood. *Experimental Cell Research* **20**, 613–6.

Morad, M., Jonasson, J. & Lindsten, J. 1973. Distribution of mitomycin C induced breaks on human chromosomes. *Hereditas* **74**, 273–82.

Morin, J., Marcollet, M., Geneix, A., Jaffray, J-Y., Malet, P. & Turchini, J-P. 1977. Marquage immunocytochimique de chromosomes humains après action d'anticorps de serum autoimmun d'origine humaine. *Comptes Rendu Academie Sciences (Paris)* D, **285**, 1211–3.

Moritz, C. 1984. The evolution of a highly variable sex chromosome in *Gehyra purpurascens* (Gekkonidae). *Chromosoma* **90**, 111–9.

Moroi, Y., Hartman, A. L., Nakane, P. K. & Tan, E. M. 1981. Distribution of kinetochore (centromere) antigen in mammalian cell nuclei. *Journal of Cell Biology* **90**, 254–9.

Moroi, Y., Peebles, C., Fritzler, M. J., Steigerwald, J. & Tan, E. M. 1980. Autoantibody to centromere (kinetochore) in scleroderma sera. *Proceedings of the National Academy of Sciences* **77**, 1627–31.

Morton, C. C., Brown, J. A., Holmes, W. M., Nance, W. E. & Wolf, B. 1983. Stain intensity of human nucleolus organizer region reflects incorporation of uridine into mature ribosomal RNA. *Experimental Cell Research* **145**, 405–13.

Moscetti, G., Aghemo, G. & Ronzoni, P. 1972. Observations in human spermatozoa by fluorescent 'Acranil' staining. *Humangenetik* **17**, 75–7.

Moscetti, G., Petriaggi, M., Barbarossa, C. G. & Tiberti, S. 1971. Fluorescence staining method for the morphological and structural study of human chromosomes. *Humangenetik* **12**, 56–8.

Moser, G. C., Fallon, R. J. & Meiss, H. K. 1981. Fluorimetric measurements and chromatin condensation patterns of nuclei from 3T3 cells throughout G1. *Journal of Cell Physiology* **106**, 293–301.

Moses, M. J. & Counce, S. J. 1974. Electron microscopy of kinetochores in whole mount spreads of mitotic chromosomes from HeLa cells. *Journal of Experimental Zoology* **189**, 115–20.

Mostacci, C., Ferraro, M., Pelliccia, F., Archidiacono, N., Rocchi, M. & de Capoa, A. 1980. Characterization of normal and rearranged human chromosomes by simultaneous Q- and R-banding with chromomycin A_3. *Cytogenetics and Cell Genetics* **28**, 3–9.

Mouras, A., Salesses, G. & Lutz, A. 1978. Sur l'utilisation des protoplastes en cytologie: amelioration d'une méthode recente en vue de l'identification des chromosomes mitotiques des genres *Nicotiana* et *Prunus*. *Carylogia* **31**, 117–27.

Müller, G., Hayman, D. & Schempp, W. 1988. Studies on early replication patterns in the marsupial *Sminthopsis crassicaudata*: no evidence for XY replication homologies. *Cytogenetics and Cell Genetics* **48**, 164–6.

Müller, H., Klinger, H. P., & Glasser, M. 1975. Chromosome polymorphism in a human newborn population. II. Potentials of polymorphic chromosome variants for characterizing the idiogram of an individual. *Cytogenetics and Cell Genetics* **15**, 239–55.

Müller, U. & Schempp, W. 1982. Homologous early replication patterns of the distal short arms of prometaphasic X and Y chromosomes. *Human Genetics* **60**, 274–5.

Müller, W., Crothers, D. M. & Waring, M. J. 1973. A non-intercalating proflavine derivative. *European Journal of Biochemistry* **39**, 223–34.

Mullinger, A. M. & Johnson, R.T. 1980. Packing DNA into chromosomes. *Journal of Cell Science* **46**, 61–86.

Mullinger, A. M. & Johnson, R. T. 1983. Units of chromosome replication and packing. *Journal of Cell Science* **64**, 179–93.

Mullinger, A. M. & Johnson, R. T. 1987. Disassembly of the mammalian metaphase chromosome into its subunits: studies with ultraviolet light and repair synthesis inhibitors. *Journal of Cell Science* **87**, 55–69.

Münke, M. & Schmiady, H. 1979. A simple one-step procedure for staining the nucleolus organizer regions. *Experientia.* **35**, 602.

Murata, M. & Orton, T. J. 1984. G-band-like differentiation in mitotic prometaphase chromosomes of celery. *Journal of Heredity* **75**, 225–8.

Murer-Orlando, M. L. & Peterson, A. C. 1985. *In situ* nick translation of human and mouse chromosomes detected with a biotinylated nucleotide. *Experimental Cell Research* **157**, 322–34.

Murtagh, C. E. 1977. A unique cytogenetic system in monotremes. *Chromosoma* **65**, 37–57.

Musich, P. R., Brown, F. L. & Maio, J. J. 1977. Subunit structure of chromatin and the organization of eukaryotic highly repetitive DNA: nucleosomal proteins associated with a highly repetitive mammalian DNA. *Proceedings of the National Academy of Sciences* **74**, 3297–301.

Nagl, W. 1974. Role of heterochromatin in the control of cell cycle duration. *Nature* **249**, 53–4.

Nagl, W. 1976. *Zellkern und Zellzyklen*. Stuttgart: E. Ulmer.

Nagl, W. & Schmitt, H-P, 1985. Transcription of repetitive DNA in condensed plant chromatin. *Molecular Biology Reports* **10**, 143–6.

Nakagome, Y. & Chiyo, H. 1976. Nonrandom distribution of exchange points in patients with structural rearrangements. *American Journal of Human Genetics* **28**, 31–41.

Nakagome, Y., Nakahori, Y., Mitani, K. & Matsumoto, M. 1986. The loss of centromeric heterochromatin from an inactivated centromere of a dicentric chromosome. *Japanese Journal of Human Genetics* **31**, 21–6.

Nakagome, Y., Teramura, F., Kataoka, K. & Hosono, F. 1976. Mental retardation, malformation syndrome and partial 7p monsomy [45, XX, tdic (7;15)(p21;p11)]. *Clinical Genetics* **9**, 621–4.

Nakagome, Y., Abe, T., Misawa, S., Takeshita, T. & Iinuma, K. 1984. The 'loss' of centromeres from chromosomes of aged women. *American Journal of Human Genetics* **36**, 398–404.

Nakai, Y. & Kohno, S. 1987. Elimination of the largest chromosome pair during differentiation into somatic cells in the Japanese hagfish, *Myxine garmani* (Cyclostomata, Agnatha). *Cytogenetics and Cell Genetics* **45**, 80–3.

Nakamura, H., Morita, T. & Yoshida, S. 1976. Enzymatic detection of X-ray-induced strand breaks in nuclear DNA on section of mouse tissue. *International Journal of Radiation Biology* **29**, 201–10.

Nardi, I., Ragghianti, M. & Mancino, G. 1973. Banding patterns in newt chromosomes by the Giemsa stain. *Chromosoma* **40**, 321–31.

Nash, W. G. & O'Brien, S. J. 1982. Conserved regions of homologous G-banded chromosomes between orders in mammalian evolution: carnivores and primates. *Proceedings of the National Academy of Sciences* **79**, 6631–5.

Natarajan, A. T. & Gropp, A. 1972. A fluorescence study of heterochromatin and nucleolar organization in the laboratory and tobacco mouse. *Exptl. Cell Res.* **74**, 245–50.

Natarajan, A. T. & Raposa, T. 1974. Repetitive DNA and constitutive heterochromatin in the chromosomes of guinea pig. *Hereditas* **76**, 145–7.

Navas-Castillo, J., Cabrero, J. & Camacho, J. P. M. 1986. Heterochromatin variants in *Baetica ustulata* (Orthoptera; Tettigoniidae) analysed by C and G banding. *Heredity* **56**, 161–5.

Nelson, M. & McClelland M. 1989. Effects of site-specific methylation on DNA modification methyltransferases and restriction endonucleases. *Nucleic Acids Research* **17**, suppl., r389–r415.

Nesbitt, M. N. & Donahue, R. P. 1972. Chromosome banding patterns in preimplantation mouse embryos. *Science* **177**, 805–6.

Neumann, H., Khalid, G., Flemans, R. J. & Hayhoe, F. G. J. 1980. A comparative study on the effect of various detergents in human chromosome G-banding prior to tryptic digestion. *Chromosoma* **77**, 105–12.

Newburger, P. E. & Latt, S. A. 1979. Improved fluorescent staining of interphase nuclei for prenatal diagnosis. *Lancet* **1**, 1144.

Niikawa, N. & Kajii, T. 1975. Sequential Q- and acridine orange-marker technique. *Humangenetik* **30**, 83–90.

Nishigaki, H., Okuda, T., Horiike, S., Tsuda, S., Taniwaki, M., Misawa, S., Inazawa, J. J. & Abe, T. 1988. DNA analysis using long-term preserved fixed cytogenetic preparations. *Japanese Journal of Human Genetics* **33**, 417–21.

Nishikai, M., Okano, Y., Yamashita, H. & Watanabe, M. 1984. Characterization of centromere (kinetochore) antigen reactive with sera of patients with a scleroderma variant (CREST syndrome). *Annals of Rheumatic Disease* **43**, 819–24.

Noda, K. & Kasha, K. J. 1978. A modified Giemsa C-banding technique for *Hordeum* species. *Stain Technology* **53**, 155–62.

Nordheim, A., Pardue, M. L., Weiner, L. M., Lowenhaupt, K., Scholten, P., Möller, A., Rich, A. & Stollar, B. D., 1986. Analysis of Z-DNA in fixed polytene chromosomes with monoclonal antibodies that show base sequence-dependent selectivity in reactions with supercoiled plasmids and polynucleotides. *Journal of Biological Chemistry* **261**, 468–76.

Nowell, P. C. 1964. Mitotic inhibition and chromosome damage by mitomycin in human leukocyte cultures. *Experimental Cell Research* **33**, 445–9.

Nussinov, R. 1987. Nucleotide quartets in the vicinity of eukaryotic transcriptional initiation sites: some DNA and chromatin structural implications. *DNA* **6**, 13–22.

Oaks, M. K., O'Malley, D. P., Kateley, J. R. & Maldonado, W. E. 1987. Detection of centromeric regions of chromosomes by immunofluorescence: procedure and application. *Journal of Medical Genetics* **24**, 498–9.

Obe, G., Natarajan, A. T. & Palitti, F. 1982. Role of DNA double-strand breaks in the formation of radiation-induced chromosomal aberrations. In *DNA repair, chromosome alterations and chromatin structure*. A. T. Natarajan (ed.), 1–9. Amsterdam: Elsevier.

O'Brien, R. L., Olenick, J. G. & Hahn, F. E. 1966. Reactions of quinine, chloroquine, and quinacrine with DNA and their effects on the DNA and RNA polymerase reactions. *Proceedings of the National Academy of Sciences* **55**, 1511–7.

O'Brien, S. J. & Seuanez, H. N. 1988. Mammalian genome organisation: an evolutionary view. *Annual Review of Genetics* **22**, 323–51.

O'Brien, S. J., Nash, W. G., Wildt, D. E., Bush, M. E. & Benveniste, R. E. 1985. A molecular solution to the riddle of the giant panda's phylogeny. *Nature* **317**, 140–4.

Ochs, R. L. & Busch, H. 1984. Further evidence that phosphoprotein C23 (110kD/pI 5.1) is the nucleolar silver staining protein. *Experimental Cell Researach* **152**, 260–5.

Ochs, R., Lischwe, M., O'Leary, P. & Busch, H. 1983. Localization of nucleolar phosphoproteins B23 and C23 during mitosis. *Experimental Cell Research* **146**, 139–49.

Ockey, C. H. 1980. Autoradiographic evidence of differential loss of BUdR-substituted DNA after UV exposure in FPG harlequin staining. *Experimental Cell Research* **125**, 511–4.

Ohno, S. 1967. *Sex chromosomes and sex-linked genes*. Berlin: Springer.

Ohno, S. 1983. Phylogeny of the X chromosome of man. In *Cytogenetics of the mammalian X chromosome*, part A, A. A. Sandberg (ed.), 1–19. New York: Alan R. Liss.

Ohno, S. & Hauschka, T. S. 1960. Allocycly of the X chromosome in tumors and normal tissues. *Cancer Research* **20**, 541–5.

Ohno, S., Beçak, W. & Beçak, M. L. 1964. X-autosome ratio and the behaviour pattern of individual X-chromosomes in placental mammals. *Chromosoma* **15**, 14–30.

Ohnuki, Y. 1965. Demonstration of the spiral structure of human chromosomes. *Nature* **208**, 916–17.

Ohyashiki, K., Yoshida, M. A., Ohyashiki, J., Koch, F., Han, T. & Sandberg, A. A. 1985. Two 14q+ chromosomes in malignant lymphoma: crucial cytogenetic changes on 14q. *Cancer Genetics and Cytogenetics* **17**, 325–31.

Okada, T. A. & Comings, D. E. 1974. Mechanisms of chromosome banding. III. Similarity between G-bands of mitotic chromosomes and chromomeres of meiotic chromosomes. *Chromosoma* **48**, 65–71.

Okada, T. A. & Comings, D. E. 1980. A search for protein cores in chromosomes: is the scaffold an artefact? *American Journal of Human Genetics* **32**, 814–32.

Okamoto, E., Miller, D. A., Erlanger, B. F. & Miller, O. J. 1981. Polymorphism of 5-methylcytosine-rich DNA in human acrocentric chromosomes. *Human Genetics* **58**, 255–9.

Olert, J. 1979. Interphase studies with a simplified method of silver staining of nucleoli. *Experientia* **35**, 283–5.

Olert, J. & Schmid, M. 1978. Comparative analysis of karyotypes in European shrew species. I. The sibling species *Sorex araneus* and *S. gemellus*: Q-bands, G-bands, and position of NORs. *Cytogenetics and Cell Genetics* **20**, 308–22.

Olert, J., Sawatzki, G., Kling, H. & Gebauer, J. 1979. Cytological and histochemical studies on the mechanism of the selective silver staining of nucleolus organizer regions (NORs). *Histochemistry* **60**, 91–9.

Olmo, E., Odierna, G. & Cobror, O. 1986. C-band variability and phylogeny of Lacertidae. *Genetica* **71**, 63–74.

Olson, M. O. J., Rivers, Z. M., Thompson, B. A., Kao, W-Y. & Case, S. T. 1983. Interaction of nucleolar phosphoprotein C23 with cloned segments of rat ribosomal deoxyribonucleic acid. *Biochemistry* **22**, 3345–51.

Olson, S. B., Magenis, R. E. & Lovrien, E. W. 1986. Human chromosome variation: the discriminatory power of Q-band heteromorphism (variant) analysis in distinguishing between individuals, with specific application to cases of questionable paternity. *American Journal of Human Genetics* **38**, 235–52.

Ord, M. G. & Stocken, L. A. 1966. Metabolic properties of histones from rat liver and thymus gland. *Biochemical Journal* **98**, 888–97.

Orkin, S. H. 1984. Wilms' tumour: molecular evidence for the role of chromosome 11. *Cancer Surveys* **3**, 465–77.

Oshimura, M. & Barrett, J. C. 1985. Double nondisjunction during karyotypic progression of chemically induced Syrian hamster cell lines. *Cancer Genetics and Cytogenetics* **18**, 131–9.

Östergren, G. 1947. Proximal heterochromatin, structure of the centromere and the mechanism of its misdivision. *Bot. Notiser* **100**, 176–7.

Overton, K. M., Magenis, R. E., Brady, T., Chamberlin, J. & Parks, M. 1976. Cytogenetic darkroom magic: now you see them, now you don't. *American Journal of Human Genetics* **28**, 417–19.

Özkinay, C. & Mitelman, F. 1979. A simple trypsin-Giemsa technique producing simultaneous G- and C-banding in human chromosomes. *Hereditas* **90**, 1–4.

Pachmann, U. & Rigler, R. 1972. Quantum yield of acridines interacting with DNA of defined base sequence. A basis for the explanation of acridine bands in chromosomes. *Experimental Cell Research* **72**, 602–8.

Page, B. M. 1973. Identification of chromosome 9 in human male meiosis. *Cytogenetics and Cell Genetics* **12**, 254–63.

Palmer, C. G. 1970. 5-bromodeoxyuridine-induced constrictions in human chromosomes. *Canadian Journal of Genetics and Cytology* **12**, 816–30.

Palmer, D. K., O'Day, K., Wener, M. H., Andrews, B. S. & Margolis, R. L. 1987. A 17-kD centromere protein (CENP-A) copurifies with nucleosome core particles and with histones. *Journal of Cell Biology* **104**, 805–15.

Pardue, M. L. & Gall, J. G. 1970. Chromosomal localization of mouse satellite DNA. *Science* **168**, 1356–8.

Passarge, E. 1979. Emil Heitz and the concept of heterochromatin: longitudinal chromosome differentiation was recognized fifty years ago. *American Journal of Human Genetics* **31**, 106–15.

Pathak, S. & Arrighi, F. E. 1973. Loss of DNA following C-banding procedures. *Cytogenetics and Cell Genetics* **12**, 414–22.

Pathak, S. & Hsu, T. C. 1979. Silver-stained structures in mammalian meiotic prophase. *Chromosoma* **70**, 195–203.

Pathak, S. & Stock, A. D. 1974. The X chromosomes of mammals: karyological homology as revealed by banding techniques. *Genetics* **78**, 703–14.

Pathak, S., Hsu, T. C. & Arrighi, F. E. 1973. Chromosomes of *Peromyscus* (Rodentia, Cricetidae). IV. The role of heterochromatin in karyotype evolution. *Cytogenetics and Cell Genetics* **12**, 315–26.

Patil, S. R. & Lubs, H. A. 1977. Classification of qh regions in human chromosomes 1, 9, and 16 by C-banding. *Human Genetics* **38**, 35–8.

Patil, S. R., Merrick, S. & Lubs, H. A. 1971. Identification of each human chromosome with a modified Giemsa stain. *Science* **173**, 821–2.

Patterson, R. M. & Petricciani, J. C. 1973. A comparison of prophase and metaphase G-bands in the Muntjak. *Journal of Heredity* **64**, 80–2.

Paulson, J. R. 1989. Scaffold morphology in histone-depleted HeLa metaphase chromosomes. *Chromosoma* **97**, 289–95.

Paulson, J. R. & Laemmli, U. K. 1977. The structure of histone-depleted metaphase chromosomes. *Cell* **12**, 817–28.

Paweletz, N. & Risueño, M. C. 1982. Transmission electron microscopic studies on the mitotic cycle of nucleolar proteins impregnated with silver. *Chromosoma* **85**, 261–73.

Pawlowitzki, I. H. & Pearson, P. L. 1972. Chromosomal aneuploidy in human spermatozoa. *Humangenetik* **16**, 119–22.

Peacock, W. J., Lohe, A. R., Gerlach, W. L., Dunsmuir, P., Dennis, E. S. & Appels, R. 1977. Fine structure and evolution of DNA in heterochromatin. *Cold Spring Harbor Symposia on Quantitative Biology* **42**, 1121–35.

Pearson, P. L. & Bobrow. M. 1970a. Definitive evidence for the short arm of the Y chromosome associating with the X chromosome during meiosis in the human male. *Nature* **226**, 959–61.

Pearson, P. L. & Bobrow. M. 1970b. Fluorescent staining of the Y chromosome in meiotic stages of the human male. *Journal of Reproduction and Fertility* **22**, 177–9.

Pearson, P. L. & van Egmond-Cowan, A. M. M. 1976. Further developments in banding techniques for mammalian chromosome identification. In *New techniques in biophysics and cell biology*, Vol. 3, R. H. Pain & B. J. Smith (eds), 213–39. London: Wiley.

Person, P. L., Bobrow, M. & Vosa, C. G. 1970. Technique for identifying Y chromosomes in human interphase nuclei. *Nature* **226**, 78–80.

Pearson, P. L., Geraedts, J. P. M. & Pawlowitzki, I. H. 1974. Chromosomal studies on human male gametes. In *Les accidents chromosomiques de la reproduction*, A. Boué & C. Thibault (eds), 219–29. Paris: INSERM.

Pearson, P. L., Bobrow, M., Vosa, C. G. & Barlow, P. W. 1971. Quinacrine fluorescence in mammalian chromosomes. *Nature* **231**, 326–9.
Pederson, D. S., Thoma, F. & Simpson, R. T. 1986. Core particle, fibre, and transcriptionally active chromatin structure. *Annual Review of Cell Biology* **2**, 117–47.
Pedrosa, M. P., Salzano, F. M., Mattevi, M. S. & Viégas, J. 1983. Quantitative analysis of C-bands in chromosomes 1, 9, 16 and Y of twins. *Acta Genet. med. Gemellol.* **32**, 257–60.
Pellicciari, C., Formenti, D., Zuccotti, M., Stanyon, R. & Manfredi Romanini, M. G. 1988. Genome size and constitutive heterochromatin in *Hylobates muelleri* and *Symphalangus syndactylus* and in their viable hybrid. *Cytogenetics and Cell Genetics* **47**, 1–4.
Pepper, D. A. & Brinkley, B. R. 1977. Localization of tubulin in the mitotic apparatus of mammalian cells by immunofluorescence and immunoelectron microscopy. *Chromosoma* **60**, 223–35.
Pera, F. 1972. Pattern of repetitive DNA of the chromosomes of *Microtus agrestis. Chromosoma* **36**, 263–71.
Peretti, D., Maraschio, P., Lambiase, S., Lo Curto, F. & Zuffardi, O. 1986. Indirect immunofluorescence of inactive centromeres as indicator of centromeric function. *Human Genetics* **73**, 12–16.
Perry, P. & Wolff, S. 1974. New Giemsa method for the differential staining of sister chromatids. *Nature* **251**, 156–8.
Peterson, J. L. & McConkey, E. H. 1976. Non-histone chromosomal proteins from HeLa cells. *Journal of Biological Chemistry* **251**, 548–54.
Petit, D., Couturier, J., Viegas-Péquignot, E., Lombard, M. & Dutrillaux, B. 1984. Tres grande similitude entre le caryotype ancestral des ecureuils (rongeurs) et celui des primates et des carnivores. *Annales de Genetique* **27**, 201–12.
Pfeifle, J., Boller, K. & Anderer, F. A. 1986. Phosphoprotein pp135 is an essential component of the nucleolus organizer region (NOR). *Experimental Cell Research* **162**, 11–22.
Phillips, R. B. 1980. New C band markers of human chromosomes: C band position variants. *Journal of Medical Genetics* **17**, 380–5.
Phillips, R. B. & Ihssen, P. E. 1986. Inheritance of Q-band chromosomal polymorphisms in lake trout. *Journal of Heredity* **77**, 93–7.
Phillips, R. B. & Zajicek, K. D. 1982. Q band chromosomal polymorphisms in lake trout (*Salvelinus namaycush*). *Genetics* **101**, 227–34.
Phillips, R. B., Pleyte, K. A. & Hartley, S. E. 1988. Stock-specific differences in the number and chromosome positions of the nucleolar organizer regions in the arctic char (*Salvelinus alpinus*). *Cytogenetics and Cell Genetics* **48**, 9–12.
Pienta, K. J. & Coffey, D. S. 1984. A structural analysis of the role of the nuclear matrix and DNA loops in the organization of the nucleus and chromosome. *Journal of Cell Science* suppl. **1**, 123–35.
Pimpinelli, S., Gatti, M. & de Marco, A. 1975. Evidence for heterogeneity in heterochromatin of *Drosophila melanogaster. Nature* **256**, 335–7.
Pimpinelli, S., Santini, G. & Gatti, M. 1976a. Characterization of *Drosophila* heterochromatin. II. C- and N-banding. *Chromosoma* **57**, 337–86.
Pimpinelli, S., Bonaccorsi, S., Gatti, M. & Sandler, L. 1986. The peculiar genetic organization of *Drosophila* heterochromatin. *Trends in Genetics* **2**, 17–20.
Pimpinelli, S., Prantera, G., Rocchi, A. & Gatti, M. 1976b. Effects of Hoechst 33258 on human leukocytes *in vitro. Cytogenetics and Cell Genetics* **17**, 114–21.
Pimpinelli, S., Sullivan, W., Prout, M. & Sandler, L. 1985. On biological functions mapping to the heterochromatin of *Drosophila melanogaster. Genetics* **109**, 701–24.
Pleyte, K. A., Phillips, R. B. & Hartley, S. E. 1989. Q-band chromosomal polymorphisms in arctic char (*Salvelinus alpinus*). *Genome* **32**. 129–33.
Ploem, J. S. 1977. Quantitative fluorescence microscopy. In *Analytical and quantitative methods in microscopy*, G. A. Meek & H. Y. Elder (eds), 55–89. Cambridge: Cambridge University Press.

Ploem, J. S. & Tanke, H. J. 1987. *Introduction to fluorescence microscopy*. Oxford: Oxford University Press.
Ploton, D., Menager, M. & Adnet, J-J. 1984. Simultaneous high resolution localization of Ag-NOR proteins and nucleoproteins in interphase and mitotic nuclei. *Histochemical Journal* **16**, 897–906.
Ploton, D., Menager, M., Jeannesson, P., Himber, G., Pigeon, F. & Adnet, J. J. 1986. Improvement in the staining and in the visualization of the argyrophilic proteins of the nucleolar organizer region at the optical level. *Histochemical Journal* **18**, 5–14.
Plumbridge, T. W. & Brown, J. R. 1979. The interaction of adriamycin and adriamycin analogues with nucleic acids in the B and A conformations. *Biochimica Biophysica Acta* **563**, 181–92.
Podugolnikova, O. A., Sushanlo, K. M., Blyumina, M. G. & Prokofieva-Belgovskays, A. A. 1984. Polymorphism of heterochromatic regions of chromosomes, 1, 9, 16, Y and mental deficiency in humans. (In Russian). *Genetika* **20**, 177–82.
Polani, P. E. 1972. Centromere localization at meiosis and the position of chiasmata in the male and female mouse. *Chromosoma* **36**, 343–74.
Polani, P. E. & Mutton, D. E. 1971. Y-fluorescence of interphase nuclei, especially circulating lymphocytes. *British Medical Journal* **1**, 138–42.
Popescu, N. C. & di Paolo, J. A. 1974. Sequential G and C chromosome banding. *Lancet* **1**, 209–10.
Pothier, L., Gallagher, J. F., Wright, C. E. & Libby, P. R. 1975. Histones in fixed cytological preparations of Chinese hamster chromosomes demonstrated by immunofluorescence. *Nature* **255**, 350–2.
Potluri, V. R., Singh, I. P. & Bhasin, M. K. 1985. Chromosomal heteromorphisms in Delhi infants. III. Qualitative analysis of C-band inversion heteromorphisms of chromosomes 1, 9 and 16. *Journal of Heredity* **76**, 55–8.
Prantera, G., Bonaccorsi, S. & Pimpinelli, S. 1979. Simultaneous production of Q and R bands after staining with chromomycin A_3 or olivomycin. *Science* **204**, 79–80.
Prantera, G., Cipriani, L., di Castro, M. & Rocchi, A. 1986. Modifications of human chromosome core after undercondensation. *Caryologia* **39**, 33–9.
Prantera, G., di Castro, M., Cipriani, L. & Rocchi, A. 1981. Inhibition of human chromosome condensation induced by DAPI as related to cell cycle. *Experimental Cell Research* **135**, 63–8.
Prantera, G., di Castro, M., Marchetti, E. & Rocchi, A. 1977. Effect of distamycin A on Chinese hamster chromosomes. *Experimental Cell Research* **109**, 459–62.
Prescott, D. M., Bostock, C. J., Hatch, F. T. & Mazrimas, J. A. 1973. Location of satellite DNAs in the chromosomes of the kangaroo rat (*Dipodomys ordii*). *Chromosoma* **42**, 205–13.
Priest, J. H. 1977. *Cytogenetics and cell culture*, 2nd ed. Philadelphia: Lea & Febiger.

Qumsiyeh, M. B. & Baker, R. J. 1988. Comparative cytogenetics and the determination of primitive karyotypes. *Cytogenetics and Cell Genetics* **47**, 100–3.

Raap, A. K., Marijnen, J. G. J., Vrolijk, J. & van der Ploeg, M. 1986. Denaturation, renaturation, and loss of DNA during in situ hybridization procedures. *Cytometry* **7**, 235–42.
Rabinovitch, P. S., Kubbies, M., Chen, Y. C., Schindler, D. & Hoehn, H. 1988. BrdU-Hoechst flow cytometry: a unique tool for quantitative cell cycle analysis. *Experimental Cell Research* **174**, 309–18.
Rabinovitch, P. S., O'Brien, K., Simpson, M., Callis, J. B. & Hoehn, H. 1981. Flow-cytogenetics. II. High-resolution ploidy measurements in human fibroblast culture. *Cytogenetics and Cell Genetics* **29**, 65–76.
Raman, R. & Sperling, K. 1981. Patterns of silver staining on NORs of prematurely condensed muntjac chromosomes following RNA inhibition. *Experimental Cell Research* **135**, 373–8.
Ranganath, H. A., Schmidt, E. R. & Hagele, K. 1982. Satellite DNA of *Drosophila nasuta nasuta* and *D. n. albomicana*: localization in polytene and metaphase chromosomes. *Chromosoma* **85**, 361–8.
Rao, P. N. & Johnson, R. T. 1974. Induction of chromosome condensation in interphase cells. *Advances in Cell and Molecular Biology* **3**, 135–89.

Raposa, T. & Natarajan, A. T. 1974. Fluorescence banding pattern of human and mouse chromosomes with a benzimidazol derivative (Hoechst 33258)., *Humangenetik* **21**, 221–6.
Rapoza, P. A., Testa, J. R., Egues, M. C. & Rowley, J. D. 1984. Evaluation of techniques for increasing mitotic yield and chromosome length in leukaemic cells. *Acta Cytologica* **28**, 740–6.
Rattner, J. B. 1986. Organisation within the mammalian kinetochore. *Chromosoma* **93**, 515–20.
Rattner, J. B. & Lin, C. C. 1985. Radial loops and helical coils coexist in metaphase chromosomes. *Cell* **42**, 291–6.
Rattner, J. B., Krystal, G. & Hamkalo, B. A. 1978. Selective digestion of mouse metaphase chromosomes. *Chromosoma* **66**, 259–68.
Ray, F. A., Bartholdi, M. F., Kraemer, P. M., & Cram, L. S. 1984. Chromosome polymorphism involving heterochromatic blocks in Chinese hamster chromosome 9. *Cytogenetics and Cell Genetics* **38**, 257–64.
Ray, M. & Hamerton, J. L. 1973. Constitutive heterochromatin in mouse chromosomes treated with trypsin. *Canadian Journal of Genetics and Cytology* **15**, 1–7.
Ray, M. & Mohandas, T. 1976. Proposed banding nomenclature for the Chinese hamster chromosomes (Cricetulus griseus). *Cytogenetics and Cell Genetics* **16**, 83–91.
Rayburn, A. L., Price, H. J., Smith, J. D. & Gold, J. R. 1985. C-band heterochromatin and DNA content in Zea mays. *American Journal of Botany* **72**, 1610–7.
Razin, A. & Cedar, H. 1984. DNA methylation in eukaryotic cells. *International Review of Cytology* **92**, 159–85.
Redi, C. A. & Capanna, E. 1988. Robertsonian heterozygotes in the house mouse and the fate of their germ cells. In *The cytogenetics of mammalian autosomal rearrangements*, A. Daniel (ed.), 315–59. New York: Alan R. Liss.
Rees, R. W., Fox, D. P. & Maher, E. P. 1976. DNA content, reiteration and satellites in Dermestes. In *Current chromosome research*, K. Jones & P. E. Brandham (eds), 33–41. Amsterdam: Elsevier/North-Holland.
Retief, A. E. & Rüchel, R. 1977. Histones removed by fixation. Their role in the mechanism of chromosomal banding. *Experimental Cell Research* **106**, 233–7.
Rhoades, M. M. 1978. Genetic effects of heterochromatin in maize. In *Maize breeding and genetics*, D. B. Walden (ed.), New York: Wiley.
Rich, A., Nordheim, A. & Wang, A. H-J. 1984. The chemistry and biology of left-handed Z-DNA. *Annual Review of Biochemistry* **53**, 791–846.
Richler, C., Uliel, E., Kerem, B-S. & Wahrman, J. 1987. Regions of active chromatin conformation in 'inactive' male meiotic sex chromosomes of the mouse. *Chromosoma* **95**, 167–70.
Rieder, C. L. 1979. Localization of ribonucleoprotein in the trilaminar kinetochore of PtK$_1$. *Journal of Ultrastructure Research* **66**, 109–19.
Rieder, C. L. 1982. The formation, structure, and composition of the mammalian kinetochore and kinetochore fibre. *International Review of Cytology* **79**, 1–58.
Rieger, R., Michaelis, A. & Green, M. M. 1968. *A glossary of genetics and cytogenetics*. London: Allen & Unwin.
Rigler, R. 1966. Microfluorometric characterization of intracellular nucleic acids and nucleoproteins by acridine orange. *Acta Physiologica Scandinavica* **67**, suppl. 267, 1–222.
Ris, H. 1975. Chromosomal structure as seen by electron microscopy. In *The structure and function of chromatin. Ciba Foundation Symposium 28*, D. W. Fitzsimmons & G. E. W. Wolstenholme (eds), 7–28. Amsterdam: Associated Scientific Publishers.
Ris, H. 1978. Higher order structures in chromosomes. *Ninth International Congress on Electron Microscopy, Toronto*, **3**, 545–56.
Ris, H. & Korenberg, J. 1979. Chromosome structure and levels of chromosome organization. In *Cell biology: a comprehensive treatise*, Vol. 2, D. M. Prescott & L. Goldstein (eds), 267–361. New York: Academic Press.
Ris, H. & Witt, P. L. 1981. Structure of the mammalian kinetochore. *Chromosoma* **82**, 153–70.
Robert-Nicoud, M., Arndt-Jovin, D. J., Zarling, D. A. & Jovin, T. M. 1984. Immunological detection of left-handed Z-DNA in isolated polytene chromosomes. Effects of ionic strength, pH, temperature and topological stress. *EMBO Journal* **3**, 721–31.

Roberts, R. J. 1990. Restriction enzymes and their isoschizomers. *Nucleic Acids Research* **18**, suppl. 2331–2365.

Robinson, J. 1971. Y chromosome fluorescence in buccal mucosa cells: a study on a newborn population. *Annals of Human Genetics* **35**, 61–5.

Robinson, J. A. & Buckton, K. E. 1971. Quinacrine fluorescence of variant and abnormal human Y chromosomes. *Chromosoma* **35**, 342–52.

Robinson, J. A., Buckton, K. E., Spowart, G., Newton, M., Jacobs, P. A., Evans, H. J. & Hill, R. 1976. The segregation of human chromosome polymorphisms. *Annals of Human Genetics* **40**, 113–21.

Rocchi, A. 1982. On the heterogeneity of heterochromatin. *Caryologia* **35**, 169–89.

Rocchi, A., di Castro, M. & Prantera, G. 1979. Effects of DAPI on human leukocytes *in vitro*. *Cytogenetics and Cell Genetics* **23**, 250–4.

Rocchi, A., di Castro, M. & Prantera, G. 1980. Effect of DAPI on Chinese hamster chromosomes. *Cytogenetics and Cell Genetics* **27**, 70–2.

Rocchi, A., Montagna, P., Lanza, V. & Prantera, G. 1986. The correlation between silver-staining and DAPI-induced chromosome undercondensation. *Genetica* **68**, 145–9.

Rocchi, A., Prantera, G., di Castro, M. & Capanna, E. 1982. Silver staining of the centromeric area of acrocentric and Robertsonian metacentric mouse chromosomes. *Genetica* **60**, 65–9.

Rocchi, A., Prantera, G., Pimpinelli, S. & di Castro, M. 1976. Effect of Hoechst 33258 on Chinese hamster chromosomes. *Chromosoma* **56**, 41–6.

Rodionov, A. V. 1985. The genetic activity of G- and R-band DNA in human mitotic chromosomes. (In Russian). *Genetika* **21**, 2057–65.

Rodman, T. C. 1974. Human chromosome banding by Feulgen stain aids in localizing classes of chromatin. *Science* **184**, 171–3.

Rodman, T. C. & Tahiliani, S. 1973. The Feulgen banded karyotype of the mouse: analysis of the mechanisms of banding. *Chromosoma* **42**, 37–56.

Rofe, R. & Hayman, D. 1985. G-banding evidence for a conserved complement in the Marsupialia. *Cytogenet. Cell Genet.* **39**, 40–50.

Röhme, D. & Heneen, W. K. 1982. Banding patterns in prematurely condensed chromosomes and the underlying structure of the chromosome. In *Premature chromosome condensation*, P. N. Rao, R. T. Johnson & K. Sperling (eds), 131–57. New York: Academic Press.

Romagnano, A. & Richer, C-L. 1985. High resolution R-bands produced in equine chromosomes after incorporation of bromodeoxyuridine. *Journal of Heredity* **76**, 377–8.

Romagnano, A., Richer, C-L., Messier, P-E. & Jean, P. 1987. Light and electron microscopy of Ag-NORs in domestic horse chromosomes identified after R-banding. *Cytobios* **49**, 23–30.

Rong, R., Chandley, A. C., Song, J., McBeath, S., Tan, P. P., Bai, Q. & Speed, R. M. 1988. A fertile mule and hinny in China. *Cytogenetics and Cell Genetics* **47**, 134–9.

Rønne, M. 1977a. *In vitro* induction of G bands in human chromosomes. *Hereditas* **85**, 81–4.

Rønne, M. 1977b. *In vitro* induction of G bands with cycloheximide. *Hereditas* **86**, 107–10.

Rønne, M. 1977c. Induction of uncoiled chromosomes with RNA'se. *Hereditas* **86**, 245–50.

Rønne, M. 1979. The effect of *in vitro* dithioerythritol exposure on human metaphase chromosome structure. *Hereditas* **91**, 105–9.

Rønne, M. 1984. Fluorouracil synchronization of human lymphocyte cultures. Induction of high resolution R-banding by simultaneous *in vitro* exposure to 5-bromodeoxyuridine/Hoechst 33258. *Hereditas* **101**, 205–8.

Rønne, M. 1985. Double synchronization of human lymphocyte cultures: selection for high-resolution banded metaphases in the first and second division. *Cytogenetics and Cell Genetics* **39**, 292–5.

Rønne, M. & Andersen, O. 1978a. Effect of 5-fluorouracil and 5-fluorouridine on metaphase chromosome structure in human lymphoid cells. *Hereditas* **88**, 127–30.

Rønne, M. & Andersen, O. 1978b. *In vitro* induction of uncoiled chromosomes with inhibitors of RNA and protein synthesis. *Cytobios* **21**, 7–21.

Rønne, M. & Sandermann, J. 1977. Simple methods to induce banding in human chromosomes. *Heriditas* **86**, 151–4.

Rønne, M., Bøye, H. A. & Sandermann, J. 1977. A mounting medium for banded chromosomes. *Hereditas* **86**, 155–8.
Rønne, M., Vang Nielsen, K. & Erlandsen, M. 1979. Effect of controlled colcemid exposure on human metaphase chromosome structure. *Hereditas* **91**, 49–52.
Rønne, M., Eldridge, F. E., Thust, R. & Andersen, O, 1982. The effect of *in vitro* distamycin A exposure on metaphase chromosome structure. *Hereditas* **96**, 269–77.
Rooney, D. E. & Czepulkowski, B. H. 1986. *Human cytogenetics: a practical approach.* Oxford: IRL Press.
Roos, U-P. 1975. Are centromere dots kinetochores? *Nature* **254**, 463.
Ross, A. & Gormley, I. P. 1973. Examination of surface topography of Giemsa-banded human chromosomes by light and electron microscopic techniques. *Experimental Cell Research* **81**, 79–86.
Rost, F. W. D. 1980. Fluorescence microscopy. In *Histochemistry, theoretical and applied*, 4th edn., Vol. 1, A. G. E. Pearse (ed.), 346–78. Edinburgh: Churchill Livingstone.
Rowley, J. D. 1973. A new consistent chromosomal abnormality in chronic myelogenous leukaemia identified by quinacrine fluorescence and Giemsa staining. *Nature* **243**, 290–3.
Rowley, J. D. 1980. Chromosome abnormalities in cancer. *Cancer Genetics and Cytogenetics* **2**, 175–98.
Rowley, J. D. & Bodmer, W. F. 1971. Relationship of centromeric heterochromatin to fluorescent banding patterns of metaphase chromosomes in the mouse. *Nature* **231**, 503–6.
Royle, N. J. 1986. New C-band polymorphism in the White Park cattle of Great Britain. *Journal of Heredity* **77**, 366–7.
Rudak, E. & Callan, H. G. 1976. Differential staining and chromatin packing of the mitotic chromosomes of the newt *Triturus cristatus*. *Chromosoma* **56**, 349–62.
Rudak, E., Jacobs, P. A. & Yanagimachi, R. 1978. Direct analysis of the chromosome constitution of human spermatozoa. *Nature* **274**, 911–3.
Rufas, J. S., Giménez-Martín, G. & Esponda, P. 1982. Presence of a chromatid core in mitotic and meiotic chromosomes of grasshoppers. *Cell Biology International Reports* **6**, 261–7.
Rufas, J. S., Gosalvez, J., Giménez-Martín, G. & Esponda, P. 1983a. Localization and development of kinetochores and a chromatid core during meiosis in grasshoppers. *Genetica* **61**, 233–8.
Rufas, J. S., Gosalvez, J., Lopez-Fernandez, C. & Cardoso, H. 1983b. Complete dependence between AgNORs and C-positive heterochromatin revealed by simultaneous Ag-NOR C-banding. *Cell Biology International Reports* **7**, 275–81.
Rumpler, Y. & Dutrillaux, B. 1979. Chromosomal evolution in Malagasy lemurs. IV. Chromosome banding studies in the genuses *Phaner, Varesia, Lemur, Microcebus,* and *Cheirogaleus. Cytogenetics and Cell Genetics* **24**, 224–32.
Rumpler, Y., Ishak, B., Dutrillaux, B., Warter, S. & Ratsirarson, J. 1986. Chromosomal evolution in Malagasy lemurs. IX. Chromosomal banding studies of *Lepilemur mustelinus, L. dorsalis,* and *L. edwardsi. Cytogenetics and Cell Genetics* **42**, 164–8.
Rumpler, Y., Warter, S., Meier, B., Preuschoft, H. & Dutrillaux, B. 1987. Chromosomal phylogeny of three Lorisidae: *Loris tardigradus, Nycticebus cousang* and *Perodicticus potto. Folia Primatologia* **48**, 216–20.
Russell, G. J., Walker, P. M. B., Elton, R. A. & Subak-Sharpe, J. H. 1976. Doublet frequency analysis of fractionated vertebrate nuclear DNA. *Journal of Molecular Biology* **108**, 1–23.
Russell, W. C., Newman, C. & Williamson, D. H. 1975. A simple cytochemical technique for demonstration of DNA in cells infected with mycoplasmas and viruses. *Nature* **253**, 461–2.
Ruzicka, V. 1899. Zur Geschichte und Kenntnis der feineren Structur der Nucleolen centraler Nervenzellen. *Anatomischer Anzeiger* **16**, 557–63.
Ryder, O. A., Epel, N. C. & Benirschke, K. 1978. Chromosome banding studies of the Equidae. *Cytogenetics and Cell Genetics* **20**, 323–50.
Ryttman, H., Tegelström, H. & Jansson, H. 1979. G- and C-banding in four related *Larus* species (Aves). *Hereditas* **91**, 143–8.

Sabaneyeva, E. V., Rautian, M. S. & Rodionov, A. V. 1984. Active nucleolar organizer regions in the macronuclei of *Tetrahymena pyriformis* (ciliates) revealed by means of Ag-NOR staining. (In Russian). *Tsitologiya* **26**, 849–52.

Sadgopal, A. & Bonner, J. 1970. Proteins of interphase and metaphase chromosomes compared. *Biochimica Biophysica Acta* **207**, 227–39.

Sahar, E. & Latt, S. A. 1978. Enhancement of banding patterns in human metaphase chromosomes by energy transfer. *Proceedings of the National Academy of Sciences* **75**, 5650–4.

Sahar, E. & Latt, S. A. 1980. Energy transfer and binding competition between dyes used to enhance staining differentiation in metaphase chromosomes. *Chromosoma* **79**, 1–28.

Sahasrabuddhe, C. G., Pathak, S. & Hsu, T. C. 1978. Responses of mammalian metaphase chromosomes to endonuclease digestion. *Chromosoma* **69**, 331–8.

Saksela, E. & Moorhead, P. S. 1962. Enhancement of secondary constrictions and the heterochromatic X in human cells. *Cytogenetics* **1**, 225–44.

Salamanca, F. & Armendares, S. 1975. Banding in old chromosome preparations. *Ann. Génét.* **18**, 139–40.

Salamanca, F., Guzman, M., Barbosa, E. & Martinez, I. 1972. A new fluorescent compound for cytogenetic studies. *Ann Génét.* **15**, 127–9.

Salinas, J., Matassi, G., Montero, L. M. & Bernardi, G. 1988. Compositional compartmentalization and compositional patterns in the nuclear genomes of plants. *Nucleic acids Res.* **16**, 4269–85.

Samols, D. & Swift, H. 1979. Genomic organization in the flesh fly *Sarcophaga bullata*. *Chromosoma* **75**, 129–43.

Sanchez, O. & Yunis, J. J. 1974. The relationship between repetitive DNA and chromosomal bands in man. *Chromosoma* **48**, 191–202.

Sanchez-Pina, M. A., Medina, F. J., Fernandez-Gomez, M. E. & Risueño, M. C. 1984. 'Ag-NOR' proteins are present when transcription is impaired. *Biology of the Cell* **50**, 199–202.

Sandberg, A. A. 1977. Cytogenetic data and prognosis in acute leukemias. In *Therapy of acute leukaemias*, F. Mandelli (ed.), 186–92. Rome: Lombardo Editore.

Sandberg, A. A. 1980. *The chromosomes in human cancer and leukaemia*. New York: Elsevier.

Sandermann, J. & Rønne, M. 1977. *In vitro* induction of G bands and uncoiling with fucidin. *Hereditas* **87**, 47–50.

Sandritter, W. & Hartlieb, J. 1955. Quantitative Untersuchungen uber den Nukleinsaureverlust des Gewebes bei Fixierung und Einbettung. *Experientia* **11**, 313–14.

San Roman, C. & Bobrow, M. 1973. The sites of radiation induced-breakage in human lymphocyte chromosomes, determined by quinacrine fluorescence. *Mutation Research* **18**, 325–31.

Santi, D. V., Norment, A. & Garrett, C. E. 1984. Covalent bond formation between a DNA-cytosine methyltransferase and DNA containing 5-azacytidine. *Proceedings of the National Academy of Sciences* **81**, 6993–7.

Santos, J., Sentis, C. & Fernandez-Piqueras, J. 1987. Pattern of nucleolar organizer region activity during male meiosis in *Callicrania seoanei* (Orthoptera) as analyzed by silver staining: evidences for a possible reactivation in the period between the two meiotic divisions. *Genome* **29**, 516–18.

Sarker, M. & Chen, F-M. 1989. Binding of mithramycin to DNA in the presence of second drugs. *Biochemistry* **28**, 6651–7.

Sarma, N. P. & Natarajan, A. T. 1973. Identification of heterochromatic regions in the chromosomes of rye. *Hereditas* **74**, 233–8.

Sasaki, M., Kodama, Y., Hayata, I. & Yoshida, M. C. 1979. Chromosome markers in 12 inbred strains of the Norway rat, *Rattus norvegicus*. *Cytogenetics and Cell Genetics* **23**, 231–40.

Sasaki, M. S. & Makino, S. 1963. The demonstration of secondary constrictions in human chromosomes by means of a new technique. *American Journal of Human Genetics* **15**, 24–33.

Sato, S. 1985. Ultrastructural localization of nucleolar material by a simple silver staining technique devised for plant cells. *Journal of Cell Science* **79**, 259–69.

Sato, S. & Sato, M. 1982. Visualization of centromeric spots in the chromosomes by UV-light exposure. *Experientia* **38**, 249–50.
Sato, S., Kuroki, Y. & Ohta, S. 1979. Two types of colour-differentiated C-banding positive segments in chromosomes of *Nothoscordum fragrans*, Liliaceae. *Cytologia* **44**, 715–25.
Sato, Y., Abe, S., Kubota, K., Sasaki, M. & Miura, Y. 1986. Silver-stained nucleolar organizer regions in bone marrow cells and peripheral blood lymphocytes of Philadelphia chromosome-positive chronic myelocytic leukaemia patients. *Cancer Genetics and Cytogenetics* **23**, 37–45.
Satoh, H. & Yoshida, M. C. 1985. Gene mapping in the Chinese hamster and conservation of syntenic groups and Q-band homologies between Chinese hamster and mouse chromosomes. *Cytogenetics and Cell Genetics* **39**, 285–91.
Satoh, K. & Busch, H. 1981. Silver staining of phosphoserine and phosphothreonine in nucleolar and other phosphoproteins. *Cell Biology International Reports* **5**, 857–66.
Satya-Prakash, K. L. & Pathak, S. 1984. Silver staining pattern of male meiosis in the house cricket. *Journal of Heredity* **75**, 319–20.
Savage, J. R. K. 1977a. Annotation: application of chromosome banding techniques to the study of primary chromosome structural changes. *Journal of Medical Genetics* **14**, 362–70.
Savage, J. R. K. 1977b. Assignment of aberration breakpoints in banded chromosomes. *Nature* **270**, 513–14.
Sawyer, J. R. & Hozier, J. C. 1986. High resolution of mouse chromosomes: banding conservation between man and mouse. *Science* **232**, 1632–5.
Scheer, U. & Rose, K. M. 1984. Localization of RNA polymerase I in interphase cells and mitotic chromosomes by light and electron microscopic immunocytochemistry. *Proceedings of the National Academy of Sciences* **81**, 1431–5.
Scheid, W. 1976. Mechanism of differential staining of BUdR-substituted *Vicia faba* chromosomes. *Experimental Cell Research* **101**, 55–8.
Scheid, W. 1979. Influence of cysteamine on differential staining of BUdR-substituted human chromosomes. *Canadian Journal of Genetics and Cytology* **21**, 145–9.
Schempp, W. & Meer, B. 1983. Cytologic evidence for three human X-chromosomal segments escaping inactivation. *Human Genetics* **63**, 171–4.
Schempp, W. & Schmid, M. 1981 Chromosome banding in Amphibia. VI. BrdU-replication patterns in Anura and demonstration of XX/XY sex chromosomes in *Rana esculenta*. *Chromosoma* **83**, 697–710.
Scheres, J. M. J. C. 1974. Production of C and T bands in human chromosomes after heat treatment at high pH and staining with 'Stains-All'. *Humangenetik* **23**, 311–14.
Scheres, J. M. J. C. 1976a. CT banding of human chromosomes. *Human Genetics* **31**, 293–307.
Scheres, J. M. J. C. 1976b. CT banding of human chromosomes. The role of cations in the alkaline pretreatment. *Human Genetics* **33**, 167–74.
Scheres, J. M. J. C. 1976c. R- and CT-banding of human chromosomes with basic fuchsin. *Differential staining of human chromosomes*. Doctoral thesis, University of Nijmegen, 65–69.
Scheres, J. M. J. C., Hustinx, T. W. J. & Merkx, G. F. M. 1980. Nomarski-optical studies of human chromosomes R-banded with barium hydroxide. *Human Genetics* **53**, 255–9.
Scheres, J. M. J. C., Merkx, G. F. M. & Hustinx, T. W. J. 1982. Prometaphase banding of human chromosomes with basic fuchsin. *Human Genetics* **61**, 8–11.
Scheres, J. M. J. C., Hustinx, T. W. J. Rutten, F. J. & Merkx, G. F. M. 1977. 'Reverse' differential staining of sister chromatids. *Experimental Cell Research* **109**, 466–8.
Schimke, R. T. 1982. *Gene amplification*. New York: Cold Spring Harbor Laboratory.
Schimke, R. T. 1984. Gene amplification in cultured animal cells. *Cell* **37**, 705–13.
Schlammadinger, J., Poulsen, H. & Mikkelsen, M. 1977. Inhibition of the development of Q-bands on human chromosomes by netropsin. *Human Genetics* **39**, 309–13.
Schlegel, R. & Gill, B. S. 1984. N-banding analysis of rye chromosomes and the relationship between N-banded and C-banded heterochromatin. *Canadian Journal of Genetics Cytology* **26**, 765–9.
Schmiady, H. & Sperling, K. 1976. Length of human C-bands in relation to the degree of chromosome condensation. *Human Genetics* **35**, 107–11.

Schmiady, H., Wegner, R-D. & Sperling, K. 1975. Relative DNA content of human euchromatin and heterochromatin after G, C and Giemsa 11 banding. *Humangenetik* **29**, 85–9.

Schmid, M. 1978. Chromosome banding in Amphibia. I. Constitutive heterochromatin and nucleolus organizer regions in *Bufo* and *Hyla*. *Chromosoma* **66**, 361–88.

Schmid, M. 1980a. Chromosome banding in Amphibia. IV. Differentiation of GC- and AT-rich chromosome regions in Anura. *Chromosoma* **77**, 83–103.

Schmid, M. 1980b. Chromosome evolution in Amphibia. In *Cytogenetics of vertebrates*, H. Muller (ed.), 4–27. Basel: Birkhauser.

Schmid, M. 1983. Evolution of sex chromosomes and heterogametic systems in Amphibia. *Differentiation* **23**, suppl., S13–22.

Schmid, M. & de Almeida, C. G. 1988. Chromosome banding in Amphibia. XII. Restriction endonuclease banding. *Chromosoma* **96**, 283–90.

Schmid, M. & Guttenbach, M. 1988. Evolutionary diversity of reverse (R) fluorescent chromosome bands in vertebrates. *Chromosoma* **97**, 101–14.

Schmid, M. & Haaf, T. 1984. Distamycin/DAPI bands and the effects of 5-azacytidine on the chromosomes of the chimpanzee, *Pan troglodytes*. *Cytogenetics and Cell Genetics* **38**, 192–9.

Schmid, M., Haaf, T. & Grunert, K. 1984a. 5-azacytidine-induced undercondensations in human chromosomes. *Human Genetics* **67**, 257–63.

Schmid, M., Olert J. & Klett, C. 1979. Chromosome banding in Amphibia. III. Sex chromosomes in *Triturus*. *Chromosoma* **71**, 29–55.

Schmid, M., Poppen, A. & Engel, W. 1981. The effects of distamycin A on gorilla-, chimpanzee- and orangutan lymphocyte cultures. *Cytogenetics and Cell Genetics* **30**, 211–21.

Schmid, M., Solleder, E. & Haaf, T. 1984b. The chromosomes of *Micromys minutus* (Rodentia, Murinae). I. Banding analysis. *Cytogenetics and Cell Genetics* **38**, 221–6.

Schmid, M., Grunert, D., Haaf, T. & Engel, W. 1983a. A direct demonstration of somatically paired heterochromatin of human chromosomes. *Cytogenetics and Cell Genetics* **36**, 554–61.

Schmid, M., Hungerford, D. A., Poppen, A. & Engel, W. 1984c. The use of distamycin A in human lymphocyte cultures. *Human Genetics* **65**, 377–84.

Schmid, M., Löser, C., Schmidtke, J. & Engel, W. 1982. Evolutionary conservation of a common pattern of activity of nucleolus organizers during spermatogenesis in vertebrates. *Chromosoma* **86**, 149–79.

Schmid, M., Müller, H., Stasch, S. & Engel, W. 1983b. Silver staining of nucleolus organizer regions during human spermatogenesis. *Human Genetics* **64**, 363–70.

Schmid, M., Haaf, T., Ott, G., Scheres, J. M. J. C. & Wensing, J. A. B. 1986. Heterochromatin in the chromosomes of the gorilla: characterization with distamycin A/DAPI, D287/170, chromomycin A_3, quinacrine, and 5-azacytidine. *Cytogenetics and Cell Genetics* **41**, 71–82.

Schmid, W. 1967. Heterochromatin in mammals. *Arch. Klaus-Stiftung Vererbungs-Forsch.* **42**, 1–60.

Schmidt, E. R. & Keyl, H-G. 1981 *In situ* binding of AT-rich repetitive DNA to the centromeric heterochromatin in polytene chromosomes of chironomids. *Chromosoma* **82**, 197–204.

Schmidt, M. 1980. Two phases of DNA replication in human cells. *Chromosoma* **76**, 101–110.

Schmidt, M. & Stolzmann, W. M. 1984. Replication variants of the human inactive X chromosome. II. Frequency and replication rate relative to the other chromosomes of the complement. *Chromosoma* **89**, 68–75.

Schmidt, M., Stolzmann, W. M. & Baranovskaya, L. I. 1982. Replication variants of the human inactive X chromosome. I. Variability within lymphocytes of single individuals. *Chromosoma* **85**, 405–12.

Schmidt, M., Wolf, S. F. & Migeon, B. R. 1985. Evidence for a relationship between DNA methylation and DNA replication from studies of the 5-azacytidine-reactivated allocyclic X chromosome. *Experimental Cell Research* **158**, 301–10.

Schmidtke, J. & Schmid, M. 1980. Regional assignment of a 2.1-kb repetitive sequence to the distal part of the human Y heterochromatin. *Human Genetics* **55**, 255–7.

Schnedl, W. 1971a. Banding pattern of human chromosomes. *Nature New Biology* **233**, 93–4.

Schnedl, W. 1971b. Fluorescenz-untersuchungen uber die Langenvariabilität des Y-Chromosoms beim Menschen. *Humangenetik* **12**, 188–94.

Schnedl, W. 1972. Giemsa banding, quinacrine fluorescence, and DNA-replication in chromosomes of cattle (*Bos taurus*). *Chromosoma* **38**, 319–28.

Schnedl, W. 1974. Der Polymorphismus des menschlichen Chromosomensatzes – eine Möglichkeit für den Vaterschaftsnachweis. *Zeitschrift fur Rechtsmedizin* **74**, 17–23.

Schnedl, W. 1978. Structure and variability of human chromosomes analysed by recent techniques. *Human Genetics* **41**, 1–9.

Schnedl, W., Dann, O & Schweizer, D. 1980. Effects of counterstaining with DNA binding drugs on fluorescent banding patterns of human mammalian chromosomes. *European Journal of Cell Biology* **20**, 290–6.

Schnedl, W., Erlanger, B. F. & Miller, O. J. 1976. 5-methylcytosine in heterochromatic regions of chromosomes in Bovidae. *Human Genetics* **31**, 21–6.

Schnedl, W., Roscher, U. & Czaker, R. 1977a. A photometric method for quantifying the polymorphisms in human acrocentric chromosomes. *Human Genetics* **35**, 185–91.

Schnedl, W., Abraham, R., Forster, M. & Schweizer, D. 1981b. Differential fluorescent staining of porcine heterochromatin by chromomycin A_3/distamycin A/DAPI and D287/170. *Cytogenetics and Cell Genetics* **31**, 249–53.

Schnedl, W., Breitenbach, M., Mikelsaar, A-V. & Stranzinger, G. 1977c. Mithramycin and DIPI: a pair of fluorochromes specific for GC- and AT-rich DNA respectively. *Human Genetics* **36**, 299–305.

Schnedl, W., Mikelsaar, A-V., Breitenbach, M. & Dann, O. 1977b. DIPI and DAPI: fluorescence banding with only negligible fading. *Human Genetics* **36**, 167–72.

Schnedl, W., Abraham, R., Dann, O., Geber, G. & Schweizer, D. 1981a. Preferential fluorescent staining of heterochromatic regions in human chromosomes 9, 15, and the Y by D287/170. *Human Genetics* **59**, 10–13.

Schnedl, W., Dev, V. G., Tantravahi, R., Miller, D. A., Erlanger, B. F. & Miller, O.J. 1975. 5-methylcytosine in heterochromatic regions of chromosomes: chimpanzee and gorilla compared to the human. *Chromosoma* **52**, 59–66.

Schollmayer, E., Schäfer, D., Frisch, B. & Schleiermacher, E. 1981. High resolution analysis and differential condensation in RBA-banded human chromosomes. *Human Genetics* **59**, 187–93.

Schonberg, S. A., Fukuyama, K., Hara, N. & Epstein, A. L. 1987. Monoclonal antibodies to human nuclear proteins as probes for human chromosome structure. *American Journal of Human Genetics* **41**, suppl., A138.

Schreck, R. R., Erlanger, B. F. & Miller, O. J. 1974. The use of antinucleoside antibodies to probe the organization of chromosomes denatured by ultraviolet irradiation. *Experimental Cell Research* **88**, 31–9.

Schreck, R. R., Erlanger, B. F. & Miller, O. J. 1977a. The structural organization of mouse metaphase chromosomes. *Chromosoma* **62**, 337–50.

Schreck, R. R., Erlanger, B. F. & Miller, O. J. 1977b. Binding of anti-nucleoside antibodies reveals different classes of DNA in the chromosomes of the kangaroo rat (*Dipodomys ordii*). *Experimental Cell Research* **108**, 403–11.

Schreck, R. R., Warburton, D., Miller, O. J., Beiser, S. M. & Erlanger, B. F. 1973. Chromosome structure as revealed by a combined chemical and immunochemical procedure. *Proceedings of the National Academy of Sciences* **70**, 804–7.

Schröder, J. 1975. Transplacental passage of blood cells. *Journal of Medical Genetics* **12**, 230–42.

Schröder, J., Suomalainen, H., van der Loo, W. & Schröder, E., 1978. Karyotypes in lymphocytes of two strains of rabbit, and two species of hare. *Hereditas* **88**, 183–8.

Schubert, I. & Rieger, R. 1979. Asymmetric banding of *Vicia faba* chromosomes after BrdU incorporation. *Chromosoma* **70**, 385–91.

Schubert, I. & Wobus, U. 1985. In situ hybridization confirms jumping nucleolus organizing regions in *Allium*. *Chromosoma* **92**. 143–8.

Schubert, I., Anastassova-Kristeva, M. & Rieger, R. 1979. Specificity of NOR staining in *Vicia faba*. *Experimental Cell Research* **120**, 433–5.

Schuh, B. E., Korf, B. R. & Salwen, M. J. 1975. Dynamic aspects of trypsin-Giemsa banding. *Humangenetik* **28**, 233–7.

Schwab, M., Ramsay, G., Alitalo, K., Varmus, H. E., Bishop, J. M., Martinsson, T., Levan, G. & Levan, A. 1985. Amplification and enhanced expression of the c-myc oncogene in mouse SEWA tumour cells. *Nature* **315**, 345–7.

Schwartz, S. & Cohen, M. M. 1985. Y chromosome length: clinical manifestations and reproductive capacity. In *The Y chromosome*, part B, A. A. Sandberg (ed.), 27–52. New York: Alan R. Liss.

Schwartz, S., Roulston, D. & Cohen, M. M. 1989. dNORs and meiotic nondisjunction. *American Journal of Human Genetics* **44**, 627–30.

Schwarzacher, H. G. 1974. Analysis of interphase nuclei. In *Methods in human cytogenetics*. H. G. Schwarzacher, U. Wolf & E. Passarge (eds), 207–34. Berlin: Springer.

Schwarzacher, H. G. & Wachtler, F. 1983. Nucleolus organizer regions and nucleoli. *Human Genetics* **63**, 89–99.

Schwarzacher, H. G., Mikelsaar, A-V. & Schnedl, W. 1978. The nature of the Ag-staining of nucleolus organizer regions. *Cytogenetics and Cell Genetics* **20**, 24–39.

Schwarzacher-Robinson, T., Cram,. L. S., Meyne, J & Moyzis, R. K. 1988. Characterization of human heterochromatin by *in situ* hybridization with satellite DNA clones. *Cytogenetics and Cell Genetics* **47**, 192–6.

Schweizer, D. 1973. Differential staining of plant chromosomes with Giemsa. *Chromosoma* **40**, 307–20.

Schweizer, D. 1976. Reverse fluorescent chromosome banding with chromomycin and DAPI. *Chromosoma* **58**, 307–24.

Schweizer, D. 1977. R-banding produced by DNase I digestion of chromomycin-stained chromosomes. *Chromosoma* **64**, 117–24.

Schweizer, D. 1980. Simultaneous fluorescent staining of R bands and specific heterochromatic regions (DA-DAPI bands) in human chromosomes. *Cytogenetics and Cell Genetics* **27**, 190–3.

Schweizer, D. 1981. Counterstain-enhanced chromosome banding. *Human Genetics* **57**, 1–14.

Schweizer, D. 1983. Distamycin-DAPI bands: properties and occurrence in species. In *Kew chromosome conference II*, P. E. Brandham & M. D. Bennett (eds) 43–51. London: Allen & Unwin.

Schweizer, D. & Ehrendorfer, F. 1976, Giemsa banded karyotypes, systematics, and evolution in *Anacyclus* (Asteraceae – Anthemideae). *Plant Syst. Evol.* **126**, 107–48.

Schweizer, D. & Ehrendorfer, F. 1983. Evolution of C-band patterns in Asteraceae – Anthemideae. *Biol. Zbl.* **102**, 637–55.

Schweizer, D. & Loidl, J. 1987. A model for heterochromatin dispersion and the evolution of C-band patterns. In *Chromosomes today*, Vol. 9, A. Stahl, J. M. Luciani & A. M. Vagner-Capodano (eds), 61–74. London: Allen & Unwin.

Schweizer, D. & Nagl, W. 1976. Heterochromatin diversity in *Cybidium*, and its relationship to differential DNA replication. *Experimental Cell Research* **98**, 411–23.

Schweizer, D., Ambros, P. & Andrle, M. 1978. Modification of DAPI banding on human chromosomes by prestaining with a DNA-binding oligopeptide antibiotic, distamycin A. *Experimental Cell Research* **111**, 327–32.

Schweizer, D., Strehl, S. & Hagemann, S. 1990. Plant repetitive DNA elements and chromosome structure. In *Chromosomes today*, Vol. 10, London: Allen & Unwin.

Schweizer, D., Mendelak, M., White, M. J. D. & Contreras, N. 1983. Cytogenetics of the parthenogenetic grasshopper *Warramaba virgo* and its bisexual relatives. X. Patterns of fluorescent banding., *Chromosoma* **88**, 227–36.

Schweizer, D., Tohidast-Akrad, M., Strehl, S. & Dann, O. 1987. Diverse fluorescent staining of human heterochromatin by the isomeric DAPI derivatives D288/45 and D288/48. *Ann. Univ. Sarav. Med.* suppl. **7**, 285–90.

Schweizer, D., Ambros, P., Andrle, M., Rett, A. & Fiedler, W. 1979. Demonstration of specific heterochromatic segments in the orangutan (*Pongo pygmaeus*) by a distamycin/DAPI double staining technique. *Cytogenetics and Cell Genetics* **24**, 7–14.

Schwemmle, S., Mehnert, K. & Vogel, W. 1989. How does inactivation change timing of replication in the human X chromosome? *Human Genetics* **83**, 26–32.

Schwinger, E., Ites, J. & Korte, B. 1976. Studies on frequency of Y chromatin in human sperm. *Human Genetics* **34**, 265–70.

Schwinger, E., Klimmeck, D. & Hansmann, M. 1978. Sex determination in nuclei of amnion fluid cells. *Clinical Genetics* **14**, 169–72.

Schwinger, E., Rakebrand, E., Müller, H. J., Bühler, E. M. & Tettenborn, U. 1971. Y-body in hair roots. *Humangenetik* **12**, 79–80.

Sciorra, L. J., Lee, M-L & Wynnyckyj, H. 1985. Study of human chromosomes. IV. Labelling of chromosomal proteins with the amino group specific fluorescent reagent fluorescamine. *Journal of Histochemistry and Cytochemistry* **33**, 1252–5.

Seabright, M. 1971. A rapid banding technique for human chromosomes. *Lancet* **2**, 971–2.

Seabright, M. 1972. The use of proteolytic enzymes for the mapping of structural rearrangements in the chromosomes of man. *Chromosoma* **36**, 204–10.

Seabright, M. 1973. Improvement of trypsin method for banding chromosomes. *Lancet* **1**, 1249–50.

Seabright, M., Cooke, P. & Wheeler, M. 1975. Variation in trypsin banding at different stages of contraction in human chromosomes and the definition, by measurement, of the 'average' karyotype. *Humangenetik* **29**, 35–40.

Searle, J. B. 1984. Three new karyotypic races of the common shrew *Sorex araneus* (Mammalia; Insectivora) and a phylogeny. *Syst. Zool.* **33**, 184–94.

Sederoff, R., Lowenstein, L., Mayer, A., Stone, J. & Birnboim. H. C. 1975. Acid treatment of *Drosophila* deoxyribonucleic acid. *Journal of Histochemistry and Cytochemistry* **23**, 482–92.

Sehested, J. 1974. A simple method for R banding of human chromosomes, showing a pH-dependent connection between R and G bands. *Humangenetik* **21**, 55–8.

Selander, R-K. 1973. Interaction of quinacrine mustard with mononucleotides and polynucleotides. *Biochemical Journal* **131**, 749–55.

Selander, R-K., & de la Chapelle, A. 1973a. Interaction of quinacrine mustard with calf thymus histone fractions. *Biochemical Journal* **131**, 757–64.

Selander, R-K., & de la Chapelle, A. 1973b. The fluorescence of quinacrine mustard with nucleic acids. *Nature New Biology* **245**, 240–4.

Seleznev, Y. V. 1973. A modified method of Giemsa staining of human chromosomes to reveal their linear differentiation. *Annales de Génétique* **16**, 139–41.

Sen, P. & Sharma, T. 1985. Characterization of G-banded chromosomes of the Indian muntjac and progression of banding patterns through different stages of condensation. *Cytogenetics and Cell Genetics* **39**, 145–9.

Sengupta, S. K., Ramsey, P. G. & Modest, E. J. 1971. Binding of quinacrine mustard and quinacrine to DNA. *Proceedings of the American Association of Cancer Research* **12**, 90.

Sentis, C. & Fernandez-Piqueras, J. 1987. Nature and distribution of heterochromatinized regions in the chromosomal races of *Pycnogaster cucullata* (Insecta, Orthoptera). *Genetica* **72**, 127–32.

Sentis, C., Santos, J. & Fernandez-Piqueras, J. 1986. C-heterochromatin polymorphism in *Baetica ustulata*: intraindividual variation and fluorescence banding patterns. *Chromosoma* **94**, 65–70.

Sentis, C., Santos, J. & Fernandez-Piqueras, J. 1989. Breaking up the chromosomes of *Baetica ustulata* by *in situ* treatments with restriction endonucleases. *Genome* **32**, 208–15.

Sessions, S. K. 1980. Evidence for a highly differentiated sex chromosome heteromorphism in the salamander *Necturus maculosus* (Rafinesque). *Chromosoma* **77**, 157–68.

Sessions, S. K. & Kezer, J. 1987. Cytogenetic evolution in the plethodontid salamander genus *Aneides*. *Chromosoma* **95**, 17–30.

Seth, P. K. & Gropp, A. 1973. Study of constitutive heterochromatin with a new and simplified fluorescence staining technique. *Genetica* **44**, 485–95.

Seth, P. K., Pera, F., Hilwig, I. & Gropp, A. 1973. Fluorescence banding pattern of the chromosomes of *Microtus agrestis* with a benzimidazol derivative. *Humangenetik* **19**, 129–34.

Seuanez, H. N. 1979. *The phylogeny of human chromosomes*. Berlin: Springer.

Seuanez, H., Fletcher, J., Evans, H. J. & Martin, D. E. 1976a. A polymorphic structural rearrangement in the chromosomes of two populations of orangutan. *Cytogenetics and Cell Genetics* **17**, 327–37.

Seuanez, H., Robinson, J., Martin, D. E. & Short, R. V. 1976b. Fluorescent (F) bodies in the spermatozoa of man and the great apes. *Cytogenetics and Cell Genetics* **17**, 317–26.

Shabtai, F., Antebi, E., Klar, D., Kimchi, D., Hart, J. & Halbrecht, I. 1985. Cytogenetic study of patients with carcinoma of the colon and rectum: particular C-band variants as possible markers for cancer proneness. *Cancer Genetics and Cytogenetics* **14**, 235–45.

Shade, M., Woodward, M. A. & Steel, C. M. 1980. Chromosome aberrations acquired *in vitro* by human B-cell lines. II. Distribution of break points. *Journal of the National Cancer Institute* **65**, 101–9.

Shafer, D. A. 1974. The intrinsic banded structure of extended chromosomes: effect of rifampicin/amphotericin B treatment in culture. *American Journal of Human Genetics* **26**, 78a.

Shafer, D. A., Selles, W. D. & Brenner, J. F. 1982. Computer image analysis of variance between human chromosome replication sequences and G-bands. *American Journal of Human Genetics* **34**, 307–21.

Shaw, D. D., Webb, G. C. & Wilkinson, P. 1976. Population cytogenetics of the genus *Caledia* (Orthoptera: Acridinae). II. Variation in the pattern of C-banding. *Chromosoma* **56**, 169–70.

Sheer, D. 1986. Chromosomes and cancer. In *Introduction to the cellular and molecular biology of cancer*, L. M. Franks & N. M. Teich (eds), 229–50. Oxford: Oxford University Press.

Sheldon, S. & Nichols, W. W. 1981a. Comparison of the patterns of chromosomal late replication. I. Human renal epithelium and lung fibroblasts *in vitro*. *Cytogenetics and Cell Genetics* **29**, 40–50.

Sheldon, S. & Nichols, W. W. 1981b. Comparison of the patterns of chromosomal late replication. II. Chick embryo lung and kidney *in vivo* and *in vitro*. *Cytogenetics and Cell Genetics* **29**, 51–9.

Sheppard, D. M., Fisher, R. A., Lawler, S. D. & Povey, S. 1982. Tetraploid conceptus with three paternal contributions. *Human Genetics* **62**, 371–4.

Sherwood, S. W. & Patton, J. L. 1982. Genome evolution in pocket gophers (genus *Thomomys*). II. Variation in cellular DNA content. *Chromosoma* **85**, 163–79.

Shettles, L. B. 1971. Use of the Y chromosome in prenatal sex determination. *Nature* **230**, 52.

Shi, L., Pathak, S. & Hsu, T. C. 1982. Demonstration of kinetochores and centrioles in spermatocytes of two species of cockroaches by silver staining. *Chromosoma* **85**, 421–6.

Shi, L., Ye, Y & Duan, X. 1980. Comparative cytogenetic studies on the red muntjac, Chinese muntjac, and their F_1 hybrids. *Cytogenetics and Cell Genetics* **26**, 22–7.

Shields, G. F. 1982. Comparative avian cytogenetics: a review. *Condor* **84**, 45–58.

Shiraishi, N., Fujihara, T., Maeda, S., Kazama, T., Takahashi, R. & Sugiyama, T. 1982. Chromosome structure of normal Long-Evans rats revealed by modified Giemsa method for DNA replication. *Acta Histochemica Cytochemica* **15**, 176–84.

Shiraishi, Y. 1970. The differential reactivity of human peripheral leucocyte chromosomes induced by low temperature. *Japanese Journal of Genetics* **45**, 429–42.

Shiraishi, Y. 1972. Differential reactivity in mammalian chromosomes. I. Relation between special segments and late-labelling patterns of DNA synthesis in human X-chromosomes. *Chromosoma* **36**, 211–20.

Shiraishi, Y. & Yosida, T. H. 1972. Banding pattern analysis of human chromosomes by use of a urea treatment technique. *Chromosoma* **37**, 75–83.

Short, R. V., Chandley, A. C., Jones, R. C. & Allen, W. R. 1974. Meiosis in interspecific equine hybrids. II. The Przewalski horse/domestic horse hybrid (*Equus przewalskii* × *E. caballus*). *Cytogenetics and Cell Genetics* **13**, 465–78.

Shows, T. B., Sakaguchi, A. Y. & Naylor, S. L. 1982. Mapping the human genome, cloned genes, DNA polymorphisms, and inherited disease. *Advances in Human Genetics* **12**, 341–452.

Sigmund, J. & Schwarz, S. 1979. Variable substructure in the secondary constriction of the human chromosome 1. *Human Genetics* **46**, 1–4.

Signoret, J. 1965. Etude des chromosomes de la blastula chez l'axolotl. *Chromosoma* **17**, 328–35.

Simeonova, M. & Raikov, Z. 1981. Chromosome banding after treatment of living cells *in vitro* with vitamins from group B. *Annales de Génétique* **24**, 126–8.

Simeonova, M. & Raikov, Z. 1982. Chromosome banding after treatment of living cells *in vitro* with planar N-heterocyclic compounds. Annales de Génétique 25, 126–8.

Simeonova, M., Manolov, G. & Levan, A. 1979. Chromosome banding after treatment of living cells *in vitro* with histamine and thymidazol. *Hereditas* **90**, 304–7.

Simi, S. & Tursi, F. 1982. Polymorphism of human chromosomes 1, 9, 16, Y: variations, segregation and mosaicism. *Human Genetics* **62**, 217–20.

Simola, K., Selander, R-K., de la Chapelle, A., Corneo, G. & Ginelli, E. 1975. Molecular basis of chromosome banding. I. The effect of mouse DNA fractions on two fluorescent dyes *in vitro*. *Chromosoma* **51**, 199–205.

Simon, M. I. & van Vunakis, H. 1962. The photodynamic reaction of methylene blue with deoxyribonucleic acid. *Journal of Molecular Biology* **4**, 488–99.

Singer, M. F. 1982. Highly repeated sequences in mammalian genomes. *International Review of Cytology* **76**, 67–112.

Singh, L., Purdom, I. F. & Jones, K. W. 1976. Satellite DNA and evolution of sex chromosomes. *Chromosoma* **59**, 43–62.

Singh, R. J. & Röbbelen, G. 1975. Comparison of somatic Giemsa banding pattern in several species of rye. *Zeitschrift Pflanzenzücht* **75**, 270–85.

Sinha, A. K., Kakati, S. & Pathak, S. 1972. Exclusive localization of C-bands within opossum sex chromosomes. *Experimental Cell Research* **75**, 265–8.

Sivak, A. & Wolman, S. R. 1974. Chromosomal proteins in fixed metaphase cells. *Histochemistry* **42**, 345–9.

Smetana, K. & Busch, H. 1974. The nucleolus and nucleolar DNA. In *The cell nucleus*, Vol. I, H. Busch (ed.), 73–147. New York: Academic Press.

Smetana, K., Ochs, R., Lischwe, M. A., Gyorkey, F., Freireich, E., Chudomel, V. & Busch, H. 1984. Immunofluorescence studies on proteins B23 and C23 in nucleoli of human lymphocytes. *Experimental Cell Research* **152**, 195–203.

Smith, B. J., Robertson, D., Birbeck, M. S. C., Goodwin, G. H. & Johns, E. W. 1978. Immunochemical studies of high mobility group non-histone chromatin proteins HMG1 and HMG2. *Experimental Cell Research* **115**, 420–3.

Smith, M. C. & Ingram, R. 1986. Heterochromatin banding in the genus *Paris*. *Genetica* **71**, 141–5.

Smith, S. G. 1965. Heterochromatin, colchicine, and karyotype. *Chromosoma* **16**, 162–5.

Sobell, H. M. 1973. The stereochemistry of actinomycin binding to DNA and its implications in molecular biology. *Progress in Nucleic Acid Research and Molecular Biology* **13**, 153–90.

Sofuni, T., Tanabe, K. & Awa, A. A. 1980. Chromosome heteromorphisms in the Japanese. II. Nucleolus organizer regions of variant chromosomes in D and G groups. *Human Genetics* **55**, 265–70.

Sofuni, T., Tanabe, K., Ohtaki, K., Shimba, H. & Awa, A. A. 1974. Two new types of C-band variants in human chromosome (6ph + and 12ph +). *Japanese Journal of Human Genetics* **19**, 251–6.

Somers, C. E. & Hsu, T. C. 1962. Chromosome damage induced by hydroxylamine in mammalian cells. *Proceedings of the National Academy of Sciences* **48**, 937–43.

Sommerville, J. 1986. Nucleolar structure and ribosome biogenesis. *Trends in Biochemical Science* **11**, 438–42.

Sommerville, L. L. & Wang, K. 1981. The ultrasensitive silver 'protein' stain also detects nanograms of nucleic acids. *Biochemical and Biophysical Research Communications* **102**, 53–8.

Soudek, D. & Laraya, P. 1976. C and Q bands in long arm of Y chromosomes; are they identical? *Human Genetics* **32**, 339–41.

Soudek, D., Langmuir, V. & Stewart, D. J. 1973. Variation in the nonfluorescent segment of long Y chromosome. *Human Genetics* **18**, 285–90.

Southern, E. M. 1970. Base sequence and evolution of guinea-pig α-satellite DNA. *Nature* **227**, 794–8.

Sozansky, O. A., Zakharov, A. F. & Benjush, V. A. 1984. Intercellular NOR-AG variability in man. I. Technical improvements and marker acrocentric chromosomes. *Human Genetics* **68**, 299–302.

Sparkes, R. S. 1984. Cytogenetics of retinoblastoma. *Cancer Surveys* **3**, 479–96.

Spector, D. L., Ochs, R. L. & Busch, H. 1984. Silver staining, immunofluorescence, and immunoelectron microscopic localization of nucleolar phosphoproteins B23 and C23. *Chromosoma* **90**, 139–48.

Sperling, K. & Rao, P. N. 1974. Mammalian cell fusion. V. Replication behaviour of heterochromatin as observed by premature chromosome condensation. *Chromosoma* **45**, 121–31.

Sperling, K. & Wiesner, R. 1972. A rapid banding technique for routine use in human and comparative cytogenetics. *Humangenetik* **15**, 349–53.

Sperling, K., Kalscheuer, V. & Neitzel, H. 1987. Transcriptional activity of constitutive heterochromatin in the mammal *Microtus agrestis* (Rodentia, Cricetidae). *Experimental Cell Research* **173**, 463–72.

Spowart, G., Forster, P., Dunn, N. & Cohen, B. B. 1985. Clinical and biochemical studies on anti-kinetochore antibody in patients with rheumatic diseases: a diagnostic marker for CREST. *Disease Markers* **3**, 103–12.

Stack, S. M. 1974. Differential Giemsa staining of kinetochores and nucleolus organizer heterochromatin in mitotic chromosomes of higher plants. *Chromosoma* **47**, 361–78.

Stack, S. M. & Clarke, C. R. 1973. Differential Giemsa staining of the telomeres of *Allium cepa* chromosomes: observations related to chromosome pairing. *Canadian Journal of Genetics and Cytology* **15**, 619–24.

Stahl, A. 1982. The nucleolus and nucleolar chromosomes. In *The nucleolus*, E. G. Jordan & C. A. Cullis (eds), 1–24. Cambridge: Cambridge University Press.

Stahl, A. & Vagner-Capodano, A-M. 1972. Etude des chromosomes du poulet (*Gallus domesticus*) par les techniques de fluorescence. *Comptes Rendu de l'Academie des Sciences (Paris)* D, **275**, 2367–70.

Stahl, A. & Vagner-Capodano, A-M. 1974. Mise en évidence des bandes des chromosomes humains en lumière ordinaire et en fluorescence après coloration par la pseudo-isocyanine. *Comptes Rendu de l'Academie des Sciences (Paris)* D, **278**, 2987–9.

Stahl, A., Luciani, J. M., Devictor, M., Capodano, A. M. & Hartung, M. 1975. Heterochromatin and nucleolar organizers during first meiotic prophase in quail oocytes. *Experimental Cell Research* **91**, 365–71.

Stanyon, R., Sineo, L., Chiarelli, B., Camperio-Ciani, A., Haimoff, A. R., Mootnick, E. H. & Sutarman, D. 1987. Banded karyotypes of the 44-chromosome gibbons. *Folia Primatologia* **48**, 56-64.

Stark, G. R. & Wahl, G. M. 1984. Gene amplification. *Annual Review of Biochemistry* **53**, 447–91.

Stark, G. R., Debatisse, M., Giulotto, E. & Wahl, G. M. 1989. Recent progress in understanding mechanisms of mammalian DNA amplification. *Cell* **57**, 901–8.

Steel, C. M., Shade, M. & Woodward, M. A. 1980. Chromosome aberrations acquired *in vitro* by human B-cell lines. I. Gains and losses of material. *Journal of the National Cancer Institute* **65**, 95–9.

Stefos, K. & Arrighi, F. E. 1974. Repetitive DNA of *Gallus domesticus* and its cytological locations. *Experimental Cell Research* **83**, 9–14.

Steiniger, G. E. & Mukherjee, A. B. 1975. Insect chromosome banding: technique for G- and Q-banding patterns in the mosquito *Aedes albopictus*. *Canadian Journal of Genetics and Cytology* **17**, 241–4.

Stephen, G. S. 1977. Mammalian chromosomes G-banded in four minutes. *Genetica* **47**, 115–6.

Stickel, S. K. & Clark, R. W. 1985. Mass characteristics of DNA obtained from chromosomes of a human carcinoma cell line. *Chromosoma* **92**, 234–41.

Stock, A. D. 1984. The occurrence of G-bands in the mitotic chromosomes of the amphibian *Xenopus muelleri*. *Genetica* **64**, 255–8.

Stock, A. D. & Mengden, G. A. 1975. Chromosome banding pattern conservation in birds and nonhomology of chromosome banding patterns between birds, turtles, snakes and amphibians. *Chromosoma* **50**, 69–77.

Stock, A. D., Arrighi, F. E. & Stefos, K. 1974. Chromosome homology in birds: banding patterns of the chromosomes of the domestic chicken, ring-necked dove, and domestic pigeon. *Cytogenetics and Cell Genetics* **13**, 410–18.

Stocker, A. J., Fresquez, C. & Lentzios, G. 1978. Banding studies on the polytene chromosomes of *Rhynchosciara hollaenderi*. *Chromosoma* **68**, 337–56.

Stokke, T. & Steen, H. B. 1986. Binding of Hoechst 33258 to chromatin *in situ*. *Cytometry* **7**, 227–34.

Stollar, B. D. 1986. Antibodies to DNA. *Critical Reviews in Biochemistry* **20**, 1–36.

Strauss, F. & Varshavsky, A. 1984. A protein binds to a satellite DNA repeat at three specific sites that would be brought into mutual proximity by DNA folding in the nucleosome. *Cell* **37**, 889–901.

Stubblefield, E. 1964. DNA synthesis and chromosomal morphology of Chinese hamster cells cultured in media containing N-deacetyl-N-methylcolchicine (Colcemid). In *Cytogenetics of cells in culture*, R. J. C. Harris (ed.), 223–48. New York: Academic Press.

Sudman, P. D. & Greenbaum, I. F. 1989. Visualization of kinetochores in mammalian meiotic preparations and observations of argentophilic differences between mitotic and meiotic kinetochores. *Genome* **32**, 380–2.

Sumner, A. T. 1972. A simple technique for demonstrating centromeric heterochromatin. *Experimental Cell Research* **75**, 304–6.

Sumner, A. T. 1974. Involvement of protein disulphides and sulphydryls in chromosome banding. *Experimental Cell Research* **83**, 438–42.

Sumner, A. T. 1976. Banding as a level of chromosome organization. In *Current chromosome research*, K. Jones & P. E. Brandham (eds), 17–22. Amsterdam: Elsevier/North-Holland.

Sumner, A. T. 1977a. Estimation of the sizes of polymorphic C-bands in man by measurement of DNA content of whole chromosomes. *Cytogenetics and Cell Genetics* **19**, 250–5.

Sumner, A. T. 1977b. Suppression of quinacrine banding of human chromosomes by mounting in organic media. *Chromosoma* **64**, 337–42.

Sumner, A. T. 1980. Dye binding mechanisms in G-banding of chromosomes. *Journal of Microscopy* **119**, 397–406.

Sumner, A. T. 1981. The nature of chromosome bands and their significance for cancer research. *Anticancer Research* **1**, 205–16.

Sumner, A. T. 1982. The nature and mechanisms of chromosome banding. *Cancer Genetics and Cytogenetics* **6**, 59–87.

Sumner, A. T. 1983. The role of protein sulphydryls and disulphides in chromosome structure and condensation. In *Kew chromosome conference II*, P. E. Brandham & M. D. Bennett (eds), 1–9. London: Allen & Unwin.

Sumner, A. T. 1984. Distribution of protein sulphydryls and disulphides in fixed mammalian chromosomes, and their relationship to banding. *Journal of Cell Science* **70**, 177–88.

Sumner, A. T. 1985. The distribution of quinacrine on chromosomes as determined by X-ray microanalysis. II. Comparison of heterochromatic and euchromatic regions of mouse chromosomes. *Chromosoma* **91**, 145–50.

Sumner, A. T. 1986a. Electron microscopy of the parameres formed by the centromeric heterochromatin of human chromosome 9 at pachytene. *Chromosoma* **94**, 199–204.

Sumner, A. T. 1986b. Mechanisms of quinacrine binding and fluorescence in nuclei and chromosomes. *Histochemistry* **84**, 566–74.

Sumner, A. T. 1987. Immunocytochemical demonstration of kinetochores in human sperm heads. *Experimental Cell Research* **171**, 250–3.

Sumner, A. T. 1989. Chromosome banding. In *Light microscopy in biology: a practical approach*, A. J. Lacey (ed.), 279–314. Oxford: IRL Press.

Sumner, A. T. & Buckland, R. A. 1976. Relative DNA contents of somatic nuclei of ox, sheep and goat. *Chromosoma* **57**, 171–5.

Sumner, A. T. & Evans, H. J. 1973. Mechanisms involved in the banding of chromosomes with quinacrine and Giemsa. II. The interaction of the dyes with the chromosomal components. *Experimental Cell Research* **81**, 223–36.

Sumner, A. T. & Robinson, J. A. 1976. A difference in dry mass between the heads of X- and Y-bearing human spermatozoa. *Journal of Reproduction and Fertility* **48**, 9–15.

Sumner, A. T. & Speed, R. M. 1987. Immunocytochemical labelling of the kinetochore of human synaptonemal complexes, and the extent of pairing of the X and Y chromosomes. *Chromosoma* **95**, 359–65.

Sumner, A. T., Carothers, A. D. & Rutovitz, D. 1981. The distribution of quinacrine on chromosomes as determined by X-ray microanalysis. I. Q-bands on CHO chromosomes. *Chromosoma* **82**, 717–34.

Sumner, A. T., Evans, H. J. & Buckland, R. A. 1971a. New technique for distinguishing between human chromosomes. *Nature New Biology* **232**, 31–2.

Sumner, A. T., Evans, H. J. & Buckland, R. A. 1973. Mechanisms involved in the banding of chromosomes with quinacrine and Giemsa. I. The effects of fixation in methanol-acetic acid. *Experimental Cell Research* **81**, 214–22.

Sumner, A. T., Robinson, J. A. & Evans, H. J. 1971b. Distinguishing between X, Y and YY-bearing human spermatozoa by fluorescence and DNA content. *Nature New Biology* **229**, 231–3.

Sumner, A. T., Taggart, M. H., Mezzanotte, R. & Ferrucci, L. 1990. Patterns of digestion of human chromosomes by restriction endonucleases demonstrated by *in situ* nick translation. *Histochemical Journal*, in press.

Sun, N. C., Chu, E. H. Y. & Chang, C. C. 1974. Staining method for the banding patterns of human mitotic chromosomes. *Caryologia* **27**, 315–24.

Sutherland, G. R. & Hecht, F. 1985. *Fragile sites on human chromosomes*. New York: Oxford University Press.

Świtoński, M., Fries, R. & Stranzinger, G. 1983. C-band variants of telocentric chromosomes in swine: evidence and inheritance studies. *Genet. Sel. Evol.* **15**, 469–78.

Szabo, P. & Ward, D. C. 1982. What's new in *in situ* hybridization? *Trends in Biochemical Science* **7**, 425–7.

Tabuchi, K., Saiga, T., Adachi, T. & Midorikawa, O. 1979. Chromosomal banding patterns produced by methyl green-pyronin staining after trypsin treatment. *Stain Technology* **54**, 125–8.

Takagi, N. 1971. A simple technique to demonstrate the centromeric 'heterochromatin' in the mouse and other animals. *Japanese Journal of Genetics* **46**, 361–3.

Takagi, N. & Sasaki, M. 1974. A phylogenetic study of bird karyotypes. *Chromosoma* **46**, 91–120.

Takayama, S. 1974. Factors for Giemsa-band formation in air-dried mammalian chromosomes. *Japanese Journal of Genetics* **49**, 189–95.

Takayama, S. 1976. Configurational changes in chromatids from helical to banded structures. *Chromosoma* **56**, 47–54.

Takayama, S. & Tachibana, K. 1980. Two opposite types of sister chromatid differential staining in BUdR-substituted chromosomes using tetrasodium salt of EDTA. *Experimental Cell Research* **126**, 498–501.

Takayama, S. & Taniguchi, T. 1986. Light and scanning electron microscopic observations on the two contrasting types of sister chromatid differential staining after ultraviolet light irradiation. *Chromosoma* **93**, 404–8.

Takehisa, S. 1968. Heterochromatic segments in *Vicia* revealed by treatment with HCl-acetic acid. *Nature* **217**, 567–8.

Takehisa, S. 1969. Positively heterochromatic segment in mitotic metaphase chromosomes of *Vicia faba*. *Experientia* **25**, 1340.
Tan, E. M. 1989. Antinuclear antibodies: diagnostic markers for autoimmune diseases and probes for cell biology. *Advances in Immunology* **44**, 93–151.
Tanaka, R. & Ohta, S. 1982. A method for differential staining of centromere regions of plant chromosomes. *Japanese Journal of Genetics* **57**, 65–73.
Tanaka, R. & Taniguchi, K. 1975. A banding method for plant chromosomes. *Japanese Journal of Genetics* **50**, 163–7.
Taniwaki, M. 1985. New chromosome banding techniques with base specific antibiotics and fluorochromes for qualified identification of marker chromosomes in leukaemia and related disorders. *Acta Haematologica Japonica* **48**, 1423–39.
Tantravahi, R., Miller, D. A. & Miller, O. J. 1977. Ag-staining of nucleolus organizer regions of chromosomes after Q-, C-, G-, or R-banding procedures. *Cytogenetics and Cell Genetics* **18**, 364–9.
Tantravahi, R., Miller, D. A., Dev, V. G. & Miller, O. J. 1976. Detection of nucleolus organizer regions in chromosomes of human, chimpanzee, gorilla, orangutan and gibbon. *Chromosoma* **56**, 15–27.
Tantravahi, R., Dev, V. G., Firschein, I. L., Miller, D. A. & Miller, O. J. 1975. Karyotype of the gibbons *Hylobates lar* and *H. moloch*. Inversion in chromosome 7. *Cytogenetics and Cell Genetics* **15**, 92–102.
Tantravahi, U., Guntaka, R. V., Erlanger, B. F. & Miller, O. J. 1981a. Amplified ribosomal RNA genes in a rat hepatoma cell line are enriched in 5-methylcytosine. *Proceedings of the National Academy of Sciences* **78**, 489–93.
Tantravahi, U., Breg, W. R., Wertelecki, V., Erlanger, B. F. & Miller, O. J. 1981b. Evidence for methylation of inactive human rRNA genes in amplified regions. *Human Genetics* **56**, 315–20.
Tartof, K. D. 1973. Unequal mitotic sister chromatid exchange and disproportionate replication as mechanisms regulating ribosomal RNA gene redundancy. *Cold Spring Harbor Symposium on Quantitative Biology* **38**, 491–500.
Taylor, E. F. & Martin-deLeon, P. A. 1980. Comparison of N-banding and silver staining of human NORs. *Human Genetics* **54**, 217–9.
Taylor, E. F. & Martin-deLeon, P. A. 1981. Familial silver staining patterns of human nucleolus organizer regions (NORs). *American Journal of Human Genetics* **33**, 67–76.
Tempelaar, M. J., de Both, M. T. J. & Versteegh, J. E. G. 1982. Measurement of SCE frequencies in plants: a simple Feulgen-staining procedure for *Vicia faba*. *Mutation Research* **103**, 321–6.
Teng, M-K., Usman, N., Frederick, C. A. & Wang, A. H-J. 1988. The molecular structure of the complex of Hoechst 33258 and the DNA dodecamer d(CGCGAATTCGCG). *Nucleic Acids Research* **16**, 2671–90.
Testa, J. R. 1984. 'High-resolution' chromosomal analysis of acute leukaemia: current assessment. *Cancer Surveys* **3**, 359–69.
Teyssier, J. R. 1989. The chromosomal analysis of human solid tumours: a triple challenge. *Cancer Genetics and Cytogenetics* **37**, 103–25.
Thiebaut, F., Rigaut, J. P. & Reith, A. 1984. Improvement in the specificity of the silver staining technique for AgNOR-associated acidic proteins in paraffin sections. *Stain Technology* **59**, 181–5.
Thiery, J-P., Macaya, G. & Bernardi, G. 1976. An analysis of eukaryotic genomes by density gradient centrifugation. *Journal of Molecular Biology* **108**, 219–35.
Third International Workshop on Chromosomes in Leukaemia. 1981. Clinical significance of chromosomal abnormalities in acute lymphoblastic leukaemia. *Cancer Genetics and Cytogenetics* **4**, 111–37.
Thomas, H. M. 1981. The Giemsa C-band karyotypes of six *Lolium* species. *Heredity* **46**, 263–7.
Thuriaux, P. 1977. Is recombination confined to structural genes on the eukaryotic genome? *Nature* **268**, 460–2.
Thust, R. & Rønne, M. 1980. Localization of AT-clusters in normal mouse chromosomes by netropsin prefixation treatment *in vitro*. *Hereditas* **93**, 321–6.

Thust, R. & Rønne, M. 1981. Structural modifications induced in Chinese hamster V79-E chromosomes by prefixation treatment *in vitro* with the AT-specific agents netropsin, distamycin A, and Hoechst 33258. *Hereditas* **94**, 209–13.
Timmis, J. N., Deumling, B. & Ingle, J. 1975. Localisation of satellite DNA sequences in nuclei and chromosomes of two plants. *Nature* **257**, 152–5.
Tishler, P. V. 1980. Genetic technology and the solution of crime. In *Genetics and the law*, Vol. 2, A. Milunsky & G. J. Annas (eds), 283–91. New York: Plenum Press.
Tishler, P. V. 1985. The human Y chromatin: an overview. In *The Y chromosome*, part A, A. A. Sandberg (ed.), 195–212. New York: Alan R. Liss.
Tjio, J. H. & Levan, A. 1950. The use of oxyquinoline in chromosome analysis. *Anal. Est. Exp. Aula Dei* **2**, 21–64.
Tjio, J. H. & Levan, A. 1956. The chromosome number of man. *Hereditas* **42**, 1–6.
Tobia, A. M., Schildkraut, C. L. & Maio, J. J. 1970. Deoxyribonucleic acid replication in synchronized cultured mammalian cells. I. Time of synthesis of molecules of different average guanine + cytosine content. *Journal of Molecular Biology* **54**, 499–515.
Tommerup, N. 1982. Specific staining of 9h in human somatic interphase cells by D287/170. *Human Genetics* **62**, 301–4.
Tommerup, N. & Vejerslev, L. O. 1985. Identification of triploidy by DA/DAPI staining of trophoblastic interphase nuclei. *Placenta* **6**, 363–7.
Traut, W., Winking, H. & Adolph, S. 1984. An extra segment in chromosome 1 of wild *Mus musculus*: a C-band positive homogeneously staining region. *Cytogenetics and Cell Genetics* **38**, 290–7.
Tsou, K. C., Giles, B. & Kohn, G. 1975. On the chemical basis of chromosome banding patterns. *Stain Technology* **50**, 293–5.
Tsukahara, M. & Kajii, T. 1985. Replication of X chromosomes in complete moles. *Human Genetics* **71**, 7–10.
Tuck-Muller, C. M., Bordson, B. L., Kane, M. M. & Hamilton, A. E. 1984. A method for combined C-banding and silver staining. *Stain Technology* **59**, 265–8.
Turner, B. M. 1982. Immunofluorescent staining of human metaphase chromosomes with monoclonal antibody to histone H2B. *Chromosoma* **87**, 345–57.
Turner, B. M. 1989. Acetylation and deacetylation of histone H4 continue through metaphase with depletion of more-acetylated isoforms and altered site usage. *Experimental Cell Research* **182**, 206–14.

Ueda, T., Irie, S. & Kato, Y. 1987. Longitudinal differentiation of metaphase chromosome of Indian muntjac as studied by restriction enzyme digestion, *in situ* hybridization with cloned DNA probes and distamycin A plus DAPI fluorescence staining. *Chromosoma* **95**, 251–7.
Ueda, T., Sato, R. & Kobayashi, J. 1988. Silver-banded karyotypes of the rainbow trout and the brook trout and their hybrids: disappearance in allotriploids of Ag-NORs originated from the brook trout. *Japanese Journal of Genetics* **63**, 219–26.
Unakul, W., Johnson, R. T., Rao, P. N. & Hsu, T. C. 1973. Giemsa banding in prematurely condensed chromosomes obtained by cell fusion. *Nature New Biology* **242**, 106–7.
Utakoji, T. 1973. Differential staining of human chromosomes treated with potassium permanganate and its blocking by organic mercurials. In *Chromosomes today*, Vol. 4, J. Wahrman & K. R. Lewis (eds), 53–9. Jerusalem: Israel Universities Press.
Utakoji, T. & Matsukuma, S. 1974. Fluorescent staining of L cell centromeres and chromocentres with 1-dimethylaminonaphthalene-5-sulphonyl chloride and G-bandings. *Experimental Cell Research* **87**, 111–19.
Utsumi, S. & Takehisa, S. 1974. Heterochromatin differentiation in *Trillium kamtschaticum* by ammoniacal silver reaction. *Experimental Cell Research* **86**, 398–401.

Valdivia, M. M. & Brinkley, B. R. 1985. Fractionation and initial characterization of the kinetochore from mammalian metaphase chromosomes. *Journal of Cell Biology* **101**, 1124–34.

Valdivia, M. M., Tousson, A. & Brinkley, B. R. 1986. Human antibodies and their use for the study of chromosome organization. *Methods and Achievements in Experimental Pathology* **12**, 200–23.

van der Ploeg, M., Bosman, F. T. & van Duijn, P. 1976. Pre-exposure of films in fluorescence photography. *Histochemical Journal* **8**, 201–4.

van der Ploeg, M., Vossepoel, A. M., Bosman, F. T. & van Duijn, P. 1977. High-resolution scanning-densitometry of photographic negatives of human chromosomes. III. Determination of fluorescence emission intensities. *Histochemistry* **51**, 269–91.

van de Sande, J. H., Lin, C. C. & Deugau, K. V. 1979. Clearly differentiated and stable chromosome bands produced by a spermine bis-acridine, a bifunctional intercalating analogue of quinacrine. *Experimental Cell Research* **120**, 439–44.

van de Sande, J. H., Lin, C. C. & Jorgenson, K. F. 1977. Reverse banding on chromosomes produced by a guanosine-cytosine specific DNA binding antibiotic: olivomycin. *Science* **195**, 400–2.

van Duijn, P., van Prooijen-Knegt, A. C. & van der Ploeg, M. 1985. The involvement of nucleosomes in Giemsa staining of chromosomes. A new hypothesis on the banding mechanism. *Histochemistry* **82**, 363–76.

van Holde, K. E. 1989. *Chromatin*. New York: Springer.

van Prooijen-Knegt, A. C., Raap, A. K., van der Burg, M. J. M., Vrolijk, J. & van der Ploeg, M. 1982. Spreading and staining of human metaphase chromosomes on aminoalkylsilane-treated glass slides. *Histochemical Journal* **14**, 333–44.

Varley, J. M. 1977. Patterns of silver staining of human chromosomes. *Chromosoma* **61**, 207–14.

Varley, J. M., Macgregor, H. C., Nardi, I., Andrews, C. & Erba, H. P. 1980. Cytological evidence of transcription of highly repeated DNA sequences during the lampbrush stage in *Triturus cristatus carnifex*. *Chromosoma* **80**, 289–307.

Vasilikaki-Baker, H. & Nishioka, Y. 1983. Immunological visualisation of 5-methylcytosine-rich regions in Indian muntjac metaphase chromosomes. *Experimental Cell Research* **147**, 226–30.

Ved Brat, S., Verma, R. S. & Dosik, H. 1979. A simplified technique for simultaneous staining of nucleolar organizer regions and kinetochores. *Stain Technology* **54**, 107–8.

Verma, R. S. & Babu, A. 1989. *Human chromosomes*. New York: Pergamon Press.

Verma, R. S. & Dosik, H. 1976. An improved method for photographing fluorescent human chromosomes. *Journal of Microscopy* **108**, 339–41.

Verma, R. S. & Dosik, H. 1980. Human chromosomal heteromorphisms: nature and clinical significance. *International Review of Cytology* **62**, 361–83.

Verma, R. S. & Lubs. H. A. 1975a. A simple R banding technique. *American Journal of Human Genetics* **27**, 110–17.

Verma, R. S. & Lubs, H. A. 1975b. Variation in human acrocentric chromosomes with acridine orange reverse banding. *Humangenetik* **30**, 225–35.

Verma, R. S. & Lubs, H. A. 1976a. Additional observations on the preparation of R-banded human chromosomes with acridine orange. *Canadian Journal of Genetics and Cytology* **18**, 45–50.

Verma, R. S. & Lubs, H. A. 1976b. Inheritance of acridine orange R variants in human acrocentric chromosomes. *Human Heredity* **26**, 315–18.

Verma, R. S. & Pandey, R. P. 1987. Non-random distribution of various sizes of human Y chromosomes in different ethnic groups. *Annals of Human Biology* **14**, 271–6.

Verma, R. S. & Rodriguez, J. 1985. Structural organization of ribosomal cistrons in human nucleolar organizing chromosomes. *Cytobios* 44, 25–8.'

Verma, R. S., Benjamin, C. & Dosik, H. 1983. Expression of nucleolus organizer regions (NORs) in human acrocentric chromosomes by NSG, QFQ and RFA techniques: are they identical? *Cytobios* **37**, 157–62.

Verma, R. S., Dosik, H. & Lubs, H. A. 1977. Frequency of RFA colour polymorphisms of human acrocentric chromosomes in caucasians: interrelationship with QFQ polymorphisms. *Annals of Human Genetics* **41**, 257–67.

Verma, R. S., Shah, J. V. & Dosik, H. 1984. Expression of ribosomal cistrons of human chromosomes at high resolution. *Stain Technology* **59**, 13–16.

Verma, R. S., Ved Brat, S. & Dosik, H. 1979. Heterochromatin of chromosomes of Indian muntjac as revealed by fluorescent banding techniques. *Journal of Heredity* **70**, 438–40.

Verma, R. S., Benjamin, C., Rodriguez, J. & Dosik, H. 1981. Population heteromorphisms of Ag-stained nucleolus organizer regions (NORs) in the acrocentric chromosomes of East Indians. *Human Genetics* **59**, 412–15.

Verma, R. S., Dosik, H., Scharf, T. & Lubs, H. A. 1978. Length heteromorphisms of fluorescent (f) and non-fluorescent (nf) segments of human Y chromosome: classification, frequencies, and incidence in normal Caucasians. *Journal of Medical Genetics* **15**, 277–81.

Verschaeve, L., Kirsch-Volders, M. & Susanne, C. 1981. Silver staining of human acrocentrics in metaphase does not reflect an induced inhibition in NOR activity. *Experimental Cell Research* **136**, 459–61.

Viegas-Péquignot, E. & Dutrillaux, B. 1976. Segmentation of human chromosomes induced by 5-ACR (5-azacytidine). *Human Genetics* **34**, 247–54.

Viegas-Péquignot, E. & Dutrillaux, B. 1978. Une methode simple pour obtenir des prophases et des prometaphases. *Annales de Génétique* **21**, 122–5.

Viegas-Péquignot, E. & Dutrillaux, B. 1981. Detection of G-C rich heterochromatin by 5-azacytidine in mammals. *Human Genetics* **57**, 134–7.

Viegas-Péquignot, E., Dutrillaux, B. & Thomas, G. 1988. Inactive X chromosome has the highest concentration of unmethylated *Hha* I sites. *Proceedings of the National Academy of Sciences* **85**, 7657–60.

Viegas-Péquignot, E., Derbin, D., Lemeunier, F. & Taillandier, E. 1982. Identification of left-handed Z-DNA by indirect immunomethods in metaphasic chromosomes of a mammal, *Gerbillus nigeriae* (Gerbillidae, Rodentia). *Annales de Génétique* **25**, 218–22.

Viegas-Péquignot, E., Dutrillaux, B., Prod'homme, M. & Petter, F. 1983b. Chromosomal phylogeny of Muridae: a study of 10 genera. *Cytogenetics and Cell Genetics* **35**, 269–78.

Viegas-Péquignot, E., Kasahara, S., Yassuda, Y. & Dutrillaux, B. 1985. Major chromosome homoeologies between Muridae and Cricetidae. *Cytogenetics and Cell Genetics* **39**, 258–61.

Viegas-Péquinot, E., Derbin, C., Malfoy, B., Taillandier, E., Leng, M & Dutrillaux, B. 1983a. Z-DNA immunoreactivity in fixed metaphase chromosomes of primates. *Proceedings of the National Academy of Sciences* **80**, 5890–4.

Vig, B. K. 1987. Sequence of centromere separation: a possible role for repetitive DNA. *Mutagenesis* **2**, 155–9.

Vig, B. K. & Zinkowski, R. P. 1986. Sequence of centromere separation: a mechanism for orderly separation of dicentrics. *Cancer Genetics and Cytogenetics* **22**, 347–59.

Viglianti, G. A. & Blumenfeld, M. 1986. Satellite DNA-correlated nucleosomal proteins in *Drosophila virilis*. *Biochemical Genetics* **24**, 79–92.

Vistorin, G., Gamperl, R. & Rosenkranz, W. 1977. Studies on sex chromosomes of four hamster species: *Cricetus cricetus, Cricetulus griseus, Mesocricetus auratus*, and *Phodopus sungorus*. *Cytogenetics and Cell Genetics* **18**, 24–32.

Vistorin, G., Gamperl, R. & Rosenkranz, W. 1978. Vergleich der Karyotypen von *Rattus rattus flavipectus* (Rodentia) und europaischen Ratten. *Zool. Anz.* **201**, 314–22.

Vogel, W. & Speit, G. 1986. Cytogenetic replication studies with short thymidine pulses in bromodeoxyuridine-substituted chromosomes of different mouse cell lines. *Human Genetics* **72**, 63–7.

Vogel, W., Autenrieth, M. & Speit, G. 1986. Detection of bromodeoxyuridine-incorporation in mammalian chromosomes by a bromodeoxyuridine-antibody. I. Demonstration of replication patterns. *Human Genetics* **72**, 129–32.

Vogel, W., Schempp, W. & Puel, V. 1978. Silver-staining specificity in metaphases after incorporation of 5-bromodeoxyuridine (BUDR). *Human Genetics* **40**, 199–203.

Vogel, W., Boldin, S., Reisacher, A. & Speit, G. 1985. Characterization of chromosome replication during S-phase with bromodeoxyuridine labelling in Chinese hamster ovary and HeLa cells. *Chromosoma* **92**, 363–8.

Vogel, W., Steinbach, P., Djalali, M., Mehnert, K., Ali, S. & Epplen, J. T. 1988. Chromosome 9 of *Ellobius lutescens* is the X chromosome. *Chromosoma* **96**, 112–18.

Voiculescu, I. 1974. A comparative study of the chromosome banding paterns of *Mesocricetus newtoni* and *Mesocricetus auratus*. *Zeitschrift Säugetierkunde* **39**, 211-19.
Voiculescu, I., Vogel, W. & Wolf, U. 1972. Karyotyp und Heterochromatinsmuster des rumanischen Hamsters (*Mesocricetus newtoni*). *Chromosoma* **39**, 215-24.
von Kalm, L. & Smyth, D. R. 1984. Ribosomal RNA genes and the substructure of nucleolar organizing regions in *Lilium*. *Canadian Journal of Genetics and Cytology* **26**, 158-66.
von Kiel, K., Hameister, H., Somssich, I. E. & Adolph, S. 1985. Early replication banding reveals a strongly conserved functional pattern in mammalian chromosomes. *Chromosoma* **93**, 69-76.
von Koskull, H. & Aula, P. 1973. Nonrandom distribution of chromosome breaks in Fanconi's anemia. *Cytogenetics and Cell Genetics* **12**, 423-34.
Vosa, C. G. 1971. The quinacrine-fluorescence patterns of the chromosomes of *Allium carinatum*. *Chromosoma* **33**, 382-5.
Vosa, C. G. 1973. Heterochromatin recognition and analysis of chromosome variation in *Scilla sibirica*. *Chromosoma* **43**, 269-78.
Vosa, C. G. 1974. The basic karyotype of rye (*Secale cereale*) analysed with Giemsa and fluorescence methods. *Heredity* **33**, 403-8.
Vosa, C. G. 1976a. Heterochromatin classification in *Vicia faba* and *Scilla sibirica*. In *Chromosomes today*, Vol. 5, P. L. Pearson & K. R. Lewis (eds), 185-92. Jerusalem: Israel Universities Press.
Vosa, C. G. 1976b. Heterochromatin banding patterns in *Allium*. II. Heterochromatin variation in species of the *paniculatum* group. *Chromosoma* **57**, 119-33.
Vosa, C. G. 1979. Heterochromatic banding patterns in the chromosomes of *Brimeura* (Liliaceae). *Plant Syst. Evol* **132**, 141-8.
Vosa, C. G. & Marchi, P. 1972a. Quinacrine fluorescence and Giemsa staining in plants. *Nature New Biology* **237**, 191-2.
Vosa, C. G. & Marchi, P. 1972b. On the quinacrine fluorescence and Giemsa staining patterns of the chromosomes of *Vicia faba*. *Giornale Bot. Ital.* **106**, 151-9.
Vosa, C. G., d'Amato, G., Capineri, R., Marchi, P. & de Dominicis, G. 1972. Quinacrine-like fluorescence of extracts from Papaveraceae and Fumariaceae. *Nature* **239**, 405-6.

Wachtler, F. & Musil, R. 1980. On the structure and polymorphism of the human chromosome no. 15. *Human Genetics* **56**, 115-18.
Wahedi, K. & Pawlowitzki, I. H. 1987. C-band polymorphisms of chromosome 9: quantification by C_e-bands. *Human Genetics* **77**, 1-5.
Wahl, G. M., Vitto, L., Padgett, R. A. & Stark, G. R. 1982. Single-copy and amplified CAD genes in Syrian hamster chromosomes localized by a highly sensitive method for *in situ* hybridization. *Molecular and Cellular Biology* **2**, 308-19.
Wahlström, J., Kyllerman, M., Hansson, A. & Taranger, J. 1985. Unequal mitotic sister chromatid exchange and different length of Y chromosomes. *Human Genetics* **70**, 186-8.
Wahrman, J., Richler, C., Neufeld, E. & Friedmann, A. 1983. The origin of multiple sex chromosomes in the gerbil *Gerbillus gerbillus* (Rodentia: Gerbillinae). *Cytogenetics and Cell Genetics* **35**, 161-80.
Wahrman, J., Goitein, R., Richler, C., Goldman, B., Akstein, E. & Chaki, R. 1976. The mongoloid phenotype in man is due to trisomy of the distal pale G-band of chromosome 21. In *Chromosomes today*, Vol. 5, P. L. Pearson & K. R. Lewis (eds), 241-8. Jerusalem: Israel Universities Press.
Wake, C. T. & Ward, O. G. 1975. The fluorescent karyotype of the tachinid fly *Voria ruralis* Fallen (Diptera). *Experientia* **31**, 291-4.
Walker, P. M. B. 1971. 'Repetitive' DNA in higher organisms. *Progress in Biophysics and Molecular Biology* **23**, 147-90.
Walker, P. R. & Sikorska, M. 1987. Chromatin structure. Evidence that the 30nm fibre is a helical coil with 12 nucleosomes/turn. *Journal of Biological Chemistry* **262**, 12223-7.
Wall, W. J. & Butler, L. J. 1989. Classification of Y chromosome polymorphisms by DNA content and C-banding. *Chromosoma* **97**, 296-300.

Wallace, A. J. & Newton, M. E. 1987. Heterochromatin diversity and cyclic responses to selective silver staining in *Aedes aegypti* (L.) *Chromosoma* **95**, 89–93.

Walther, J-U., Stengel-Rutkowski, S. & Murken, J-D. 1974. Observations with G-banding of human chromosomes. Reduction of dye concentration in Soerensen buffered solutions is sufficient for demonstrating G bands. *Humangenetik* **25**, 49–51.

Wandall, A. 1989. Kinetochore development in two dicentric chromosomes in man. A light and electron microscope study. *Human Genetics* **82**, 137–41.

Wang, H. C. & Fedoroff, S. 1972. Banding in human chromosomes treated with trypsin. *Nature New Biology* **235**, 52–4.

Wang, H. C. & Juurlink, B. H. J. 1979. Nucleolar organizer regions (NORs) in Chinese hamster chromosomes as visualized by Coomassie brilliant blue. *Chromosoma* **75**, 327–32.

Wang, H. C. & Kao, K. N. 1988. G-banding in plant chromosomes. *Genome* **30**, 48–51.

Warburton, D. & Henderson, A. S. 1979. Sequential silver staining and hybridization *in situ* on nucleolus organizing regions in human cells. *Cytogenetics and Cell Genetics* **24**, 168–75.

Ward, D. C., Reich, E. & Goldberg, I. H. 1965. Base specificity in the interaction of polynucleotides with antibiotic drugs. *Science* **149**, 1259–63.

Waring, M. 1970. Variation of the supercoils in closed circular DNA by binding of antibiotics and drugs: evidence for molecular models involving intercalation. *Journal of Molecular Biology* **54**, 247–79.

Warley, A., Stephen, J., Hockaday, A. & Appleton, T. C. 1983. X-ray microanalysis of HeLa S3 cells. II. Analysis of elemental levels during the cell cycle. *Journal of Cell Science* **62**, 339–50.

Wayne, R. K., Nash, W. G. & O'Brien, S. J. 1987a. Chromosomal evolution of the Canidae. I. Species with high diploid numbers. *Cytogenetics and Cell Genetics* **44**, 123–33.

Wayne, R. K., Nash, W. G. & O'Brien, S. J. 1987b. Chromosomal evolution of the Canidae. II. Divergence from the primitive carnivore karyotype. *Cytogenetics and Cell Genetics* **44**, 134–41.

Webb, G. C. 1976. Chromosome organization in the Australian plague locust, *Chortoicetes terminifera*. I. Banding relationships of the normal and supernumerary chromosomes. *Chromosoma* **55**, 229–46.

Webb, G. C. & Neuhaus, P. 1979. Chromosome organization in the Australian plague locust, *Chortoicetes terminifera*. II. Banding variants of the B-chromosome. *Chromosoma* **70**, 205–38.

Webb, G. C., White, M. J. D., Contreras, N. & Cheney, J. 1978. Cytogenetics of the parthenogenetic grasshopper *Warramaba* (formerly *Maraba*) *virgo* and its bisexual relatives. IV. Chromosome banding studies. *Chromosoma* **67**, 309–39.

Webber, L. M., Brasch, J. M. & Smyth, D. R. 1981. DNA extraction during Giemsa differentiation of chromatids singly and doubly substituted with BrdU. *Chromosoma* **81**, 691–700.

Weber, B., Schempp, W. & Wiesner, H. 1986. An evolutionarily conserved early replicating segment on the sex chromosomes of man and the great apes. *Cytogenetics and Cell Genetics* **43**, 72–8.

Weide, L. G., Dev, V. G. & Rupert, C. S. 1979. Activity of both mouse and Chinese hamster ribosomal RNA genes in somatic cell hybrids. *Experimental Cell Research* **123**, 424–9.

Weimarck, A. 1975. Heterochromatin polymorphism in the rye karyotype as detected by the Giemsa C-banding technique. *Hereditas* **79**, 293–300.

Weintraub, H. & Groudine, M. 1976. Chromosomal subunits in active genes have an altered conformation. *Science* **193**, 848–56.

Weisblum, B. 1973. Why centric regions of quinacrine-treated mouse chromosomes show diminished fluorescence. *Nature* **246**, 150–1.

Weisblum, B. & de Haseth, P. L. 1972. Quinacrine, a chromosome stain specific for deoxyadenylate-deoxythymidylate-rich regions in DNA. *Proceedings of the National Academy of Sciences* **69**, 629–32.

Weisblum, B. & Haenssler, E. 1974. Fluorometric properties of the bibenzimidazole derivative Hoechst 33258, a fluorescent probe specific for AT concentration in chromosomal DNA. *Chromosoma* **46**, 255–60.

Weisbrod, S. & Weintraub, H. 1979. Isolation of a subclass of nuclear proteins responsible for conferring a DNase-I-sensitive structure on globin chromatin. *Proceedings of the National Academy of Sciences* **76**, 630–4.

Weith, A. 1983. Mg^{2+}-dependent compactness of heterochromatic chromosome segments. *Experimental Cell Research* **146**, 199–203.

Weltens, R., Kirsch-Volders, M., Hens, L., Defrise-Gussenhoven, E. & Susanne, C. 1985. NOR variability in twins. *Acta Genet. med. Gemellol.* **34**, 141–51.

Welter, D. A. & Hodge, L. D. 1985. A scanning electron microscopic technique for three-dimensional visualization of the spatial arrangement of metaphase, anaphase and telophase chromatids. *Scanning Electron Microscopy* **2**, 879–88.

Westerman, M. & Woolley, P. A. 1990. Cytogenetics of New Guinean dasyurids and genome evolution in the Dasyuridae (Marsupialia). *Australian Journal of Zoology* **37**, 521–31.

Wheeler, L. L., Arrighi, F., Cordeiro-Stone, M. & Lee, C. S. 1978. Localization of *Drosophila nasutoides* satellite DNAs in metaphase chromosomes. *Chromosoma* **70**, 41–50.

White, M. J. D. 1973. *Animal cytology and evolution*, 3rd edn. Cambridge: Cambridge University Press.

Wiberg, U. H. 1983. Sex determination in the European eel (*Anguilla anguilla*, L). *Cytogenetics and Cell Genetics* **36**, 589–98.

Widom, J. & Klug, A. 1985. Structure of the 300Å chromatin filament: X-ray diffraction from oriented samples. *Cell* **43**, 207–13.

Wienberg, J. & Stanyon, R. 1988. DA/DAPI fluorescent bands in the chromosomes of *Pan paniscus*. *American Journal of Primatology* **14**, 91–6.

Will, H. & Bautz, E. K. F. 1980. Immunological identification of a chromocentre-associated protein in polytene chromosomes of *Drosophila*. *Experimental Cell Research* **125**, 401–10.

Willard, H. F. 1977. Tissue-specific heterogeneity in DNA replication patterns of human X chromosomes. *Chromosoma* **61**, 61–73.

Williams, M. A., Kleinschmidt, J. A., Krohne, G. & Franke, W. W. 1982. Argyrophilic nuclear and nucleolar proteins of *Xenopus laevis* oocytes identified by gel electrophoresis. *Experimental Cell Research* **137**, 341–51.

Wilson, G. B. & Boothroyd, E. R. 1944. Temperature-induced differential contraction in the somatic chromosomes of *Trillium erectum* L. *Canadian Journal of Research* C, **22**, 105–19.

Wilson, G. N. 1982. The structure and organization of human ribosomal genes. In *The cell nucleus*, Vol. X, H. Busch & L. Rothblum (eds), 287–318. New York: Academic Press.

Wittekind, D. H. 1983. On the nature of Romanowsky-Giemsa staining and its significance for cytochemistry and histochemistry: an overall view. *Histochemical Journal* **15**, 1029–47.

Wittekind, D. H., Kretschmer, V. & Sohmer, I. 1982. Azure B-eosin Y stain as the standard Romanowsky-Giemsa stain. *British Journal of Haematology* **51**, 391–3.

Wojcik, J. M. & Seale, J. B. 1988. The chromosome complement of *Sorex granarius* – the ancestral karyotype of the common shrew (*Sorex araneus*)? *Heredity* **61**, 225–9.

Wolff, S. 1982. *Sister chromatid exchange*. New York: Wiley.

Wolff, S. & Bodycote, J. 1977. The production of harlequin chromosomes by chemical and physical agents that disrupt protein structure. In *Molecular human cytogenetics*, R. S. Sparkes & C. F. Fox (eds), 335–40. New York: Academic Press.

Wong, A. K. C. & Rattner, J. B. 1988. Sequence organization and cytological localization of the minor satellite of mouse. *Nucleic Acids Research* **16**, 11645–61.

Woodard, J. & Swift, H. 1964. The DNA content of cold-treated chromosomes. *Experimental Cell Research* **34**, 131–7.

Worton, R. G., Ho, C. C. & Duff, C. 1977. Chromosome stability in CHO cells. *Somat. Cell Genet.* **3**, 27–45.

Wray, W. & Stefos, K. 1976. Quinacrine bands in isolated chromosomes. *Cytologia* **41**, 729–32.

Wrigley, J. M. & Graves, J. A. M. 1988. Sex chromosome homology and incomplete, tissue-specific X-inactivation suggest that monotremes represent an intermediate stage of mammalian sex chromosome evolution. *Journal of Heredity* **79**, 115–18.

Wu, C-I., True, J. R. & Johnson, N. 1989. Fitness reduction associated with the deletion of a satellite DNA array. *Nature* **341**, 248–51.

Wu, R. S., Panusz, H. T., Hatch, C. L. & Bonner, W. M. 1986. Histones and their modifications. *CRC Critical Reviews in Biochemistry* **20**, 201–63.
Wurster, D. H. & Atkin, N. B. 1972. Muntjac chromosomes: a new karyotype for *Muntiacus muntjak*. *Experientia* **28**, 972–3.
Wurster-Hill, D. H. & Bush, M. 1980. The interrelationship of chromosome banding patterns in the giant panda (*Ailuropoda melanoleuca*), hybrid bear (*Ursus middendorfi* × *Thalarctos maritimus*), and other carnivores. *Cytogenetics and Cell Genetics* **27**, 147–54.
Wurster-Hill, D. H. & Centerwall, W. R. 1982. The interrelationships of chromosome banding patterns in canids, mustelids, hyena, and felids. *Cytogenetics and Cell Genetics* **34**, 178–92.
Wurster-Hill, D. H. & Gray, C. W. 1975. The interrelationships of chromosome banding patterns in procyonids, viverrids, and felids. *Cytogenetics and Cell Genetics* **15**, 306–31.
Wyandt, H. E., Anderson, R. S., Patil, S. R. & Hecht, F. 1980. Mechanisms of Giemsa banding. II. Giemsa components and other variables in G-banding. *Human Genetics* **53**, 211–15.
Wyandt, H. E., Vlietinck, R. F., Magenis, R. E. & Hecht, F. 1974. Coloured reverse-banding of human chromosomes with acridine orange following alkaline/formalin treatment: densitometric validation and applications. *Humangenetik* **23**, 119–30.
Wyandt, H. E., Wysham, D. G., Minden, S. K., Anderson, R. S. & Hecht, F. 1976. Mechanisms of Giemsa banding. I. Giemsa–11 banding with azure and eosin. *Experimental Cell Research* **102**, 85–94.

Yamada, K. 1985. Relation of stature to length and heterochromatin of the human Y chromosome. In *The Y chromosome*, part B, A. A. Sandberg (ed.), 1–13. New York: Alan R. Liss.
Yamamoto, M. 1979. Cytological studies of heterochromatin function in the *Drosophila melanogaster* male: autosomal meiotic pairing. *Chromosoma* **72**, 293–328.
Yamamoto, M. & Miklos, G. L. G. 1978. Genetic studies on heterochromatin in *Drosophila melanogaster* and their implications for the functions of satellite DNA. *Chromosoma* **66**, 71–98.
Yamasaki, N. 1956. Differentielle Färbung der somatischen Metaphasechromosomen von *Cypripedium debile*. I. *Chromosoma* **7**, 620–6.
Yamasaki, N. 1959. Differentielle Färbung der Chromosomen der ersten meiotischen Metaphase von *Cypripedium debile*. *Chromosoma* **10**, 454–60.
Yamasaki, N. 1961. Differentielle Färbbarkeit der somatischen und meiotischen Metaphasechromosomen von *Cypripedium debile* nach DNase-Behandlung. *Chromosoma* **11**, 479–83.
Yamasaki, N. 1965. Differentielle Färbbarkeit der somatischen Metaphasechromosomen von *Cypripedium debile* durch die Methylgrun-Pyronin-Methode. *Chromosoma* **16**, 411–14.
Yamasaki, N. 1971. Karyotypanalyse an Hand des Färbungsmusters der Metaphasechromosomen von *Cypripedium debile* und *Trillium kamtschaticum*. *Chromosoma* **33**, 372–81.
Yamasaki, N. 1973. Differentielle Darstellung der Metaphasechromosomen von *Cypripedium debile* mit Chinacrin- und Giemsa-Färbung. *Chromosoma* **41**, 403–12.
Yaniv, M. & Cereghini, S. 1986. Structure of transcriptionally active chromatin. *CRC Critical Reviews in Biochemistry* **21**, 1–26.
Yonenaga-Yassuda, Y., Kasahara, S., Chu, T. H. & Rodrigues, M. T. 1988. High-resolution RBG-banding pattern in the genus *Tropidurus* (Sauria, Iguanidae). *Cytogenetics and Cell Genetics* **48**, 68–71.
Yoshida, M. A., Takagi, N. & Sasaki, M. 1983. Karyotypic kinship between the blue fox (*Alopex lagopus* Linn.) and the silver fox (*Vulpes fulva* Desm.). *Cytogenetics and Cell Genetics* **35**, 190–4.
Yoshida, M. C., Ikeuchi, T. & Sasaki, M. 1975. Differential staining of parental chromosomes in interspecific cell hybrids with a combined quinacrine and 33258 Hoechst technique. *Proceedings of the Japanese Academy* **51**, 184–7.
Yosida, T. H. 1975. Diminution of heterochromatic C-bands in relation to the differentiation of *Rattus* species. *Proceedings of the Japanese Academy* **51**, 659–63.
Yosida, T. H. & Sagai, T. 1972. Banding pattern analysis of polymorphic karyotypes in the black rat by a new differential staining technique. *Chromosoma* **37**, 387–94.

Yosida, T. H. & Sagai, T. 1975. Variation of C-bands in the chromosomes of several subspecies of *Rattus rattus*. *Chromosoma* **50**, 283–300.
Yosida, T. H., Moriwaki, K. & Sagai, T. 1974. Oceanian type black rats (*Rattus rattus*) with a subtelocentric M_2 chromosome and C-type transferrin obtained from North America. *Experientia* **30**, 742–4.
Yunis, J. J. 1974. *Human chromosome methodology*, 2nd edn. New York: Academic Press.
Yunis, J. J. 1976. High resolution of human chromosomes. *Science* **191**, 1268–70.
Yunis, J. J. 1981. Mid-prophase human chromosomes. The attainment of 2000 bands. *Human Genetics* **56**, 293–8.
Yunis, J. J. 1982. Comparative analysis of high-resolution chromosome techniques for leukaemic bone marrows. *Cancer Genetics and Cytogenetics* **7**, 43–50.
Yunis, J. J. 1983. The chromosomal basis of human neoplasia. *Science* **221**, 227–36.
Yunis, J. J. & Bahr, G. F. 1979. Chromatin fibre organization of human interphase and prophase chromosomes. *Experimental Cell Research* **122**, 63–72.
Yunis, J. J. & Prakash, O. 1982. The origin of man: a chromosomal pictorial legacy. *Science* **215**, 1525–30.
Yunis, J. J. & Sanchez, O. 1973. G-banding and chromosome structure. *Chromosoma* **44**, 15–23.
Yunis, J. J. & Sanchez, O. 1975. The G-banded prophase chromosomes of man. *Humangenetik* **27**, 167–72.
Yunis, J. J. & Tsai, M. Y. 1978. Mapping of polysomal messenger RNA and heterogeneous nuclear RNA to the lightly staining G-bands of human chromosomes. *Cytogenetics and Cell Genetics* **22**, 364–7.
Yunis, J. J. & Yasmineh, W. G. 1971. Heterochromatin, satellite DNA, and cell function. *Science* **174**, 1200–9.
Yunis, J. J., Kuo, M. T. & Saunders, G. F. 1977. Localization of sequences specifying messenger RNA to light-staining G-bands of human chromosomes. *Chromosoma* **61**, 335–44.
Yunis, J. J., Sawyer, J. R. & Ball, D. W. 1978. The characterization of high-resolution G-banded chromosomes of man. *Chromosoma* **67**, 293–307.
Yunis, J. J., Sawyer, J. R. & Dunham, K. 1980. The striking resemblance of high-resolution G-banded chromosomes of man and chimpanzee. *Science* **208**, 1145–8.
Yunis, J. J., Roldan, L., Yasmineh, W. G. & Lee, J. C. 1971. Staining of satellite DNA in metaphase chromosomes. *Nature* **231**, 532–3.
Yurov, Y. B., Mitkevich, S. P. & Alexandrov, I. A. 1987. Application of cloned satellite DNA sequences to molecular-cytogenetic analysis of constitutive heterochromatin heteromorphisms in man. *Human Genetics* **76**, 157–64.

Zakharov, A. F. & Egolina, N. A. 1972. Differential spiralization along mammalian mitotic chromosomes. I. BUdR-revealed differentiation in Chinese hamster chromosomes. *Chromosoma* **38**, 341–65.
Zakharov, A. F., Davudov, A. Z., Benjusch, V. A. & Egolina, N. A. 1982. Polymorphisms of Ag-stained nucleolar organizer regions in man. *Human Genetics* **60**, 334–9.
Zakharov, A. F., Baranovskaya, L. I., Ibraimov, A. I., Benjusch, V. A., Demintseva, V. S. & Oblapenko, N. G. 1974. Differential spiralization along mammalian mitotic chromosomes. II. 5-bromodeoxyuridine and 5-bromodeoxycytidine-revealed differentiation in human chromosomes. *Chromosoma* **44**, 343–59.
Zanenga, R., Mattevi, M. S. & Erdtmann, B. 1984. Smaller autosomal C band sizes in blacks than in Caucasoids. *Human Genetics* **66**, 286.
Zankl, H. & Bernhardt, S. 1977. Combined silver staining of the nucleolus organizing regions and Giemsa banding in human chromosomes. *Human Genetics* **37**, 79–80.
Zatsepina, O. V., Polyakov, V. Y. & Chentsov, Y. S. 1988. Differential decondensation of mitotic chromosomes during a hypotonic treatment of cells as a possible cause of G-banding. (In Russian). *Tsitologiya* **30**, 1172–9.
Zech, L. 1969. Investigation of metaphase chromosomes with DNA-binding fluorochromes. *Experimental Cell Research* **58**, 463.

Zech, L., Haglund, V., Nilsson, K. & Klein, G. 1976. Characteristic chromosomal abnormalities in biopsies and lymphoid-cell lines from patients with Burkitt and non-Burkitt lymphomas. *International Journal of Cancer* **17**, 47–56.

Zelenin, M. G., Zakharov, A. F., Zatsepina, O. V., Polijakov, V. Y. & Chentsov, Y. S. 1982. Reversible differential decondensation of unfixed Chinese hamster chromosomes induced by change in calcium ion concentration of the medium. *Chromosoma* **84**, 729–36.

Zhang, A., Lin, M. S. & Wilson, M. G. 1987. Effect of C-banded heterochromatin on centromere separation. *Human Heredity* **37**, 285–9.

Zinkowski, R. P., Vig, B. K. & Broccoli, D. 1986. Characterization of kinetochores in multicentric chromosomes. *Chromosoma* **94**, 243–8.

Zuelzer, R. T., Ottenbreit, M., Inoue, S. & Zuelzer, W. W. 1973. Banding in old chromosome preparations. *Lancet* **2**, 270.

Zuffardi, O., Tiepolo, L., Dolfini, S., Barigozzi, C. & Fraccaro, M. 1971. Changes in the fluorescence patterns of translocated Y chromosome segments in *Drosophila melanogaster*. *Chromosoma* **34**, 274–80.

Additional references

Finaz, C. & de Grouchy, J. 1971. Le caryotype humain après traitement par l'α-chymotrypsine. *Annales de Génétique* **14**, 309–11.

Holmgren, P., Johansson, T., Lambertsson, A. & Rasmuson, B. 1985. Content of histone H1 and histone phosphorylation in relation to the higher order structures of chromatin in *Drosophila*. *Chromosoma* **93**, 123–31.

Hsu, T. C., Cooper, J. E. K., Mace, M. L. & Brinkley, B. R. 1971. Arrangement of centromeres in mouse cells. *Chromosoma* **34**, 73–87.

Kozak, C. A., Lawrence, J. B. & Ruddle, F. H. 1977. A sequential staining technique for the chromosomal analysis of interspecific mouse/hamster and mouse/human somatic cell hybrids. *Experimental Cell Research* **105**, 109–17.

Lee, M-L., Lazzarini, A. M. & Sciorra, L. J. 1978. Chromosome banding with thrombin. *American Journal of Human Genetics* **30**, 87A.

Müller, W. & Rosenkranz, W. 1972. Rapid banding technique for human and mammalian chromosomes. *Lancet* **1**, 898.

Polisky, B., Greene, P., Garfin, D. E., McCarthy, B. J., Goodman, H. M. & Boyer, H. W. 1975. Specificity of substrate recognition by the Eco RI restriction endonuclease. *Proceedings of the National Academy of Sciences* **72**, 3310–14.

Trusler, S. 1975. G bands produced by collagenase. *Lancet* **1**, 44–5.

Vosa, C. G. 1970. Heterochromatin recognition with fluorochromes. *Chromosoma* **30**, 366–72.

Index

References to figures are printed in *italics*

Acranil 124
  structure *7.1*
acridine orange
  binding to DNA 115
  fluorescence properties 116
  heteromorphisms **6.5.3**, 114, 293–4, 298
  R-banding **6.2.2**, 105
  replication banding 235
actinomycin D
  inhibition of chromosome condensation 223
  inhibition of rRNA transcription 199
adenosine, antibodies to 273
adriamycin
  banding patterns 166
  properties Table 8.1
  structure *8.2*
Ag-AS method for NORs 194–5
Ag-I method for NORs 195
Ag-NOR staining **9.4.1**
  active NORs 198–9
  ageing 203
  applications **9.4.3**
  heteromorphisms 204, 298
  in gametogenesis 203
  in hybrids 203
  mechanism 200–2
  neoplastic cells 203
  satellite associations 205
  sites of ribosomal genes 198
Ailuropoda melanoleuca (giant panda)
  chromosomal evolution 323
Allium spp.
  effect of heterochromatin on chiasmata 52
  heterogeneity of heterochromatin 51
  heteromorphism 308
  interphase arrangement of chromosomes 58
  Q-banding 132, 149
  replication bands 242
alphoid satellite DNA
  heteromorphisms 290, 291
  kinetochores 208
7-amino-actinomycin D **8.4.2**
  banding patterns 174
  binding to chromatin 174

  properties 174, Table 8.1
  structure *8.4*
Amphibia
  cold treatment of chromosomes 224–5
  DNA in C-bands Table 4.1
  heterochromatin variation between species 314–16
  hybrid species 326
  Q-banding 136, 149
  replication banding 242, 318
  sex chromosomes 327–8
Anemone
  base-specific banding 180
  heterochromatin variation between species 316
  heteromorphism 308
AS-SAT method for NORs 194
autoradiography 3, 233
5-azacytidine
  effect on Ag-NOR staining 199
  effect on transcription 199
  inhibition of chromosome condensation 221, **11.2.3**
  inhibition of methylation 199, 232
  structure *11.6*

band
  definition 12
  number of genes in 6
banding
  definition **1.1**
  history **1.2**
  homology between distantly related groups 324–5
  with combinations of fluorochromes **8.5**, Table 8.5
banding methods
  classification **2.3**
  three-letter code **2.3.3**
banding patterns, evolutionary conservation **15.2.1**
banding profiles 131–2
bands
  nomenclature Table 2.2, Table 2.3
  numbering of **2.4**
birds

chromosomal evolution 313, 325
DNA in C-bands Table 4.1
heteromorphisms 305
sex chromosomes 328
bone marrow transplants
  distinguishing donor and recipient cells 150
Bos taurus (ox)
  DNA in C-bands Table 4.1
  heteromorphism 303–4
  inhibition of chromosome condensation 228, 230
  5-methylcytosine in heterochromatin 277
  Q negative heterochromatin 133, Table 7.1
  satellite DNA 133
Bovidae
  chromosomal evolution 314, 323
  heteromorphism 303
  Robertsonian fusion 319
bromodeoxyuridine (BUdR)
  antibodies to 272
  inhibition of chromosome condensation 223, 233, 245
  lateral asymmetry **11.3.5**
  light sensitivity 235
  photolysis of DNA 238–9
  quenching of fluorescence 238
  sister chromatid exchanges 6, 232
  structure *11.8*
  study of replication 6, **11.3**

cancer
  Ag-NOR staining 203
  applications of banding **5.5.4**
  chromosome heteromorphisms 300
  use of high resolution banding 103
Capra hircus (goat)
  DNA type in C-bands Table 4.1
  heteromorphism 303
  5-methylcytosine in heterochromatin 277
Carnivora, chromosomal evolution 313, 322–3
C-banding
  applications **4.2.7**
  BSG method **4.2.2**
  combined with Ag-NOR staining 196–8
  definition 12, 40, Table 2.1
  DNA denaturation 53
  DNA extraction 54
  introduction 5, 40
  mechanism **4.2.5**, 278
  methods **4.2.1**
  protein extraction 54
  using DNase 42, 55, 253
C-bands
  absence in chromosomes 57–8
  differences between species 51, 59
  DNA types in 48, Table 4.1
  effect on chiasmata 52
  electron microscopy **4.2.6**, 287

functions **4.2.4**
genes in 50
heterochromatin and 47–8
heterogeneity 50–1
heteromorphism 51, 58
in interphase nuclei 58
protein composition 49
subdivision 287
time of DNA replication 50
transcription 50
C_d-banding 13, 292
  method 209, **10.2.1**
  of inactive centromeres 210–11
C_e-bands 287
cell cycle synchronization
  for high-resolution banding 79, 244–5
centromere
  definition 206
  inactive 210–11, 215
Cetacea
  chromosomal evolution 316, 323
  DNA type in C-bands Table 4.1
  enlarged sex chromosomes 329
chiasmata, effects of heterochromatin 52, 334
Chinese hamster
  DNA type in C-bands Table 4.1
  heteromorphism 304–5
  inhibition of chromosome condensation 227–8
  low content of repetitive DNA 48
  replication bands 234, 240
Chiroptera (bats),
  chromosomal evolution 313, 324
chorionic villus cells,
  exclusion of maternal contamination 150
chromomeres
  and euchromatic bands 32, 83, 180, 340
  universal distribution 342, Table 16.1
chromomycin A_3, *see* chromomycinone antibiotics
chromomycinone antibiotics **8.4.1**
  banding patterns 169–71
  binding to DNA 172
  combination with DA/DAPI 182
  combination with DAPI **8.5.1**
  fading of fluorescence 172
  properties 172, Table 8.1
  staining method 171–2
  staining of NORs 191
  structure *8.4*
chromosomal evolution
  bats 324
  birds 325
  Bovidae 323
  Carnivora 322–3
  Cetacea 316, 323
  Equidae 324
  giant panda 323
  great apes 321

Marsupialia 323–4
  plants 316
  primates 321
  reptiles 314, 316, 325–6
  rodents 321–2
chromosome breakpoints 97
  identification with banding **5.5.2**
  identification with high-resolution banding 103
chromosome condensation **3.4**
  and G-banding 82–3
chromosome condensation, inhibition 220–1, Table 11.1
  5-azacytidine **11.2.3**
  cold treatment **11.2.1**
  DAPI 226
  distamycin 226
  for high-resolution banding 79–80
  Hoechst 33258 160, 211, **11.2.2**
chromosome coils, and G-bands 88, 341–2
chromosome preparations
  denaturation of DNA 278
  effects of ageing on banding 110, 114
  storage for banding 110
chromosome sheath 189
chromosome structure **3.3**
  coiling 28, 30, 33
  domains 30
  'folded-fibre' model 28–9
  higher order packing **3.3.3**
  nucleosomes **3.3.1**
  'radial-loop' model 30
  scaffold (core) 29–31, 67, 194
  solenoids (30 mm fibre) **3.3.2**
'circumferential grooves' 341
colchicine, inhibition of chromosome condensation 223
cold treatment of chromosomes 3, **11.2.1**
condensation banding **11.2**
Coomassie brilliant blue, staining of NORs 191
counterstain-enhanced chromosome banding 178
CREST sera
  identification of kinetochore proteins 218–19, Table 10.1
  kinetochore labelling 208, **10.2.3**
  labelling of inactive centromeres 215
  labelling of pre-kinetochores 215–127
  staining method 213–15
crossing-over
  sites of 345–6
  unequal 285
CT-banding 111–12
cytidine, antibodies to 273
cytosine, methylation 23

D287/170 **8.3.5**
D288/45

chromosome banding 168–9
properties Table 8.1
structure *8.2*
D288/48
  chromosome banding 168–9
  properties Table 8.1
  structure *8.2*
DAPI
  applications 163
  banding patterns 163–4
  combination with chromomycinone antibiotics **8.5.1**
  combination with other fluorochromes 163
  inhibition of chromosome condensation 226–8
  properties 164, Table 8.1
  resistance to fading 163
  staining method 164
  staining meiotic chromosomes 181
  structure *8.2*
daunomycin
  banding patterns 166
  binding to DNA 166–7
  properties 166, Table 8.1
  structure *8.2*
D-bands
  DNase hypersensitivity 256
  using daunomycin 166
'dénaturation ménagée' 105
deoxyribonucleases
  banding 42, 55, 253–4
  hypersensitivity 253, **12.2.1**
  resistance of NORs 191
deoxyribonuclease sensitivity
  active genes **12.2.1**
  euchromatic bands 85, 256–7
  inactive X chromosome 256
  meiotic XY bivalents 256, 258
  patterns on chromosomes 255–9
dibutyl-proflavine
  banding patterns 165
  binding to DNA 165
  properties Table 8.1
  structure *8.2*
dicentric chromosomes, inactive centromeres 210–11, 215
'differential reactivity' 223
DIPI
  properties Table 8.1
  resistance to fading 163
  staining method 164
  structure *8.2*
Dipodomys
  DNA type in C-bands Table 4.1
  lateral asymmetry 247–8
  quinacrine fluorescence of heterochromatin 144
  satellite DNA 144
Diptera
  chromosomal evolution 316

heteromorphism 307
light-induced banding 184
distamycin, inhibition of chromosome condensation 226–8
distamycin/DAPI **8.5.2**
  combination with chromomycin 182
  fluorescence of heterochromatin 181
  meiotic chromosomes 181
  specificity 183
  staining method 182
  taxonomic distribution of bands 181, 302–3, Table 8.7
DNA
  antibodies to **13.2.3**
  distribution along chromosome 31–2
  genic DNA 345–6
  highly repetitive 23
  in C-bands Table 4.1
  intermediate repetitive 23
  inverted repeats 23
  'junk' 336
  methylation 23
  nucleoskeletal DNA 345–6
  photolysis 128, 144, 172, 184, 211, 238–9, 273, 275
  satellite 22–3
  'selfish' 336
  structure 22
  unique sequences 23
DNA denaturation 22
  during C-banding 53
  during R-banding 115–18
  effect of base composition, 22, 117
  for immunocytochemical labelling 273
  in chromosome preparations 278
DNA, double-stranded, antibodies to 278
DNA, loss during banding
  C-banding 54–5
  G-banding 88
  R-banding 115
DNA, nuclear amount (C value)
  correlation with amount of heterochromatin 314
  variation in spite of constant G-band patterns 323
DNA-protein cross-linking, BrdU substitution 239
DNA replication,
  and euchromatic bands 83–5, 240, 337
  in chromosomes 221, 233, 240–3
  in heterochromatin 50
  timing 243
  and transcription 346–7
DNA, single-stranded, antibodies to 277–8
DNA, triple-stranded
  antibodies to **13.2.5**
  functions 280
DNase-sensitivity, *see* deoxyribonuclease sensitivity
double minutes 101

absence of C-bands 58
gene amplification 99
Drosophila spp.
  DNA type in C-bands Table 4.1
  effect of heterochromatin on chiasmata 52
  genes in heterochromatin 50, 333
  heterochromatin variation between species 317
  heterogeneity of heterochromatin 51, 317
  heteromorphism 306–7
  Hoechst banding 157, 162
  inhibition of chromosome condensation 226, 228
  lateral asymmetry 248
  light-induced banding 184
  N-banding 64
  proteins in heterochromatin 49
  Q-banding 123, 132, 141–2, 144, 149, 184, Table 7.1
  restriction endonuclease banding 264
  sex chromosomes 330–1
  Y chromosome 334
'dynamic banding' 252

early replication patterns 235, 240, 243
Ellobius lutescens, sex chromosomes 330
energy transfer, in chromosome banding 177–8, Table 8.6
Equidae
  chromosomal evolution 324
  hybrids 324
ethidium bromide,
  banding 156–7
  inhibition of rRNA transcription 199
  properties 155–6, Table 8.1
  structure *8.1*
euchromatic bands
  chromosome breakage 346
  circumferential grooves 341
  coiling 33, 341
  condensation 33, 82, 97
  constancy 27, 82, 122–3
  definition **2.2.2**
  DNA base composition 84, 337
  DNA replication 83–5, 123, 240, 337
  gene distribution 85, 337–8
  intermediate repetitive sequences 85, 338–9
  meiotic recombination 346
  number 81, 336
  restriction to higher vertebrates 70, 81, 135–6, **16.3.3**
  size 336
euchromatin, 'ontogenetic' 11
'euchromatin transformation' 309, 315
Eutheria, sex chromosomes 329–30

facultative heterochromatin 333
  staining 10
F-bodies, human 150
Feulgen banding 65–7, 74

fish chromosomes
  banding patterns Table 16.1
  heteromorphism 305–6
  replication bands 242
  sex chromosomes 327
fixation of chromosomes **3.5**
  effects on DNA 35, 37
  extraction of proteins 36, 38, 280–1
FPG method 235, 237–8
fragile sites 220

gametogenesis, variation of Ag-NOR staining 203
G-banding
  applications **5.5**
  ASG method **5.2.2**
  and base-specific banding 347
  chromosome identification **5.5.1**
  combination with Ag-NOR staining 197
  deoxyribonuclease 253
  DNA loss 88
  euchromatic bands 11
  gene mapping **5.5.6**
  high-resolution banding **5.2.4**
  homogeneously staining regions **5.5.5**
  inorganic ions 93
  introduction 5
  mechanisms **5.4**
  meiotic chromosomes 83, 95
  methods **5.2.1**
  proteases 73, **5.2.3**, Table 5.1
  protein loss 88
  restriction endonucleases 263
  staining mechanism 89–91
  stains 73, 74, 76, 92
G-bands
  characteristics **5.3**
  chromosomal proteins 86, 281–2
  chromosome coils 88
  definition Table 2.1, **5.1**
  fusion during chromosome condensation 82, 84, 347
  pachytene chromomeres 32, 83, 340
  replication patterns 83–4, 240
  taxonomic distribution 70, 344, Table 16.1
G-11 banding 37, **4.3**
  applications **4.3.3**
  mechanism **4.3.2**
  method **4.3.1**
  taxonomic distribution 59
gene mapping 288, 302
  use of G-banding 102
  use of heteromorphisms 59, 302
genes, chromosomal distribution 85, 338
Giemsa, *see* Romanowsky dyes
grasshopper chromosomes
  DNA in C-bands Table 4.1
  effect of heterochromatin on chiasmata 52, 306

heterochromatin variation between species 51
  heteromorphism 306
  hybrid species 326
  sex chromosomes 330–1
great apes
  chromosomal phylogeny 321
  distamycin/DAPI fluorescence 302–3
  DNA type in C-bands Table 4.1
  heteromorphisms 302–3, **14.5**
  inhibition of chromosome condensation 230
  5-methylcytosine in heterochromatin 277
  replication bands 240
  restriction endonuclease banding 263
guanine, photo-oxidation 273
guanosine, antibodies to 273

heterochromatic bands **16.2**
  definition **2.2.1**
  heterogeneity 50–1, 149, 288–9, 309
  silver staining **4.6**, 194
  types of DNA Table 4.1
heterochromatin
  C-banding 47–8
  early observations 2–3
  effect on chiasmata 52, 306, 334
  effect on disjunction 334
  elimination 52, 334
  equilocal distribution 317–18
  'functional' 11
  functions **4.2.4**, **16.2**
  genes in 333–4
  inertness 333
  'intercalary' 11
  heterogeneity 50–1, 149, 288–9, 309
  inhibition of condensation **11.2.1**, **11.2.2**
  lateral asymmetry **11.3.5**
  5-methylcytosine in 275, 278
  non-C-banded 10, 40, 47–8, 65
  'nucleotypic' effect 335
  transcription 50, 334
  under-replication 52, 334
heteromorphism
  applications **14.4.11**
  classification **2.4.2**
  definition 283
  formation **14.2**
  human chromosomes **14.4**
  inheritance 58–9, 150, 204, 284, 296, 305, **14.4.8**
  measurement 132, **14.3**
  NORs 204
  paternity testing 59, 150, 301–2
  phenotypic effects 59
  restriction endonucleases 263
high-resolution banding 6–7
  applications **5.5.7**
  cancer chromosomes 103
  cell cycle synchronization 79, 244–5

G-banding **5.2.4**
'microcytogenetics' 103
nomenclature 80
replication banding **11.3.4**
histones **3.2.3**
   antibodies to **13.3.1**
   extraction by fixation 36, 280–1
   modifications in heterochromatin 49
HMG proteins 25
   antibodies to 281–2
Hoechst 33258
   applications 160
   banding **8.3.1**
   fluorescence of human heterochromatin Table 8.3
   fluorescence quenching by BrdU 238
   inhibition of chromosome condensation 160, 211, **11.2.2**
   properties 160, 162, Table 8.1
   replication banding 235
   staining method 160
   structure *8.2*
homogeneously staining regions (HSRs) **5.5.5**
human chromosomes
   banding with chromomycinone antibiotics 169
   banding with D287/170 168
   banding with daunomycin and adriamycin 166
   banding with dibutyl-proflavine 165
   banding with Hoechst 33258 166, Table 8.3
   distamycin/DAPI fluorescence 181
   DNA type in C-bands Table 4.1
   effect of cold treatment 225
   high-resolution banding *2.2*
   inhibition of condensation 227–30
   lateral asymmetry 246–7, Table 11.3
   5-methylcytosine in heterochromatin 277
   number of bands 6–7, 336
   Q-band heteromorphism 135, 149–50
   quinacrine fluorescence in interphase **7.2.3**
   quinacrine positive heterochromatin 127, 132, Table 8.3
   replication patterns 241
   standard karyotype *2.1*, *2.2*, 147
   Y chromosome 132, **7.5.3**
human chromosome heteromorphisms **14.4**
   chromosome 1 **14.4.1**
   chromosome 3 **14.4.2**
   chromosome 4 **14.4.2**
   chromosome 6 **14.4.3**
   chromosome 9 62, **14.4.4**
   chromosome 16 **14.4.6**
   D group chromosomes **14.4.5**
   G group chromosomes **14.4.5**
   inheritance **14.4.8**
   phenotypic effects **14.4.10**

population studies **14.4.9**
racial differences 59, 300
Y chromosome **14.4.7**
Hy-banding 13, 65, 67
immunocytochemical methods
   bromodeoxyuridine 237
   DNA **13.2**
   histones **13.3.1**
   kinetochores **10.2.3**
   5-methylcytosine **13.2.2**
   non-histone chromosomal proteins **13.3.2**
   NORs 192
   nucleosides **13.2.1**
   replication banding 237
'immunocytogenetics' 272
inactive X chromosome
   DNase sensitivity 256, 258
   methylation 267, 277
   replication patterns 241
   restriction endonuclease banding 267
inhibition of chromosome condensation **3.4**
   5-azacytidine **11.2.3**
   cold treatment **11.2.1**
   DAPI 226–8
   distamycin 226–8
   high-resolution banding 79–80
   Hoechst 33258 160, 211, **11.2.2**
   mechanisms 228–9, 232
   netropsin 226
   silver staining of uncondensed regions 228
inorganic ions **3.2.5**
in situ hybridization 5
   chromomycin banding 170
   Q-banding 148
intercalation 116
   acridine orange 116
   quinacrine 139–40
   quinacrine mustard 137
interphase nuclei
   C-banding 58
   orientation of chromosomes 58
isochores 84, 143, 344
   euchromatic bands 84, 142–3, 337, 344
   origin 347–8
   plants 344
   taxonomic distribution 344, Table 16.1

kinetochores
   C_d-banding 5, 13, **10.2.1**
   composition 208
   CREST staining **10.2.3**
   definition 206
   early observations 2, 5–6
   Giemsa staining 209–11
   immunocytochemical labelling 5–6, **10.2.3**
   meiotic chromosomes 212
   pre-kinetochores 215–17
   proteins 218–19, Table 10.1, Table 10.2

silver staining 5, **10.2.2**
size variation 215
staining methods **10.2**
structure **10.1.2**
lateral asymmetry **11.3.5**
  compound 247
  Drosophila 248
  inheritance 247
  mammals 247–8
  mechanism 245–6
  Robertsonian fusion 246
  Vicia faba 249
leukaemias, alteration of replication pattern 241
light-induced banding **8.6**
Lilium spp., Q-bands throughout chromosomes 136, 344
LINEs, chromosomal distribution 85, 338–9

marsupials
  chromosomal evolution 313, 323–4
  DNA types in C-bands Table 4.1
  inhibition of chromosome condensation 226–7
meiotic chromosomes
  distamycin/DAPI banding 181
  identification by C-banding 58
  kinetochores 212
  Q-banding 148
  Y chromosome 122
mental retardation, and chromosome heteromorphisms 59, 299
5-methylcytosine 293–4
  antibodies to **13.2.2**
  heteromorphism 277
  in heterochromatin 275, 277
methylation of DNA
  inactive X chromosome 267
  inhibition of restriction endonucleases 261
  repression of transcription 277
  suppression by 5-azacytidine 199
'microcytogenetics' 103
mithramycin **8.4.1**
  see also chromomycinone antibiotics
Monotremata, sex chromosomes 328–30
mountants
  banded chromosomes 47
  G-banding 76–7
  pH for Q-banding 125–6
  to inhibit fading of fluorescence 160
mouse chromosomes
  DNA type in C-bands Table 4.1
  heterochromatin variation between species 317
  heteromorphism 304
  Hoechst banding 157, 160
  homogeneously staining regions 99
  light-induced banding 184
  5-methylcytosine in heterochromatin 275
  quinacrine fluorescence 133, 143–4, 185

restriction endonuclease banding 263, 266
Robertsonian fusion 97, 318–19
satellite DNA 133
mule, fertility 324
Muntiacus muntjak (Indian muntjac)
  kinetochores 216
  5-methylcytosine in heterochromatin 277
  'muntjac scandal' 320
Myopus schisticolor (wood lemming)
  sex chromosomes 329–30

N-banding
  definition 64
  mechanism **9.3.2**
  methods 63, 64, **9.3.1**
  NORs 13, 191, **9.3**
  specificity 192–3
  staining of heterochromatin 64
netropsin, inhibition of chromosome condensation 226
nick translation of chromosomes
  DNase 255–6
  restriction endonucleases **12.3.2**
non-disjunction, and chromosome heteromorphisms 300
non-histone chromosomal proteins **3.2.4**
  antibodies **13.3.2**
NORs
  applications of silver staining **9.4.3**
  Coomassie blue staining 191
  definition 12, 189
  Giemsa staining 191
  heteromorphisms 204, 294, 296, 298
  immunofluorescence 192
  mechanisms of staining **9.3.2**, **9.4.2**
  methylation of inactive NORs 277
  N-banding 191, **9.3**
  resistance to DNase 191, 253
  ribosomal genes 189
  sites 192, 202
  staining methods **9.2**, **9.4.1**
  staining with chromomycinone antibiotics 191
  staining with Coomassie blue 191
NORs, silver staining 2, 5, 191, **9.4**
  active NORs 198–9, 203–4
  combined with other banding methods 196–8
  electron microscopy 195
  methods **9.4.1**
  suppression in hybrids 203
'nucleic acid starvation' 3
nucleolar proteins Table 9.1
  Giemsa staining 193
  silver staining 193, 200–2
nucleolar organisers, *see* NORs
nucleoli
  proteins Table 9.1
  silver staining, early observations 2

structure 189, *9.2*
nucleosomes **3.3.1**
nucleotypic effects 335
olivomycin **8.4.1**
  see also chromomycinone antibiotics
Ovis aries (sheep)
  DNA type in C-bands Table 4.1
  heteromorphism 303–4
  5-methylcytosine in heterochromatin 277

packing ratio 26
parameres 292
paternity testing, using heteromorphisms 59, 150, 288, 301–2
Peromyscus spp.
  chromosomal evolution 59, 322
  DNA types in C-bands Table 4.1
  heteromorphisms 304
Philadelphia chromosome 98
photography
  acridine orange R-bands 110
  Q-bands 131–2
plant chromosomes
  absence of euchromatic bands 344
  chromosomal evolution 316
  cold treatment 224
  DAPI banding 163
  heteromorphism 307–8
  Q-banding 149, 316
  replication bands 242, Table 16.1
polymorphism, definition 282
pre-kinetochores
  interphase distribution 215–17
  in spermatozoa 217
primates
  chromosomal evolution 313, 319, 321
  DNA types in C-bands Table 4.1
  lateral asymmetry 247–8
  replication bands 240
  structural polymorphisms 303
propyl quinacrine mustard 123–4
  structure *7.1*

Q-banding
  applications 122–3, **7.5**
  base specificity 133, 135, 137, 184, **7.4**, Table 7.1
  combination with other methods 197, **7.2.4**
  counterstaining 128
  effect of chromosome condensation 136
  fluorochromes 123–4
  methods **7.2.1**
  mounting 125–7, 130
  pachytene chromosomes 148
  replication patterns and 135, 240–1
  sperm chromosomes 148
  spermine bis-acridine **7.2.2**
Q-bands
  definition 122, Table 2.1
  euchromatic bands 135–7
  heterochromatic 132–5, **7.5.2**
  heteromorphisms 149–50, 290, 293–4
  measurement 132, 288
  introduction 122
  observation **7.2.5**
  photography 130–1
  taxonomic distribution Table 16.1
quinacrine
  base specificity of fluorescence 133–5, **7.4.2**
  binding to chromosomes 139–40
  binding to DNA 139–40
  differential excitation of fluorescence **7.4.2**
  fading 126
  fluorescence of human heterochromatin Table 8.3
  structure *7.1*
quinacrine mustard
  binding to DNA 138–40
  Q-banding 4
  structure *7.1*

R-banding
  acridine orange 105, **6.2.2**
  acridine orange heteromorphisms 114, **6.5.3**
  ageing of chromosome preparations 110, 114
  alternative dyes 108
  applications **6.5**
  combination with other methods 197, **6.5.1**
  definition 105, Table 2.1
  'dénaturation ménagée' 105
  DNA loss 115
  DNase hypersensitivity 256–7, 259
  electron microscopy 115
  euchromatic bands 1
  Giemsa **6.2.1**
  introduction 5, 105
  mechanisms **6.4**
  photography 110
  protein loss 115
  replication patterns and 240–1
replication banding 6, 79, **11.3**
  applications **11.3.3**
  G-banding and 123, 240–1
  high-resolution banding **11.3.4**
  homology of X & Y chromosomes 241, 330
  lower organisms 242, Table 16.1
  mechanisms of staining **11.3.2**
  methods **11.3.1**
  plants 242, Table 16.1
  principle of differential labelling *11.7*
  recombination and 346
  reverse staining 236, 239
  tissue-specific patterns 241, 347
  universal distribution 344, Table 16.1
  use of BrdU 70, *11.3*
replicons 337
reproductive failure, and chromosomal heteromorphisms 299

reptiles
    chromosomal evolution 314, 316, 325–6
    heteromorphisms 305
    sex chromosomes 328
restriction endonucleases
    banding heteromorphisms 263, 289, 290, 295
    banding methods 261
    banding patterns 262–5
    band nomenclature **12.5**
    effect of DNA methylation 261, 263, 265, 267
    enzyme nomenclature 259
    isoschizomers 259
    mechanism of action on chromosomes **12.3.3**
    nick translation of chromosomes **12.3.2**
    use for banding 6, **12.3**, Table 12.1, Table 12.2
retroposons (mobile elements) 339
ribosomal RNA 187–8
    genes for 187–8, 293, *9.1*
RNA **3.2.2**
Robertsonian fusion
    Bovidae 319
    DNA polarity 246
    heteromorphism 304
    in evolution 318–19
    lateral asymmetry and 246
    loss of heterochromatin 59, 314
    Mus musculus 97, 318–19
    number of centromeres 211
    sheep 907
    Sorex (shrews) 319
rodents
    chromosomal evolution 314, 317, 319, 321–2
    DNA in C-bands Table 4.1
    enlarged sex chromosomes 329
    heteromorphism 304
    replication bands 240
    transcription of heterochromatin 50
Romanowsky dyes 61, 89–92

satellite associations, silver staining 205
satellite DNA 22–3
    C-bands 48–9, Table 4.1
    human 48–9
    'library' hypothesis 317
    methylation 232
    restriction endonuclease digestion 264, 266
Scilla spp.
    DNA in C-bands Table 4.1
    ethidium banding 156
    heterochromatin variation between species 51, 316
    heteromorphism 308, 316
    Q-banding 132, Table 7.1
Secale spp. (rye)

DNA in C-bands Table 4.1
    heterochromatin variation between species 316
    heteromorphisms 307
secondary constrictions 3
    induction 221
    NORs 189
sex chromosomes
    evolution **15.4**
    homology of pairing segments 241, 330
silver staining
    active NORs 198–9, 203
    chromosomal proteins 67, 194
    chromosome cores 67, 194, 196
    DNA 67, 200
    heterochromatin 194, 196, 212, **4.6**
    histones 67, 194
    kinetochores 5, 194, **10.2.2**
    NORs 5, 196–8, **9.4**
    nucleoli 2
    satellite associations 205
    sites in NORs 199
    substrate in NORs 200–2
    synaptonemal complexes 194
    uncondensed chromatin 228
SINEs, chromosomal distribution 85, 338–9
sister chromatid exchange 6, 232, 234, 238, 240
    principle of differential labelling *11.7*
solenoids **3.3.2**
spermatozoa
    aneuploidy 63
    pre-kinetochores 217
    Y chromosome 152–4
spermine bis-acridine 125
    method for Q-banding **7.2.2**
    structure *7.1*
standard karyotypes Table 2.4
Sus scrofa (pig)
    D287/170 banding 168
    heteromorphism 304
    standard karyotype Table 2.4
synaptonemal complexes
    CREST labelling of kinetochore 217
    silver staining 194

T antigen, distribution on chromosomes 282
T-bands
    DNA composition 112, 114
    introduction 105
    mechanisms 118
    methods **6.2.3**
    telomeres and 112
Tetrahymena pyriformis, silver staining of NORs 202
thymidine, antibodies to 273
topoisomerase I, in mouse centromeres 282
Trillium spp.
    cold treatment of chromosomes 224

ethidium banding 156
Q-banding 132
triploidy, origin 150, 301
trisomy, origin 150, 301
Triticale, effect of heteromorphism 307
Triturus spp.
   cold treatment of chromosomes 224
   heterochromatin variation between species 316
   heteromorphism 305
turtles, 200-million-year-old chromosomes 325–6

unequal crossing-over 285

variant, definition 283–4
Vicia faba
   banding with hydrochloric acid 65
   daunomycin bands 166
   DNA in C-bands Table 4.1
   heterogeneity of heterochromatin 51

inhibition of chromosome condensation 229
lateral asymmetry 249
Q-banding 132, 149
replication bands 242
transcription of repetitive DNA 50

X chromosome, replication patterns 241

Y chromosome, human
   heteromorphism 150, 288, 298–300, **14.4.7**
   homology with X 241
   interphase 150–2
   meoisis 122
   quinacrine fluorescence **7.5.3**
   racial differences 298
   spermatozoa 122, 152–4

Z-DNA 23–4, 278
   antibodies to **13.2.4**
   specificity of antibodies 279–80